Energy

Its Use and the Environment

third edition

Roger A. Hinrichs

State University of New York, College at Oswego

Merlin Kleinbach

State University of New York, College at Oswego

HARCOURT COLLEGE PUBLISHERS

Fort Worth Philadelphia San Diego New York Orlando Austin San Antonio Toronto Montreal London Sydney Tokyo

Publisher: Emily Barrosse
Publisher/Acquisitions Editor: John Vondeling
Marketing Strategist: Kathleen Sharp McLellan
Developmental Editor: Peter McGahey
Project Editors: Robin Bonner, Dana L. Passek
Production Manager: Alicia Jackson
Art Director and Text and Cover Designer: Jacqueline LeFranc

Cover Credit: Windmills on hill. *(©Dale Sanders/Masterfile)*

Energy: Its Use and the Environment, Third Edition
ISBN: 0-03-031834-3
Library of Congress Catalog Card Number: 00-111713

Address for domestic orders:
Harcourt College Publishers, 6277 Sea Harbor Drive, Orlando, FL 32887-6777
1-800-782-4479
e-mail collegesales@harcourt.com

Address for international orders:
International Customer Service, Harcourt, Inc.
6277 Sea Harbor Drive, Orlando, FL 32887-6777
(407) 345-3800
Fax (407) 345-4060
e-mail hbintl@harcourt.com

Address for editorial correspondence:
Harcourt College Publishers, Public Ledger Building, Suite 1250, 150 S. Independence Mall West, Philadelphia, PA 19106-3412

Web Site Address
http://www.harcourtcollege.com

Printed in the United States of Ameria

123456789 039 10 98765432

THIS BOOK IS PRINTED ON **ACID-FREE, RECYCLED** PAPER

Contents

Preface

Introduction to the Third Edition

Energy is an introductory textbook that emphasizes the physical principles behind energy and its effects on our environment. It can be used in physics, technology, physical science, and environmental science courses for nonscience majors. Much of the pedagogy used herein comes from a one-semester general education course I (R. H.) have taught at State University of New York, College at Oswego, over the past 20 years. No math or other science prerequisite is necessary.

For this third edition, a second author has been added. Lin Kleinbach, as Emeritas Professor of Technology Education at SUNY-Oswego, brings a wealth of information on current energy technologies and a very practical approach to energy education. Both of us have been involved in conducting energy education workshops for New York State teachers for the last 15 years.

This third edition provides updated data on energy resources, energy use, and energy technologies. Since the publication of the second edition (only 5 years ago), we have seen deregulation and increased competition in the electric utility industry, escalating oil prices, and increased worldwide commitment to renewable energy. These trends are emphasized throughout the book. Because the field of energy changes so quickly, we have added for this edition a website that provides an up-to-date list of Internet-related material for each chapter.

We continue to emphasize the environmental impact of fossil-fuel consumption, in part by placing chapters that discuss fossil-fuel use, air pollution, and global warming material early in the text. More examples dealing with energy use in other countries have been added. Additional "hands-on" Activities have been placed within the chapters when a topic is introduced, and more have been added at the end of the chapters.

Goals of this Book

Because the subject of energy is multifaceted, this book has several purposes. First and foremost, it seeks to explain the basic physical principles behind the use of energy, including the study of mechanics, electricity and magnetism, thermodynamics, and atomic and nuclear physics. At the same time, it covers crucial environmental questions that currently are receiving much public attention, such as global warming, radioactive waste, and municipal solid waste. The best way to understand the consequences of present and future energy alternatives, and the environmental, social, and economic trade-offs that must be made, is to understand the scientific principles involved. These principles are presented with a minimal amount of math and with the help of everyday examples. Each chapter has a set of problems (mainly nonmathematical) that seeks to emphasize the principles and apply them to situations dealing with energy and the environment. Many of the standard topics found in introductory physics textbooks are included. As a consequence, this book can be used as the text in a conceptual physics course with energy as the central theme, as has been my practice.

Second, the book examines different aspects of each energy resource, including the principles involved and the environmental and economic consequences of its use. Renewable energy is treated early on, after heat transfer, and everything from solar heating systems to photovoltaic devices to wind-turbine generators is covered. The principles of nuclear power and the current debate over radioactive waste disposal are covered in depth. At the heart of most energy conversion schemes is the heat engine, so general principles in this area are developed early (Chapter 4). The environmental aspects of generating electricity and using it are also an underlying theme. Air pollution and global warming are treated early on, as are some ways in which their impacts can be lessened.

Third, the textbook seeks to integrate the complex questions of energy policy and possible energy strategies. There are no simple answers or single alternatives that can provide all of our energy needs, preserve our economic prosperity, and protect our environment. As a consequence, questions (many unanswered) are brought up throughout the book to cause readers to critically think ahead and maybe even begin to develop their own solutions. The important theme of energy conservation is emphasized with examples *throughout* the textbook, rather than in a separate chapter, because the efficient use of energy should be of utmost concern in every area—from energy mechanics to heat transfer to electricity use. Practical suggestions are given that we hope will cause the readers to evaluate their roles in using energy more effectively.

Many books have an energy theme, but few seek both to teach general physical principles and to cover the many alternatives of energy supply and conservation. These themes are presented in such a way that they will appeal to the growing audience of nonscience majors, as well as provide information relevant to those interested in science and engineering.

Coverage

While the material in this textbook is organized to follow a somewhat traditional sequence presented in a physics course, we have tried to order the topics in a way that touches the current interests of students. Solar heating is covered early (Chapter 6), but only after concepts of heat transfer (Chapter 4) and residential energy conservation (Chapter 5) have been discussed. Chapters 8 and 9 deal with the environmental aspects of energy use, namely air pollution (from both stationary and mobile sources), global warming, and thermal pollution. These topics follow Chapter 7 on fossil fuels. The physics of oil exploration and recovery is covered in Chapter 7 as a Special Topic. Photovoltaics and the rapidly growing area of wind energy follow Chapters 10 and 11 on electrical energy production. New automotive technologies incorporating fuel cells and hybrids and electric vehicles are treated within these two chapters. Nuclear fission is covered in Chapter 14, directly after a chapter on the fundamentals of atomic and nuclear physics. Chapter 17 covers the broad topic of biomass, from wood combustion to municipal solid waste.

Features

Many features are designed to encourage students to understand the crucial role of energy in our society and the implications stemming from their own consumption of energy. The textbook is designed to make physics and its applications relevant and interesting to nonscience majors. Features include:

- **Hands-On Activities,** which provide opportunities for students to engage in experiments that will reinforce the material covered. These are both integrated into the chapters and placed at their ends. Some examples are the construction of a simple shoe box model to examine solar heating, an experiment to study heat transfer from insulated soda pop cans, and investigation of atmospheric particulates. All of the activities have been tested, many of them in the summer workshops that we direct for secondary school science and technology teachers.
- **Residential use of energy,** which is emphasized through many examples that are relevant to the home. Analysis of heat losses in the home and ways to reduce them are covered in Chapter 5. A home energy audit is provided in the Appendices. New energy-efficient lighting fixtures are discussed in Chapter 10 on electricity. Recycling and questions related to solid waste management are covered in Chapter 17.
- **International perspectives on energy,** which are emphasized throughout the textbook. Although the book primarily focuses on the United States' use of energy, we are part of an interdependent global village. As we well know, political events in other parts of the world can have important effects on our own economy. Attention also is focused on those people who live in developing countries, where energy plays a critical role. Several new Focus-On boxes, which highlight interesting applications or unusual facts about energy, have

been added in this area. Both of us have had a number of years experience working in developing countries.

- **Internet references,** which are up-to-date, are provided on a website for each chapter. Some general sites on energy use are found on the inside front cover.

Other features include appendices that provide current information on energy use in the United States and the world, end-of-chapter summaries, questions and problems, worked examples, a glossary, and a (separate) Instructor's Manual.

Teaching Options

Because of the many facets of energy use and our environment, some sections and chapters could be omitted without any loss in continuity. For example, the section on heat engines in Chapter 4 could be omitted (especially for non-science majors); the Carnot efficiency could be introduced later for that part of Chapter 9 on thermal pollution. For a shorter course, you might wish to skip Chapter 15 on the biological effects of radiation and Chapter 16 on fusion, and concentrate instead on biomass conversion and municipal solid waste in Chapter 17. The nontechnical style of the writing allows the instructor to assign sections in each chapter as outside reading, making it possible to cover more of the book than could be covered in class discussion alone.

Acknowledgments

Many people were of great help in the construction and revision of this textbook. The contributions of the following greatly refined this third edition: Manuscript reviewers—Patrick Gleeson, *Delaware State University* and Daryl Prigmore, *University of Colorado, Colorado Springs*. Prerevision reviewers—Terry Carlton, *Oberlin College;* Laurent Hodges, *Iowa State University;* Jack Pinnix, *Chicago State University;* Robert Poel, *Western Michigan University;* Don Reeder, *University of Wisconsin, Madison;* Karin Shen, *Vanderbilt University.* Reviewers from early editions included David Appenbrink, *University of Chicago;* Joseph Katz, *Johns Hopkins University;* Philip Krasicky, *Hamilton College;* Wesley Lingren, *Seattle Pacific University;* Robert Poel, *Western Michigan University;* Ljubisa R. Radovic, *Pennsylvania State University;* Don Reeder, *University of Wisconsin;* Peter Schroeder, *Michigan State University;* Carl Voiles, *Michigan State University;* Thomas Weber, *Iowa State University.* We offer our sincere thanks to them all.

At Harcourt College Publishers, we have appreciated the encouragement and professional advice of Peter McGahey, Developmental Editor; Alicia Jackson, Production Manager; Robin Bonner and Dana L. Passek, Project Editors; Jacqueline LeFranc, Art Director; and Kathleen Sharp McLellan, Marketing Strategist. In these days of electronic communication, it's a shame we can't thank them in a more personal way.

Roger Hinrichs and Merlin Kleinbach
Oswego, New York
December, 2000

1

Introduction

A. Energy: An Initial Definition

Energy is one of the major building blocks of modern society. Energy is needed to create goods from natural resources and to provide many of the services we have come to take for granted. Economic development and improved standards of living are complex processes that share a common denominator: the availability of an adequate and reliable supply of energy. The modernization of the West from a rural society to an affluent urban one was made possible through the employment of modern technology based on a multitude of scientific advances—all of which are energized by fossil fuels. Political events, beginning with an oil embargo in 1973 and continuing through the Iranian revolution of 1979 and the Persian Gulf War of 1991, made many people aware of how crucial energy is to the everyday functioning of our society. Long gasoline lines and cold winters with natural gas shortages in the 1970s are still unhappy memories for some people. The energy crises of the 1970s were almost forgotten by the 1980s. However, that decade brought an increased awareness of our environment. Concerns about global warming, acid rain, and radioactive waste are still very much with us today, and each of these topics is related to our use of energy.

While an interest in being energy self-sufficient and producing one's own power was a strong desire of some in the 1970s and 1980s, during the second half of the 1990s the entire public began to have another choice—that of being able to select their own provider of electricity. The electric power industry moved away from a traditional, highly regulated industry to one of deregulation and competition. Beginning in 1997, customers were given the chance to shop for their own supplier—and the bottom line (cost) was not the only criteria. Many people decided to buy from the producer who polluted least, so-called "green power" alternatives.

Energy pervades all sectors of society—economics, labor, environment, international relations—in addition to our own personal lives—housing, food, transportation, recreation, and more. The use of energy resources has relieved us from many drudgeries and made our efforts more productive. Humans once had to depend on their own muscles to provide the energy necessary to do work. Today our muscles supply less than 1% of the work done in the industrialized world.

Energy supplies are key limiting factors to economic growth. We have become a very interdependent world, and access to adequate and reliable energy resources is central for economic growth. About 40% of the world's energy comes from oil, much of which is imported by the industrialized nations and much of which comes from the Persian Gulf. From this region, Japan imports two thirds of its oil, the United States imports 20% of its oil, while one third of France's total oil needs comes from there. If industrialized nations encounter any significant restriction to their sources of oil, through either reduced supplies or large price increases, their economies would suffer considerable damage.

Your own picture of energy might be colored in many ways by your experiences. You might think of the "energy" (or the lack of it) that a particular person possesses, or the kinetic energy that a stone gains as it drops, or the energy responsible for the movement of automobiles, or the energy used in the production of heat and light. One dictionary defines energy as the "capacity for vigorous action; inherent power; potential forces." Energy is found in many forms, and one purpose of this book will be to identify them and study how they can be used. Energy is found in such forms as wind and flowing water, and stored in matter such as fossil fuels—oil, coal, natural gas—where it can be burned for "vigorous action."

Energy might best be described in terms of what it can do. We cannot "see" energy, only its effects; we cannot make it, only use it; and we cannot destroy it, only waste it (that is, use it inefficiently). Unlike food and housing, energy is not valued in itself but for what can be done with it.

Energy is not an end in itself (notes Richard Balzhiser, former president of Electric Power Research Institute). *The fundamental goals we should have in mind are a healthy economy and a healthy environment. We have to tailor our energy policy as a means to those ends, not just for this country but in global terms as well.*

Energy is a basic concept in all the sciences and engineering disciplines. As we will discuss in the next chapter, a very important principle is that energy is a conserved quantity, that is, the total amount of energy in the universe is a constant. Energy is not created or destroyed but just converted or redistributed from one form to another, such as from wind energy into electrical energy, or from chemical energy into heat. We will study the various forms of energy—chemical, nuclear, solar, thermal, mechanical, electrical—and the useful work that energy is capable of doing for us. We will explore both energy resources and energy conversion processes.

Understanding energy means understanding energy resources and their limitations, as well as the environmental consequences of their use. Energy and environment and economic development are closely linked. Over the past two decades, global energy consumption has increased by about 25%, while U.S. consumption increased by 15%. Much of this global growth has been in less-developed countries. (In the next two decades, estimates are that energy consumption will rise by over 100% in developing nations.) With this growth has been a decline in urban air quality as well as serious land and water degradation. Since fossil fuels represent almost 90% of our consumption, we continue to increase the emissions of carbon dioxide, which may alter the earth's climate irreversibly. The proper use of energy requires consideration of social issues as well as technological ones. Indeed, sustained economic growth in this century, together with improvements in the quality of everyone's lives, may be possible only by the well-planned and efficient use of limited energy resources and the development of new energy technologies.

B. Energy Use and the Environment

We live in an age of environmental awareness. Politicians would have a hard time getting elected if they did not at least state they had a concern for the environment. The 20th anniversary of Earth Day on April 22, 1990, became the focus of attention for millions of people who wanted to launch a decade of environmental activism. Many changes in the environment have occurred in the 30 years since the first Earth Day and some are listed in Focus On 1.1, "Our Earth—Then and Now."

The 25th anniversary of Earth Day in 1995 focused on the progress made to improve our air and water quality. In air pollution, smog has declined nationally by about one third since 1970. In 1999, Los Angeles did not record one ozone reading high enough to trigger a smog alert; 20 years earlier there were

Focus On 1.1

OUR EARTH—THEN AND NOW

	1970	1997
World population	3.3 billion	5.8 billion
10^3 Tons of lead emitted, United States	204	4
Tons of waste recycled	8 million	49 million
U.S. homes using solar energy	35,000	2 million
Tons of garbage generated annually in United States	121 million	217 million
Percentage of oil imported to United States	23%	56%
Percentage of federal budget spent for environment	3%	1.5%
Atmospheric CO_2 concentration (ppM)	325	367
World CO_2 emissions, 10^9 tons/yr	14	23

120 smog alerts in a year. New cars in 1995 emitted about 1% of the pollution per mile of 1970 model cars! Sulfur dioxide emissions, the primary cause of acid rain, have fallen by one third since 1970. In 1970, only about one quarter of our rivers met federal standards for fishing and swimming; in 1995, about 60% did. These accomplishments did not come about without great efforts. Federal and state expenditures for pollution abatement and control have risen sharply since 1970 (to $100 billion per year). However, concerns over federal spending, the national debt, and the role of the federal government continue to prompt legislative drives for drastic environmental law reforms and modifications in regulations affecting clean air and water, toxic waste, pesticides, endangered species, etc.

The use of our energy resources is one of the major factors affecting the environment. (Our use of chemicals is another.) Increased use of fossil fuels since the beginning of the industrial age has increased the carbon dioxide concentration in the atmosphere by 30%, and has probably also increased the earth's temperature (Fig. 1.1). Warmer global temperatures can lead to a melting of the polar ice caps and higher ocean levels, which will force a movement of population away from low-lying land near the seas. It can also mean a shift of agricultural areas as precipitation patterns move northward.

Getting rid of our garbage is also an increasingly serious environmental problem. Americans dispose of almost 4 pounds of garbage per person per day—that's about 3 tons per family per year, and twice the rate of disposal by Europeans. We're running out of acceptable places to bury our garbage. We have gone from 14,000 landfills in 1970 to about 3000 today, for more people.

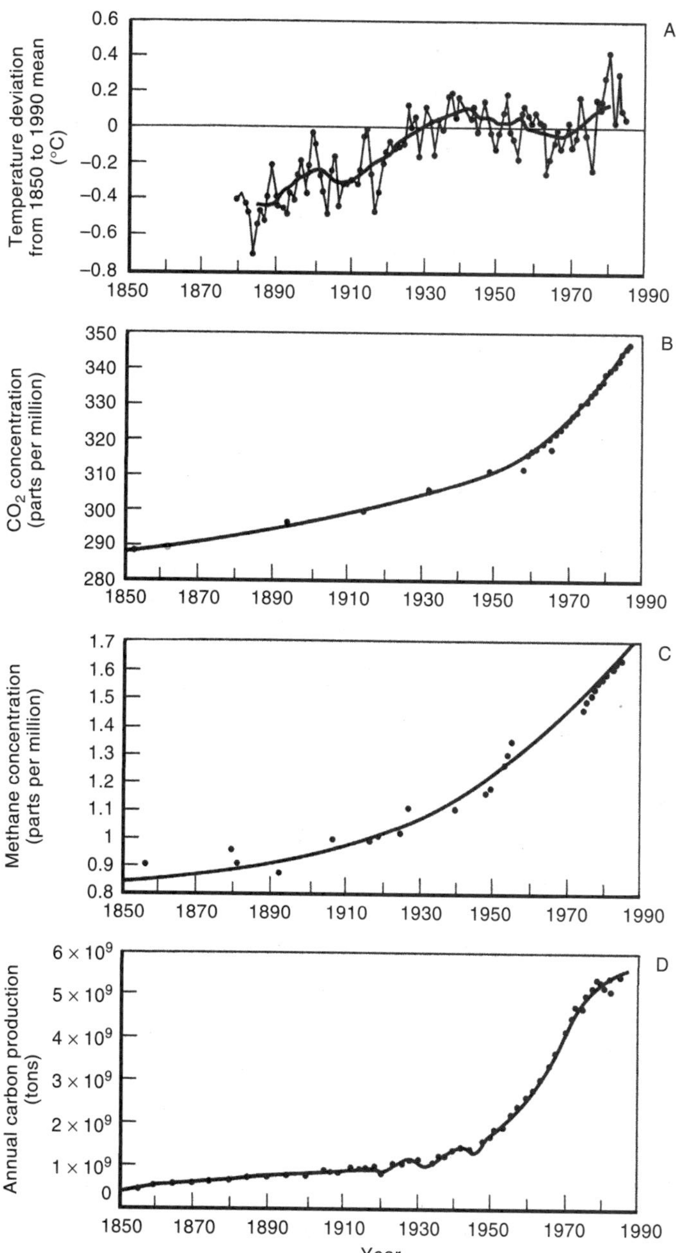

FIGURE 1.1

Correlation among global temperature change (A), atmospheric carbon dioxide and methane concentrations (B, C), and annual carbon production from fossil-fuel burning (D), displayed in order (Houghton, R. A., and G. M. Woodwell. 1989. Global Climatic Change. *Scientific American,* April). A more convincing correlation between carbon dioxide concentrations and the earth's temperature over the past 160,000 years is shown in Figure 9.1.

Do we deal with solid waste by incinerating it (and using the heat for industrial purposes, to generate electricity, or both) and put only the ash into landfills? There is a lot of opposition to this approach because of possible air, water, and thermal pollution. How much of this problem can be solved by recycling, by reduced packaging, or by other means?

In each of these examples, tough choices have to be made. If we want to reduce the quantity of fossil fuels burned because we're concerned about global warming, what substitutes should be made? More solar or nuclear energy? At what point do we say we're confident enough to endorse a method for burying the radioactive wastes generated at nuclear power plants? What can take the place of gasoline in our cherished autos? Is ethanol made from grain an energy-efficient substitute? (Today, 10% of the gasoline sold in the United States contains some ethanol, usually made from corn.) Should food be used for fuel, when people are malnourished? Should solar energy be subsidized to compete with the less expensive fossil fuels, since we know that our fossil fuel supplies are finite and that their use causes damage to the environment?

> *There is an often poorly understood link between ethical choices that seem quite small in scale and those whose apparent consequences are very large, and that a conscious effort to adhere to just principles in all our choices—however small—is a choice in favor of justice in the world. Both in our personal lives and in our political decisions, we have an ethical duty to pay attention, resist distraction, be honest with one another and accept responsibility for what we do—whether as individuals or together . . . We can believe in the future and work to achieve it and preserve it, or we can whirl blindly on, behaving as if one day there will be no children to inherit our legacy. The choice is ours; the earth is in the balance.* Al Gore

C. Energy Use Patterns

Until the 1980s, energy consumption in the world—especially the United States—had been increasing annually at a rapid rate. Figure 1.2 shows the energy consumption in the United States over the last 200 years, by fuel used. Between 1850 and 2000, the use of commercial fuels increased by a factor of 100. In the late 1940s and 1950s, an average of 2.9% more energy was used each year in the United States than the previous year. In the 1960s and early 1970s, the rate of growth was higher still: 4.5% per year. Such a growth rate would result in a doubling of energy consumption in only 15 years. In the late 1970s, the growth rate of U.S. energy consumption slowed to 3%, and in the early 1980s actually declined: in 1983 the United States used 11% less energy than it did in 1979 even with an increase in population. During the latter 1980s, U.S. energy consumption did increase modestly, although at a smaller rate than the Gross Domestic Product (GDP), indicating a continuing trend toward greater energy

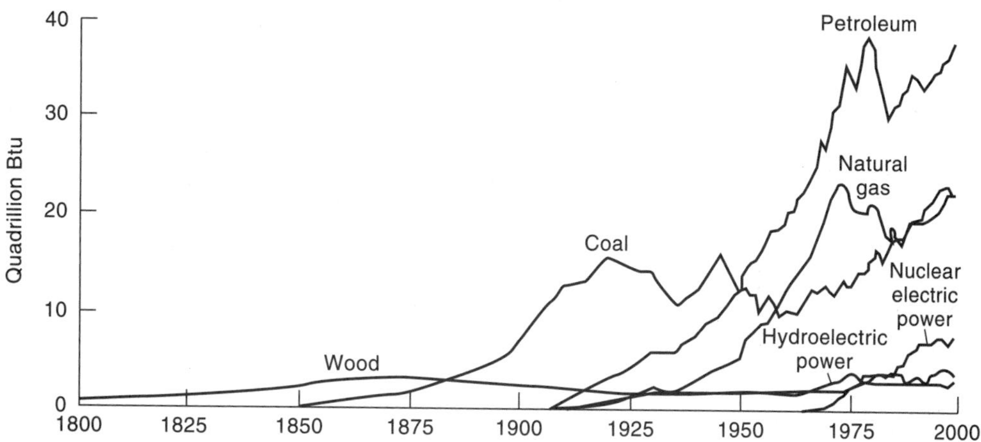

FIGURE 1.2

Energy consumption in the United States over the last 200 years, by fuel used. A Btu is a unit of energy. A quadrillion Btu (or Quad) is 10^{15} Btu.
(United States Energy Information Administration, USEIA)

efficiency. In the 1990s, energy consumption continued to increase, but at a slightly greater pace than in the 1980s as the nation recovered economically. Between 1978 and 1998, energy consumption was up by 17%, but the GDP rose by 67%.

Global demand for energy has tripled in the past 50 years and might triple again in the next 30 years. The majority of this increased demand in the past was in the industrialized countries, and 90% of this demand was met with fossil fuels. However, in the years ahead, most of the increased energy demand will come from the developing countries as they seek to meet development goals and as they experience population increases much larger than the industrialized

Morning rush hour, Canton, China.
(Terry Qing/FPG International, LLC)

countries. It is projected that energy consumption in the industrialized countries will rise by only 1% per year in the next several decades, while the growth rate in energy consumption in the developing countries will be about 4% per year. If such projections come to pass, the developing countries will be using more energy than the industrialized countries by the year 2020. Figure 1.3 shows such projections of energy use to the year 2020. Also shown is a more detailed breakdown of world energy consumption by region for 1996.

The United States, with only 4.6% of the world's population, consumes about one fourth of the energy used in the world today (Fig. 1.4). We have the dubious distinction of having one of the highest per capita rates of energy consumption of any country, the equivalent of using 7 gallons of oil (or about 70 pounds of coal) per person per day. This is about five times the world's average! If the developing countries were to increase their per capita rates to that of the developed world, world energy consumption would increase threefold.

The principal **sources of energy** used in the United States and the world are illustrated in Figure 1.5. Note that about 85% of our energy comes from fossil fuels. (For the world, if traditional noncommercial fuels are included—such as wood and dung—then renewable energy accounts for about 20% of the world's total.) The fuel mix has certainly changed over the years. Originally, humans added to their own muscles by using animals, water, and wind to do work. Preindustrial society drew only on **renewable** forms of energy, that is, those sources that cannot be used up, such as water, wind, solar,

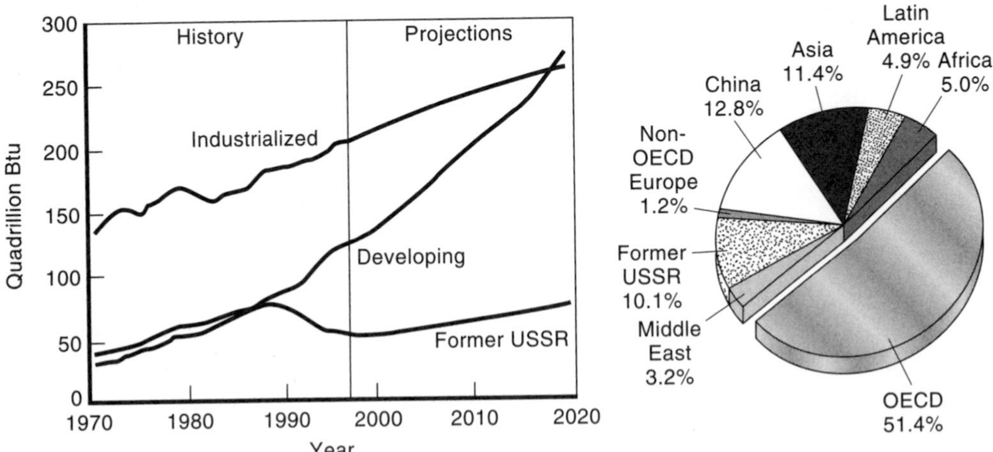

FIGURE 1.3
World energy consumption, 1970–2020 for industrialized countries, developing countries, and Eastern Europe/Former Soviet Union (EE/FSU). Also shown are regional shares of total final consumption for 1996. (OECD is the Organization for Economic Cooperation and Development.) (UNITED STATES ENERGY INFORMATION ADMINISTRATION, USEIA)

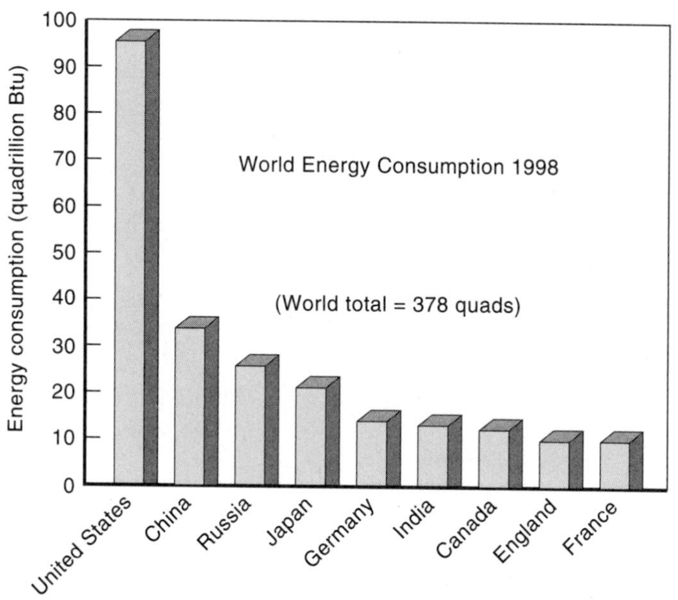

FIGURE 1.4

World energy consumption by country: 1998. (UNITED STATES ENERGY INFORMATION ADMINISTRATION, USEIA)

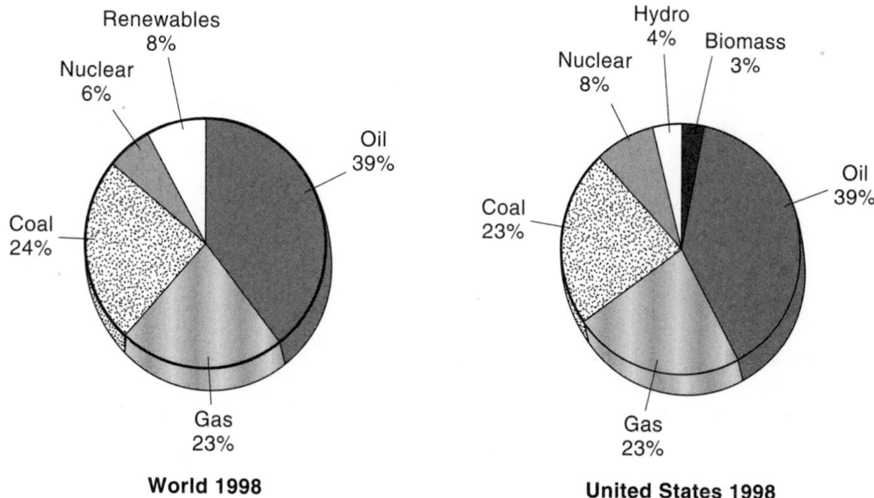

FIGURE 1.5

Energy consumption by source for the world and for the United States: 1998. (UNITED STATES ENERGY INFORMATION ADMINISTRATION, USEIA)

Oil fields in Texas in the 1920s. (AMERICAN PETROLEUM INSTITUTE)

and biomass. A shift to **nonrenewable** resources began in the 18th century, as an increasingly industrialized society started to burn fossil fuels to make steam for steam engines (invented in 1763) and to smelt iron.

The first modern oil well was drilled in Pennsylvania in 1859, and oil found increased use after the invention of the internal combustion engine in the 1870s. As both the number of engines and the availability of petroleum increased, the contribution of oil rose rapidly after 1920. Its relatively clean-burning features were desirable for environmental reasons. Eventually, coal was replaced by oil in industries and power utilities. Today oil accounts for about 40% of the U.S. and the world's fuel consumption.

The use of natural gas in the United States was small and localized until the discovery of large deposits in Texas and Louisiana and the construction of a network of long-distance pipelines to the north. Today natural gas accounts for 23% of the U.S. energy consumption, primarily for home heating and industrial operations. Because of increased discoveries and electricity deregulation, the percentage contribution by natural gas to the total energy economy in the United States and the world has been rapidly increasing.

In the history of humanity, the fossil-fuel age will be a small interval of time. Figure 1.6 shows the percentage contribution of each major energy resource in the United States over the last century. Note the large decrease in the percentage contributions from wood and coal and the rapid increase in oil and natural gas since World War II. Until the 1940s, the United States produced nearly all the oil it needed. However, increasing energy demand and declining production forced the United States to import petroleum beginning in the late 1950s. Production reached its highest level in 1970 (at 11 million barrels per day, abbreviated MBPD). Production was augmented in the late 1970s by oil from Alaska, but this resource started to fall in 1988. Today, the total U.S. output has dipped to less than 8 MBPD. Figure 1.7 shows petroleum production and con-

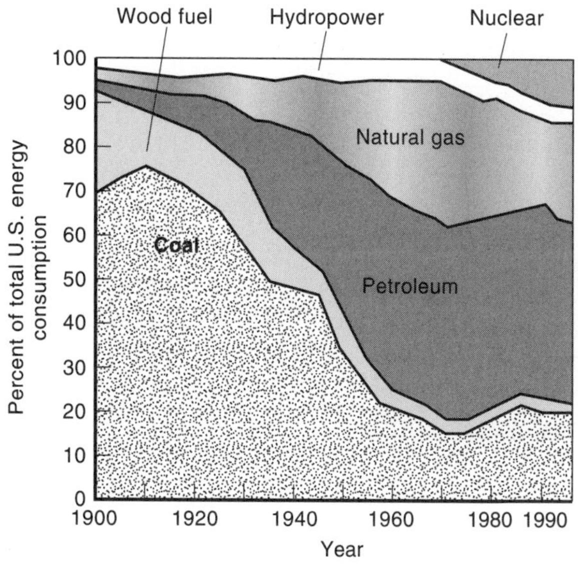

FIGURE 1.6
United States energy consumption by fuel share for the last century. (UNITED STATES ENERGY INFORMATION ADMINISTRATION, USEIA)

sumption for the last half of the 20th century. After 1992, imports exceeded production—doubling between 1985 and 1997. The cost of these imports is about $60 billion per year. The five leading suppliers of petroleum to the United States in 1999 were Venezuela, Canada, Saudi Arabia, Mexico, and Nigeria.

Renewable energy sources include hydroelectric, biomass (wood and wood products), wind, photovoltaics, and radiant solar energy for heating, cooling, and the production of electricity. Although they contributed less than 10%

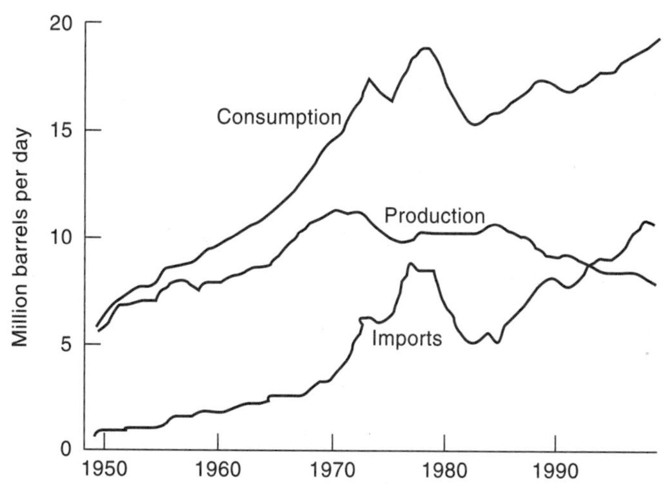

FIGURE 1.7
United States petroleum production and imports: 1949–1999. (Petroleum includes crude oil and natural gas plant liquids.) (UNITED STATES ENERGY INFORMATION ADMINISTRATION, USEIA)

A 300-kW photovoltaic power plant at Gun Hill Bus Depot in New York City. Solar cells supplement the terminal's electrical energy needs. (NEW YORK POWER AUTHORITY)

toward meeting the total U.S. energy demand, some of these technologies are growing rapidly. Wind energy, in particular, is the world's fastest growing energy source. Although presently supplying but 0.2% of our total energy, its rate of growth is about 10% per year, and an astounding 37% per year in Europe. Denmark currently supplies 8% of its electricity using wind turbines.

Recall from the beginning of this chapter that energy is not an end in itself, but is valued for what can be done with it. Consequently, it is important to examine where energy is used. The **end uses of energy** are traditionally broken down into four sectors: transportation, industrial, residential (single and multifamily dwellings), and commercial (offices, stores, schools, etc.). Figure 1.8 shows these uses in the United States in 1998. Figure 1.9 illustrates

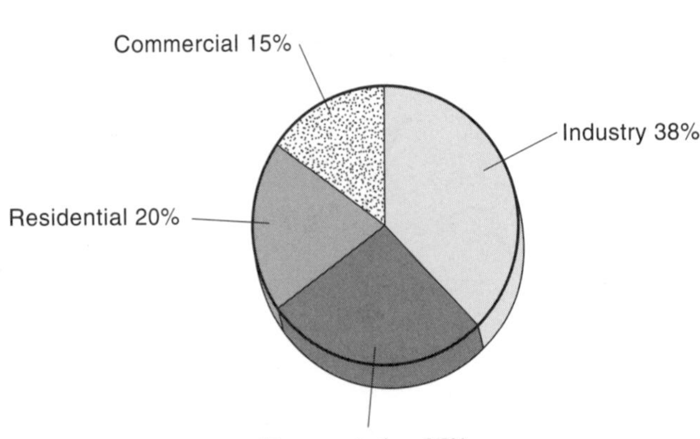

FIGURE 1.8
United States end uses of energy by sector: 1998. (UNITED STATES ENERGY INFORMATION ADMINISTRATION, USEIA)

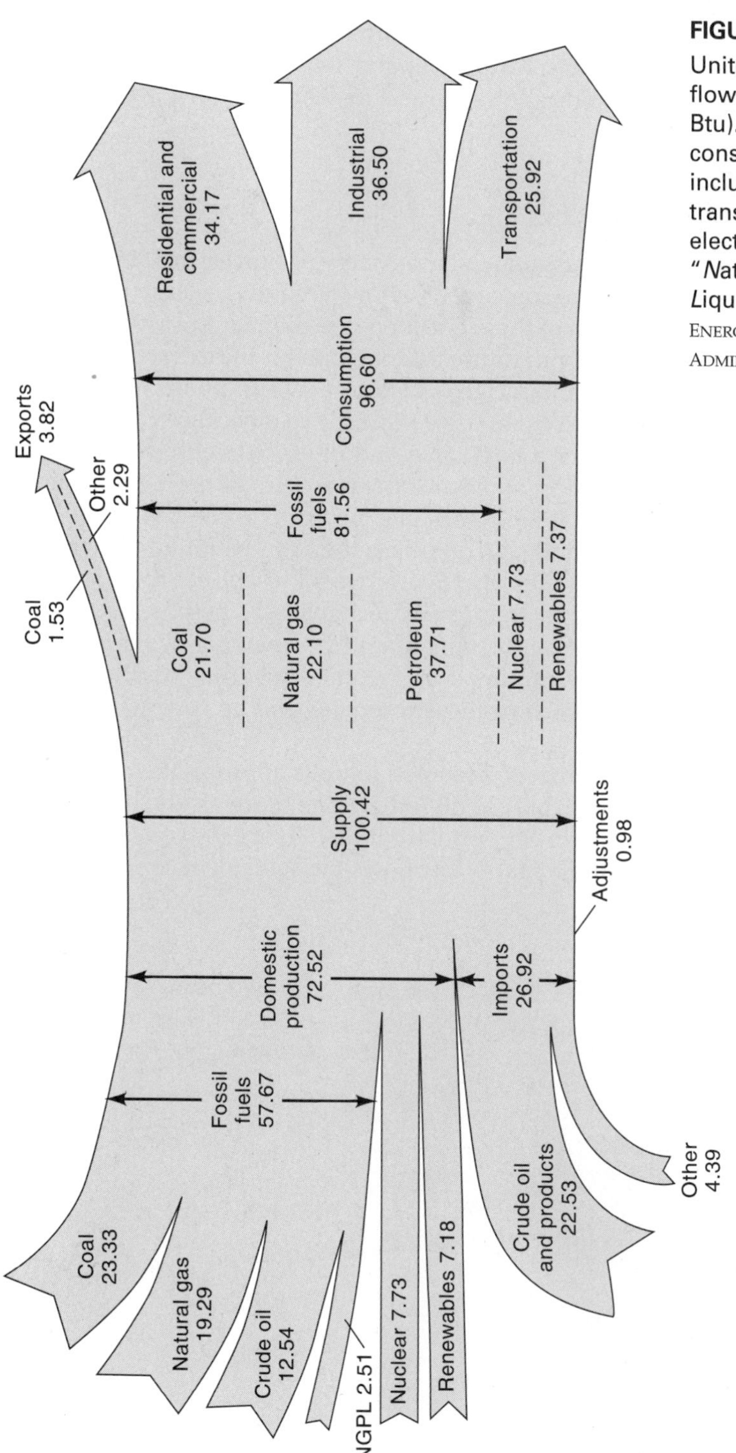

FIGURE 1.9

United States total energy flow in 1999 (Quadrillions of Btu). Total energy consumed—96.6 Quads—includes conversion and transmission losses of electric utilities. (*NGPL* = "*N*atural *G*as *P*lant *L*iquids.) (UNITED STATES ENERGY INFORMATION ADMINISTRATION, USEIA)

the complexity of the flow of energy from source to end use. On the left side of the figure are the energy inputs, by amount and source, including oil and natural gas imports. The right side shows the sectors that consume the energy.

D. Energy Resources

To understand energy, one must understand energy resources, their limitations, and their uses. One must have some idea of how large each energy resource is and how long it will last. Both of these questions are difficult to answer because assumptions must be made about future technologies for the extraction of these resources, future fuel prices, and the growth rate of consumption.

Estimates of fossil-fuel resources are easiest for coal because coal deposits occur in extensive seams over a large area and often crop out on the earth's surface. Estimates of petroleum and natural gas resources are more difficult because deposits occur at scattered sites and lie underground at depths between several hundred feet and several miles; they can be found only by exploration. Table 1.1 lists estimates of U.S. and world fossil-fuel resources that can be recovered profitably with present technology. These resources are called **reserves.** Reserves are not a static quantity—they are being added to every year through discovery and improved methods of economically extracting the particular resource. Each of these resources will be covered in a later chapter.

Each type of energy resource is measured in units appropriate to its physical form: tons of coal, barrels (bbl) of oil (where one barrel is equal to 42 U.S. gallons), and trillion cubic feet (tcf) of natural gas. To enable you to compare apples and oranges, so to speak, Table 1.1 shows the equivalent of each reserve

Caribou near the Trans-Alaska Pipeline in Alaska. (AMERICAN PETROLEUM INSTITUTE)

Table 1.1 WORLD AND UNITED STATES PROVEN RESERVES: 1998

Resource	World	United States	Lifetime*
Oil	1020×10^9 bbl 5.9×10^{18} Btu	21×10^9 bbl 0.11×10^{18} Btu	8 yr
Natural gas	5090×10^{12} ft^3 5×10^{18} Btu	165×10^{12} ft^3 0.17×10^{18} Btu	9 yr
Coal	1.09×10^{12} tons 27×10^{18} Btu	0.58×10^{12} tons 14×10^{18} Btu	500 yr
Tar sands	300×10^9 bbl 1.7×10^{18} Btu	22×10^9 bbl 0.12×10^{18} Btu	8 yr

*Ratio of U.S. reserves to 1998 U.S. production rate

Source: U.S. Energy Information Administration

in a common energy unit, the British thermal unit (Btu). This unit is defined as the amount of energy needed to raise the temperature of 1 lb of water by 1°F. (1 Btu is approximately the energy released by burning a wooden kitchen match.)

EXAMPLE

U.S. oil reserves are estimated at 21 billion bbl, and we currently produce about 8 MBPD. How long will these reserves last at this production rate?

Solution

The yearly production is

$$8{,}000{,}000 \text{ bbl/d} \times 365 \text{ d/yr} = 2{,}920{,}000{,}000 \text{ bbl/yr}$$

The lifetime will be

$$\frac{21{,}000{,}000{,}000 \text{ bbl}}{2{,}920{,}000{,}000 \text{ bbl/yr}} = 7.2 \text{ yr}$$

To Americans, most of all, it is still difficult to understand that we are running short of the fuels that propelled the United States into the position of global economic leadership that it occupies. The nation progressed by recognizing no limits, by making the most of our citizens' ingenuity, and by taking chances. The economy was built on a price of \$3 per barrel of oil. That is no

Focus On 1.2

ENERGY IN CHINA

Although about 20% of the world's people live in China, the Chinese accounted for less than 10% of the world's total energy consumption in 1997. The per capita energy consumption was less than one tenth that of the United States and one third the global average. However, China's GDP grew by almost 8% per year in the 1990s. In 1982, 3% of Beijing's households had refrigerators. In 1995, 81% did (using three times as much electricity as U.S. models). Unlike the pattern in most Western countries, coal dominates the commercial energy resources of China, accounting for 71% of the country's energy consumption (Fig. 1.10). China is the world's largest producer and consumer of coal. However, the pictures for urban and rural energy resource consumption are quite different. Of China's one billion people, 80% live in rural areas and consume only 40% of the total energy. Of the rural consumption, 90% is from plant and animal sources (called biomass), and 4.5 million anaerobic digesters produce natural gas from animal waste for cooking and lighting.

Energy is becoming a key constraint in China's economic growth. It is estimated that 20% of potential industrial output is lost because of a shortage of electricity. Hydropower produces about 30% of China's electricity and is rapidly expanding, primarily through the construction of small-scale units. More than 100,000 hydroelectric plants have been built over the past 20 years. China is now building what will be the world's largest dam—the Three Gorges, along the

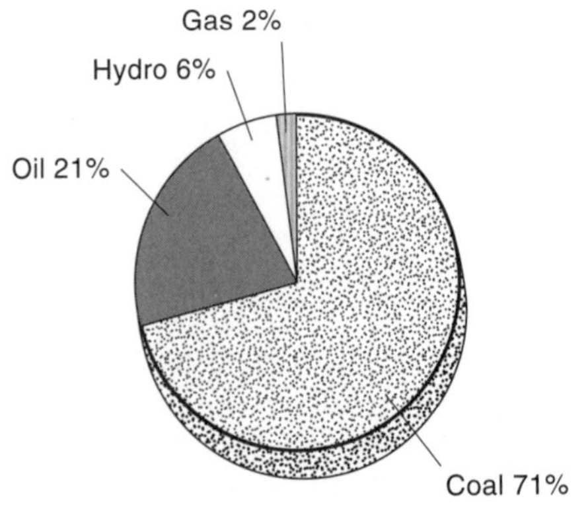

FIGURE 1.10
Energy resources used in China: 1997. (UNITED STATES ENERGY INFORMATION ADMINISTRATION, USEIA)

Yangtze River near Wuhan (see map). When completed in 2009, the dam will have an output of 18,600 MW. It is so large that many consider it to be the eighth wonder of the world. The dam will be 2 km across and create a reservoir 600 km long. A quarter of a million people will be displaced. China is the world's fifth largest oil producer, yet imports 25% of its needs.

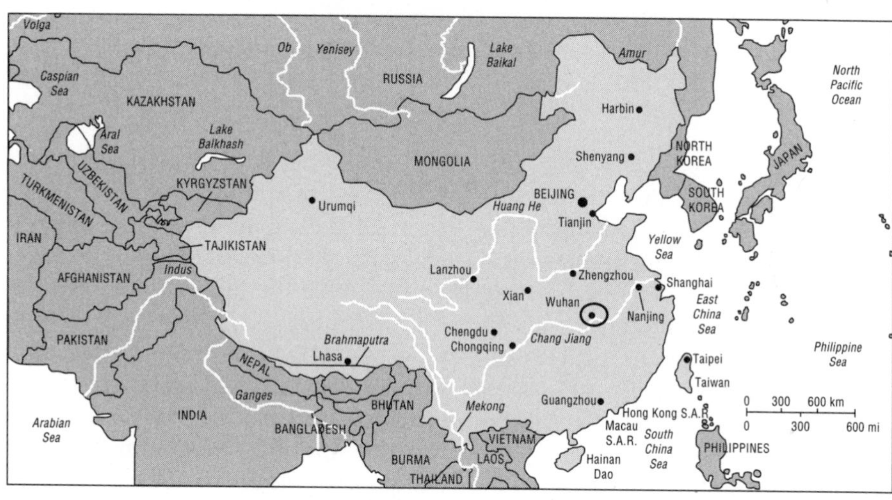

longer the case. To remain strong economically, we must acknowledge the limits of our resources. Failure to recognize this finiteness is certainly one element responsible for the energy crises of the past. The question of resource depletion is addressed in the next section.

E. Exponential Growth and Resource Depletion

An important factor in estimating the lifetimes of energy resources is the **growth rate** of consumption. Earlier figures in this chapter displayed such data. For example, between 1960 and 1970, U.S. energy consumption grew at an average rate of 4.5% per year. It is useless to state the lifetime of a resource if nothing is said about how fast the use of that resource is increasing (or decreasing).

Exponential growth illustration in southern India.

There are many types of growth, but one that is of particular interest to us is **exponential growth.** If a quantity is growing at the same percentage rate each year, then we say its growth is exponential. Said another way, a quantity that is growing exponentially will always increase in size by the same factor in a given period of time—the time required to double the quantity will be the same regardless of the starting amount.

Consider the growth of money, say \$1000, in a savings account that is earning a simple 10% annual interest rate. Table 1.2 shows the amount of money in the bank at the end of each year, assuming that no withdrawals are made. Each year the amount grows by 10% of the amount that is in the account at the beginning of the year. By the end of the seventh year, the \$1000 investment has grown to \$1948, or almost double. By the 14th year, this amount has almost doubled again, to about \$3800. In the 22nd year, \$8000 is in the bank, double the amount available 7 years earlier. This quantity is growing exponentially because the amount in the bank is increasing at a fixed percentage rate, and the time required to double your money is constant—about 7 years.

A useful approximate relationship between the doubling time (in years) and the percentage growth rate is

$$\text{Doubling time} \approx \frac{70 \text{ yr}}{\% \text{ growth rate}}$$

If we had a growth rate of 7% per year for electrical energy, the amount of electrical energy consumed would double in about $70/7 = 10$ years. In other words, the number of electrical power plants needed would double in 10 years, and *quadruple* in 20 years. To specify the lifetime of a resource, you must also specify the expected rate of growth in its use. At a zero growth rate for coal production, U.S. coal resources will last about 500 years. However, if the growth rate for coal production were 5% per year, this lifetime would drop to less than 70 years!

Clearly, the use of a particular resource will not continue to grow exponentially until we have exhausted that fuel, and then suddenly stop. The pattern of

Table 1.2 MONEY IN THE BANK—AN EXAMPLE OF 10% ANNUAL EXPONENTIAL GROWTH

Year End	Amount	Year End	Amount
0	$1000	12	$3138
1	1100	13	3453
2	1210	14	3798
3	1331	15	4178
4	1464	16	4596
5	1610	17	5056
6	1771	18	5562
7	1948	19	6118
8	2143	20	6730
9	2357	21	7403
10	2594	22	8143
11	2854		

growth and decline in resource use has been analyzed by M. K. Hubbert of the U.S. Geological Survey. In general, the use or exploitation of a resource shows an initial period of increase. As the high-quality deposits run out, production reaches a maximum and then declines, eventually going to zero at the exhaustion of the resource. The production curve will be bell-shaped, as depicted in Figures 1.11, 1.12, and 1.13. As a resource begins to be depleted, discovery and production become more difficult, prices rise, and other resources begin to take the place of the original fuel. If one graphs the annual production versus time, the total area under the curve represents the total amount of the resource known to be recoverable. The amount used so far is the area under the curve up to the present year.

These bell-shaped production curves allow an estimate of the time until the complete depletion of a resource; they also provide an estimate of when maximum production will occur. Figure 1.11 shows a curve for the production of coal in the world. The graph implies that coal resources are large enough to last more than 500 years, and that the peak in production will not occur for almost 200 years. The situation is considerably different for oil and natural gas.

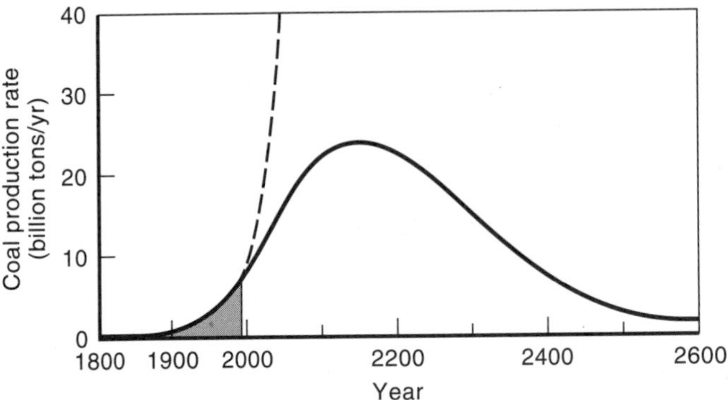

FIGURE 1.11
World coal production cycle. The probable exploitation of a fossil fuel (coal in this case) can be characterized by the solid curve. Production initially increases exponentially (as shown by the dashed line), but its rate of increase eventually decreases. Production then declines as extraction becomes more difficult and the rate of discovery decreases. Knowing the amount of fuel initially present, we can use this pattern to determine the lifetime of a resource; in this example, the lifetime of coal reserves is 400 to 600 years. (The amount of coal used so far is shown by the shaded area.) (CURVES BY M. K. HUBBERT, U.S. GEOLOGICAL SURVEY. ADAPTED FROM *AMERICAN JOURNAL OF PHYSICS*, NOVEMBER 1981)

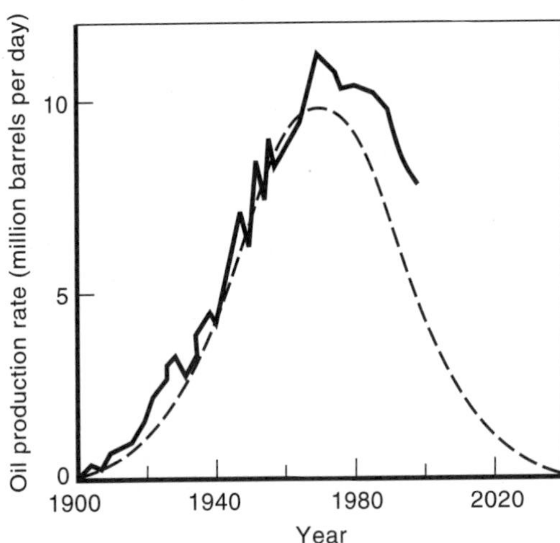

FIGURE 1.12
United States oil production. Comparison of estimated (Hubbert) production curve *(dashed line)* and actual production *(solid line)*.

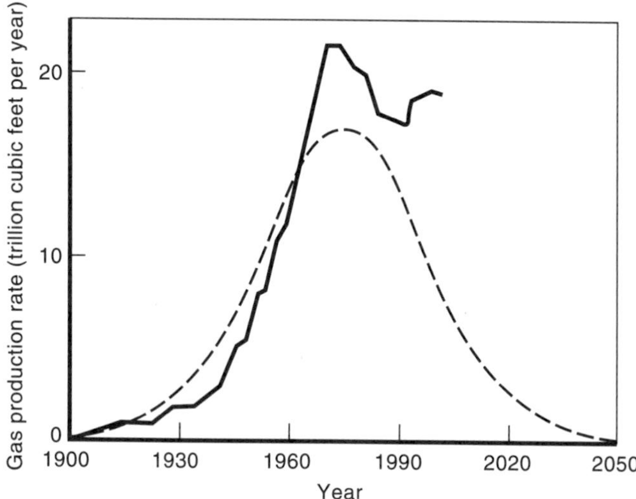

FIGURE 1.13
United States natural gas production. Comparison of estimated (Hubbert) production curve *(dashed line)* and actual production *(solid line)*.

Figure 1.12 shows U.S. production for oil. It suggests that within 20 years the production rate of oil in the United States will be one third what it is today. It also indicates that the peak of U.S. oil production should have occurred about 1970, and this indeed did happen. The same conclusions can be drawn from Figure 1.13 for natural gas; U.S. gas production peaked in 1973. The production rate for natural gas has not fallen off as fast as the Hubbert curve would indicate. Advanced drilling techniques, offshore deposits, and increased demand from electric utilities and industry have pushed natural gas production up from these predictions. However, consumption outpaces production, and imports have been rising steadily to where they now account for almost one fifth of natural gas used.

F. Oil: A Critical Resource

Oil has fueled most of the increase in global energy consumption since World War II. In 1950, oil accounted for less than one third of world energy use, and today it accounts for almost half. The low cost of oil and its adaptability to many uses—from space heating to transportation to electric energy production—made it the fuel of choice for an expanding economy. The rapid growth rate of U.S. oil consumption, about 5% per year, is reflected in Figure 1.2.

The last three decades have been extremely volatile for the world energy picture and for world economics, and an examination of oil prices over time reflects

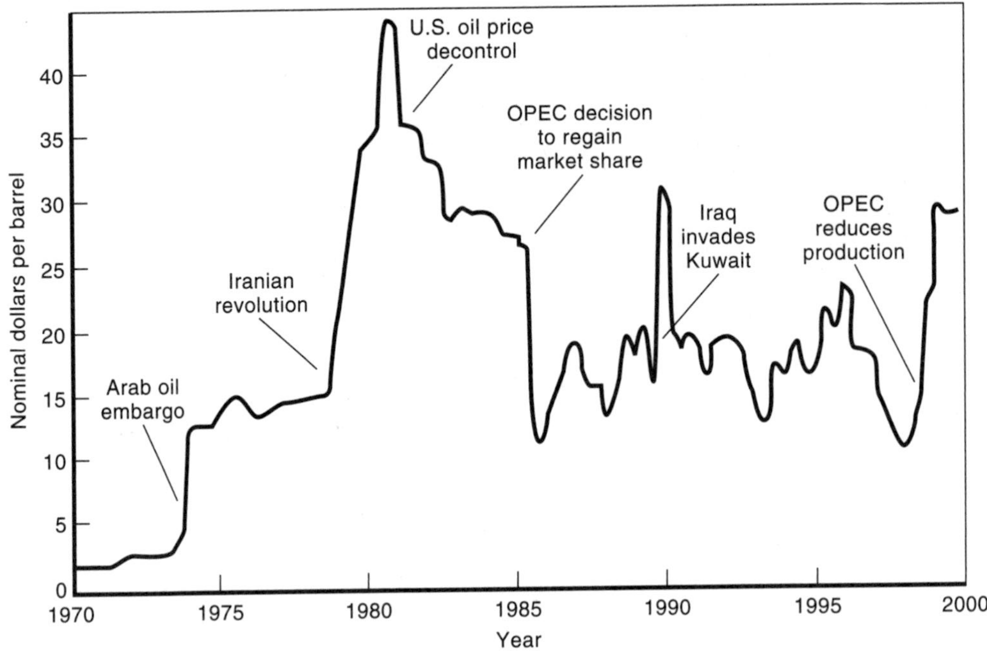

FIGURE 1.14
World oil prices: 1970–2000. Oil prices reflect international events. (UNITED STATES ENERGY INFORMATION ADMINISTRATION, USEIA)

these international events (Fig. 1.14). In constant dollars, the real price of oil declined during the 1950s and 1960s, encouraging a rapid increase in its rate of use. During the early part of this expansion, most oil production was controlled by large multinational companies. However, producing countries pushed for more control over oil operations. A cartel of oil-producing states called the Organization of Petroleum Exporting Countries (OPEC)* was formed in 1960, and its influence increased because of political changes and an increasing worldwide demand for oil. As the OPEC countries increased their market share of oil sales in the early 1970s, they began to set their own prices for their exports and to take control of oil away from foreign companies. Several events in the 1970s and early 1980s brought about a series of sudden increases in oil prices, which tended to remain in effect long after the political situations changed.

1. At the outbreak of the Arab-Israeli war in October 1973, the Arab member countries of OPEC imposed an oil embargo against selected Western countries, including the United States, and cut back

*Member states of OPEC are Algeria, Indonesia, Iran, Iraq, Kuwait, Libya, Nigeria, Qatar, Saudi Arabia, Venezuela, and the United Arab Emirates.

production. This supply disruption caused prices in the world market to triple, from about $8 per barrel to more than $25 per barrel (in 1985 dollars).

2. The Iranian revolution in 1978 and 1979 disrupted that country's production of almost 6 million barrels per day. Even though other countries stepped up production and took up some of the slack, the net effect was a loss in the world oil market of about 2 MBPD. During these events, prices doubled from about $22 per barrel to $44 per barrel.
3. The response of the world energy economy to high oil prices was to reduce energy consumption, use energy more efficiently, and develop alternative energy resources. In the United States, President Ronald Reagan decontrolled oil prices in 1981. Domestic production increased and the drilling rate reached all-time records. As a result of

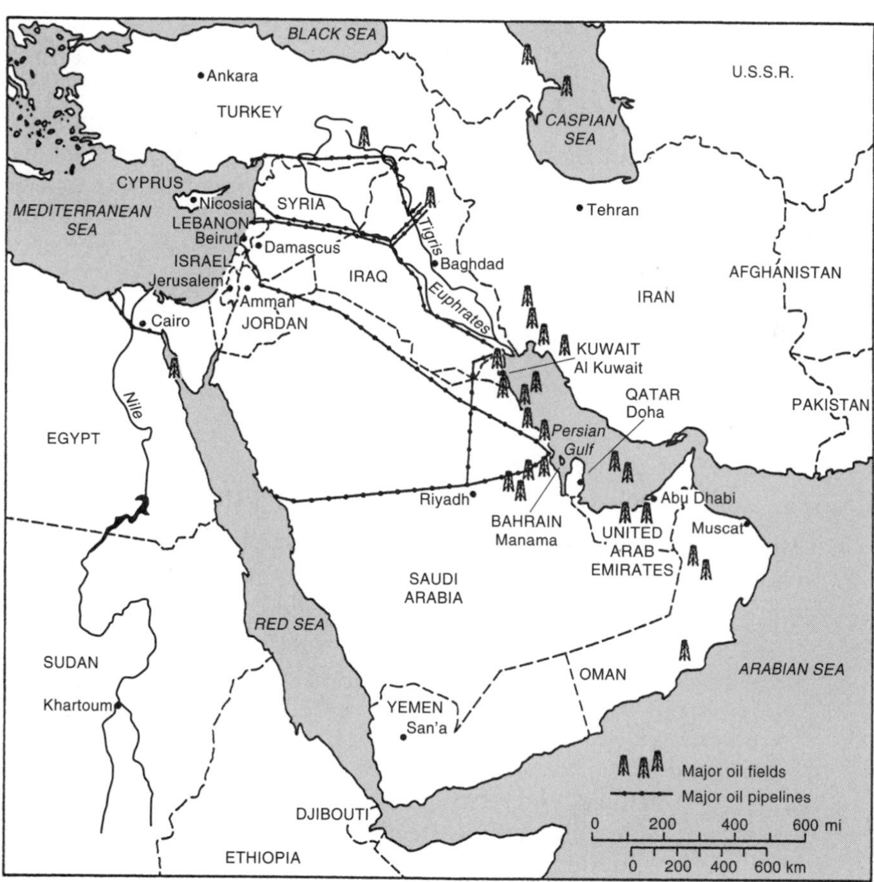

A map of the Middle East.

these market responses to higher oil prices, the world's dependence on OPEC declined from about 28 MBPD in 1980 to about 17 MBPD in 1985. World oil consumption fell by 14% during that period.

4. Oil prices began a drop in 1981. In 1986, prices dropped sharply by almost a factor of three as OPEC tried to regain its share of a shrinking world oil market by increasing production and lowering prices. In less than a year Saudi Arabia tripled its production rate to almost 6 MBPD.
5. Iraq's invasion of Kuwait in August 1990 prompted a sudden increase in the price of oil on the world market, and it reached its highest point in eight years. As other countries such as Saudi Arabia began to replace Kuwait's production, prices dropped again. The Allied liberation of Kuwait in January 1991 caused another sharp drop in prices.
6. Worldwide, oil imports are increasing, setting the stage for future energy crises. World oil prices are difficult to predict. While oil prices in 1994 were at their lowest point since 1988 because of an overabundance of oil on the world market, prices at the beginning of the 21st century rose to their highest since 1990 (at over $30/bbl) as OPEC cut back production and most countries were experiencing increased demand. High gasoline prices in 2000 might have aggravated many drivers but did not seem to dampen driving habits in a robust economy. In the years ahead, most growth in demand will probably come from Eastern Europe and China, while most growth in supply will be from Saudi Arabia, Kuwait, and the United Arab Emirates.

G. Energy Conservation

The total energy consumed in any activity can be thought of as the product of two factors:

Total energy consumption =
energy required for the activity (intensity) × frequency of activity

The factor we call intensity of use is the amount of energy required to do the task once; the level of activity is the number of times the task is done—the frequency. For instance, if your car uses one gallon of gasoline to go between your home and your job (the activity) and you drive ten times per week (the frequency), the energy consumption for this activity is 10 gallons per week.

We can represent these two factors in a graph (Fig. 1.15) as quantities along the x and y axes. Their product, the total energy consumption for that activity, is represented by the area of the rectangle. The figure shows two rectangles, both with the same area representing the same amount of total energy consumed. For rectangle (a), a high activity frequency was possible because the intensity of use (energy required for the activity) was low. In rectangle (b), the same amount of energy was consumed but with a greater intensity (more energy required per activity), so there needed to be a reduced frequency for that activity.

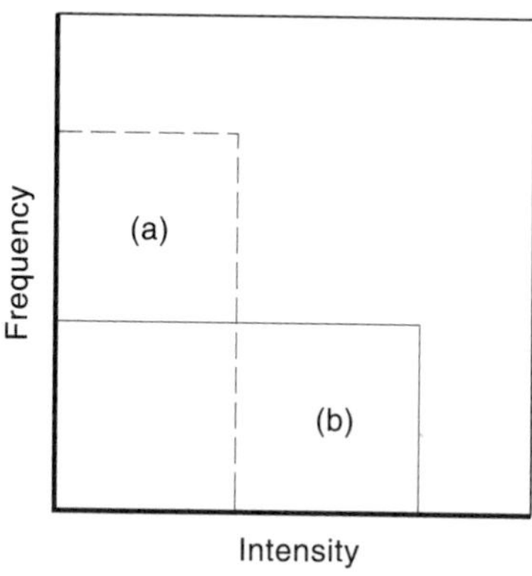

FIGURE 1.15
Characterization of total energy use as a function of intensity of use and frequency of activity.

Efforts of **energy conservation** usually concentrate on one or the other of these factors. In the context of Figure 1.15, energy conservation strives to reduce the size of the rectangle representing the total energy used. The two approaches are

1. The "technical fix"; this consists of using fuel more efficiently to perform the same task, such as driving a car with a more efficient engine (reducing the energy required for that activity).

2. The "lifestyle change"; this means consciously using less fuel by such behaviors as turning down the thermostat or driving fewer miles (thereby reducing the frequency of the activity).

The maximum possible success of technical fixes for energy conservation is limited by the laws of physics (the first and second laws of thermodynamics, discussed in Chapter 4). However, there is still a lot of room for improvement in this approach for energy conservation, especially in the efficiency of energy use for particular tasks. For example, a 20-watt fluorescent bulb gives the same light output as a 75-watt incandescent bulb and lasts ten times as long. The initial cost of a fluorescent bulb is higher, but the savings in electricity costs during average use over one year will pay back the investment. If we replaced incandescent bulbs with fluorescent ones, fewer electric power plants would be needed. The investment made in constructing an industry to build energy-efficient light bulbs will be much less than that needed to build an electrical power plant. This type of economics is of prime importance in developing countries.

In energy conservation, the issues are more than just technological, because energy consumption also depends on the "frequency of the activity." There are many barriers to adopting the measures we will discuss in this book, such as market restraints (e.g., the initial cost of home insulation). There is also a general reluctance to move toward what are envisioned as "lifestyle changes" (such as changes in comfort control or in preferences for certain materials).

Many people assert that energy prices should reflect what it will cost to *replace* the dwindling supplies of nonrenewable fuels such as oil and natural gas, rather than just what it costs to obtain them. Societies will not switch to renewable energy technologies and to more efficient equipment if fossil fuels are priced as if they were almost free. One of the main forces behind our per capita reduction in energy use during the early 1980s was higher oil prices (Fig. 1.16). Until that time we had seen a steady rise in energy use per person. Between 1900 and 1980, per capita energy use rose from 80 million to 320 million Btu per year. Did the quality of life improve that much? Do you believe that you are *four times* better off than your great grandparents? The per capita use of electricity was 6 times higher in 1998 than 1950.

Increased emphasis on energy conservation is based on some convincing arguments:

1. Conservation technologies are cost-effective alternatives to the development of additional supply technologies. That is, in most cases it will cost less to save a barrel of oil than to develop a new barrel of an oil substitute. "Investment in energy conservation provides a better return than investment in energy supply," stated the International Energy Agency in 1987.
2. Conservation will stretch the earth's limited energy resources, not only for the United States but for other countries as well. Today more than half of the less-developed countries rely on imported oil for 75% or more of their commercial energy needs. Conservation will gain time for the possible development of inexhaustible resources such as solar energy and nuclear fusion.
3. Conservation will reduce the pollution of our environment. If we use less energy, there will be less air and water pollution, less radiation and thermal pollution, and less global warming and acid rain.
4. Conservation technologies can be put to use more quickly than we can increase supplies. It takes 2 to 4 years to open a new coal mine,

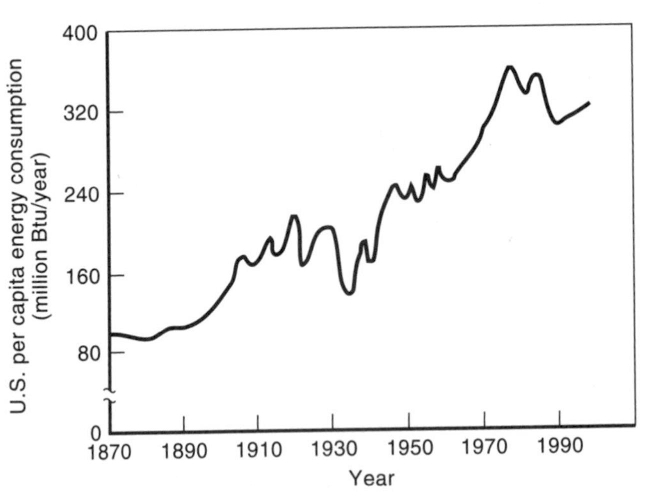

FIGURE 1.16
United States per capita energy consumption over the past 130 years. (UNITED STATES ENERGY INFORMATION ADMINISTRATION, USEIA)

2 to 3 years to build a gas-turbine power plant, 5 to 7 years to build a coal-fired electric generating plant, and 9 to 11 years to construct a nuclear power plant. Many conservation practices can begin immediately because the technology is available and simple, such as better insulation of buildings. The money needed for such energy-saving measures is less than that necessary for the capital-intensive supply technologies.

5. Conservation of fossil fuel resources is particularly crucial for the future, since their use as the raw materials for chemical industries (such as pharmaceuticals and plastics) is far more important than their use as fuels in power generation.
6. Conservation measures can be readily practiced in some way by each individual, with the incentive of saving money as well as energy. Such practices can also contribute to our own health; for example, bicycle riding provides more exercise than driving a car.

H. Economic and Environmental Considerations

The belief was strongly held some years ago that economic growth always meant increasing the amount of energy used. Since it takes energy to produce a given output, one might expect a constant relationship between the GDP and energy consumption. This relationship *was* steady until the early 1980s; then higher oil prices mandated energy conservation and increased efficiency, and this caused a significant decrease in per capita energy use (Fig. 1.17). From 1980 to 1998, energy consumption increased by only an average of 1.0%

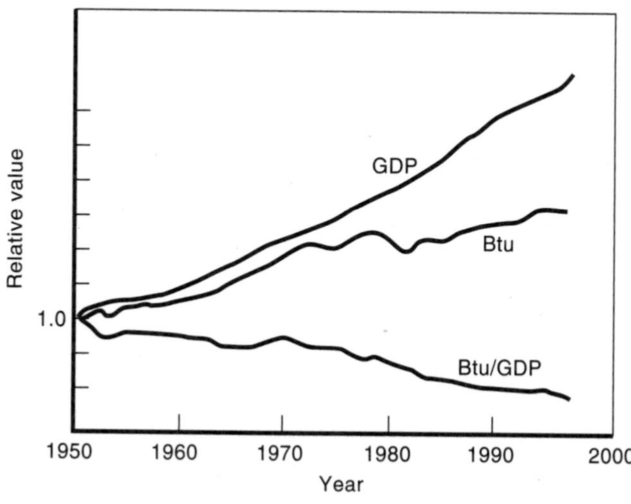

FIGURE 1.17
United States energy use (Btu) compared to GDP over time, and their ratio. (UNITED STATES ENERGY INFORMATION ADMINISTRATION, USEIA)

Focus On 1.3

THE KYOTO PROTOCOL ON CLIMATE CHANGE

In December 1997, 167 nations met in Japan, under the auspices of the United Nations, to forge what is known as the "Kyoto Protocol." This document is the first international attempt to place legally binding limits on greenhouse gas emissions from developed countries. The Protocol set as a goal to cut by 2008–2012 the combined emissions of greenhouse gases from developed countries by about 5% from their 1990 levels. However, the Kyoto Protocol does not set any binding limits on emissions from developing countries. The United States, which emits the most CO_2 of any country, and which also has the highest per capita CO_2 emissions, will need to meet a reduction goal of 7% less than 1990. The U.S. Department of Energy estimates that U.S. carbon emissions by 2010 will increase by 34% in the absence of any change in energy policy or consumer behavior.

Since about 83% of greenhouse gas emissions in 1990 were CO_2 released by the use of fossil fuels, actions taken to reduce these emissions are likely to have a significant impact on energy markets. Increases in energy prices may be required to meet these limits. It is expected that the cost of electricity generated by coal will rise and the share of electricity generated by natural gas and renewables will increase.

annually, while the GDP (in constant dollars) grew by 3.3% per year. A report by the Office of Technology Assessment stated that two thirds of this improvement in energy use was a result of conservation and increased efficiency, while the other third came from a shift toward a more service-oriented economy. Energy use has risen more rapidly since 1986, but the trend toward energy conservation continues as one can see by the continued decrease in the Btu/GDP ratio.

Energy policy should be concerned not only with finding new sources and reducing energy consumption, but also with weighing the effects of new technologies and energy-related lifestyles on our lives and on our planet. Energy policy should be shaped by an awareness of long-term constraints as well as short-term predicaments. What do we give up, and for what? Do we sacrifice the tundra in Alaska's Arctic National Wildlife Reserve in return for an additional ten years' supply of oil? Was the 1989 *Exxon Valdez* Alaskan oil spill an acceptable part of our effort to provide stable oil resources? Can we cope with increased emissions from fossil-fuel power plants and automobiles? Are potential radiation dangers too severe to continue the use of nuclear energy? Even though many of the following chapters will deal with the details of supply technologies, we will also cover environmental constraints and measures for more efficient fuel use.

Understanding energy use means also understanding the environmental consequences of its use. A major concern in the burning of fossil fuels is the possibility of large global climatic changes caused by increased levels of carbon dioxide and other greenhouse gases in the upper atmosphere. More than 5 billion tons of carbon are added to our atmosphere each year by fossil-fuel combustion. The average global temperature has already risen by ½°C since 1900 (see Fig. 1.1). Increased global temperatures could affect agricultural production, local temperatures, severe weather patterns, and sea level heights. Chapter 9 will focus on this topic in greater detail.

Acid rain caused by the emissions of coal-fired power plants harms trees, crops, and animals. About 20% of Europe's forests have been damaged by acid rain, while hundreds of lakes in the United States and Canada have become empty of fish. The effects of increased emissions of sulfur oxides, nitrogen oxides, and hydrocarbons have led to severe health problems around the world.

Nuclear power has its own set of environmental constraints, including the need for the permanent disposal of radioactive wastes and the assurance of safety during operation.

I. Future Scenarios

The energy situation today is dramatically different from that in the early 1970s, but this is not something about which to be complacent. Lower oil prices in the 1990s brought increased oil consumption and discouraged energy conservation and the development of alternative energy resources. However, the economic environment has changed in a way that may make it easier to handle future supply disruptions or shortages:

1. The United States depends less today on oil for its fuel mix and more on coal, natural gas, nuclear, and renewable technologies than we did a decade ago.
2. Oil production today is more dispersed among non-OPEC nations than it was in 1973 when 56% of the world supply was produced by OPEC members.
3. Thanks in part to the high oil prices of the 1970s, we have learned both to conserve energy in the residential and industrial sectors and

to build more energy-efficient machinery. The fuel efficiency of new cars today is 62% greater than in the middle 1970s (up from 17.5 miles per gallon to 28.5 mpg today). New refrigerators are 300% more energy efficient today than in 1973.

4. There is now a Strategic Petroleum Reserve that provides a backup oil supply of approximately 60 days in the event that all oil imports to the United States are cut off. It was used in 2000 to help lower rising fuel prices.
5. Renewable energy, all but unheard of in 1973 (except hydropower), has been growing steadily in both developed and developing countries.

However, future energy crises certainly can, and probably will, occur. We still have a limited resource base, especially of oil and natural gas. No act of Congress can increase our fossil-fuel reserves. The world's strong dependence on oil will continue to be a factor in limiting economic growth, especially in developing countries, and the oil supply will still be vulnerable to the political situation in the Middle East. Lower oil prices benefit the economy and the consumer in the short run, while cutting the federal deficit. However, lower oil prices provide less incentive to invest in energy-efficient equipment, discourage domestic exploratory drilling, and reduce research and development efforts on alternative technologies. Finally, the higher economic growth rate caused by low oil prices brings with it increased environmental pollution.

Oil prices are quite volatile. While oil prices dropped to $11 per barrel in the late 1990s, they were about $30 per barrel at the start of the year 2000. In December 1998, depressed oil prices were predicted to last for ten more years. Such uncertainties make forecasts of energy demand very difficult. Many other things can also change to alter predictions. These include new technologies, new laws and regulations, as well as economic growth or downturn. Figure 1.3 includes a prediction for world energy demand through 2020. World energy consumption is predicted to increase by 50% in this time period, with the largest increases coming in Asia and Central and South America. But experts have been wrong before.

Our growth as an industrialized society has been fueled by cheap, abundant resources. Progress was achieved by developments in science and technology, and energy resources were available to do the work. The situation today in our global marketplace is somewhat reversed in that resource availability will dictate our progress and our style of living much more than it has in the past. The time scale for change will be much shorter—decades rather than centuries. We have left the era of cheap energy, and we will have to make lifestyle changes regardless of what path we take.

Although this book emphasizes energy use and energy technology, one must keep in mind that *energy is just a means to an end.* Human conditions and values can be damaged as much by having too much energy, too soon, as by having too little, too late.

Focus On 1.4

THE GREEN GAMES 2000

The Sydney 2000 Olympic Games boasted many renewable energy projects that set environmental standards for the future. The Olympic Village, dubbed the world's largest solar suburb, used rooftop-mounted photovoltaic panels. Energy-efficient buildings and passive cooling designs (air circulation without the use of fans) cut energy use by 50%. Electricity from renewable energy sources provided all the electricity for the Superdome. Many of the buses used for spectator transport were powered by compressed natural gas. The pace car for the marathon was run with a fuel cell powered by liquefied hydrogen. With these games, the environment has become the third pillar of the Olympics, along with sport and culture.

Sydney Opera House (DONALD MIRALLE/LIAISOM AGENCY)

Internet References

For an up-to-date list of Internet resources related to the material in this chapter, go to the Harcourt College Publishers website at **http://www.harcourtcollege.com.** The links are in the *Energy: Its Use and the Environment* site on the Physics page. General energy related sites and some guidelines for using the World Wide Web in your class are on the inside front cover of this book.

References

Chapter 1

Ausubel, J., and H. Sladovich, eds. 1989. *Technology and Environment.* Washington, D.C., National Academy Press.

Bates, R. W. 1993. The Impact of Economic Policy on Energy and the Environment in Developing Countries. *Annual Review of Energy,* 18.

Brown, L., et al. Annual. *State of the World.* New York, W. W. Norton.

Brown, L., et al. Annual. *Vital Signs.* New York, W. W. Norton.

Cassedy, E., and P. Grossman. 1990. *An Introduction to Energy: Resources, Technology, and Society.* Cambridge, Cambridge University Press.

Commoner, B. 1977. *The Poverty of Power: Energy and the Economic Crisis.* New York, Bantam.

Commoner, B. 1992. *Making Peace with the Planet.* New York, W. W. Norton.

Darmstadter, J., and R. W. Fri. 1992. Interconnections Between Energy and the Environment. *Annual Review of Energy,* 17.

Economic Development. 1980. *Scientific American,* 243 (September).

Energy. 1981. *National Geographic* (February).

Energy for Planet Earth. 1990. *Scientific American,* 263 (September).

Fowler, J. 1975. *Energy and the Environment.* New York, McGraw-Hill.

Freeman, D. 1974. *Energy: The New Era.* New York, Vintage.

Harrison, P. 1993. *Inside the Third World: The Anatomy of Poverty.* London, Penguin.

Helm, J. L., ed. 1990. *Energy: Production, Consumption, and Consequences.* Washington, D.C., National Academy Press.

Houghton, R. A., and G. M. Woodwell. 1989. Global Climatic Change. *Scientific American* (April).

Hubbert, M. K. 1969. Energy Resources. *Resources and Man.* San Francisco, W. H. Freeman.

Hubbert, M. K. 1971. Energy Sources of the Earth. *Scientific American,* 224 (September).

Levine, M. D., F. Liu, and J. E. Sinton. 1992. China's Energy System. *Annual Review of Energy,* 17.

Lovins, A. 1977. *Soft Energy Paths.* New York, Harper.

Murota, Y., and Y. Yano. 1993. Japan's Policy on Energy and the Environment. *Annual Review of Energy,* 18.

1990. Managing Planet Earth [Readings from *Scientific American*]. New York, W. H. Freeman.

Ramachandran, A., and J. Gururaja. 1977. Perspectives on Energy in India. *Annual Review of Energy,* 2.

Ross, M., and R. Williams. 1981. *Our Energy: Regaining Control.* New York, McGraw-Hill.

Sathaye, J., A. Ghirardi, and L. Schipper. 1987. Energy Demand in Developing Countries: A Sectoral Analysis of Recent Trends. *Annual Review of Energy,* 12.

Sathaye, J., and S. Tyler. 1991. Transitions in Household Energy Use in Urban China, India, the Philippines, Thailand, and Hong Kong. *Annual Review of Energy,* 16.

Schipper, L. 1976. Raising the Productivity of Energy Utilization. *Annual Review of Energy,* 1.

Schipper, L., R. B. Howarth, and H. Geller. 1990. U.S. Energy Use from 1973 to 1987: The Impacts of Improved Efficiency. *Annual Review of Energy,* 15.

Schramm, G., and J. Warford, eds. 1989. *Environmental Management and Economic Development.* Washington, D.C., World Bank.

Stobaugh, R., and D. Yergin, eds. 1979. *Energy Future—Report of the Energy Project at Harvard Business School.* New York, Ballantine.

U.S. Department of Energy. *Annual Energy Review.* Washington, D.C., U.S. Government Printing Office.

World Resources Institute, United Nations Environment Program, The World Bank. 1992, 1994, 1996. *World Resources.* New York, Oxford University Press.

Yergin, D. 1993. *The Prize: Epic Quest for Oil, Money, and Power.* New York, Simon & Schuster.

QUESTIONS

1. Identify the principle energy sources used worldwide and classify them as either renewable or nonrenewable.
2. What energy source has seen the most rapid growth in the past 50 years and why?
3. If the world use of oil is about 66 MBPD, how long would you expect this resource to last at this consumption rate? (See Table 1.1.)
4. (a) What is exponential growth?
   (b) Today the United States has the equivalent of 400 standard-sized 1000 MW power plants. If electrical energy consumption continues to rise at the present rate of 2% per year, how many additional power plants will be needed in 35 years to meet those needs?
5. If the world's population is increasing at an annual rate of 1.3%, and there were 5 billion people in the year 1986, then in what year will the world's population be 10 billion?
6. List reasons why the U.S. per capita consumption of energy over the past three generations has risen by almost a factor of four.
7. Discuss the statement that the price of energy should reflect the true cost of replacing it.
8. The amount of energy used depends on what two factors? Give some additional examples of each.
9. What is meant by the statement that energy is but a means to an end, not the end itself.
10. What has been the impact of the 1991 Persian Gulf War on the availability of oil worldwide?
11. Find the price of oil today in terms of dollars per barrel.
12. Why is the energy use per capita in the world rising?
13. Some of the advantages of energy conservation were given in this chapter. What are some disadvantages?
14. Should environmental impacts always be given first concern when it comes to the use of energy?
15. Even though the subject has not been covered thoroughly yet, list alternative or substitute fuels that would begin to reduce the carbon dioxide emissions of fossil-fueled power plants that contribute to global warming. What problems might these alternatives present?

16. What changes would you have to make personally if the amount of energy that you use per year was mandated for a 25% reduction?
17. If the world's population rises by a factor of two in the next 60 years (as projected), does that mean we will have twice as much pollution and/or twice as much energy consumption? Elaborate.
18. Investigate what choices are available in your community for choosing a supplier of your electricity.
19. What problems are there with the continued growth in energy demand in developing countries?
20. If an American consumes the energy equivalent of 7 gal of oil per day, then how much oil does an Indian consume daily? (See Figure 1.4.)
21. Why has the industrial world's per capita use of energy remained about the same in the past decade, even though the world's economy has increased by 50%?
22. In his book, *Earth in the Balance,* Al Gore argues that "research in lieu of action is unconscionable. A choice to do nothing in response to the mounting evidence [on global warming] is actually a choice to continue and even accelerate the environmental destruction that is creating the catastrophe at hand." Comment.
23. There is always a significant time period between book publication and available data for such things as energy use. From current data on the Web, find the present (or most recent) numbers for world and U.S. consumption of energy. Cite your URLs.
24. Determine what energy resources are used to provide energy in your state and their percentage contributions. Cite two URLs, at least one of which must be a government one (state or federal). Data must be no older than two years.

2

Energy Mechanics

A. Introduction

Not many years ago an invention came on to the market that claimed to put out more energy than was put into it. Some demonstrations of the device were given and a good deal of stock in the company was sold. The entrepreneur claimed this device would solve the energy crisis. It *did* solve his own financial crisis, but proved to be nothing but a clever trick. Had the investors known some simple physical laws about energy, they could have saved their money.

This chapter will serve to introduce some of the basic principles needed for an understanding of energy conversion devices and energy technologies. Much of this background comes from a study of basic physics. Physics is an experimental science concerned with understanding the natural world. The term "physics" is derived from the Greek word *"physike,"* meaning science or knowledge of nature; one dictionary defines physics as "the science dealing with the properties, changes, interactions, etc., of matter and energy." Some significant technological advances will have occurred since the writing of this book in the areas of energy conservation, resource utilization, and power production, but the basic concepts explained here and in the following chapters should provide a key to understanding many of them.

Science as a way of knowing is devoted to discovering the general principles that govern our world. A good part of this process is **observing** things around us in order to understand how they work. For example, our knowledge of ozone depletion and air pollution and their effects on humans have come from observations over the past several decades. After many observations, a **hypothesis** is usually proposed that tries to generalize the observations. The observation that collies and poodles have four legs might lead us to hypothesize that all dogs have four legs. But before we can be sure, we have to **test** our hypothesis by observing more dogs. A hypothesis that has been supported by a large body of observations and experiments becomes a **theory.** A good theory grows or is revised as new facts and observations arise. A scientific theory must be organic and open to change. Present-day theories of global warming are not fixed in stone. There are a variety of opinions within the scientific community on such issues. At the end of each of the chapters to come there are "Further Activities" suggested that will allow you to try your hand in the scientific process.

As we study energy, we must be aware of the limitations of science. Many things of great concern to us cannot be studied by a scientific approach; it is not the only road to knowledge. Science can address the questions of *how* something happened but not *why* something happened. As issues of energy policy emerge in the chapters ahead, we shall consider many questions or problems that lie outside the domain of science. There is clearly a need for more research in many areas of energy technology, but when and how these discoveries are applied will probably depend as much on the social, political, and economic atmosphere as it will on science and engineering. For example, some politicians feel that requiring more fuel-efficient cars would be an unwarranted federal intrusion into the private sector. Likewise, the selection of a site for the disposal of high-level radioactive waste is as much a political decision as a geophysical one.

B. Forms of Energy and Energy Conversions

As mentioned in Chapter 1, too often the term "energy" brings to mind only a vague picture of an electrical generating plant or a person bounding up from the breakfast table "full of energy." We more correctly might think of energy as something that makes automobiles move or that provides us with light and heat. We would have a better definition of energy if we thought of it as related to the capacity of certain materials in certain situations to perform useful tasks. To be consistent in our use of this term for the remainder of this book, we need a more rigorous definition. We will begin by identifying various forms of energy and the transformations of energy from one form to another.

One of the basic types of energy is the energy associated with an object's *motion*. We call this **kinetic energy.** A moving car or a rotating shaft has kinetic en-

With a total output of 4000 MWe, Korea Electric Power's Seoinchon is the world's largest and most efficient (57%) combined cycle facility (gas turbine and steam turbine system). (GENERAL ELECTRIC POWER SYSTEMS)

ergy. There is also energy associated with an object's *position*, which is called **potential energy.** A stretched spring or a ball positioned above a table has potential energy. Kinetic and potential energy can be classified as forms of what we call **mechanical energy.**

Other forms of energy are **chemical energy, nuclear energy, thermal energy, light (or radiant) energy,** and **electrical energy.** The fossil fuels as well as food possess chemical energy. The energy found within the atomic nucleus is nuclear energy. A hot object possesses thermal energy (a function of its mass and its temperature). Radiant energy is also called electromagnetic radiation, and covers everything from radio and television waves to infrared radiation to visible light to X-rays. The electromagnetic radiation received from the sun is usually referred to as solar energy. Electrical energy is produced at an electrical power plant or from the batteries in your "walkman."

All these kinds of energy on a microscopic level are examples of kinetic energy or potential energy. The chemical energy stored within oil may be considered as potential energy associated with molecular bonds, which are changed or broken during combustion. Radiant energy and electrical energy may be loosely thought of as related to the kinetic energy of light or electrons, respectively. The thermal energy of an object consists primarily of the sum of the kinetic energy of all the molecules of that object. We can categorize the primary energy sources introduced in Chapter 1 into chemical, nuclear, or radiant energy. The "end uses" of energy—the ways in which we see energy being used—include light, heat, motion, electricity, and some chemical reactions. Table 2.1 summarizes the forms, sources, and end uses of energy.

The transformation of energy from primary sources to end uses usually occurs through one or more *energy conversion processes.* Electrical energy is not a primary energy source but is the result of a conversion process that began with chemical, nuclear, or solar energy sources. For example, the chemical energy contained in oil is converted into other forms (thermal, electrical, and/or mechanical energy) beginning with combustion. The heat energy released by

Table 2.1 FORMS OF ENERGY

Chemical
Nuclear
Radiant
Thermal
Electrical
Mechanical (kinetic, potential)

Primary Sources		End Uses
Coal	Chemical	Heat
Oil	Chemical	Light
Natural gas	Chemical	Motion
Uranium—nuclear		Electricity
Sun—radiant/solar		Chemical processes

burning oil in a boiler turns water into steam, which drives a turbine that is connected to a generator to produce electrical energy.

Another example of energy conversion occurs in a solar cell. Sunlight impinging on a solar cell (Fig. 2.1) produces electricity, which in turn can be used to run an electric motor. Energy is converted from the primary source of solar energy into electrical energy and then into mechanical energy.

Table 2.2 lists a variety of devices to illustrate conversions from one energy form to another. For example, a toaster illustrates the conversion of electrical energy to thermal energy; a battery converts chemical energy into electrical energy. Mechanical energy (the kinetic energy part) of a car is converted into heat when the brakes are applied.

Let's further discuss the two different types of mechanical energy **(ME)**. Kinetic energy is the energy associated with the motion of an object. Examples of

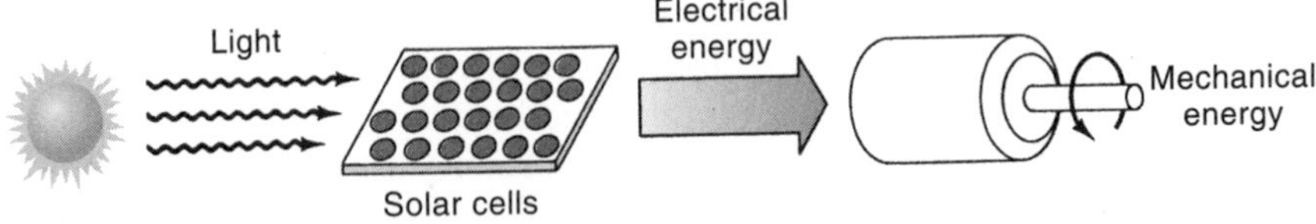

FIGURE 2.1
Illustration of conversions between different forms of energy. Here, solar energy is converted into electrical energy by a solar cell, which is used to run a motor.

Table 2.2 ENERGY CONVERSIONS

	To Chemical	To Electrical	To Heat	To Light	To Mechanical
From Chemical	food plants	battery fuel cell	fire food	candle phosphores-cence	rocket animal muscle
From Electrical	battery electrolysis electroplating	transistor transformer	toaster heat lamp spark plug	flourescent lamp light-emitting diode	electric motor relay
From Heat	gasification vaporization	thermo-couple	heat pump heat exchanger	fire	turbine gas engine steam engine
From Light	plant photosynthesis camera film	solar cell	heat lamp radiant solar	laser	photoelectric door opener
From Mechanical	heat cell (–crystal-lization)	generator alternator	friction brake	flint spark	flywheel pendulum water wheel

objects having **kinetic energy (KE)** include a moving stream of water, a bug flying through the air, a spinning flywheel, and the wind. Moving water has kinetic energy by virtue of its motion; this can be converted into useful work as it hits the blades of a waterwheel (Fig. 2.2). As moving air interacts with the blades of a wind turbine, the shaft rotates. The kinetic energy of the wind is converted into the kinetic energy of the shaft and then into electrical energy through a generator.

The other form of mechanical energy is associated with the relative position of an object. It is stored energy. The water at the top of a dam has gravitational **potential energy (PE)** by virtue of its position relative to the bottom of the dam. The amount of gravitational potential energy depends on the amount of water and the height of the water behind the dam wall. There is also potential energy

ACTIVITY 2.1

Provide other examples of devices to illustrate the energy conversion processes found in Table 2.2. This activity works best in small groups.

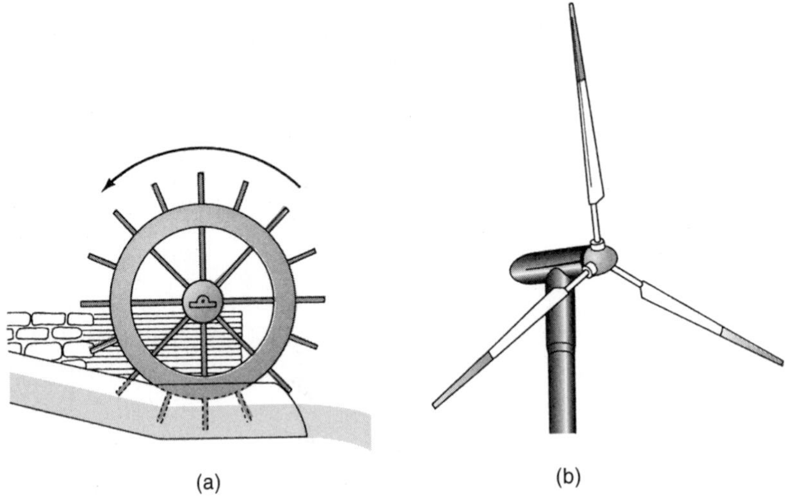

FIGURE 2.2
Two examples illustrating the conversion of kinetic energy (KE) of water or air into the motion of a waterwheel or a blade, which can be used to grind grain or generate electricity, respectively. *(a)* An undershot waterwheel. *(b)* A horizontal-axis, three-bladed, wind-powered generator.

associated with a compressed spring. The potential energy of an object attached to the spring is proportional to the displacement of the spring from its equilibrium (uncompressed) position (Fig. 2.3). You can probably remember the wind-up toys you had as a child; potential energy stored in the spring could be released to turn the toy's wheels, giving the toy kinetic energy.

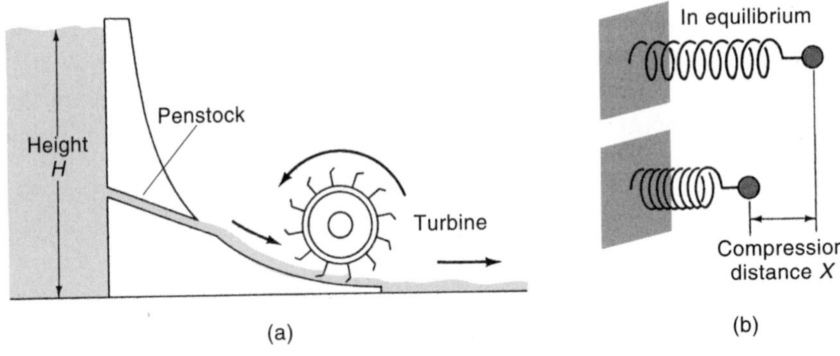

FIGURE 2.3
Examples of potential energy. *(a)* The gravitational potential energy of the water in the reservoir behind the dam is equal to the weight of the water times its height above the turbine. *(b)* The potential energy of the compressed spring is proportional to the square of the displacement of the spring from equilibrium X.

C. Motion

Before attempting a more rigorous definition of "energy," and in order to appreciate the subject of energy from a physics perspective, you should be familiar with motion and its causes. We will briefly discuss this in this section. These topics are covered in more detail at the end of this chapter in the Special Topic "Newton's Laws of Motion."

One of the most basic terms in the description of motion is **speed.** The speed of an object is equal to the distance it has traveled divided by the time taken to travel that distance. Commonly used units for speed are meters per second (m/s), kilometers per hour (km/h), feet per second (ft/s), and miles per hour (mph). **Velocity** provides additional information about motion, namely its direction; our velocity while walking briskly across campus might be one meter per second to the northeast.

In our everyday experience, it is more common to observe objects speeding up or slowing down than to see them moving at a constant velocity. These objects are accelerating; **acceleration** is the change in velocity divided by the time elapsed during that change. If an object's velocity changes at a constant rate, such as would occur to a coin dropped from your desk, its acceleration is a constant. The International System of Units (SI) for acceleration are m/s^2, pronounced meters per second per second.

What causes the velocity of an object to change (that is, to accelerate) is a **force,** specifically a **net** (or unbalanced) **force.** A force can be defined as the interaction of an object with other objects in its environment, and usually takes the form of a push or a pull. The net force is the (vector) sum of all the forces acting on the object. **Newton's second law of motion** states the mathematical relationship between net force and acceleration. It says that the acceleration of an object is directly proportional to the net force acting on it and inversely proportional to the object's mass *m*:

$$a = \frac{F_{net}}{m} \text{ or } F_{\text{net}} = ma$$

The SI unit of force is the **newton** (N). An object with a mass of 1 kilogram (kg) will be accelerated 1 m/s^2 by the application of a net force of 1 N. In the English system of measurement, the unit of force is the pound (lb). One pound is equal to about 4 N.

EXAMPLE

A 6-kg meteor is moving in space. If a 3-N force is applied to it, what will be its acceleration?

Solution

Newton's second law gives us $a = \frac{F}{m} = \frac{3N}{6\text{ kg}} = 0.5\text{ m}/s^2$

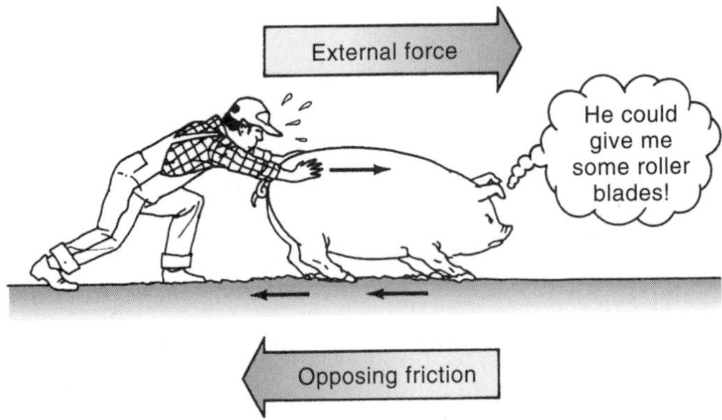

FIGURE 2.4
Friction enters into almost every situation in the real world. In order to accelerate the object, the force of the person's push must exceed the force of friction.

Acceleration only occurs if the object is acted on by a net force, that is, if the sum of all the forces acting on an object is not zero. One of the most common forces in nature is the force of friction, which always acts to *oppose* the motion taking place (Fig. 2.4). If a cart is pushed across the ground at a constant velocity, the net force on the cart must be 0 (Fig. 2.5). If your force pushing the cart is, say 100 N, this force is balanced or opposed by a force of

FIGURE 2.5
Pushing a cart at a constant velocity means that the net force on the cart (the person's force minus the force of friction on the tires minus the force of gravity down the hill) must be zero. The acceleration is zero. (P. HINRICHS)

ACTIVITY 2.2

You can study forces and Newton's second law with the following activity:

(a) Attach a rubber band to a shoe box or a Styrofoam bowl or an aluminum tart pan so that it can be pulled along a tabletop. Measure how far the rubber band will stretch (final length minus initial length) to move the box at a constant velocity with a 1-lb weight in it. Make this measurement when the box is moving, not just when it starts. (Why should the velocity be kept constant?) Add weights to the box and repeat the experiment. Propose a relationship between the stretch distance and the weight of the box.

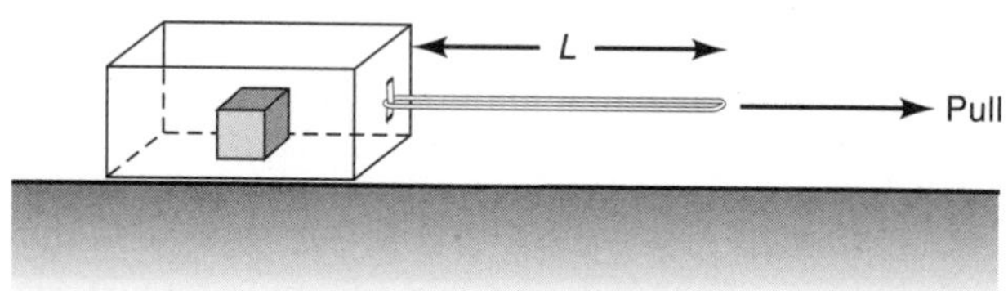

(b) Put some round pencils or dowels under the box and repeat the experiment. Compare these results with your results in part (a).

friction of 100 N, so that the net force is 0. (Note that a constant velocity implies that the acceleration is 0.)

As an example of Newton's laws of motion, consider the following: One of the environmental problems associated with the burning of fossil fuels is the emission of particulates (very small fly-ash particles) from the stack. These particles (from about one millionth to a hundred millionth of a meter in size) have been known to travel many hundreds of kilometers before landing, depending on wind velocity. This mobility is a problem because of the health effect these particles will have on those who inhale them. Their motion is possible if the net

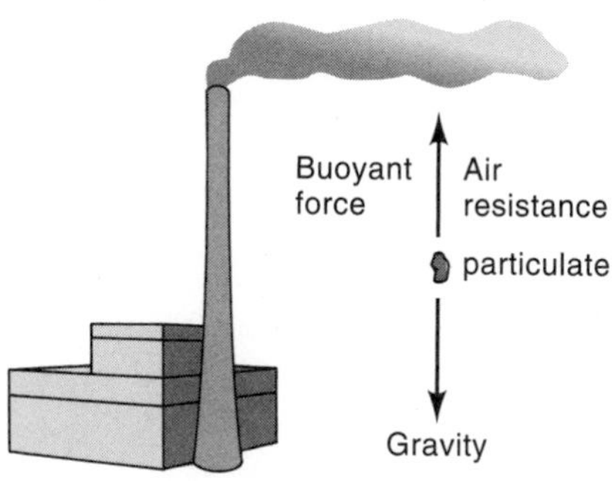

Focus On 2.1

ENERGY LOSSES IN A CAR

Overall automobile fuel efficiency is a function of two factors: engine efficiency (called thermal efficiency—how much of the chemical energy of the fuel is converted into work moving the pistons) and mechanical efficiency—the fraction of that work delivered by the engine that actually goes to move the vehicle. This includes aerodynamic losses and frictional losses within the engine. Thermal efficiency (see Chapter 4) for standard gasoline engines today is about 38%. Mechanical efficiency at cruising speeds is about 50%.

The net force on a moving automobile is equal to the difference between the force delivered by the engine and the friction forces due to air drag, rolling resistance of the tires, and friction within the engine. This can be written as

$$F_{net} = F_{engine} - F_{friction} = ma$$

When a car is cruising on level ground at a constant speed, F_{net} is equal to 0, since the acceleration is 0.

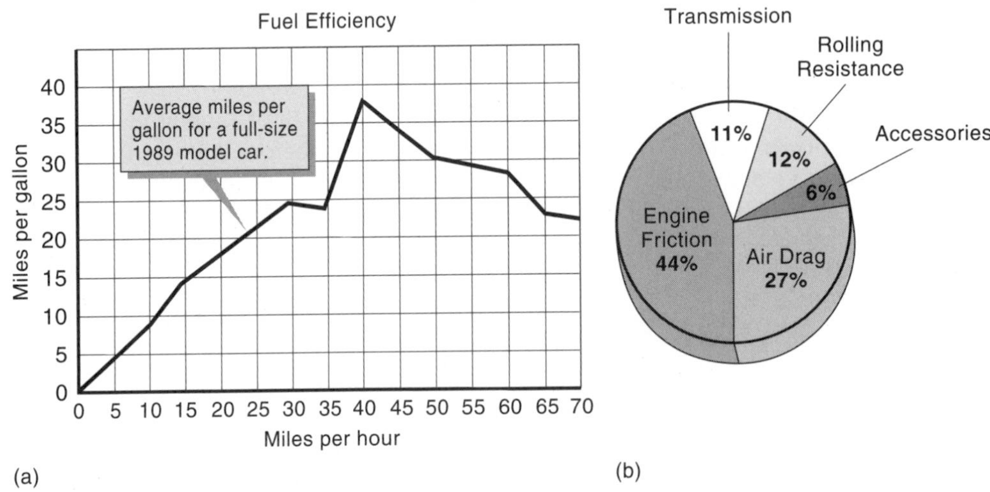

(a) Gas mileage as a function of the car's speed. (INSURANCE INSTITUTE FOR HIGHWAY SAFETY); *(b)* Energy losses in an average sales–weighted car at cruising speeds. (Engine efficiency is not included.) (*ANNUAL REVIEW OF ENERGY*, VOL. *19*)

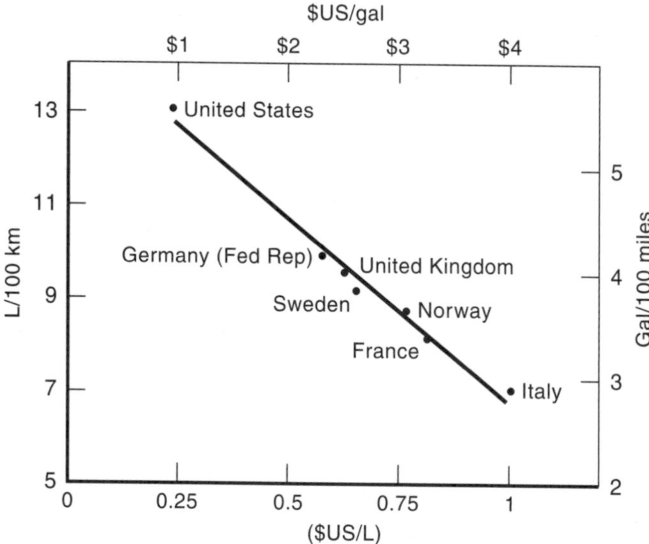

Frictional losses within the engine are much larger at lower speeds, while air drag increases as the square of the velocity, that is, air drag will be four times as much at 60 mph than at 30 mph. The figure to the left shows fuel economy in mpg as a function of the car's speed. The most efficient cruise speed is about 40 mph. Fuel economy drops off at higher speeds due to the air drag.

About two thirds of all the oil the United States uses is for transportation (see Fig. 7.7). The fuel efficiency of new cars rose steadily until the late 1980s and leveled off through the 1990s (see Table 2.3). However, the total amount of oil used for transportation has been increasing because there are more cars and more miles traveled per car than in the past.

One of the reasons for the greater use of gasoline per person in the United States than in other countries is that fuel prices are comparatively low. The figure above shows the interesting correlation between fuel prices and average automobile fuel economy (measured in gallons per 100 miles or liters per 100 km) in different countries.

Table 2.3 NEW PASSENGER CAR FLEET: AVERAGE CHARACTERISTICS

Characteristics	1978	1988	1998
Weight (lb)	3349	2831	3071
Horsepower	136	116	129
Engine size (in.^{3} displacement)	260	161	164
Miles per gallon (combined city/highway)	19.9	28.6	28.6

Source: U.S. Department of Transportation

vertical force on the particle is zero, or close to it, and so there is very little acceleration toward the ground. The force of gravity downward on the particle is balanced out by the upward buoyant force of the air and by air resistance; thus the particles can drift with the wind great distances.

Newton's second law says that the acceleration of an object depends on both the net force acting on it and its own mass. For example, if identical engines were placed into a Cadillac and a Dodge Neon, the acceleration of the Neon would be greater than that of the Cadillac because the mass of the Neon is much less, even though the force accelerating the car is the same in both cases.

Driving in a city with stop-and-go traffic, we burn more fuel than we do driving comparable distances in the country. In city driving we quite frequently have to accelerate from rest, which requires a net force acting on the car. The gas mileage of new cars has improved substantially in recent years as a result of many factors. The most important change has been because of the reduction in the mass of a car, not more efficient engines or better aerodynamics. Focus On 2.1 "Energy Losses in a Car" explains some of the factors that determine fuel efficiency.

D. Energy and Work

From this introduction to the forms of energy and to motion, let us now proceed to a definition of energy. **Energy is defined as the "capacity to do work."** While the word "work" might bring to mind various pictures in such fields as literature and biology, in physics **work** is defined as *the product of a force times the distance through which that force acts.* If you push this book across the desk, a force is being applied. The work done on an object (such as the book) equals the

Table 2.4 UNITS IN MECHANICS

Quantity	SI	English	Conversions
Velocity	m/s	ft/s	1 ft/s = 0.305 m/s
Acceleration	m/s^2	ft/s^2	1 ft/s^2 = 0.305 m/s^2
Force	newton (N)	lb	1 lb = 4.45 N
Energy	joule (J)	ft-lb	1 ft-lb = 1.356 J
Power	watt (W)	ft-lb/s, hp	550 ft-lb/s = 1 hp = 746 W

applied force times the distance through which that object moves in the direction of the force. This may be expressed by the formula

$$\text{Work} = \text{force} \times \text{distance}$$
$$W = F \times d$$

According to this definition, no work is done if the object on which you are applying a force does not move, no matter how hard you push or pull it.

To look at energy and work in another way, we can say that the **consequence of doing work on an object is to give the object energy.** If you apply a force to a cart and move it a certain distance on a flat surface, work has been done and the cart has gained kinetic energy. When an object is raised to a certain height, work has been done to increase the object's gravitational potential energy. The force in this case is the one required to lift the object against the gravitational force on it. The distance through which the force acts is the height through which the object was raised.

The units of both energy and work are those of force times distance. In the SI system, this is expressed as newtons times meters, or joules (J). (One newton times one meter equals one joule. One joule is approximately equivalent to the potential energy of one apple one meter above the floor.) In the English system, the units are pounds times feet, or foot-pounds (ft-lb); 1 ft-lb ≈ 1.4 joules. See Table 2.4 for a summary of units in mechanics.

EXAMPLE

A man pushes a box across the floor by exerting a force of 150 N on it in the direction of motion. If the box is moved 3 m, how much work (W) did he do?

Solution

$$W = F \times d = 150\ \text{N} \times 3\ \text{m} = 450\ \text{J}$$

Work is one way of transferring energy to an object. If we push an object up a hill from rest, we are doing work to give it both kinetic energy and gravitational potential energy. The work has been used to increase the object's mechanical energy. We can write this as an equation

$$W = \Delta(KE + PE)$$

The Δ (delta) in this expression means a "change in." The mechanical energy—kinetic plus potential energy—of the object (our system) has become larger by the amount of work done.

Let's return to other forms of energy. Recall from the section "Forms of Energy and Energy Conversions" that the thermal energy (TE) of an object is a function of its temperature. TE is the internal energy of an object and is equal to the total of all the microscopic energies of the molecules that make up that object. We can change the thermal energy of a body by rubbing it on a rough surface. If you move a block of wood back and forth on a flat surface, the kinetic energy and potential energy of the block do not change (the speed remains about the same) but work has been done. What is occurring is that the block's temperature is increasing and so its TE is changing. As a result, we need to add a thermal energy term to our expression for work:

$$W = \Delta(KE + PE + TE)$$

In the previous paragraph, if there was friction on the hill, some of the work done on the object would go to an increase in its thermal energy.

Another way of transferring energy to a system is by the addition of heat. (This will be covered in more detail in Chapter 4.) The block of wood in the preceding paragraph could have its thermal energy increased by placing it against a very hot object: heat will flow from the hot object to the cooler block. **Heat is the energy transferred as a result of a temperature difference between two objects.** Note the distinction between heat and thermal energy: Heat is never contained within an object; an object contains thermal energy.

Putting these two ideas together, we say that the thermal energy of an object can be changed by doing work W as well as by adding heat Q. Our basic energy equation now becomes:

$$W + Q = \Delta(KE + PE + TE)$$

(Work or heat can also change the electrical energy or chemical energy of a system, so these terms could also be added to this energy equation, but we will ignore them now for simplicity.)

Summarizing this equation, the total energy of a system can be increased by doing work on it or by adding heat. This relationship is known as the **first law of thermodynamics.**

E. Examples of Work and Energy

Let's bring together the concepts of energy and work that we have discussed so far. Work is done by the application of a force over a distance. Doing work on an object gives it more energy. Conversely, energy is the capacity for doing work. Let us consider two examples of work and mechanical energy:

1. A moving object has KE. A moving body can exert a force on another object and cause it to move, thereby doing work. For example, a moving bullet can embed itself in a block of wood and move the wood. Energy is the capacity for doing work.
2. If we are pushing a block on a horizontal surface and it is moving at a constant velocity, the kinetic energy remains the same. Since there is no acceleration, the net force on the block must be zero. This must mean that the force we are applying is balanced by the force of friction. However, *we* are doing work as *we* are applying a force over a distance. Our work generates heat as a result of the friction between the block and the table.

As introduced earlier, gravitational potential energy (PE_G) is energy as a result of the relative height of an object. It is stored energy. When an object is raised to a certain height, work has been done to give it gravitational potential energy. Force is required to lift the object against the gravitational force on it.

Greer's Ferry Dam on the Little Red River in north central Arkansas. (USDOE)

Since the force of gravity on an object is equal to the weight of the object, mg, where g is the acceleration due to gravity, 9.8 m/s^2, the object has gained an amount of gravitational potential energy equal to its weight times the vertical height h through which the object was raised.

$$PE_G = \text{weight} \times \text{height} = mgh$$

Note that the "height" h in this expression is not an absolute number (such as the elevation above sea level) but a vertical distance measured from a selected reference point. For example, the water at the top of a dam possesses a certain amount of potential energy (stored energy) relative to the water level at the bottom of the dam. The height h is the vertical distance of the water behind the dam measured from the bottom of the dam.

EXAMPLE

How much potential energy is possessed by 10,000 kg of water (about 10 m^3 or 2600 gal) behind a dam if the distance the water will fall before it hits the blades of a turbine is 20 m?

Solution

$PE_G = \text{weight} \times \text{height} = 10^4 \text{ kg} \times 9.8 \text{ m/s}^2 \times 20 \text{ m} = 196 \times 10^4 \text{ J}$.
(This is equivalent to the energy contained in about 1⁄60th of a gallon of gasoline.)

Energy associated with motion is called **kinetic energy**. An object at rest has no kinetic energy. The expression for kinetic energy (KE) of an object in motion is

$$KE = \frac{1}{2} mv^2$$

where m is the mass of the object and v is its velocity.

EXAMPLE

What is the kinetic energy of 1 kg of air (about 1 m^3) moving at 15 m/s (about 32 mph)?

Solution

The expression for kinetic energy is ½ mv^2, so

$$KE = \frac{1}{2} \times 1 \text{ kg} \times (15 \text{ m/s})^2 = 112 \text{ J}$$

(One of the problems with generating electricity with the wind is the low density (mass per volume) of air. An equivalent volume of water with the same velocity will have about 1000 times as much energy.)

F. Power

Another basic concept of energy mechanics is "power." **Power** is the rate of doing work or the rate at which energy is used, produced, or transferred.

$$\text{Power} = \frac{\text{work done}}{\text{time taken}} = \frac{\text{energy used}}{\text{time taken}}$$

As you lift a block up to a table from the ground, work is being done by you on the block as you apply a force over a distance. The same amount of work is done on the block whether it took one second or one hour to do the task; however, the power supplied by you is different if the work was done in different time intervals. More power is required for the short time interval job.

The unit of power is the unit of energy divided by the unit for time. In SI units this is joule/second, which is given the name watt (abbreviated W).

$$1 \text{ watt} = \frac{1 \text{ joule}}{1 \text{ second}}$$

Since the watt is a relatively small unit of power, we commonly use the kilowatt, where 1 kilowatt (kW) = 1000 watts (W). In English units, the unit for power is ft-lb/s. Similarly, a larger unit called the horsepower (hp) is often used in the English system, where 1 hp = 550 ft-lb/s. Note that 1 hp = 746 W (Table 2.4).

EXAMPLE

If it takes 2 s to raise an 8-kg block a vertical height of 1 m, what is the power output?

Solution

$$\text{Power} = \frac{\text{work}}{\text{time}} = \frac{\text{weight} \times \text{height}}{\text{time}} = \frac{8 \text{ kg} \times 9.8 \text{ m/s}^2 \times 1 \text{ m}}{2 \text{ s}} = 39.2 \text{ W}$$

ACTIVITY 2.3

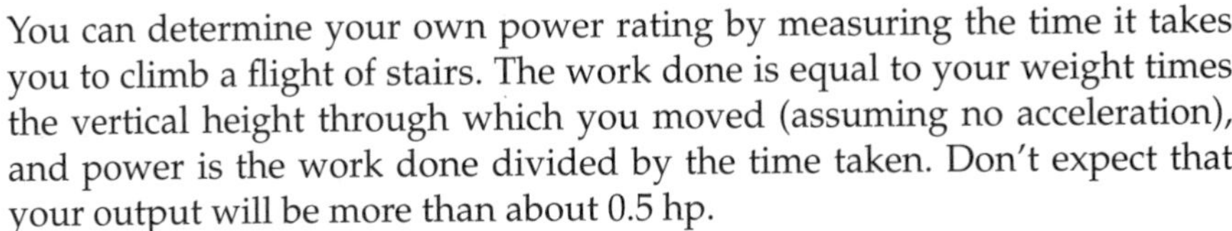

You can determine your own power rating by measuring the time it takes you to climb a flight of stairs. The work done is equal to your weight times the vertical height through which you moved (assuming no acceleration), and power is the work done divided by the time taken. Don't expect that your output will be more than about 0.5 hp.

We can rearrange the equation for power, to yield

$$\text{Energy used} = \text{power} \times \text{time of use}$$

This is especially useful when one wants to find the energy used in a particular conversion when the power expended is known. Your electrical bill is a charge for the amount of energy you have used, not power. To figure the cost of using an electrical appliance, you must know the time the appliance is operating and the power rating of the appliance. Electrical energy is usually expressed in units of kilowatts × hours, abbreviated kWh. The cost of running an appliance is equal to the energy used times the cost per kilowatt-hour.

ACTIVITY 2.4

Locate an electric bill for your house or apartment. Notice the total bill and the price per kWh. Do the rates vary at different times of the day and night? Does the bill divide the cost into distribution and generation? How much electricity did you use last month per person? What item do you think is the biggest contributor to your total electric bill?

EXAMPLE

An electric heater has a power rating of 1500 W (1.5 kW). If the heater is run for 6 hours and electricity costs \$.12/kWh, how much will it cost to run the heater for this time interval?

Solution

Since Energy = power × time of use, the energy used = 1.5 kW × 6 h = 9 kWh. The cost is equal to 9 kWh × \$.12/kWh = \$1.08.

Focus On 2.2

ENERGY USE IN INDIA

In recent years, economic reforms in India, such as privatization of many industries, have helped to double the GDP growth rate to approximately 4% per year. The commercial and industrial use of energy has been rising at a rate of about 5% per year, the highest of any major country, yet the per capita use of energy is only about one eighth the world's average. The main energy resources used today are biomass (wood, dung) and coal; oil provides about one fourth of the energy mix, over half of which is imported. Increasing use of energy and a population growth rate of 1.9% per year continue to bring changes in the environment. Access to clean water and sanitation in

continued

Street scene, New Delhi, India. (HAROLDO DE FAR CASTRO/FPG INTERNATIONAL, LLC)

both rural and urban areas and increasing levels of air pollution in the large cities are serious concerns. About 70% of India's electricity is generated by highly polluting coal. The key energy challenge facing India today is to prevent bottlenecks in the energy supply from constraining economic growth.

A few interesting points:

- Biomass fuels provide about one third of India's total energy needs
- Electricity use has been rising about 8% per year since 1970. Annual additions are not able to keep up with demand, leading to power shortages
- Three fourths of Indians live in rural areas
- India has the largest solar-cooking program in the world
- More than 8000 villages use solar cells for electricity
- India is the third largest producer of electricity from wind energy
- India has more than tripled food production since 1950 (mainly through the use of high-yield grains), outpacing population growth.

The average power expended per person in the United States is about 12 kW. This is calculated as follows: The annual consumption of energy in the United States is about 95×10^{15} Btu/yr $= 100 \times 10^{18}$ J/yr. The average per capita energy consumption is thus

$$\frac{100 \times 10^{18}\ \text{J/yr}}{275{,}000{,}000\ \text{people}} = 3.64 \times 10^{11}\ \text{J/person/yr}$$

Since 1 year $= 3.16 \times 10^7$ s, the average per capita power expenditure in the United States is

$$\frac{3.64 \times 10^{11}\ \text{J/person/yr}}{3.16 \times 10^{7}\ \text{s/yr}} = 12 \times 10^3\ \text{W} = 12\ \text{kW/person}$$

This number comes not only from an individual's personal use of energy but also from a share of cooling shopping malls, making steel and aluminum, lighting offices, and so forth. The United States has one of the highest energy uses per capita in the world, but not the highest standard of living. In most countries, higher standards of living, as measured by Gross Domestic Product (GDP), are matched by higher levels of energy consumption. This is seen in Figure 2.6. The average power consumption per person is 6 kW for Switzerland

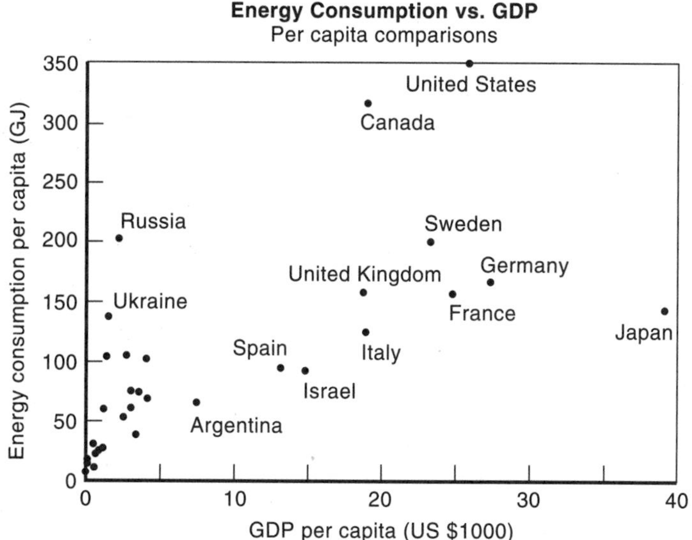

FIGURE 2.6

Comparison of energy use per capita per year versus GDP per capita for various countries. 1 GJ = 10^9 J. 320 GJ/yr = 10 kW. (WORLD RESOURCES INSTITUTE, 1998. *WORLD RESOURCES*. NEW YORK, OXFORD UNIVERSITY PRESS.)

and 5 kW for Japan, countries that have approximately the same standard of living as the United States. The power consumption per capita in India is about 0.2 kW. (An enlargement of this figure for developing countries is shown in the questions at the end of the chapter.)

The rate of producing electricity by an electrical generating plant is electrical power, expressed in units of watts. A modern-size plant produces electrical energy at a rate of about 1 billion watts, or 1000 Megawatts (1000 MWe, where the "e" signifies electrical power output, not thermal power output). Note that this is not expressed as MW/s, since watts already has time included as J/s.

An example of electrical power obtained from a renewable source is hydropower. It now provides about 9% of the U.S. total electrical energy needs. Hydropower in the United States generates as much electricity as about 80 large-sized (1000 MWe) coal power plants. Because of its zero fuel cost, it is used by utilities whenever available. The largest hydroelectric dam in the United States is the Grand Coulee Dam in the state of Washington, with a capacity of 6500 MWe and a height of 168 m. The largest operating hydroelectric plant in the world is in Venezuela, with a capacity of 10,000 MWe. Hydroelectric power will be discussed in more detail in Chapter 12.

G. Summary

In this chapter we have defined and illustrated the fundamental concepts of work and energy. Work is defined as the product of an applied force times the distance through which that force acts. Energy can be found in many forms (mechanical, thermal, electrical, radiant, chemical, nuclear) and is the capacity to do work. Mechanical energy is the sum of an object's kinetic energy and its potential energy. The study of energy includes a study of its transformations from one form to another, that is, from mechanical to electrical to thermal energy. This will be examined in more detail in the next chapter.

Internet Sites

For an up-to-date list of Internet resources related to the material in this chapter, go to the Harcourt College Publishers website at **http://www.harcourtcollege.com.** The links are in the *Energy: Its Use and the Environment* site on the Physics page. General energy-related sites and some guidelines for using the World Wide Web in your class are on the inside front cover of this book.

References

Chapter 2

Cutnell, J., and K. Johnson. 2001. *Physics.* 5th ed. New York, John Wiley & Sons.

DeCicco, J., and M. Ross. 1994. Improving Automobile Efficiency. *Scientific American,* 271 (December).

Gartrell, J. 1989. *Methods of Motion: An Introduction to Mechanics.* Washington, D.C., National Science Teachers Association.

Hewitt, P. 1998. *Conceptual Physics.* 8th ed. Reading, MA, Addison-Wesley.

Hobson, A. 1999. *Physics: Concepts and Connections.* 2d ed. Englewood Cliffs, NJ, Prentice-Hall.

Levine, M. D., F. Liu, and J. E. Sinton. 1992. China's Energy System. *Annual Review of Energy,* 17.

Ostdiek, V., and D. Bond. 1995. *Inquiry into Physics.* 3rd ed. Minneapolis, West.

Ross, M. 1994. Automobile Fuel Consumption and Emissions: Effect of Vehicle and Driving Characteristics. *Annual Review of Energy,* 19.

Sathaye, J., and S. Tyler. 1991. Transitions in Household Energy Use in Urban China, India, the Philippines, Thailand, and Hong Kong. *Annual Review of Energy,* 16.

Schafer, A., and D. Victor. 1997. The Past and Future of Global Mobility, *Scientific American,* 277 (October).

Schipper, L. 1995. Determinants of Automobile Use and Energy Consumption in OECD Countries. *Annual Review of Energy,* 20.

Serway, R., and J. Faughn. 1999. *College Physics.* 5th ed. Philadelphia, Saunders College Publishing.

Shonle, J. 1975. *Environmental Applications of General Physics.* Boston, Addison-Wesley.

Starr, C. 1971. Energy and Power. *Scientific American,* 225 (September).

VanCleave, J. 1991. *Physics for Every Kid.* New York, John Wiley & Sons.

Walsh, M. 1990. Global Trends in Motor Vehicle Use and Emissions. *Annual Review of Energy,* 15.

World Resources Institute, 1998. *World Resources.* New York, Oxford University Press.
Zubrowski, B. 1986. *Wheels at Work: Building and Experimenting with Models of Machines.* New York, Beech Tree.

QUESTIONS

1. List five different forms or types of energy. Give one example of a conversion from each of these forms to another form.
2. Discuss the types of energy transformations that are involved in the following devices or events:
   (a) Striking a match
   (b) Windmill
   (c) Ball rolling off tabletop and bouncing on floor until it stops
   (d) Microphone
   (e) Flashlight
3. List the energy conversions that occur when (a) riding a bicycle and (b) using a windmill to pump water.
4. Discuss the transformation of the potential energy of water behind a dam as it flows through a pipe at the bottom to turn a turbine generator.
5. What happens to the kinetic energy of a car as it rolls up an incline and stops?
6. Refer to the discussion of energy losses in a car. What are some ways to increase the fuel economy of a car?
7. Why is there a large difference in the per capita use of energy between the United States (or Canada) and European countries (see Fig. 2.6). Answer the same question for countries of Eastern Europe and other developing countries. (See Fig. 2.7 on page 58.)
8. If a constant net force is applied to an object, what can you say about the velocity and the acceleration of the object?
9. If the acceleration of an object is not zero, can the velocity of the object be zero?
10. A car accelerates from 30 mph to 40 mph. Give an expression for the net force causing this acceleration in terms of the forces acting on the car. How does this expression change when the car is moving at a constant velocity of 40 mph?
11. Work is expressed as force times distance. There is a gain in energy when work is done. In terms of energy, what happens to the work done in pushing an object across a level floor with a constant force?
12. Distinguish between work done in completing a task and the power expended.
13. Categorize the following units as those of work or power: Btu, joules, watts, kilowatt-hours, ft-lb, calories, Btu/h, ft-lb/min.
14. What factors determine the amount of electrical power that can be produced by a stream or river?

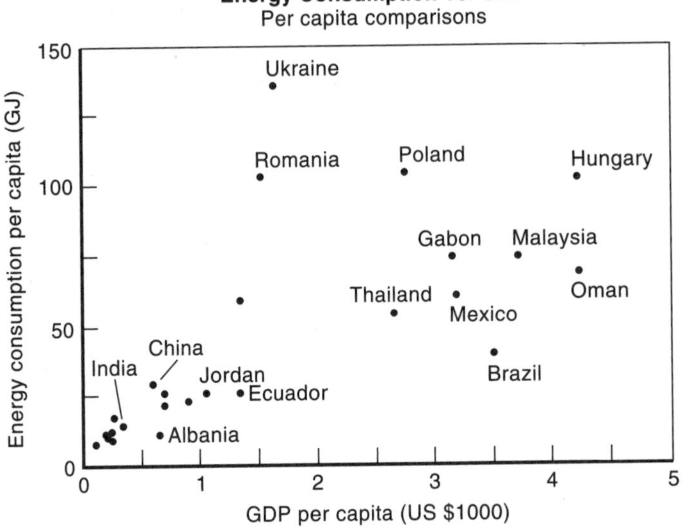

FIGURE 2.7

Comparison of energy use per capita versus GDP per capita for various developing countries. Included in the energy consumption numbers are both commercial fuels and traditional ones (dung, biomass). (WORLD RESOURCES INSTITUTE, 1998. *WORLD RESOURCES*. NEW YORK, OXFORD UNIVERSITY PRESS.)

PROBLEMS

1. A 50-kg meteor is moving in outer space. If a 12-N force is applied opposite the direction of motion, what will be its deceleration?
2. Suppose you apply a force of 40 N to a box of mass 2 kg. The force of friction opposing the motion of the box is 15 N. What is the acceleration of the box?
3. What is the potential energy of a 70-kg person sitting on a ladder 2 m off the ground?
4. Calculate the work done in lifting a 4-lb book to a height of 8 ft.
5. How much work do you expend to move a 450-lb refrigerator 6 ft across the floor if you are exerting a force of 90 lb?
6. What is the kinetic energy of a 3000-lb car traveling at 60 mph? (Remember that pounds are not units of mass.)
7. To what height would a 2-kg object have to be raised such that its gravitational potential energy would equal the kinetic energy it would possess if it were moving at 3 m/s?

8. A cart of mass 20 kg, initially at rest, is pushed with a net force of 40 N on a flat surface. If the cart is pushed 5 m, what is the final KE? What is the final velocity? (Assume no heat losses.)
9. A ball of mass 0.5 kg is dropped from a height of 5 m. What is the greatest velocity it will have just before hitting the ground?
10. What is the potential energy of the water in a lake of surface area 10 square miles, average depth 40 ft, and elevation above the electrical generator of 600 ft? (Note that 1 square mile is about 28 million square feet and that 1 ft^3 of water weighs 62 lb.)
11. The per capita rural consumption of household energy in some Bangladesh villages is 7×10^9 J/yr. This is what fraction of the U.S. average for residential per capita energy use?
12. An engine performs 4000 J of work in 10 s. What is its power output in kilowatts and in horsepower?
13. In one day, the average food intake for a person in the United States is about 2500 kilocalories (1 cal = 4.2 J). Assuming half of this energy goes into work (the other half goes into heat radiated by the body), what is the average power output (in watts) of a person for 24 hours?
14. If you lifted a 50-kg mass to a height of 0.7 m in 1.2 s, how much work did you do? What was the power exerted in watts? (Express this in hp also.) If this is done 20 times, compare the total work done with the food intake for one day. (See Problem 13.)
15. A 100-W light bulb is accidentally left on for two days in a basement. If electricity costs 12¢/kWh, how much did this oversight cost?
16. A small stream flowing at a rate of 8 L/s has a vertical drop of 1.5 m. What is the maximum power one can obtain from this stream? (One liter of water has a mass of 1 kg.)

FURTHER ACTIVITIES

1. Be a meter reader. Record the numbers on your electric meter at the beginning of one day and the beginning of the next. How many kWh of electric energy were used in one day? (The meter below is reading 4831.)

Kilowatt-hours

2. Measure the speed (in m/s) of one or more of the following:
   (a) The water in a stream or river
   (b) An ant or other insect moving across the floor
   (c) Yourself, as you go to your first morning class
3. Why do heavy cars use more gasoline than light cars? Check this out (in your library or on the Internet) by comparing EPA mileage ratings with car weight.
4. Find (or remember) a self-propelled toy you used to play with. Describe the energy transformations that take place. A toy that can sometimes be found at an outdoor market is a "rubber band racer," in which a stretched rubber band powers the rear wheels of the car. If you have access to one of these, an interesting study can be made of the number of turns of the rear wheels (thereby stretching the rubber band) versus the distance the car travels.
5. You can study acceleration and the forces of friction and air resistance on an automobile with the following investigations. While moving on an open, level highway (with little traffic), shift your car into neutral and measure the time it takes for the velocity to decrease from:
   (a) 55 to 45 mph
   (b) 45 to 35 mph
   (c) 35 to 25 mph

FIGURE 2.8
Dropping eggs from the third floor doesn't always lead to disaster. (R. BROWN)

Find the deceleration of the car (and consequently, the frictional drag) as a function of velocity. Estimate the mass of the car, or use 2000 kg. How does this compare with the statements made in the Focus On 2.1 box "Energy Losses in a Car" earlier in this chapter?

6. A somewhat classic and "eggciting" illustration of Newton's laws of motion is the egg-drop experiment. The object here is to drop a raw egg onto a hard surface from a height of several meters or more (Fig. 2.8). For the egg *not* to break, the force on its shell should be minimized as it hits. One can achieve this by minimizing the deceleration (see Fig. 2.11). One can carry out this project *either* by using any materials available *or* (as the author does) restricting each participant to the same materials, such as two Styrofoam cups, two paper towels, and a small handful of toothpicks. (It might be wise to put the egg into a plastic bag.) These materials can be modified as needed. The second approach can be done in class.
7. Demonstrate several types of energy conversions by making a working model of a steam turbine.
   (a) Draw the pattern below and cut along the dashed lines.

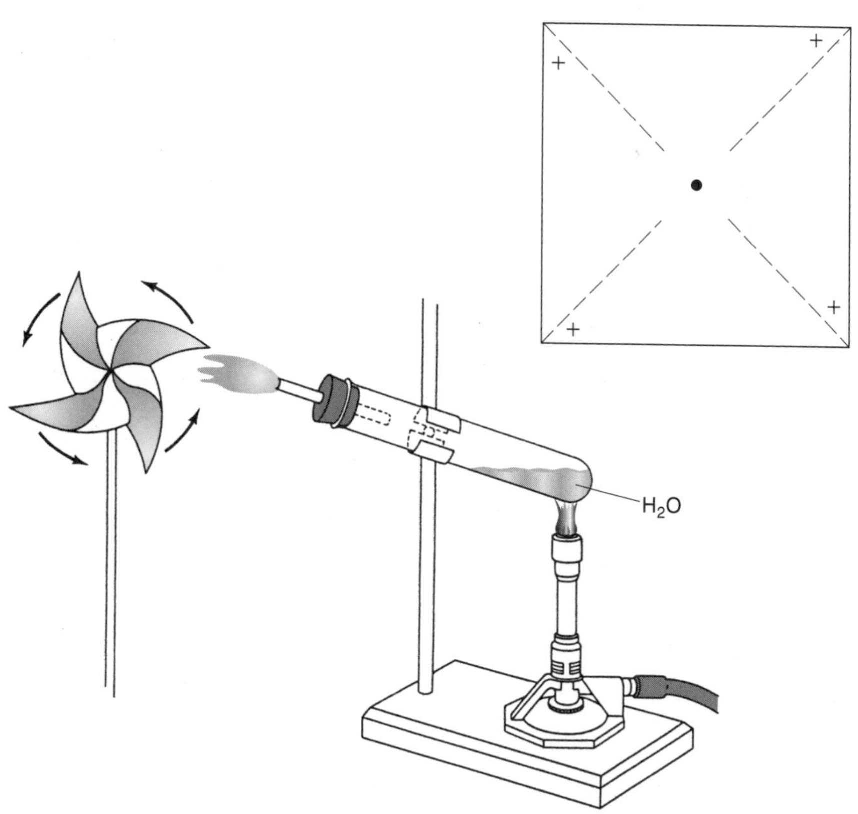

(b) Pin the starred corners through the center to the eraser of a pencil.
(c) For the steam generator, put a little water into a test tube and attach the tube to a support stand, or use a tea kettle with a metal tube inserted into the hole in the lid.
(d) For the test-tube nozzle, use the glass tube from a medicine bottle (with the rubber bulb removed). Insert it into a one-hole stopper. (Be careful when inserting glass into a stopper; use a lubricant such as glycerine.)
(e) Heat the water in the test tube or the kettle on a hot plate. Allow the steam to hit the paper blades of the turbine. (*Warning:* Steam can cause a severe burn.)
(f) What types of energy transformations are involved here?

8. Determine the number of light bulbs in a restroom in your school or office building and their wattage. (Assume 40 W if you cannot get this number.) If the lights are accidentally left on over the weekend (6 PM Friday to 8 AM Monday), how much energy was wasted (in kilowatt-hours)? What is the *annual* cost of leaving lights on in *all* the restrooms in this building every weekend? Determine the cost of purchasing and installing a motion sensor for this room. What is the payback time?

SPECIAL TOPIC

Newton's Laws of Motion

This section can serve as a review of the terms and concepts of energy mechanics as presented earlier in the chapter. It amplifies some of the material presented in the section on motion.

Speed, Velocity, Acceleration

One of the most basic terms in the description of motion is speed. *Speed* is equal to the ratio of distance traveled to the time taken to travel that distance. Mathematically,

$$\text{Speed} = \frac{\text{distance traveled}}{\text{elapsed time}}$$

If you travel a distance of 20 kilometers (km) in 30 minutes, your average speed over that interval is 20 km per 30 minutes, or 40 km/hr. Each car's starting position for the Indianapolis 500 race is determined by measuring the time that car takes to drive one lap around the track. The average speed is the total distance traveled divided by the total time taken. Average speeds in excess of 320 km/hr (200 mph) are recorded for this race. However, the speed at any time on the track varies, depending on whether the race car is on a curve or a straightaway. The speed at a particular time is called the instantaneous speed.

Many times we speak of "velocity" when we mean speed. *Velocity* provides additional information about the motion of an object, namely its direction. Velocity indicates not only that your speed is 40 km/hr, but also that you are going in a particular direction, for example, west. Two cars that leave a parking lot and proceed with the same average speed can end up at entirely different places because their directions of travel were different; they had different velocities.

The equation for speed can be arranged another way to yield:

$$\text{Distance traveled} = \text{average speed} \times \text{time}$$

We'll use this expression in the following example.

EXAMPLE

A train travels in a straight line between two cities in 40 min at an average speed of 30 m/s. How far apart are the cities?

Solution

We use the formula in the form distance = speed × time, and remember to convert minutes to seconds:

$$d = vt = 30\ \text{m/s} \times 40\ \text{min} \times 60\ \text{s/min} = 72{,}000\ \text{m} = 72\ \text{km}$$

EXAMPLE

The speed of light is approximately 300,000 km/s, and the distance from the earth to the sun is 155,000,000 km. How much time does it take light to travel to earth?

Solution

Since

$$\text{Speed} = \frac{\text{distance traveled}}{\text{time}}$$

then

$$\text{Time} = \frac{\text{distance}}{\text{speed}} = \frac{155{,}000{,}000\ \text{km}}{300{,}000\ \text{km/s}} = 517\ \text{s} = 8.6\ \text{min}$$

Recall the illustration in the section on motion on particulate emissions from a power plant. With some assumptions on velocity, we can be quantitative on the distance a particulate will travel, as in the next example.

EXAMPLE

An average sized fly-ash particle has a (constant) settling velocity of 0.3 m/s. If these particulates are emitted from a 200-m-high stack and there is a 15-km/h wind, how far away from the stack will the particle land?

Solution

The time it will take for the particle to reach the ground is

$$\text{Time} = \frac{\text{distance}}{\text{vertical velocity}} = \frac{200\ \text{m}}{0.3\ \text{m/s}} = 667\ \text{s} = 0.19\ \text{h}$$

In this time it will have covered a horizontal distance of $d = vt = 15\ \text{km/h} \times 0.19\ \text{h} = 2.8\ \text{km}$.

The change in velocity in a certain amount of time is called the acceleration, written as

$$\text{Acceleration} = \frac{\text{change in velocity}}{\text{time elapsed}}$$

Acceleration is the rate of change in either an object's speed, or its direction, or both. If a car goes around a corner at a constant speed, its direction of motion is changing, and so we can say that the car is accelerating. When a car slows down, it is decelerating or experiencing a negative acceleration.

If an object's velocity changes at a constant rate, its acceleration is constant. When any object is dropped to the ground near sea level, its velocity changes 9.8 meters per second for each second of travel. At the end of the first second its velocity will be 9.8 m/s, after 2 s it will be 19.6 m/s, after 3 s it will be 29.4 m/s, and so on. This is the "acceleration due to gravity" of 9.8 m/s^2. In English units, it is 32 ft/s^2.

EXAMPLE

A car accelerates from rest to a speed of 20 m/s (45 mph) in 2 s. What is its acceleration?

Solution

$$\text{Acceleration} = \frac{\text{change in velocity}}{\text{time elapsed}} = \frac{(20\ \text{m/s} - 0\ \text{m/s})}{2\ \text{s}} = 10\ \text{m/s}^2$$

(This is close to the acceleration due to gravity.)

Figure 2.9 shows the vertical positions of an object at 1-s intervals after it has been dropped and is experiencing an acceleration due to gravity. The distance traveled for any 1-s interval is greater than during any previous 1-s interval, since the velocity is constantly changing. The same figure shows a graph of velocity versus time; note that the velocity is changing but the acceleration (the rate of change of velocity) remains constant (at 9.8 m/s^2).

Force and Newton's Laws of Motion

FIRST LAW The part of mechanics that deals with the causes of motion is called **dynamics,** and it is based on three laws that are called Newton's laws of motion, after Isaac Newton (1642–1727). The first of these laws refers to the "natural motion" of an object:

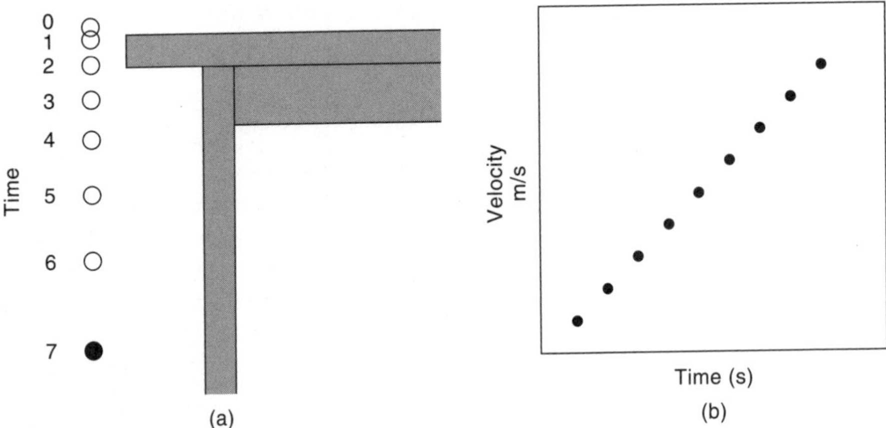

FIGURE 2.9
A freely falling object. *(a)* Positions of a ball at equally spaced time intervals after it was dropped from a tabletop. *(b)* Graph of velocity of the ball versus time. Even though the velocity is changing, the rate of change of velocity (the acceleration) remains constant. (Air resistance is assumed to be zero.)

> I. A body continues in a state of rest, or motion with a constant velocity (i.e., with a constant speed in a straight line), unless compelled to change by an unbalanced (net) outside force.

This statement was formulated not by Newton but by the Italian scientist Galileo Galilei (1564–1642) on the basis of experiment, and many consider it to be the first successful application of the scientific method. It is also called the law of inertia. **Inertia** is the tendency of all objects to resist changes in their motion. A body continues to move because of its inertia. The measure of an object's inertia is its mass. It is more difficult to stop a heavy car than a light car, both moving at the same speed, because of the greater inertia of the heavy car. The dining-room trick of pulling a tablecloth out from under a setting of dishes without disturbing them is possible because of the inertia of the dishes and silverware (Fig. 2.10).

Newton's first law is certainly not self-evident and appears to be contradicted by our everyday experience. If you take your foot off the gas pedal of your car, you will slow down. Give this book a push across the table and it will come to rest in a short distance. In fact, most scientists before Galileo, dating back at least to Aristotle (~300 BC), felt that the natural state of all objects was to be at rest. They thought that a push or a pull was required to keep an object moving.

However, objects such as the car and book stop moving because of an external resistive force acting on them, known as the force of friction. Friction is a result of the interaction of two surfaces in contact and always acts to oppose the

FIGURE 2.10
Because of their inertia, the dishes should stay on the table after the tablecloth is quickly pulled out.

motion that is taking place. The amount of friction depends on the type of surfaces in contact and the mass of the moving object. According to the law of inertia, the friction force causes the motion of the object to change. We don't find many situations in nature in which the force of friction is absent (or very small). One example is motion on an icy surface. If we set a hockey puck in motion across a frictionless pond of ice, it will continue to move at a constant velocity until it reaches the far side; the resistive force on the puck is near zero. Another situation occurs in outer space, in which a spacecraft can continue moving indefinitely after the engines are turned off.

As our experience also indicates, the application of a force to an object will not necessarily cause the velocity of the object to change. A person pushing on a car will not usually cause it to move. Several forces can act on an object in such a way as to produce no change in its motion. What changes the motion is the application of a *net* or *unbalanced* force, as the first law states. In the example of the car, the force exerted by one person is balanced by the force of friction. If the net force on an object is zero, then we say that the object is in equilibrium; its velocity is not changing; the acceleration is zero. Note that the velocity can be nonzero and the object can still be in equilibrium (i.e., there is zero acceleration).

SECOND LAW As the first law states, the application of a net force to an object results in a change in its velocity, that is, an acceleration. The second law of motion quantifies this relationship between net force and acceleration:

> II. The acceleration of an object is directly proportional to the net force acting upon it and inversely proportional to its mass. The direction of the acceleration will be the same direction as the net force.

Mathematically,

$$a = \frac{F_{\text{net}}}{m} \quad \text{or} \quad F_{\text{net}} = ma$$

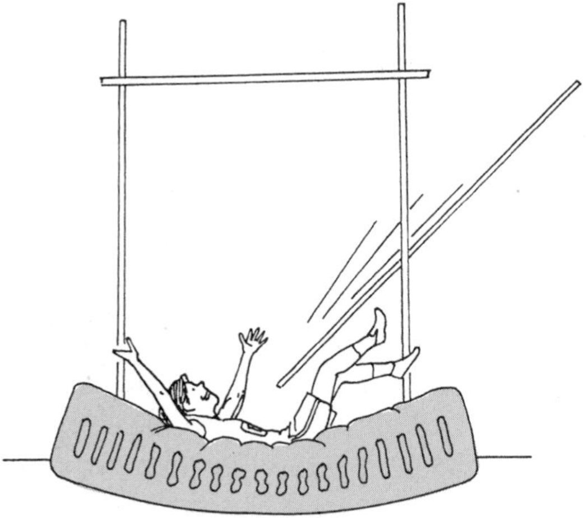

FIGURE 2.11
Foam rubber slows your landing. The small deceleration of the pole-vaulter while landing on the foam rubber makes the force he experiences also small, since $F = ma$.

A force can cause deceleration as well as acceleration. If a car smashes into a brick wall, its rate of change of velocity is very large, that is, its deceleration is large. If the deceleration were smaller, as would occur if the car smashed into the barrels of water that guard some exits from freeways, then the force on the car and its occupants would be smaller. A pole-vaulter would rather land on foam rubber than concrete because his or her deceleration will not be as large on the foam rubber, and therefore the force on the athlete will be smaller (Fig. 2.11).

EXAMPLE

A cart of mass 15 kg has a *net* force applied on it of 120 N. What will be its velocity after 5 s if it starts from rest?

Solution

We need to first find the acceleration of the car.
Newton's second law says that $a = F/m = 120\text{ N}/15\text{ kg} = 8\text{ m/s}^2$.
Now acceleration a = change in velocity/time = $(v - 0)/5\text{ s}$.
Therefore,

$$v = at = (8\text{ m/s}^2)(5\text{ s}) = 40\text{ m/s}$$

Newton's second law also says that the acceleration of an object will be in the direction of the net force. If two people pull on a box in opposite directions, one with a force of 10 lb and the other with a force of 20 lb, the box will accelerate in the direction of the 20-lb force. Note that the result of a net force is the acceleration of the object, such as, a change of the velocity it had before the (net) force was applied.

THIRD LAW Have you ever stopped to think what is responsible for the acceleration of a rocket in outer space? There certainly is no air on which the rocket can push, yet it *can* accelerate. The explanation for this phenomenon is the same as that for many common, everyday experiences: "forces act in pairs." If you push on a wall with a force (called the "action force"), then the wall pushes back on you with a force (called the "reaction force"), equal in magnitude but opposite in direction to the action force. It is as if a person were standing on the other side of the wall pushing on you. (However, you don't move because friction keeps your feet from sliding.) Newton saw the importance of this relationship and stated it as the third law of motion:

> III. For every action force there is an equal and opposite reaction force.

A good illustration of this law occurs in the interaction between two skaters on a frozen pond. If one pushes on the other, both skaters will move, the one who initially pushed being subject to the reaction force of the other skater. If both skaters have the same mass, then they will both experience the same acceleration, but in opposite directions (Fig. 2.12).

Note that the action and reaction forces always act on different objects. In the acceleration of a space shuttle into outer space, the rockets force the gas out of the ship (action force); the reaction force is the force of the exiting gas on the rocket, thus propelling the ship forward (Fig. 2.13).

WEIGHT VERSUS MASS In physics a distinction must be made between the weight of an object and its mass. Weight is a force and is a measure of the force of gravity on an object. Mass is an intrinsic property of a substance. Mass is a measure of the "inertia" of an object; it never changes. However, an object's weight depends on its position in a gravitational field. On the moon, the weight of an object is one

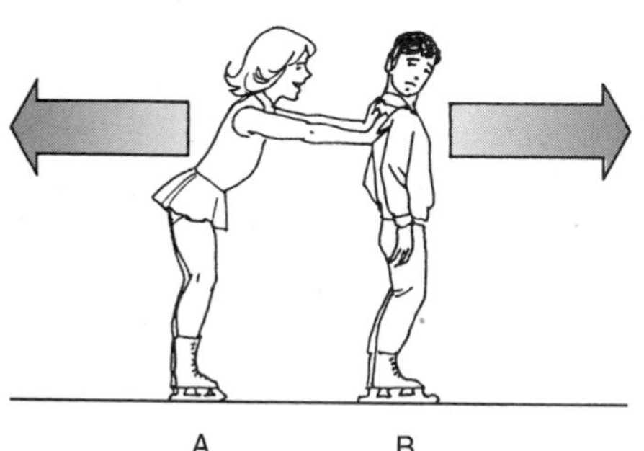

FIGURE 2.12

Skater A experiences a force equal in magnitude but opposite in direction to the force she exerts on skater B.

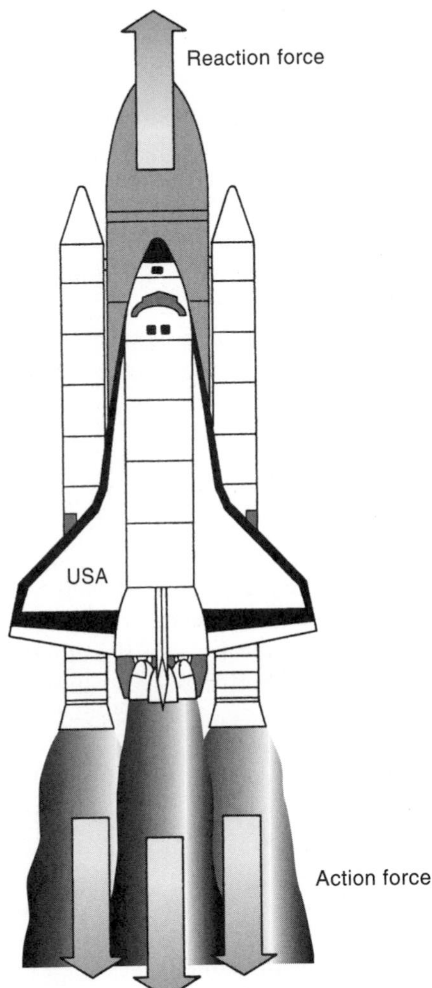

FIGURE 2.13
The reaction force of the exiting gases on the rocket accelerates the *Columbia*.

sixth of that when it is on earth, but its mass is the same. From Newton's second law, $F = ma$, we know that the weight w of an object is equal to its mass m times the acceleration due to gravity g (9.8 m/s^2 or 32 ft/s^2, at sea level): $w = mg$. For example, the weight of a 1 kg mass at sea level is equal to 1 kg $\times$ 9.8 m/s^2 = 9.8 N.

In the English system of measurement, the unit of mass is defined as the mass of a body whose acceleration is 1 ft/s^2 when the net force on the body is 1 lb. This unit of mass is called a slug. Therefore 1 lb = 1 slug $\times$ 1 ft/s^2. In everyday usage the pound is used to refer to a quantity of matter, but it really is a unit of force or weight. A useful fact is that an object with a mass of 1 kg will have a weight (at sea level) of 2.2 lb. (This won't be true on the moon.) An object with a mass of 1 kg does not weigh 1 kg—a common mistake.

Work and Energy and Units

Recall that the kinetic energy of an object is given by $KE = ½\,mv^2$. From this expression, we can see that the units of energy are kg-m^2/s^2, which is defined as a joule (J). Note that since $W = F \times d$ and $F = ma$, the units for work are kg-m/s^2 $\times\ m$, which is also a joule. The units for the terms of motion and energy are given in Table 2.4. Conversion factors between the English system and SI are also listed. A more complete set of conversions is found in Table 3.4.

Since KE and PE are both forms of energy, we can compare the amount of energy found in each of these forms, as the following example illustrates.

EXAMPLE

A cart with a mass of 10 kg is originally moving at a speed of 5 m/s. To what height above the ground would it have to be raised so that its potential energy there would have the same value as its KE?

Solution

The cart's original kinetic energy is $KE = ½\,mv^2 = ½ \times 10$ kg $\times$ (5 m/s)^2 = 125 J. Now, the weight of the cart ($w = mg$) is 10 kg $\times$ 9.8 m/s^2 = 98 N. We want $PE = KE$, so 98 N $\times$ height = KE = 125 J. Therefore, the height is 125 J/98 N = 1.28 m.

Simple Machines

A machine is a device that multiplies a force at the expense of a distance. Energy is still conserved for these devices, but machines can reduce the force needed to perform a task. One of the simplest machines is the **lever,** which is a rigid bar pivoted at a point called the fulcrum (Fig. 2.14). A force F applied at one end of the bar lifts a load A (the output force) at the other end. The ratio A/F is called the "mechanical advantage" of the machine and is usually greater than 1. Levers have been used since prehistoric times to move heavy stones, lift water, and help in building construction.

Another example of a lever (also shown in the figure) is one in which the load is between the fulcrum and the applied force, as in a wheelbarrow. To lift a heavy load, you can apply a smaller force at the end of the handle; since energy is conserved, this force will be applied over a greater distance than the distance the load is raised.

Another very simple machine is the **inclined plane.** Pushing a cart up a plane is easier than carrying the same cart straight up to the top by a ladder, because the applied force is less; however, the work done in both cases is the same. Inclined planes or ramps were probably used during the construction of the Egyptian pyramids to move the large blocks of stone to the top.

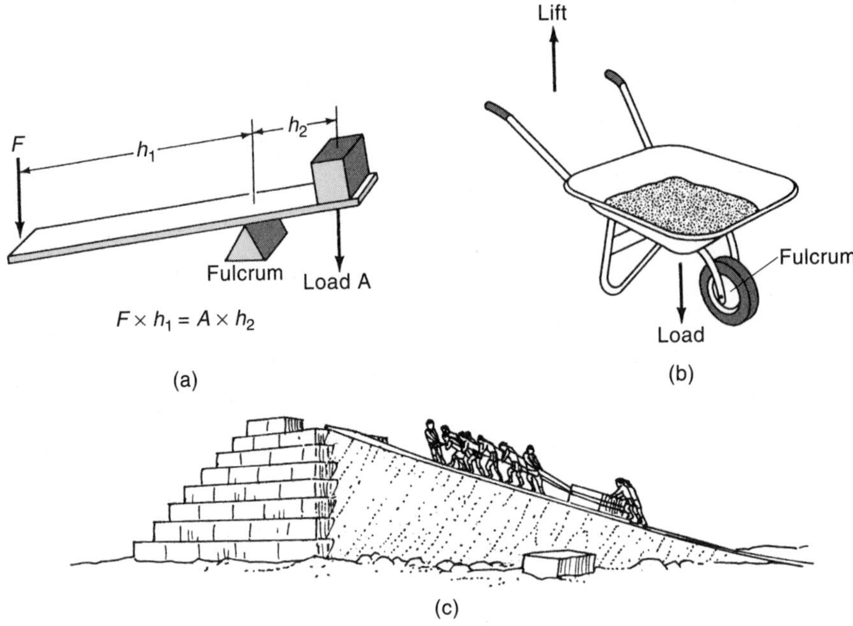

FIGURE 2.14
Three types of simple machines: *(a)* lever; *(b)* wheelbarrow; *(c)* inclined plane (pyramid construction).

PROBLEMS

1. How long does it take a car to travel 1 mile if its speed is 40 mph?
2. If you shout across a canyon and the echoes return in 2 s, how far away is the other side? (Take the velocity of sound to be 330 m/s.)
3. In a television tube, an electron moves at a speed of 4×10^7 m/s. How long does it take to hit the screen 0.5 m away?
4. A ship can maintain an average velocity of 30 km/h on an ocean voyage. How far will the ship have traveled in 4 days?
5. A bird takes 2 min to travel 350 m. What distance would this bird cover in 10 s, assuming uniform motion?
6. A car accelerates at a constant rate of 4 m/s^2 from rest. Find its velocity after 6 s.
7. A net force of 10 N is applied to a 3-kg block at rest on a smooth, level surface. Find the block's velocity after 9 s. (*Hint:* Find the acceleration first.)
8. A 1300-kg car has a net force of 3900 N applied. After 100 m, what is the car's kinetic energy and velocity?

3

Conservation of Energy

A. Introduction

The past two decades have seen a minor revolution in our concept and understanding of the role of energy in diverse societies, especially in the industrialized world. This has been through the employment, implementation, and acceptance of energy conservation. The developed nations to a large extent have come to realize that sustainable development can take place without irreversibly bringing about damage to our environment and without over using our resources. Energy conservation, once considered a "weak sister" to resource deployment, is now *the* approach of choice when it comes to meeting the needs of a growing population and a growing economy. Although there is much room for improvement, the U.S. situation has demonstrated that a society can grow economically without a commensurate increase in energy consumption. In the past 20 years, the U.S. Gross Domestic Product (GDP) rose 66% (in constant dollars) while energy use only rose 18%. What is as interesting is the 40% *rise* in per capita GDP versus the 3% *decrease* in per capita energy use. In contrast to this, developing countries are seeing a rapid increase in per capita energy consumption. While they account for more than 80% of the world's population, they consume only about one third of the world's energy. However, this share is expected to grow to nearly 40% in the next decade due to rapid industrial expansion, high population growth and urbanization, and rising incomes that enable families to purchase energy-consuming appliances and cars.

Chapter 1 provided some good arguments for energy conservation. The technical side of this approach is emphasized throughout the book, so there is no separate chapter on energy conservation. Recall that one approach to energy conservation was the "technical fix," in which we employ more efficient technologies to allow us to do a particular task with less energy. For example, illuminating a room with fluorescent bulbs rather than incandescent bulbs provides the same light for one quarter of the energy. Increased efficiency for a conversion process is a centerpiece of energy conservation—that is, doing the same task with a smaller expenditure of energy.

B. Conservation of Energy Principle

Recall from Chapter 2 that two ways the energy of a system can be changed is by doing work on it or by adding heat to it. (Mass can also be added.) Of special interest in the equation of the first law of thermodynamics is the case in which no work and no heat is added to the system: $W + Q = 0$. This is an example of what is called an "isolated" or **closed** system. No forces act on an object in the system from the outside. In an isolated system, the *change* in the total energy will always be zero: $0 = \Delta(KE + PE + TE)$. Stated in another way:

> The total amount of energy in an isolated system will always remain constant.

We say that the energy in the isolated system is conserved. Energy does not appear from nowhere. If our isolated or closed system is the universe itself, then the law of conservation of energy states that the *total energy in the universe is a constant and will remain so.* Within this isolated system we can certainly have transformations or conversions of energy from one form to another, as from potential energy into kinetic energy. For example, X units of chemical energy might be converted (via combustion) into Y units of thermal energy plus Z units of mechanical energy. Conservation of energy then states that $X = Y + Z$.

Note that the principle of the conservation of energy is different than "energy conservation" discussed in the previous section. The latter concept has to do with reducing the amount of energy used through reduced activity (turning down thermostats, driving fewer miles) and/or increased efficiency in the performance of a particular task (more efficient furnaces, automobiles, etc.).

Conservation of energy is a very important and useful principle in describing physical processes, and will be used many times throughout this book. It is a principle that cannot be proven from basic principles, but it is a very good bookkeeping device and no exceptions have ever been found. When all forms of energy are included in the previous equation, the total energy of an isolated system is conserved.

As an example of energy transformations and the principle of the conservation of energy, consider the "nosecracker" seen in Figure 3.1. This is a ball suspended by a rope from the ceiling. (This is much like a simple pendulum.) In the raised position, the ball is at rest and has gravitational potential energy only. Consider the potential energy of the ball to be the energy it possesses relative to the lowest point of its swing. The potential energy is at a maximum when the ball is at the top of its swing. After the ball is released, the potential energy decreases and the kinetic energy increases as the ball swings closer to the bottom. The ball's kinetic energy reaches a maximum at the bottom of the swing, where the potential energy is zero. In the absence of frictional forces, the maximum kinetic energy at the bottom of the swing is *exactly equal* to the initial potential energy of the ball at the top of its swing.

The main point of the nosecracker example is that the total mechanical energy ($PE + KE$) of the ball is conserved; no energy is transferred into the system by work or heat, so $\Delta(KE + PE) = 0$. (The earth must be included in our isolated system since it exerts a gravitational force on the ball.) Any loss in PE goes into KE acquired by the ball. That is, the mechanical energy ($KE + PE$) initially equals the mechanical energy ($KE + PE$) after. If the ball began with 100 J of PE at the top of the swing, the sum of ($KE + PE$) would always remain at 100 J. Energy is transformed from one form to another—from potential at the top of the swing to kinetic at the bottom; in between these two positions, the ball has both KE and PE. In everyday life there are other forms into which the original

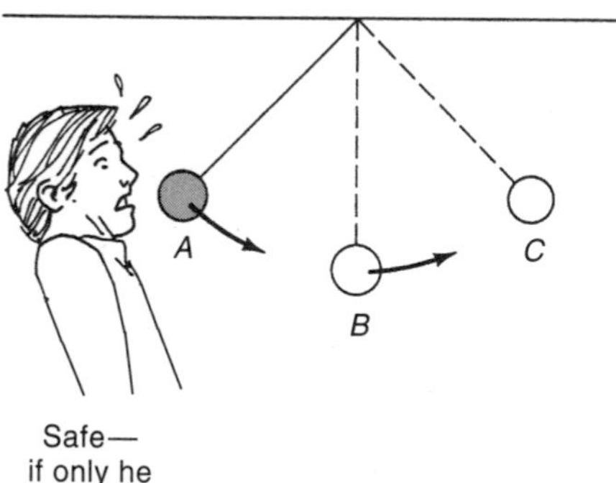

FIGURE 3.1
If the ball is released carefully from Point *A* to swing through Points *A–B–C–B–A*, it will not go higher than *A* on its return (and so will not break the person's nose). The mass possesses a fixed amount of energy at Point *A*, which is in the form of gravitational potential energy. At Point *B* all the energy of the mass is in kinetic form, and it has its maximum velocity at this point. The mass cannot return higher than Point *A* because to do so it would need to gain energy, which is impossible in the absence of external forces.

ACTIVITY 3.1

One can study the conversion of PE into KE with the following activity. On a smooth, level surface, place a ruler and a book to make an incline. Place a marble at the 10-cm position on the ruler and let it roll down the ruler. When it hits the level surface, measure the time it takes the marble to roll 1 m. Place the marble at the 20-cm and the 30-cm positions and again measure the times on the level surface. Find the velocity of the marble for all three positions. Plot velocity squared versus distance up the ruler. What is the shape of the plot? If it is a straight line, it shows that the marble's KE at the bottom is proportional to the PE of the marble at the release point. (Recall that KE is proportional to the square of the velocity.)

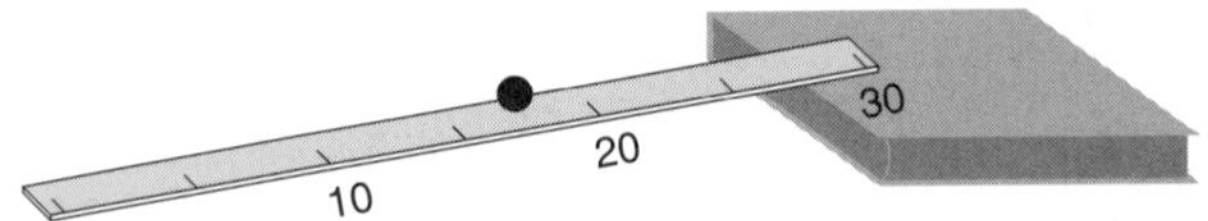

energy is transformed. One of the most common forms (and ultimately the final form) to which the original energy goes is thermal energy. In the example of the nosecracker, it was assumed that none of the mechanical energy was lost to heat, that is, there was no friction at the socket in the ceiling to which the rope was attached and no air resistance. The transformation of mechanical energy to thermal energy in the socket and the air over many swings is why the ball eventually stops swinging.

Since energy in a closed system is not destroyed or created or generated, one might then wonder why we need be so concerned about our energy resources, since energy is a conserved quantity. The problem is that the final result of most energy transformations is the production of **waste heat** that is transferred to the environment and no longer useful for doing work. To state it in another way, the potential for energy to produce useful work has been "degraded" in the energy transformation. (We will elaborate on this very important point in Chapter 4.) Because of the use of our resources, we are beset with problems of thermal pollution, air pollution, and possible global climatic changes.

C. Energy Conversion Examples

Let us now examine a system capable of interacting with the outside world and consider the transfer of energy into and out of the entire system. This is no longer an isolated system, but energy is still a conserved quantity. The *net* energy added

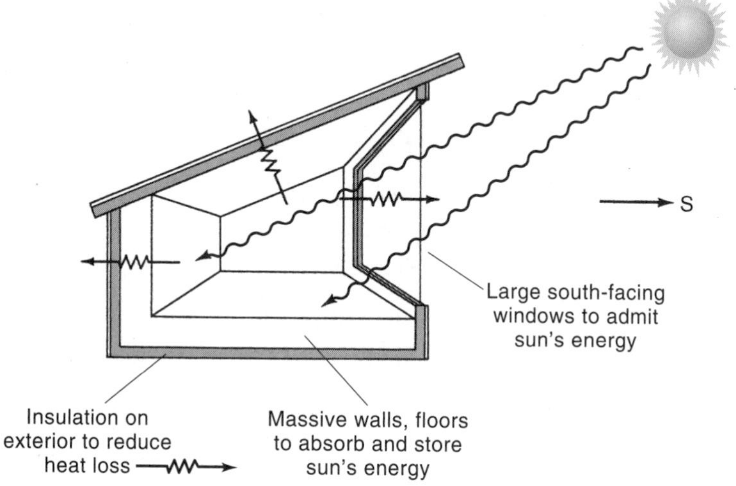

FIGURE 3.2

A passive solar energy house. Energy in = energy out + energy stored.

(the energy in minus the energy out) is equal to the change in the energy of the system. The principle of the conservation of energy can now be stated as:

> The energy into a system equals the energy out plus the increase of energy stored in the system.

This law of the conservation of energy can be illustrated with a passive solar house (Fig. 3.2), as will be covered in detail in Chapter 6. Such a house acts as a collector of solar energy. The solar energy coming through the south-facing glass windows (input) is equal to the energy that leaves the

Passive solar home, Oswego, New York. Thirty-five percent of the heating needs are provided by solar energy. (C. SALVAGIN)

house (as heat transferred out through the walls and windows and ceiling to the outside) plus the increase of energy stored in the material within the house.

Another example (one in which the energy storage term can be ignored) is a fossil-fueled steam power plant for the generation of electrical energy (Fig. 3.3). Although many of the details will have to be developed later, the overall functions are straightforward. Fuel (oil, coal, natural gas) enters the boiler unit of the steam plant where combustion takes place using air from the atmosphere. Combustion of the fuel generates heat that converts the water in the boiler into steam. (The water gains thermal energy.) The high-temperature, high-pressure steam is directed through nozzles onto the blades of a turbine. The rotating blades of the turbine turn a shaft that is connected to a generator for the production of electricity. The steam, at a reduced temperature and pressure, leaves the turbine and passes through a condenser, where it becomes a liquid. The condenser takes cold water (from a lake or other body of water) and passes it through heat exchanger coils to condense the steam; this cold water becomes warmer and is discharged into

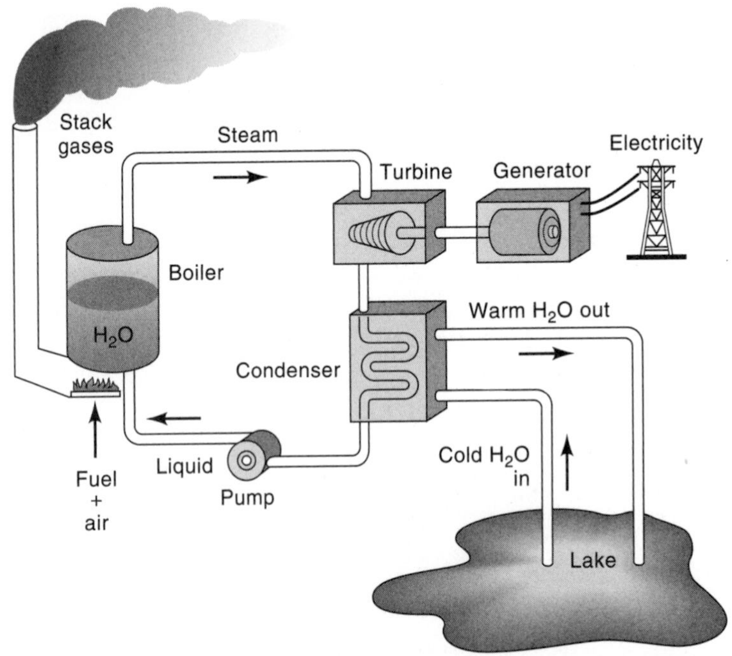

FIGURE 3.3

Block diagram of a fossil-fuel-powered electric generating station. Energy in = energy out, because no energy is stored.

the lake, resulting in thermal pollution. The condensed water is then pumped back to the boiler for recycling through the entire process. The total energy input to this system is found in the chemical energy of the fuel and the air for combustion, and in the thermal energy of the cooling water for the condenser. The energy output is contained in the electrical energy generated and leaving the plant, in the thermal energy of the hot water leaving the condenser, and in the combustion gases emitted from the smoke stack. No energy is stored, as the water returns to the boiler with the same thermal energy as when the process originally started. Using the notation E_x as the energy associated with item x, we can write the equation for the conservation of energy for the power plant as

$$E_{\text{fuel}} + E_{\text{air}} + E_{\text{water in}} = E_{\text{electricity}} + E_{\text{water out}} + E_{\text{combustion gases}}$$

D. Energy Conversion Efficiencies

Even though energy is conserved in an energy conversion process, the output of *useful* energy or *useful* work will be less than the energy input. The **efficiency** of an energy conversion process is defined as

$$\text{Efficiency} = \frac{\text{useful energy or work output}}{\text{total energy input or energy stored}} \times 100\%$$

The energy input that does not go into useful work goes into unusable energy forms (such as waste heat). In the power plant example of the preceding section, only a fraction of the energy input (primarily the chemical energy of the fuel) to the generating plant is transformed into electrical energy. The percentage of the fuel energy converted into electrical energy is the plant's efficiency and is about 35% for a conventional fossil-fuel-powered plant. Using the previous notation,

$$\text{Efficiency} = \frac{E_{\text{electricity}}}{E_{\text{fuel}}} = 0.35 \times 100\% = 35\%$$

While essentially all of the fuel's chemical energy is converted into heat during combustion, 65% of this heat is transferred to the water leaving the condenser and gases going up the smokestack. This heat ends up in a lake or river as increased thermal energy of the water (or in the atmosphere when cooling towers are used).

The efficiencies of other energy conversion devices are given in Table 3.1. Efficiency values range from about 5% for the incandescent light bulb (converting electrical energy into light) to 95% for the best electrical motor (converting electrical energy into mechanical). As we shall see in Chapter 4, the laws of

Table 3.1 EFFICIENCIES OF SOME ENERGY CONVERSION DEVICES AND SYSTEMS

Device	Efficiency
Electric generators (mechanical → electrical)	70–99%
Electric motor (electrical → mechanical)	50–95%
Gas furnace (chemical → thermal)	70–95%
Wind turbine (mechanical → electrical)	35–50%
Fossil-fuel power plant (chemical → thermal → mechanical → electrical)	30–40%
Nuclear power plant (nuclear → thermal → mechanical → electrical)	30–35%
Automobile engine (chemical → thermal → mechanical)	20–30%
Fluorescent lamp (electrical → light)	20%
Incandescent lamp (electrical → light)	5%
Solar cell (light → electrical)	5–28%

thermodynamics place a limit on the efficiencies that can be attained in some of these conversions, notably those involving heat to mechanical energy. Increased efficiency in the use of energy means that the same job can be done with a smaller amount of energy.

If we have a multistep process, the **overall efficiency** is equal to the *product* of the individual efficiencies. For example (Fig. 3.4), if the efficiency for the conversion of the chemical energy in coal into electricity is 35%, the efficiency for the transmission of electricity through high voltage lines is 90%, and the conversion of electricity into light in an incandescent bulb is 5%, then the overall efficiency for the conversion of chemical energy into visible light is $= 0.35 \times 0.90 \times 0.05 = 0.016 = 1.6\%$. (Note that the overall efficiency will never be greater than that of the step with the lowest efficiency in the process.)

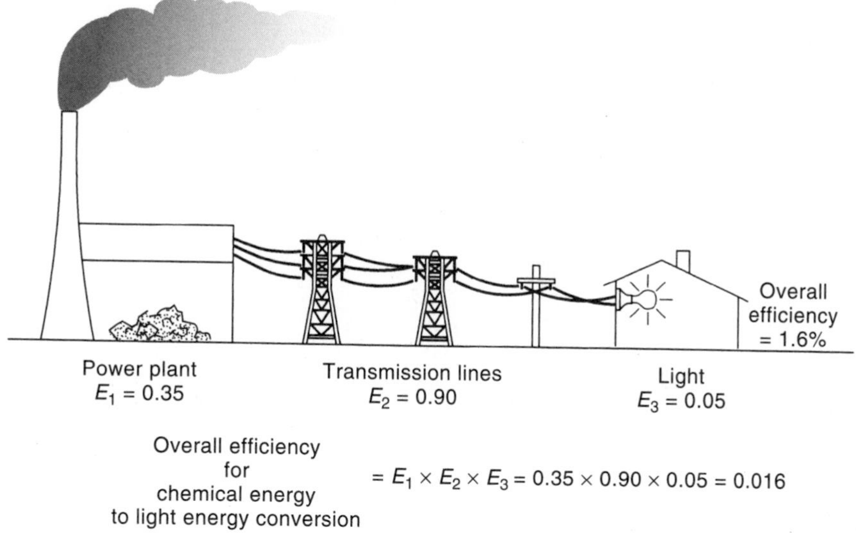

FIGURE 3.4

Calculation of the overall efficiency for a multistep process involves multiplying the efficiencies of the individual steps.

E. Energy Use in Developing Countries

In many developing countries, human power still provides a large share of the energy consumption per person, at least when traditional fuels (wood, biomass) are excluded. (See Fig. 3.5.) We must go back far in history to find societies that used only human power. As fire began to be used for cooking, the consumption of energy rose. With the domestication of animals, additional mechanical energy became available. It was not until the industrial revolution and the invention of the steam engine (in the 18th century) that useful energy on a large scale began to be available.

Developing countries today comprise almost three quarters of the world's population but consume only one quarter of the energy. There is a large disparity in per capita energy use among the developed countries and the developing countries. (See Figures 2.6 and 2.7.) The average American consumes about three times as much commercial energy as a person from France, 30 times as much as an Indian, and 300 times as much as a person from Haiti or Nepal. (These latter comparisons are misleading, since a large majority of energy used in developing countries comes not from commercial fuels such as oil but from traditional fuels of firewood, plant residues, and dung.) Since 1960, developing countries have quadrupled their energy use while tripling their per capita use.

FIGURE 3.5
An interesting method of pumping water in Burkina Faso, West Africa. The boy is doing work using his PE, gained by jumping in the air. (W. BENNETT)

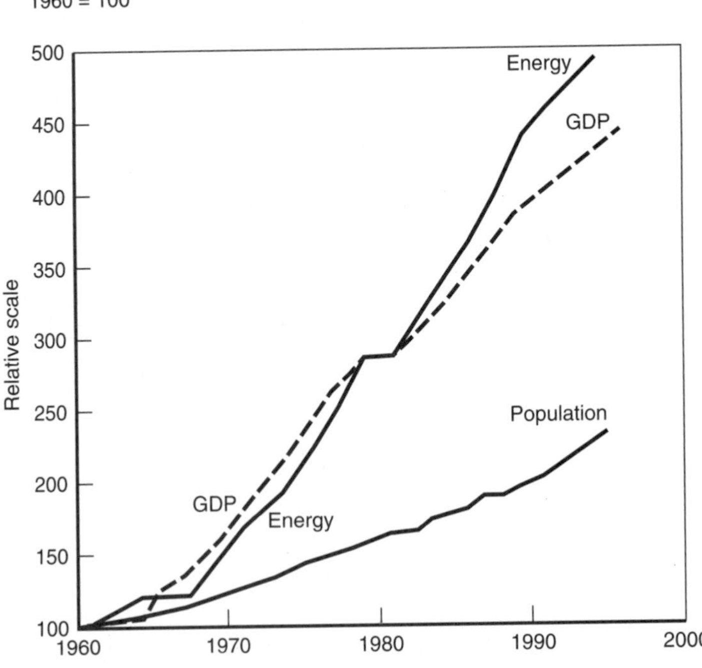

FIGURE 3.6
Growth in energy use, GDP, and population in developing countries: 1960–1995. (SOURCES: BRITISH PETROLEUM, UNITED NATIONS, WORLD BANK, POPULATION REFERENCE BUREAU)

A rapidly growing economy in China has led to increased use of coal combustion. About 75% of its electricity is generated with coal. (FORREST ANDERSON/GAMMA LIAISON)

(See Fig. 3.6 and Table 3.2.) However, such rapid growth has caused staggering problems in some (but not all) developing countries as a result of massive increases in foreign debt from oil imports and serious health and environmental problems from a decline in air quality and water degradation. Three quarters of developing countries are oil importers, and of the poorest countries, three quarters import more than 70% of their commercial energy.

Table 3.2 WORLD COMMERCIAL ENERGY CONSUMPTION: 1970 AND 1999

	1970		1999	
Region	**Energy Consumption**	**Per capita**	**Energy Consumption**	**Per capita**
	(10^{18} J)	**(10^9 J/person)**	**(10^{18} J)**	**(10^9 J/person)**
Developing countries	31	12	137	34
Latin America	8	26	16	49
Asia	19	10	110	34
Africa	4	10	11	22
Industrial countries	129	180	183	221
Centrally planned economies	44	120	38	128
WORLD	203	55	358	70

Sources: British Petroleum, *BP Amoco Statistical Review of World Energy* (London: 2000); United Nations, *World Population Prospects 1990* (New York: 1991).

Table 3.3 ENERGY USE IN DEVELOPING COUNTRIES

Source	Percent
Biomass	35
Oil	26
Coal	25
Natural gas	8
Other renewables	6
Nuclear	<1

Source: Worldwatch Institute, (www.worldwatch.org), "State of the World, 1993."

The developing countries' energy plans seem to be trying to follow the example of the industrial nations, which grew economically through the use of technologies fueled especially by fossil fuels and electricity. Oil and coal already provide more than 50% of the energy used in developing countries (Table 3.3). In the next 20 years, the developing countries' share of world CO_2 emissions might increase from 28% to 44%. Coal, which produces the highest CO_2 emissions of any fossil fuel, generates 70% of the electricity in China and India. Electricity demand is projected to continue to rise by 7% per year—a doubling time of 10 years. Already, industrial and residential demand has outstripped the growth in power-generating capacity of some nations, leading to power shortages in large cities. India's electricity shortfall averages 9%, while China's shortfall idled one quarter of the country's industrial capacity in 1987. In the 1990s, China's strong industrial growth required the *addition* of 12,000 MWe of electrical power each year over the decade. That's equivalent to building about one large power plant each month!

To deal with these problems, developing countries will need to invest in more energy-efficient technologies as well as to develop alternatives to oil and coal. Desired economic development and population growth will make this difficult. In the next 30 years, global population is projected to grow 40%, from 6 billion to 8.5 billion. Almost all of this growth will be in developing countries, primarily in urban areas.

F. A Barrel, a Calorie, a Btu? Energy Equivalencies

In discussions of energy demand and supply, we must use a consistent set of units when talking about different quantities. A gallon of crude oil has the capacity to heat a certain amount of water by so many degrees. It could also be

Focus On 3.1

SUSTAINABLE DEVELOPMENT

Both economic development and population growth exert tremendous pressures on our natural resources and systems. For example, land and water degradation make the expansion of food production extremely difficult. Many of the environmental problems associated with the increased use of energy to fuel this growth have already been introduced. "Sustainable development" is that type of development that meets the needs of the present without compromising the ability of future generations to meet their own needs. It necessitates protecting the natural systems needed for food and fuel while expanding production to meet the needs of a growing population.

The approach to meet this goal will be different for developed and developing countries. Industrial countries have a special responsibility for leadership in sustainable development because their past and present consumption of natural resources is disproportionately large. On a per capita basis, developed countries use many times more of the world's resources than do people from developing countries (see Fig. 2.6). The industrialized countries also have the financial and technical resources to develop cleaner, less resource-intensive technologies.

Building a house from handmade mud and straw bricks.
(HABITAT FOR HUMANITY, KENYA)

For developing countries, sustainable development is the use of resources for the improvement of their standard of living. One fifth of the world's population have annual per capita GDPs of less than $500. They are also faced with severe health problems. The citizens of poor countries usually have limited access to safe drinking water and sanitation, are malnourished, and have the lowest levels of education. One half of the population in developing countries are illiterate; average life expectancy is 55 years (76 years in industrialized countries); and infant mortality is 90 deaths per 1000 live births (8 per 1000 in industrial countries). These countries must provide for basic human necessities, stabilize population, and alleviate poverty while at the same time conserve natural resources essential to economic growth—a difficult task.

refined into gasoline and be used to run a car for so many miles. One barrel of oil (42 gallons) converted into gasoline can fuel the family automobile for about 1300 km (780 miles) or heat the average home for 4 days to 20°C when the outside temperature is 0°C. However, much of the chemical energy inherently contained within the gasoline will not be converted into mechanical work because of the low efficiency of the automobile engine, and engines vary widely. Because of the various efficiencies involved in different machines, it is better to speak of energy quantities in terms of heating values rather than mechanical work done. The "heating value" of a fuel is the amount of heat the fuel could provide if completely burned. The heating values of different fuels are usually given in the United States in Btu's. One Btu is the energy needed to raise the temperature of one pound of water by 1°F. (Two Btu's of heat energy will raise the temperature of one pound of water by 2°F, or two pounds of water by 1°F.) A similar unit in the metric system is the calorie; one calorie is the energy needed to raise the temperature of one gram of water by 1°C. One Btu = 252 calories (cal) = 1055 joules (J). A food Calorie is equal to 1000 calories, or 1 kilocalorie, the energy necessary to raise the temperature of 1 kilogram of water by 1°C. Our food intake is about 2000 to 3000 kilocalories per day. A list of heating values for different types of fuels is given in Table 3.4. For example, 1 gallon of gasoline has a heating value equivalent to about 125,000 Btu; if burned in an electric power plant, it would provide about 10 kWh of electrical energy.

It is also convenient to express the energy stored in different fuels in another way. The heating values of coal, uranium, and natural gas, for example, can be equated to the heating value of so many barrels of oil. In terms of energy consumption, it is convenient to express the total annual consumption of other fuels as being equivalent to using so many million barrels per day of oil (MBPD) for an entire year. For example, burning 500 million tons of coal per year would provide the same energy output as burning 6 MBPD of oil for a year. The rate of consumption of crude oil in the United States is about 6 billion barrels per year, or 16 MBPD. Data for conversions between different units of energy and power, as well as heating values of different fuels, appear in Table 3.4, as well as Appendix B and the inside back cover.

Table 3.4 CONVERSIONS AND EQUIVALENCIES

Energy Units	1 Btu = 1055 J = 778 ft-lb = 252 cal 1 ft-lb = 1.356 J = 0.33 cal 1 calorie = 4.184 J 1 food Calorie = 1000 cal = 1 kcal 1 hp-hr = 2.68×10^6 J = 0.746 kWh 1kWh = 3.61×10^6 J = 3413 Btu = 2.65×10^6 ft-lb 1 Quad = 10^{15} Btu 1 GJ = 10^9 J = 948,000 Btu
Power Units	1 watt = 1 J/s = 3.41 Btu/h 1 hp = 550 ft-lb/s = 2545 Btu/h = 746 W
Fuel Relationships	1 barrel (bbl) crude oil = 42 gal = 5.8×10^6 Btu = 6.12×10^9 J 1 standard ft^3 natural gas (SCF) = 1000 Btu 1 therm = 100,000 Btu 1 gal gasoline = 1.24×10^5 Btu 10^6 ft^3 natural gas = 172 bbl crude oil 1 ton bituminous coal = 25×10^6 Btu 1 ton ^{235}U = 70×10^{12} Btu 1000 bbl/day of oil = 2.117×10^{12} Btu/yr 1 million bbl/day of oil (1 MBPD) = 5.8×10^{12} Btu/day = 80 million tons per year of coal = ⅕ ton per year of uranium oxide
Fuel requirements for a 1000-MWe Power Plant (2.4×10^{11} Btu/d Input)	Coal: 9000 tons/d or 1 unit train load (100–90 ton cars)/d Oil: 40,000 bbl/d or 1 tanker /wk Natural gas: 2.4×10^8 SCF/d Uranium (as ^{235}U): 3 kg/d
Energy Needs	U.S. Total Energy Consumption (1999) = 97×10^{15} Btu (97 Quads) = 45 MBPD oil equivalent = 102×10^9 GJ

continued

Table 3.4 CONVERSIONS AND EQUIVALENCIES *continued*

Everyday Usage and Energy Equivalencies	1 barrel (bbl) of oil = driving 1400 km (840 mi) in average car Electricity for city of 100,000 takes 4000 bbl/d of oil State of California energy needs for 8 h = 1 million bbl oil 1 gal gasoline = 11 kWh electricity (@ 30% generation efficiency) = 5 hours of operation of standard air conditioner = 200 days for electric clock = 48 hours for color TV = average summer days' solar energy incident on 2 m^2 (22 ft^2)
One Million Btu Equals Approximately	90 lb of coal 125 lb of oven-dried wood 8 gal of motor gasoline 10 therms of natural gas 1 day energy consumption per capita in the United States 100 kWh of electricity produced at a power plant
Power Data	1000 MWe utility, at 60% load factor, generates 5.3×10^9 kWh/yr, enough for a city of about 1 million people. U.S. per capita power use = 12 kW Human, sitting = 60 W Human, running = 400 W Automobile at 65 mph = 33 kW

EXAMPLE

If one ton of bituminous coal is burned to generate electricity, how many kWh could be produced if the efficiency of this conversion is 35%?

Solution

From Table 3.4, one ton of bituminous coal = 25×10^6 Btu. Thirty-five percent of this energy will become electricity, or $(25 \times 10^6)(0.35) = 8.9 \times 10^6$ Btu. Now 1 kWh = 3413 Btu, so the amount of electricity produced will be 8.9×10^6 Btu/3413 Btu/kWh = 2560 kWh.

G. Summary

In this chapter we saw that doing work on an object (or adding heat) causes a change in the object's total energy; for example, lifting this book to the top of your dresser gives the book more potential energy. This was expressed as the first law of thermodynamics: $W + Q = \Delta(KE + PE + TE)$. The energy of a closed system (in which no work or heat is added) is conserved. The conservation of energy also states that the energy put into a system = the energy output + the energy stored in the system. The study of energy includes a study of its transformations from one form to another, such as, from mechanical to electrical to thermal energy. We are especially interested in the efficiencies of such transformations. Efficiency is the ratio of useful energy or work output to the total energy input.

Internet References

For an up-to-date list of Internet resources related to the material in this chapter, go to the Harcourt College Publishers website at **http://www.harcourtcollege.com**. The links are in the *Energy: Its Use and the Environment* site on the Physics page. General energy-related sites and some guidelines for using the World Wide Web in your class are on the inside front cover of this book.

References

Chapter 3

Barnes, D., and W. Floor. 1996. Rural Energy in Developing Countries. *Annual Review of Energy*, 21.

Cutnell, J., and K. Johnson. 2001. *Physics.* 5th ed. New York, John Wiley & Sons.

"Economic Development." 1980. *Scientific American*, 243 (September).

Gartrell, J., and L. Schaefer. 1990. *Evidence of Energy.* Washington, D.C., National Science Teachers Association.

Hobson, A. 1999. *Physics: Concepts and Connections.* 2d ed. Englewood Cliffs, NJ, Prentice-Hall.

LeBel, P. 1982. *Energy Economics and Technology.* Baltimore, Johns Hopkins.

Lovins, A. B. 1978. Soft Energy Technologies. *Annual Review of Energy*, 3.

Ostdiek, V., and D. Bond. 1995. *Inquiry into Physics.* 3d ed. Minneapolis, West.

Ross, M., and R. Williams. 1981. *Our Energy: Regaining Control.* New York, McGraw-Hill.

Schipper, L., R. B. Howard, and H. Geller. 1990. U.S. Energy Use from 1973 to 1987: The Impacts of Improved Efficiency. *Annual Review of Energy*, 15.

Summers, C. 1971. The Conversion of Energy. *Scientific American*, 225 (September).

QUESTIONS

1. Since energy is a conserved quantity, what happens to the kinetic energy of a car after you take your foot off the accelerator?
2. For a tea kettle on a hot electric stove element, record the forms of energy going into the system (the kettle). Equate this to the energy stored plus the energy flowing out of the system.
3. The pendulum of a clock swings back and forth. At what position will its kinetic energy be the greatest? At what position will its kinetic energy equal its potential energy?
4. What is the difference between energy conservation and the principle of the conservation of energy? Give some examples of each.
5. Write an expression (using words and not numbers) for the efficiency of an automobile engine; be specific about what energy forms you are referring to.

6. If the overall efficiency for the conversion of the chemical energy of coal into the useful work of an electric motor is 29%, and the efficiency for the production of electricity is 35% while the efficiency for its transmission to the motor is 85%, then what is the efficiency of the motor itself for conversion from electrical to mechanical energy?
7. Table 3.5 examines the conversion processes in an internal combustion engine (ICE) and an electric vehicle. For both power systems, calculate the overall efficiencies. (See the following example, and then complete the second half of the table by filling in the dashes.)

Table 3.5 ENERGY CONVERSION EFFICIENCIES

Efficiency for a process is the product of the efficiencies for the individual steps.
Example of lighting:

Process	Efficiency of Step	Overall Efficiency
Production of coal	96%	96%
Transportation of coal	97%	93%
Generation of electricity	33%	31%
Transmission of electricity	85%	26%
Lighting		
Incandescent bulb	5%	1.3%
Fluorescent bulb	20%	5.2%

Example of automobile:

Automobile Power Units	Gasoline (ICE)		Electric Car	
	Step Efficiency	Overall Efficiency	Step Efficiency	Overall Efficiency
Raw fuel production	83%	83%	96%	96%
Generation of electricity			33%	______
Transmission of electricity			90%	______
Battery			80%	______
Engine	25%	______	90%	______
Mechanical	70%	______		
Transmission system	70%	______	90%	______

8. If the efficiency of a coal-fired electrical generating plant is 35%, then what do we mean when we say that energy is a conserved quantity?
9. About how many Btu's of waste heat are dumped into the environment by a fossil-fueled electrical generating plant that uses 10,000 Btu's of chemical energy during combustion?
10. The efficiency of a light bulb can be written as the ratio of what two quantities?
11. What do we mean when we say that 50 million tons of coal could replace the use of 0.6 MBPD (million barrels per day) of oil used over a year?
12. What is meant by "sustainable development"? Give some examples of development that might occur in industrial or developing countries that are not "sustainable."

PROBLEMS

1. A skateboard with a mass of 3 kg is moving at a speed of 5 m/s on a level surface. It encounters a hill and rolls up until it stops. To what vertical height does it rise? (Ignore friction.) (*Hint:* See equations in Chapter 2.)
2. How many pounds of coal has the same heating value as 20 gal of gasoline?
3. A household furnace has an output of 100,000 Btu/h. What size electrical heating unit (in kW) would be needed to replace this?
4. If a coal-burning electrical generating plant burns 2 tons of coal to generate 6000 kWh of electricity, calculate the efficiency of the plant as the ratio of electricity output to fuel energy input (see Table 3.4).
5. If a 1000-MWe power plant is shut down for one day, what will be the loss in revenue if the utility could sell its electricity at 12¢/kWh?
6. If 60% of U.S. petroleum use goes for transportation, and 1% of this goes for buses, how much oil could be saved per year if all buses were electrified (assuming the electrical energy was generated from nonpetroleum resources)? What is this in MBPD?
7. Suppose you left a 100-W light bulb on continuously for one month. If the electricity generation and transmission efficiency is 30%, how much chemical energy (in joules) was wasted *at the power plant* for this oversight? If the fuel consumption for one meal in China using a kerosene wick stove is 6 MJ (1 MJ = 1,000,000 J), how many equivalent meals could be obtained with this wasted energy?
8. Refer back to Problem 10 of Chapter 2. Using the conversions of Table 3.4, to how many gallons of oil is the potential energy of the water in the lake equivalent?

FURTHER ACTIVITIES

1. Drop a rubber ball or tennis ball and measure the heights to which it rebounds for 10 bounces. What energy conversions have taken place? Graph the height (y-axis) versus the number of bounces (x-axis). Does it rebound the same percentage each bounce? Explain how energy is conserved.
2. This activity is similar to the one earlier in this chapter. Use a ruler and book to make an incline. Place an inverted Styrofoam cup or margarine cup, with a "doorway" cut out of it, at the bottom of the ruler. Place the marble at the 10-cm position on the ruler and let it roll down the incline into the cup. Measure the distance the cup moves. (Do this two or three times to get an average.) Repeat with the marble at the 20-cm and the 30-cm position. Did the cup move twice as far when the marble was twice as far up the ruler? Try the same things for a steel ball and compare the results.
3. Energy in Developing Countries—A Web Assignment

   Purpose: To use the World Wide Web to identify and explore the use of energy in another country and its effect on the economic and political situation.

   Country:

   Capital:

   Date:

   Population:

   % Urban:

   Growth rate:

   Per capita income:

   % Literacy:

   Inflation:

   Per capita energy use:

   Unemployment:

   Principal exports:

   Principal imports:

   Primary energy fuels used:

   Natural energy resources:

   Potential fuels for meeting future energy needs:

   Describe this country's economic and food situations. Discuss how energy resources play a role in these issues. Mention environmental problems that are particularly troublesome. What changes in the use of energy and economic situation have there been in the past 10 to 20 years? List the URLs used for this assignment.

4

Heat and Work

A. Introduction

Approximately one quarter of the energy used in this country goes into the heating and cooling of buildings. In the residential sector, an average of 50% of this energy is used for space heating (Fig. 4.1). It is in this sector that you and I can have a significant impact, both in our own dwellings and in the community in which we live. As discussed in Chapter 1, energy conservation—or

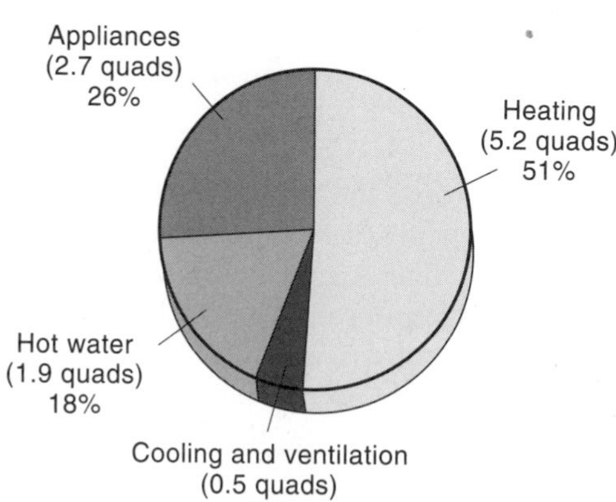

FIGURE 4.1
United States household energy consumption by end use.
1 Quad = 10^{15} Btu. (UNITED STATES ENERGY INFORMATION ADMINISTRATION, 1997)

more precisely, increased efficiency in the use of energy—should be the first step in dealing with environmental impacts of energy use, taking priority even over the increased employment of renewable resources.

The fundamental concepts involved in this chapter and the next are some of the most important ones in this book. In the following sections we will study the area of science and engineering that deals with heat and work, called "thermodynamics." We will elaborate on the laws of energy introduced in Chapters 2 and 3, study the methods by which heat is transferred, and detail steps that can be taken to reduce energy consumption in the areas of space heating and cooling.

B. Heat and Work and the First Law of Thermodynamics

Chapter 2 introduced the different forms of energy. We can group many of these together into what is called an object's "total energy *E*," which is the sum of the object's kinetic energy (KE), potential energy (PE), thermal energy (TE), chemical energy, and electrical energy.

$$E = KE + PE + TE + \text{chemical energy} + \text{electrical energy}$$

Recall that heat and work are the only two ways in which energy can be added to an object to change its total energy if no mass is added. Chapter 2 dealt mainly with the work done on an object to change its energy. The same changes can also be brought about by the transfer of heat energy. However, this was not always thought to be the case.

One of the important discoveries of the 18th century was that heat is just the transfer of energy between two bodies as a result of a temperature difference. Previously, heat was mistakenly thought to be a material fluid, called "caloric," that would flow from a hot body into a cold one, causing an increase in both the body's temperature and its mass. One of the outstanding experiments of the 19th century was the determination of the equivalence between mechanical work and heat. The English physicist James Joule, with a device such as the one shown in Figure 4.2, measured the increase in temperature of a water bath when a paddle wheel was turned in it. He observed that the *same effect* (a rise in the water's temperature) occurred *either* with the performance of work *or* by the addition of heat.

The units for heat could be expressed in joules or foot-pounds because heat is a form of energy. However, historically, the separation between heat and other forms of energy led to the establishment of different units for each. The unit of heat was originally the *calorie,* which is the amount of heat that must be added to one gram of water to raise its temperature by 1°C. To heat 250 g (about one cup) of water from 20°C to 100°C requires 250 g × 80°C = 20,000 calories of energy (if no heat is lost to the surroundings). In the English system, the unit of heat energy is the *Btu,* which is the amount of heat that must be added to one pound of water to raise its temperature by 1°F.

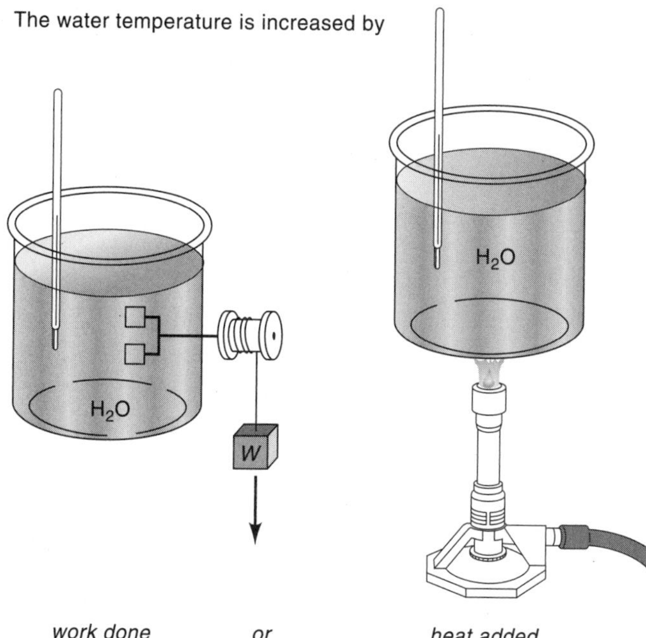

FIGURE 4.2
Relationship between work and heat. A temperature change in the water can be caused either by letting the weight drop (causing the blades to rotate) or by adding heat from the gas burner.

Heat is not contained in a body but is manifest only as the interaction of that body with its surroundings. Heat is a "happening." It is nonmaterial but nonetheless very real in what it can do. Heat addition will usually result in an increase in the temperature of a body. The body's thermal energy has increased. Conversely, in a heat engine, heat is converted in part into mechanical energy. Steam turbines and automobile engines are heat engines.

The interchange between work, heat, and total energy E can be summarized in the **first law of thermodynamics** (as given in Chapter 2):

$$W_{on} + Q_{to} = \Delta(KE + PE + TE) = \Delta E$$

That is, **the work done *on* a system plus the heat added *to* it is equal to the change in total energy of that system.** This law is nothing more than a statement of the conservation of energy: The energy put into a system is equal to the energy output plus the energy stored.

Work done *on* a system is the negative of work done *by* the fluid, that is, $W_{on} = -W_{by}$, so another expression for the first law is

$$Q_{to} = \Delta E + W_{by}$$

This states that the (net) heat Q added *to* a system is equal to the change in the total energy ΔE of that system plus the work W_{by} done *by* the system. For example, if heat from the sun's radiant energy is added to a flexible air-filled balloon,

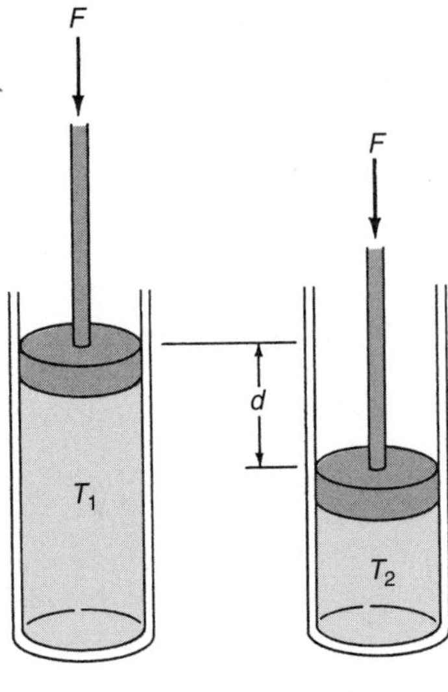

FIGURE 4.3
Bicycle pump. Work done on the air in pushing the handle down results in an increase in the air's thermal energy. $W_{on} = F \times d = \Delta TE$.

the temperature of the air in the balloon will increase (its TE will increase) and the balloon will expand, doing work on the surroundings.

A bicycle pump can also serve to illustrate this law (Fig. 4.3). If the pump is well insulated and the handle is pushed down, work is done on the system, resulting in an increase in the air's temperature (and so in its thermal energy):

$$W_{on} = \Delta E = \Delta TE.$$

If the pump is left standing, the air in the pump will eventually cool off; heat has flowed to the cooler environment and there is a decrease in the air's thermal energy.

C. Temperature and Heat

Many of our experiences with temperature and heat are found in our sensations of hot and cold. When we think of temperature, we might think of how hot the roast is in the oven or of how cold the air is outside. **Temperature** is a property of an object, much like color or shape. When you measure the temperature of a substance, the temperature is the same whether you measure the temperature of a small piece or of the entire substance. (We are assuming the substance has a uniform temperature.) Temperature cannot tell us the amount of energy contained in the substance, since it is independent of the mass. You can quench a candle with your fingers, but you would not put your hand in a pot of boiling water, even though the candle is hundreds of degrees hotter than the water. From a microscopic viewpoint, temperature is proportional to the average kinetic energy of the atoms or molecules of that body. The higher the temperature, the more energetic are the atoms or molecules.

To make the concept of "temperature" scientifically useful, we need a temperature scale. We can define a scale by using reference points related to the temperatures of specific states of a substance. Two such points are the freezing and boiling points of pure water. On the **Celsius** scale, the freezing point of water (at atmospheric pressure) is 0°C and the boiling point is 100°C (at atmospheric pressure). (The temperatures for these points on the **Fahrenheit** scale are 32°F and 212°F.) Since temperature is related to the average kinetic energy of the molecules of a substance, it is convenient to define an absolute temperature scale such that at 0 degrees absolute (0 K) the molecule motion will be at a minimum. This is the **Kelvin** scale, in which 0 K corresponds to –273°C (–460°F); K = °C + 273, so 100°C corresponds to 373 K. Table 4.1 relates the reference temperatures of several substances in these three scales.

When heat is added to a substance, we usually find an increase in its temperature. There is a simple relationship between a temperature change ΔT and the amount of heat added (or removed) Q:

$$Q = mc\Delta T \quad \text{or} \quad \Delta T = \frac{Q}{mc}$$

In this equation, m is the mass of the substance and c is the **specific heat** of that substance. Different materials will absorb different amounts of heat in undergoing the same temperature increase; conversely, the same amount of heat added to two

Table 4.1 TEMPERATURES OF SOME COMMON PHENOMENA*

	°C	°F	K
Water, ice point	0	32	273
Water, boiling point	100	212	373
Absolute zero	–273	–460	0
Liquid nitrogen, boiling point	–196	–319	77
Liquid helium, boiling point	–269	–454	4
Zinc, melting point	420	787	693
Gold, melting point	1063	1945	1336
Solid CO_2 (dry ice) sublimation†	–78	–109	195

*At atmospheric pressure

†Process of going from a solid directly to a gas phase

Table 4.2 SPECIFIC HEATS OF COMMON SUBSTANCES

Material	Specific Heat (J/kg-°C)	Specific Heat (Btu/lb-°F)
Water	4186	1.00
Aluminum	900	0.22
Iron	448	0.12
Copper	387	0.093
Concrete	960	0.23
Glass	840	0.20
White pine	2800	0.67
Ice	2090	0.50
Air	1004	0.24
Rock	840	0.20

different materials can result in significantly different temperature increases. The specific heat is the amount of heat added (or removed) per unit mass per degree of temperature increase (or decrease). In metric units, the specific heat is the number of calories needed to raise the temperature of one gram of material by 1°C. In English units, the specific heat of a substance is the number of Btu's needed to raise the temperature of one pound of material by 1°F. Table 4.2 lists specific heats

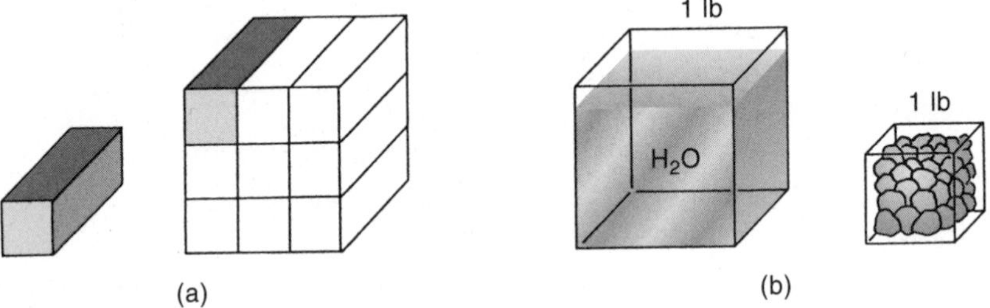

FIGURE 4.4

Thermal energy. (*a*) If both brick assemblies are heated in a kiln for several hours, they will have the same temperature, but the larger array will store nine times as much thermal energy as the smaller one. (*b*) A 1-lb sample of water stores about five times as much thermal energy as 1 lb of rocks at the same temperature, because water has a higher specific heat.

ACTIVITY 4.1

We usually estimate that an average person uses about 10 gal of hot water each day. How many gallons of water do you use for a shower? A bath? Measure the amount by using a bucket. Assuming that the water temperature must be raised from 50° to 120°F, how many Btu's of energy are used? What is the cost for fuel (at $10 per MBtu) for your shower or bath?

of several materials. For example, the specific heat of iron is 0.12 Btu/lb-°F, meaning that it requires 0.12 Btu to raise the temperature of 1 lb of iron by 1°F. In contrast, 1.00 Btu is needed to raise the temperature of 1 lb of water by 1°F.

An object with a large specific heat will release more heat energy Q to the surroundings in undergoing a temperature *decrease* ΔT than an object of the same mass but lower specific heat (that undergoes the same temperature change); see Figure 4.4. We say that the object with the greater specific heat has a larger thermal storage capacity. A hot object with a high specific heat will cool down slower than an object with a low specific heat. For example, the aluminum foil on a heated TV dinner cools down much faster than the contents, which are mainly water. A hot apple pie from your local fast-food restaurant usually comes with a warning label: "Caution: Filling may be hot" compared to the outer crust. One must remember that freshly served coffee will stay hot for a long time (Fig. 4.5).

FIGURE 4.5
Water has a high specific heat. (McDONALDS CORP.)

EXAMPLE

How much heat is required to raise the temperature of 4 gal of water from 60°F to 100°F?

Solution

The temperature of the water must be raised by 40°F. Since it takes 1 Btu of heat to raise the temperature of 1 lb of water by 1°F, it will take 40 Btu to raise the temperature of each pound of water by 40°F. Since there are 8.3 lb/gal, the total amount of heat energy that must be added is 4 gal × 8.3 lb/gal × 40°F × 1 Btu/lb-°F = 1328 Btu.

When heat is added to an object, its temperature will not necessarily increase; there might be a **change of phase.** Matter exists in three phases: solid, liquid, and gas.* Transitions between different phases are accompanied by the absorption or liberation of heat. It takes a good deal of heat to change water from the liquid phase into the gaseous phase called steam. The quantity of heat that must be added to a liquid per unit mass at its boiling point to convert it entirely to a gas *at the same temperature* is called the **heat of vaporization** of the material. For water, this value is 540 calories per gram or 540 kilocalories per kg or 2260 kJ per kg or 970 Btu per lb. The quantity of heat that must be added per unit mass at the solid's melting point to convert it entirely to a liquid at the same temperature is called the **heat of fusion** of the material. For water, this value is 80 calories per gram or 140 Btu per lb. (*Note:* The quantity of heat necessary to cause a phase change can also be called the **latent heat of fusion** or **latent heat of vaporization.**)

If the change of phase proceeds in the opposite direction (as from steam to liquid), then heat is liberated in the process. This is the phenomenon behind the operation of a "steam radiator": the condensing steam liberates 540 cal/g. Similarly, if a liquid is cooled to the temperature at which it melted, it will give up an amount of heat equal to the heat of fusion as it solidifies. Phase change materials for storing solar energy make use of this; certain salts are initially placed in the sun, where they absorb solar energy and undergo a phase change from a solid to a liquid. At night, the salts cool, giving off heat. They undergo a phase change back to a solid when they reach their melting temperature, providing heat to the room. This will be discussed in Chapter 6.

Figure 4.6 shows the phase change phenomena for water. The figure is a graph of water temperature versus heat added for 1 kg of water. If we start at –50°C with 1 kg of ice and begin to add heat Q, the ice temperature increases. At 0°C the ice begins to melt and the temperature remains constant until all the ice has melted, which requires a total of 80 kcal. When all the ice has melted, the

---

*Some physicists consider plasma, which consists of free electrically charged particles, to be a fourth phase of matter. See Chapter 16.

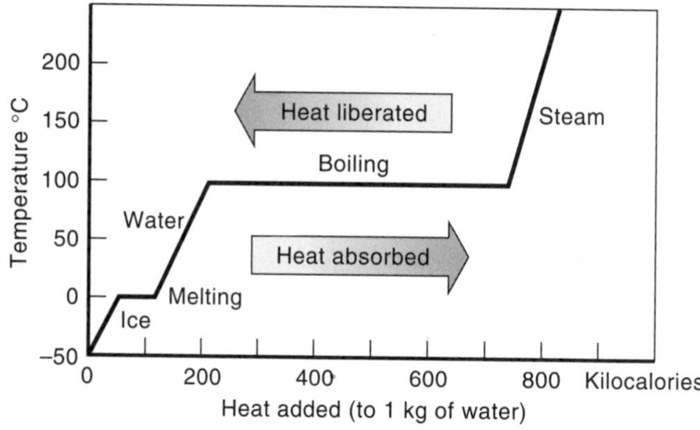

FIGURE 4.6
Changes of phase for water.

water temperature begins to increase again as more heat is added. When 100°C is reached, the water begins to turn to steam. The temperature remains at 100°C until all the water has become steam, which requires a total of 540 kcal. Above this point, the steam rises in temperature as heat is added. Conversely, if we start with steam at a temperature above 100°C and remove heat, the steam temperature decreases and we proceed from the right side of Figure 4.6 to the left; heat energy is liberated rather than absorbed in this direction. As a consequence, materials with phase-change temperatures in the range of common room temperatures are used to store or release substantial amounts of energy.

EXAMPLE

How much heat is required to bring 1 L (1 kg) of water to a boil from 20°C? How much heat is required to boil away this amount of water?

Solution

The amount of heat needed to bring the water to 100°C is

$$Q = mc\Delta T = (1\text{ kg}) \times (4186\text{ J/kg-°C})(100\text{°C} - 20\text{°C}) = 334{,}000\text{ J} = 334\text{ kJ}$$

To bring about the change of phase from liquid to vapor at 100°C, we need 2260 kJ per kg, the heat of vaporization of water.

D. Heat Transfer Principles

If a cup of hot coffee and a glass of iced tea are placed in a room (which is at 20°C), one will cool off and the other will warm up until both reach room temperature. In both cases there has been a transfer of heat that has caused a change in the temperature of both objects. For the hot coffee, energy has been transferred to the room through the flow of heat. In the case of the iced tea, heat has been added to the tea. On a cold winter day, if the glass of iced tea is placed outdoors, the opposite situation will occur; the tea will become colder. Heat will flow from one object to another only when there is a temperature difference between the objects, and then only from a hot object to a colder one (Fig. 4.7). This point will be very important when we study the direction in which physical processes can proceed.

There are three ways by which heat can be transferred from a hot body to a cold body: conduction, convection, and radiation.

> Heat Transfer
>
> - One of two ways in which energy can be transferred to a body (work is the other)
> - Occurs only when there is a temperature difference between two bodies
> - Occurs through the processes of conduction, convection, and radiation.

1. CONDUCTION If one end of a copper rod is placed into a fire, the other end will become hot as heat is conducted through the rod. A spoon handle in a bowl of soup similarly becomes hot (Fig. 4.8). In the process of **conduction,** heat is transferred through molecular collisions from a hot object to a colder one. One molecule transfers energy by "bumping" into another, which in turn collides with another and so on, and heat is transferred through the material.

Materials differ markedly in their ability to conduct heat. Have you felt the difference between a block of wood and a piece of metal that are sitting on a table? The metal feels colder, although both are at the same temperature. This is

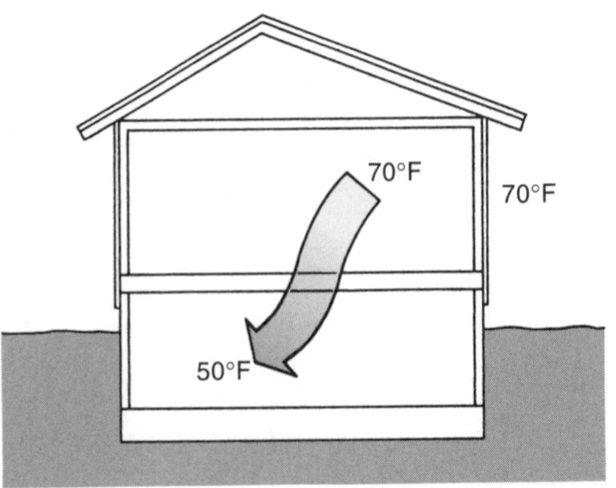

FIGURE 4.7
Heat flows when there is a temperature difference ΔT. In this case, $\Delta T = 70° - 50° = 20°F$.

FIGURE 4.8
Heat is transferred by conduction through the metal spoon from the hot coffee to the colder fingers.

because metal is a better conductor than wood and is more effective in transferring heat from your body (37°C) to itself (21°C).

How rapidly heat is conducted through a piece of material depends on several factors. The greater the temperature difference across the material, the greater is the rate of heat flow. More heat will be transferred through a wall if the area of the wall is larger. If the wall is thin, more heat will be transferred through it than through a thicker wall. The type of material through which the heat is transferred is also important, as we already discussed. The expression for the **rate of heat transfer by conduction** (abbreviated as Q_c/t, where t is time) is

$$\frac{Q_c}{t} = \frac{\text{heat transferred via conduction}}{\text{elapsed time}} = \frac{k \times A \times (T_2 - T_1)}{\delta}$$

In this equation, A is the surface area and δ is the thickness of the material through which the heat flows, T_1 and T_2 are the temperatures on either side of the material, and k is the **thermal conductivity** of the material (see Fig. 4.9).

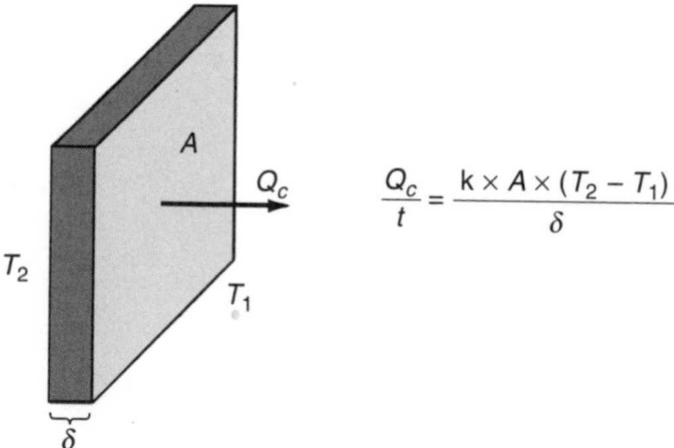

FIGURE 4.9
Heat flow through a wall by conduction. The rate Q_c/t at which heat energy is transferred through the material depends on the temperatures on either side (T_1 and T_2), the wall's area A, its thickness δ, and its thermal conductivity k.

This rate is usually expressed in units of Btu per hour or watts. Although the rate of heat transfer for a given substance becomes smaller with increasing thickness, don't be misled by claims that insulating material is good based solely on its thickness because thermal conductivity varies enormously from material to material.

An illustration of heat transfer via conduction is the heat flow through the exterior surfaces of a house. Although opportunities for reducing heat loss (or heat gain in the summer) with increased insulation will be discussed in the next chapter, two examples are interesting here. The preceding equation shows that a reduction in the interior temperature T_2 during the winter can reduce heat loss, since the temperature difference $(T_2 - T_1)$ between the inside and outside will be smaller. As an example of such a reduction, suppose that you turn down the thermostat from 72°F to 66°F when the outside temperature is 20°F. Before you do this, the rate of conductive heat loss is $Q_c/t = \text{constant} \times (72 - 20) = \text{constant} \times 52$, where the constant depends upon the thermal conductivity, the area, and the thickness of the surface. After you lower the thermostat, the rate becomes $Q_c/t = \text{constant} \times (66 - 20) = \text{constant} \times 46$. The percentage savings achieved by your action is $(52 - 46)/52 = 0.12 \times 100\% = 12\%$, which could translate into savings of \$10 to \$20 per month. Figure 4.10 shows the percentage savings that can be achieved in several cities representing various climates, as a function of the thermostat setting. The percentage saving is higher in cities with milder winter temperatures, but the total amount of fuel and money saved is greater in colder climates.

Another illustration of the previous equation deals with the surface area A. A two-story house will lose less heat than an equivalent one-story house of the

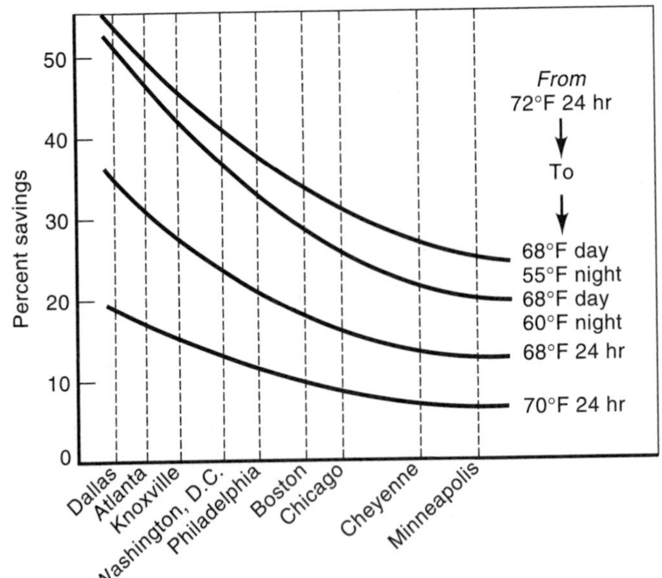

FIGURE 4.10
Percentage of energy saved by lowering the thermostat from 72°F to the values shown on the curved lines, for the time periods shown.

same floor space, because the total surface area of the exterior surfaces is smaller for the two-story house. In general the dwelling with the smallest ratio of surface area to interior volume will lose the least amount of heat per cubic foot of living space. The shape with the smallest surface area for a given volume is a sphere. A dome house uses this idea. As the absolute size of the dome increases, the surface-area-to-volume ratio goes down. As an illustration of this principle, note that 10 lb of ice cubes melts much faster than a 10-lb block of ice. The block has the same volume as the ice cubes but the block has a smaller surface area than the total surface area of the ice cubes. Large animals lose their heat more slowly than smaller animals, which have a relatively large surface-area-to-volume ratio. A small warm-blooded animal like a mouse must eat almost continually in a cold climate to maintain its body temperature; pound for pound, a mouse eats 17 times as much food as a human.

2. CONVECTION In a gas or liquid, the molecules are too far apart for heat to be transferred effectively by conduction. Heat is transferred through fluids primarily by the motion of the gas or liquid, a process called **convection.** The density of a fluid is less when it is warm than when it is cold, and so density differences cause the warmer fluid to rise, setting up convection currents. A pot of water is heated primarily in this way. Heat is conducted through the metal from the stove element into the water on the bottom of the pot. The warm water rises to the top because its density is less than that of the surrounding cold water. A convection current is set up, as shown in Figure 4.11.

Convection processes can occur naturally as a result of density differences or by forced convection through the use of a fan or the presence of wind. Storm windows are very effective in reducing heat loss by forced convection, because

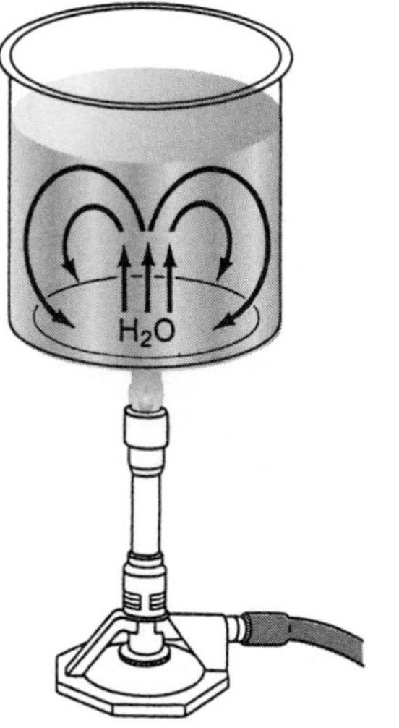

FIGURE 4.11
Convection currents in water.

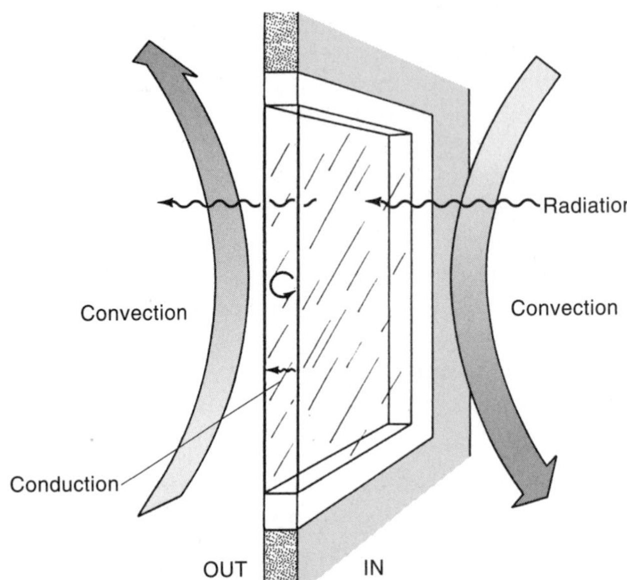

FIGURE 4.12
Heat transfer through a double-pane window.

they establish an insulating layer of still air next to the glass pane (Fig. 4.12). However, some heat transfer by natural convection (as well as conduction) will occur between the glass panes because there is an air gap and a temperature difference across the gap.

Convection currents are important to the operation of some types of solar heating systems. A solar air heater that can be placed in a south-facing window is shown in Figure 4.13. Solar radiation transmitted through the glass

ACTIVITY 4.2

A colorful activity to illustrate natural convection uses water at different temperature in two setups.

1. Fill a mixing bowl or measuring beaker with cold water (preferably ice water). Fill a small vial (like a shot glass) with hot water and add several drops of food coloring. Carefully place the vial in the bowl and observe what happens to the colored water.
2. Repeat this step but place ice water in the small vial (with food coloring added) and hot water in the bowl. Place the vial in the bowl and observe what happens. Explain the results in terms of heat transfer principles.

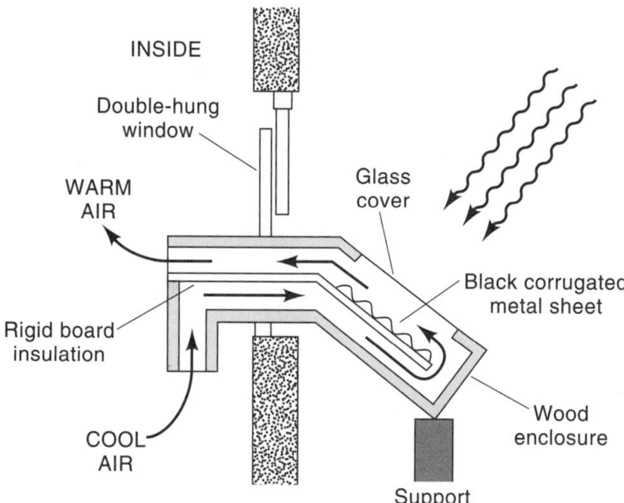

FIGURE 4.13
Solar air heater to use in a window.

cover is absorbed by the corrugated black metal sheet. Air under the glass and in contact with the metal is heated and rises into the house because it is less dense. Cold air from the room is drawn into this collector to replace the warmer rising air, setting up convection currents. This window box heater can raise the temperature of a small room on a sunny day by 10°F to 20°F. (Figure 6.22 shows a "thermosiphoning" solar water heater, which circulates water through its coils without the aid of a pump, because of the lower density of the warmer water.)

3. RADIATION A third and very important means of heat transfer is **radiation.** Unlike conduction and convection, heat transfer by this means does not need a medium in which to propagate. Radiation is emitted from an object in the form of electromagnetic waves, which consist of electric and magnetic fields whose amplitudes vary with time. These waves propagate out from the object at the speed of light. Types of electromagnetic radiation include many familiar waves: visible light, radio waves, microwaves, X-rays, and infrared waves.

To understand radiation, it is instructive to discuss waves on a rope. If you tie one end of a long rope to a post, stretch it out to its full length, and move the other end up and down with your hand, you will generate a series of waves that move toward the post (Fig. 4.14). These waves are characterized by their amplitude A, or maximum displacement from the equilibrium position of the stretched rope, and their wavelength λ. The waves will move with a velocity v (which is a function of the tension and the mass per unit length of the rope). The rate or frequency f at which the free end of the rope is moved up and down by your hand determines the number of waves that pass a given

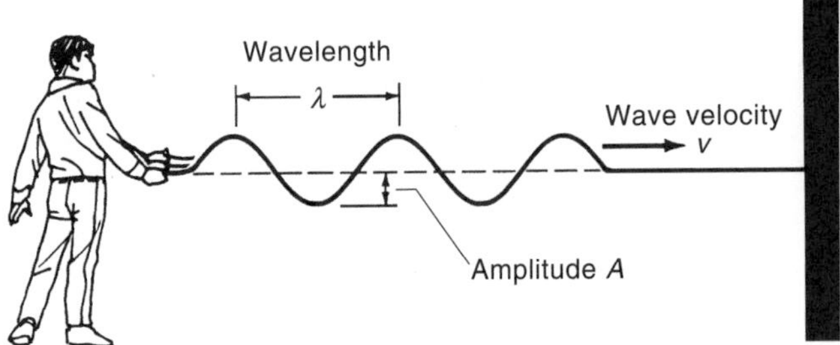

FIGURE 4.14
Waves generated by moving the end of a rope.

point on the rope per unit time. The velocity v, wavelength λ, and frequency f of the waves are related by the expression (with units in parentheses):

$$v\ (\mathrm{m/s}) = \lambda\ (\mathrm{m}) \times f\ (/\mathrm{s})$$

The different types of electromagnetic radiation all share one property: They all have the same velocity v in a vacuum—the speed of light, 3.0×10^8 m/s (1.1 billion km/h or 186,000 mi/s). The difference between these waves is their frequency and their wavelength. X-rays have very high frequencies (10^{18} Hertz [Hz], where 1 Hz = one cycle per second), and very short wavelengths (10^{-10} meters). Radio waves have relatively low frequencies (such as 600 kHz for AM

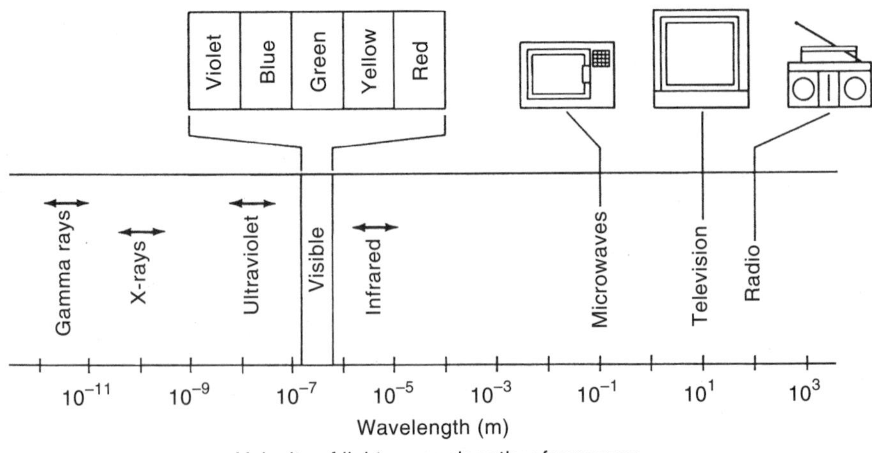

FIGURE 4.15
The electromagnetic spectrum, shown as a function of wavelength.

and 100 MHz for FM) but long wavelengths (from several meters to hundreds of meters). The electromagnetic "spectrum" (range of wavelengths) is shown in Figure 4.15, in which the different types of radiation are categorized according to their wavelength. Visible light is that group of wavelengths that is visible to the human eye and occupies only a very small region of this spectrum, which is shown expanded in the figure. The visible spectrum's wavelengths go from violet light at about 4×10^{-7} m to red light at 7×10^{-7} m.

EXAMPLE

What is the wavelength of electromagnetic radiation emitted from a radio station broadcasting at 1500 kHz?

Solution

$$\text{Wavelength} \times \text{frequency} = \text{velocity of light}$$

So

$$\text{Wavelength } \lambda = \frac{3 \times 10^8 \text{ m/s}}{1500 \times 10^3 \text{ Hz}} = 200 \text{ m}$$

All objects whose temperatures are above absolute zero emit electromagnetic radiation. At temperatures below 1000°C, predominantly microwave or infrared radiation is emitted. As the temperature of an object increases above 1000°C, some of the radiation moves into the visible range, as evidenced by the red glow of hot charcoal in the barbeque. As the temperature increases, the colors shift from red to violet, making the object appear bluish-white. Our sun, whose surface temperature is about 6000°C, emits a spectrum of radiation (Fig. 4.16) that is centered in the visible region, mainly around the color yellow. However, there are also strong components of infrared (longer wavelength) and ultraviolet (shorter wavelength) light (50% and 9%, respectively). The

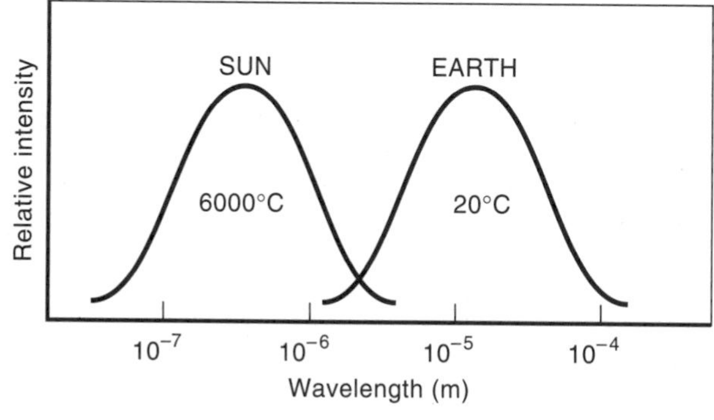

FIGURE 4.16
The spectrum of radiation emitted from the sun and the earth. The vertical scales for each spectrum are different, as the intensity of radiation from the sun is many times greater than that from the earth.

amount of radiation emitted from an object every second is proportional to its temperature. The higher the temperature, the greater is the rate of emission of radiation.

Our own bodies, at relatively low temperatures, emit radiation in the infrared region. This can be "seen" at night by some animals, such as snakes and cats. Heat losses through poorly insulated parts of a house, or the location of enemy troops in the jungle at night, can be seen with the help of infrared detectors such as photoconductors (photosensitive semiconductors). Infrared film is used in studying objects with higher temperatures.

Since a body is always losing energy by radiation, exactly as much energy must be added as is radiated if that body is to maintain the same temperature (Fig. 4.17). An inanimate object achieves an equilibrium temperature when it receives as much heat energy (from the sun, the surrounding atmosphere, and the earth) as it loses. At night a body continues to radiate and, unless it is in good contact with a source of heat such as the earth, its temperature will decrease. Radiative cooling of the earth at night, especially when there are no clouds, can lead to cool temperatures.

The human body maintains an equilibrium internal temperature of 37°C (98.6°F). In cold weather, heat is lost from the body via radiation into the surrounding atmosphere and by convection. This is about 100 W (350 Btu/h) for a sitting person. If you stand near a window or exterior wall on a cold day, you will lose more heat energy because of the lower temperature of the window, and thus your body will feel cold. The net heat loss due to radiation will be more since the cold window radiates less, and heat loss due to conduction and convection will increase because of a larger T. On days when the air temperature is about 38°C (100°F), the body keeps cool mainly by evaporation. Heat is removed from the body as it evaporates perspiration during a change of phase from liquid to vapor. Such cooling is less effective on "muggy" days when the water content (the humidity) of the surrounding air is very high and therefore evaporation is inhibited.

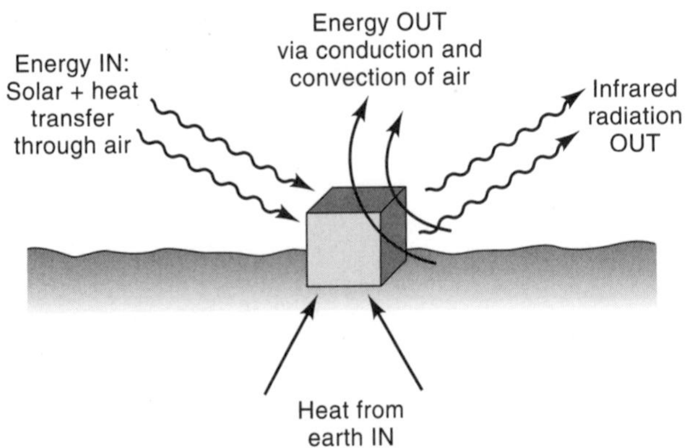

FIGURE 4.17
The equilibrium temperature of an object is maintained if the energy input is equal to the energy output.

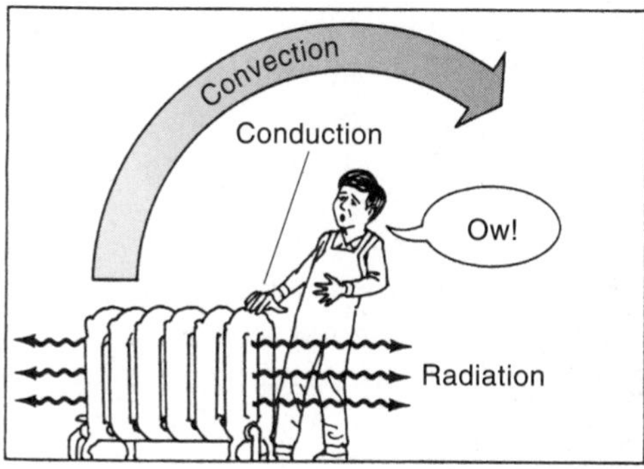

FIGURE 4.18
A hot-water radiator as an illustration of heat transfer via conduction, convection, and radiation.

A hot-water radiator illustrates all three heat transfer processes (Fig. 4.18). Heat energy is conducted through the metal from the hot water to the air in contact with the radiator (or your hand if you touch it); the motion of the warmer air via convection currents carries the heat around the room. Heat energy is also transferred to objects in the room by the process of direct radiation from the radiator.

E. Heat Engines*

In previous chapters we learned that energy appears in many forms (mechanical, electrical, nuclear, solar, etc.) and that it can be converted from one form to another. In such conversions, the total energy of an isolated system must be conserved; energy cannot be created or destroyed, only transformed. In the following sections we will expand our discussion of energy conversion by considering a class of machines called **"heat engines."** This group includes all devices in which heat is converted into useful work, and therefore the heat engine is a central part of many of our energy conversion devices. An automobile engine converts the chemical energy of gasoline (through combustion) into the mechanical energy of a piston and crankshaft; the turbine in an electrical generating plant converts heat into shaft work to run a generator.

Since heat is the energy transferred from one substance to another when there is a temperature difference between them, we need a source of heat. This usually comes from a fuel that is burned, although it can also be solar or nuclear in origin. The flow of heat proceeds through a fluid medium such as a gas or a liquid. This medium is called the "working fluid." For example, the combustion of wood can heat air, which can be used to turn a turbine. Heat has flowed from a hot source (the wood fire) to the medium of the combustion gases that, in turn, do work by driving the turbine.

---

*For those interested mainly in practical applications of heat transfer, this section and the next can be skipped in a first reading.

Figure 4.19 shows the energy flow for a heat engine. Heat Q_H flows from a hot "source" at temperature T_H to a cold "sink" at temperature T_C. Some of this heat is transformed into work W. Since energy is conserved, the heat Q_H leaving the source is equal to the heat Q_C entering the sink plus the work done by the engine: $Q_H = Q_C + W$. (No energy is stored.) The lower the sink temperature T_C or the higher the source temperature T_H, the more work is available from the heat engine. The energy available for work comes from a decrease in the total energy of the working fluid. The greater the temperature change, the greater the decrease in the energy of the working fluid, and hence there is a greater quantity of energy available to do work.

After doing work, the working fluid can either be exhausted into the environment or sent back to the heat source to start the cycle over. The first case is called an "open cycle"; the second case is a "closed cycle." If the fluid is returned to its initial state, there has been no change in its total energy and so $\Delta E = 0$. Consequently, from the first law of thermodynamics, the total work done by the system is just equal to the net heat added (heat in minus heat out):

$$W = Q_H - Q_C$$

A common example of a heat engine is the *steam turbine*, such as the ones used for electricity generation (see Fig. 3.3). The working fluid in this closed cycle system is water, in both liquid and vapor states. Heat is transferred from the burning fuel to the water in the boiler, increasing its total energy and turning it into steam. The steam turns the blades of the turbine, giving up some of its total energy to turn the shaft. In the condenser, steam is condensed to the liquid phase as some of its energy is transferred to the cooling water and passed on to the environment. The water is finally pumped to a higher pressure and re-

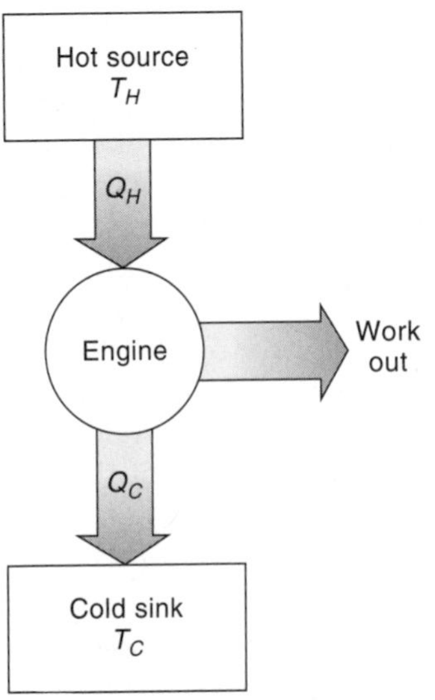

FIGURE 4.19
A heat engine transforms heat into work.

turned to the boiler. (The work necessary to operate this pump comes from the turbine generator.) For the *entire* generating system, the energy balance relationship follows from the first law of thermodynamics:

Q_H	=	W	+	Q_C
Heat into plant	=	net work out	+	net heat out
(fuel combustion)		(to generate electricity)		(of condenser)

The work done by the system comes from a decrease ΔE in the energy of the steam; the greater the change ΔE achieved, the more work is possible from the turbine. This is the main reason for the condenser: It provides a low-temperature region in which the steam will condense to liquid at a low pressure (less than atmospheric pressure), thereby providing a large change in ΔE for the water. The liquid is also much easier (requiring less energy) to pump back into the boiler than it would be as a gas.

Only a fraction of the energy contained within the fuel of a heat engine goes into useful work. We shall expand on the presentation in Chapter 3 of energy conversion efficiencies, to see that heat engines have theoretical limits to their efficiencies. This results in "thermal pollution": the addition of unwanted heat to the environment, in particular to natural waters. As noted in Chapter 3, about two thirds of the chemical energy of the fuel input to a fossil-fueled electrical generating plant ends up as nonproductive waste heat.

There are several general types of heat engines, some of which are listed in Table 4.3. They are characterized by the type of cycle the working fluid carries out. A cycle in which the working fluid changes state, as in a steam turbine loop, is called a **vapor cycle** or **Rankine cycle.** A cycle in which the fluid is a gas at every point is called a **gas cycle.** (This gas is usually primarily hot air and is not to be confused with the fuel natural gas.) Gas cycle engines can have either external or internal combustion, depending on whether the fuel is burned outside or inside the chamber in which the power is produced. Rankine cycle engines are all external combustion devices. External combustion gas turbine engines are widely used in aircraft and marine engines and in load-peaking units for electrical generating stations. With the current apparent surplus of natural gas, their uses are sure to increase. The most familiar example of the internal combustion engine is the spark-ignition reciprocating engine of the automobile, which uses the "Otto" cycle.

Table 4.3 EXAMPLES OF HEAT ENGINES

Vapor or Rankine Cycle	Steam engine (electrical power plant, old train locomotive) Refrigerator, heat pump (using Freon)
Gas Cycle	Internal combustion: Otto, diesel cycles (automobiles, trucks) External combustion: gas turbine (airplanes), Stirling cycle

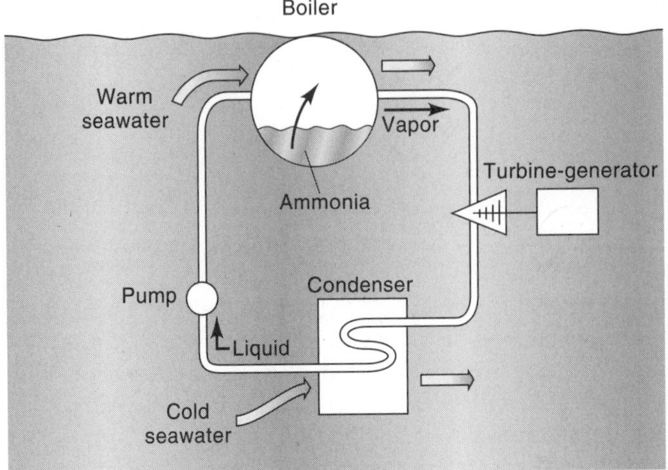

FIGURE 4.20
Ocean Thermal Energy Conversion (OTEC). The temperature difference between waters on the top of the water and down deep allows one to construct a heat engine.

Another example of a heat engine is a device that generates electricity by using the temperature differences between the warm tropical surface waters and those waters 1000 m below. This is called **Ocean Thermal Energy Conversion,** or **OTEC.** One type of cycle (Fig. 4.20) uses the warm surface waters to boil ammonia. The ammonia vapor then drives a turbine generator. The cold deep waters condense the ammonia to a liquid, which is then pumped back to the boiler. The temperature difference in this plant might be 25°C – 5°C = 20°C. With a small ΔT (compared to 500°C in a steam power plant), the overall efficiency of such a plant is very small (3–4%). Thus to produce much power, the plant must move a lot of water, and so much of its output goes to run the pumps. A demonstration plant in Hawaii generated 250 kW, with a net output of 100 kW, from 1979 to 1999. OTEC has much potential, but cost and corrosion from saltwater are two impediments.

F. The Second Law of Thermodynamics*

Direction and Disorder

In our discussion and analysis of energy conversion processes, we frequently have used the principle of energy conservation, or the first law of thermodynamics. However, observation shows us that there must be other physical prin-

*For those interested mainly in practical applications of heat transfer, this section and the previous can be skipped in a first reading.

ciples that govern energy processes. Have you ever seen a book that is lying on a table move on its own accord? To do so, the book would have to take thermal energy from the table and convert it into kinetic energy (Fig. 4.21a). The table would be at a lower temperature afterward. Energy conservation does not prevent this from happening, as this is nothing but a transformation of heat into mechanical energy, like a heat engine. Yet we don't see things occurring this way. Another illustration comes from the vast supply of thermal energy that is stored in the oceans. Why don't we extract energy from the warm oceans and use it for a heat engine (Fig. 4.21b)? (This is different than OTEC where a temperature difference between the top and bottom layers of water is used.)

Another physical law, called the **second law of thermodynamics,** must be used to explain these situations. The second law speaks about the *direction* of physical processes: why a process can go in one direction but not in the other. For example, we say that heat flows from a hot body to a cold one, but we never see it flow the other way *by itself.* (But, you say, what about a refrigerator? Yes, heat does pass from a cold source to a hot source, but *only* with the help of the external electrical energy supplied to the refrigerator's compressor.)

If you put a drop of black ink into a glass of water, the ink will disperse and make the water cloudy and gray; *the system has become more disordered.* We never see grayish water like this clear up and leave a spot of black ink in the middle. A pendulum held to one side has all of its energy in the form of potential energy. As it swings back and forth, its amplitude decreases as some of its mechanical energy turns into heat. It finally stops; its total energy, initially in one form, has been distributed to many other molecules—the system has become more disordered. We never see the opposite of this take place, in which a pendulum starts itself by taking heat from the surrounding air; although this process would not violate the principle of conservation of energy, it is forbidden by the second law.

A quantity that is used to measure the disorder of a system is called **entropy.** Entropy is a property of the system. Remember that we can change the

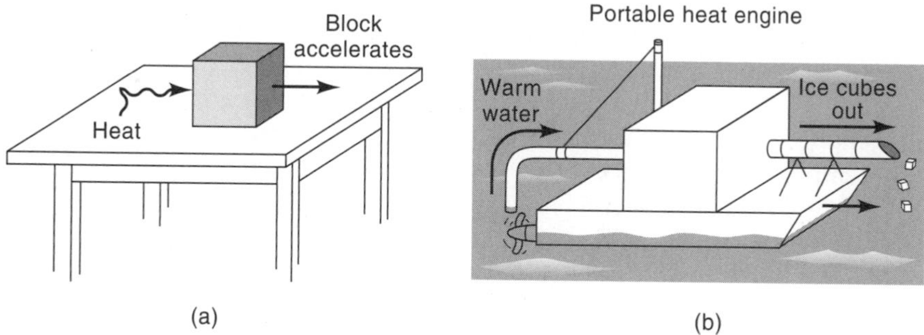

FIGURE 4.21

Impossibilities according to the second law of thermodynamics. (*a*) Heat withdrawn from the table is converted into mechanical energy—the kinetic energy of the block. (*b*) Heat from seawater is converted into electrical energy (the resulting ice cubes are discarded).

energy of a system by doing work on the system or adding or subtracting heat. When heat is added, the disorder of that system increases and so the entropy increases. If heat flows out, the disorder decreases and entropy decreases.

There also can be an entropy change in an isolated system in which no heat is added. **The second law states that for any spontaneous process, the entropy of an isolated system can only increase or stay the same, but never decrease.** If we take our isolated system to be the entire world, then the amount of disorder in the world is continually increasing. Entropy provides a kind of "arrow of time," establishing a direction in which natural processes are allowed to occur. The ink-in-water example and the pendulum are cases in which entropy increases. To proceed in the opposite direction would lead to a decrease in entropy, as we are moving toward a system with increased order. (The order of a system can be increased, but only by doing external work, such as rearranging the pieces of a broken dish that has dropped off the table or sorting marbles by their colors.)

There are two important statements that follow from the second law. They are:

Statements of the Second Law

1. Heat can flow spontaneously (by itself) *only* from a hot source to a cold sink.
2. No heat engine can be constructed in which heat from a hot source is converted *entirely* to work. Some heat has to be discharged to a sink at a lower temperature.

The first statement we have dealt with already. A refrigerator works only with the aid of external work delivered to the compressor.

The second statement says that we need a hot source *and* a cold sink for there to be the flow of heat and the extraction of useful work. This argues against our hypothetical example of a book that moves by extracting energy from the table: We don't have a cold sink to which heat can flow or be exchanged. The same reasoning holds true for the example of extracting work from the vast thermal energy contained in our oceans. The second law states that for a heat engine to function there must be some heat discharged to a cold sink—such as our environment. We must have a ΔT (a temperature difference).

Let us look at the second law in more detail. Chapter 3 defined the percentage efficiency of a device as the ratio

$$\text{Efficiency} = \left(\frac{\text{useful work out}}{\text{energy input}}\right) \times 100\%$$

The principle of conservation of energy says that the work output equals the heat input into a system minus the heat transferred out. Therefore,

$$\text{Efficiency} = \left(\frac{\text{heat in} - \text{heat out}}{\text{heat in}}\right) \times 100\% = \left(1 - \frac{\text{heat out}}{\text{heat in}}\right) \times 100\%$$

If some heat is transferred out to a cold sink, as the second law states it must, then this expression says we can *never* have a system that is 100% efficient. The value for the efficiency must always be less than 100%. Therefore we will *never* have perpetual motion machines. People have worked on such devices for countless years, hoping to find a machine that will run forever, one in which energy does not have to be continually supplied to sustain the operation of the machine, but they haven't been successful. Even in the absence of friction in the machine, some heat will be transferred to a cold sink and the efficiency will be less than 100%, causing the machine eventually to slow down and stop.

Maximum Efficiency—Less than Perfect

If some heat has to be discarded to the environment, what is the *best* we can do? What is the highest efficiency we can obtain if we start with a heat source at temperature T_H and a cold sink at temperature T_C? This question was of real concern in the development of the steam engine in the early 19th century. Sadi Carnot, a French engineer, showed that for an ideal engine (as we will define later in this section), the **maximum possible efficiency** for a heat engine operating between a boiler temperature of T_H and a cold sink or condenser temperature T_C is given by

$$\text{Maximum (Carnot) efficiency} = \left(1 - \frac{T_c}{T_H}\right) \times 100\% = \left(\frac{T_H - T_c}{T_H}\right) \times 100\%$$

(For this equation to be correct the temperatures must be expressed using the Kelvin or absolute scale. Remember that K = °C + 273.) We can never do better than this, and in fact we always do worse. In the real world, most heat engines operate at one half to two thirds of the Carnot efficiency.

EXAMPLE

For a conventional vapor-cycle power plant, the temperature of the steam entering the turbine is 540°C or 813 K. The temperature of the cold reservoir (the cooling water) is 20°C or 293 K. What is the *maximum* possible efficiency of this heat engine?

Solution

$$\text{Maximum efficiency} = \left(\frac{T_H - T_C}{T_H}\right) \times 100\%$$

$$= \left(\frac{813 - 293}{813}\right) \times 100\% = 64\%$$

If the actual plant efficiency is 35% (a typical value), then it is operating at 0.35/0.64 = 55% of the Carnot efficiency.

From the expression for efficiency we see that we want to use the largest possible temperature extremes for the operation of a heat engine. One of the biggest advances in steam engine development was the addition of the low-pressure condenser by James Watt. Before this development, the steam was thrown away after it passed through the turbine. Unfortunately, the steam leaving at normal atmospheric pressure and 100°C was quite energetic. If a condenser was added, which was operating at a reduced pressure, the condensation would occur at a lower temperature T_C (as low as the temperature of the environment) and so the steam engine's efficiency would be increased by almost a factor of 2. (Water will boil and condense at temperatures lower than 100°C for pressures less than those at sea level.)

The Ideal Case: Reversibility

The maximum efficiency of an energy conversion process is achieved when the process is completely **reversible,** that is, the reverse process is possible according to the second law. (A process in which this is not possible is called **irreversible.**) In a reversible process, everything is restorable to its original state—there are no net effects on the surrounding environment. For example, a mass on a frictionless spring oscillating back and forth is a reversible process. No heat is lost to the surroundings, so every part of the process can be repeated. In reality, temperature increases do occur in the parts, so there will be irreversibility with a normal spring; eventually the spring will come to rest, having converted all its mechanical energy into heat energy (transferred to itself, the attached mass, and the environment). An ideal process is a reversible one, but we don't live in an ideal world and so we don't observe such processes, although we can try to maximize the degree of reversibility. *In a **reversible process,** the entropy remains constant: There is no net increase of disorder in the system.* The mechanical energy of the ideal spring is a combination of kinetic and potential energies, and is known at every position. In reality, disorder increases as heat energy is transferred to the molecules of the spring and the environment.

Available Energy

Another way to examine the consequences of the second law of thermodynamics is through the concept of "available energy," or availability. A hot reservoir has the potential of being the source of energy for a heat engine to do work. However, if heat had been transferred from the hot reservoir to a cold sink *without* the production of useful work, then there would have been a *loss in available energy,* as the cold sink could not be used to do work unless there were another sink at a lower temperature to which the heat could flow. Even if the cold sink could be used to do work, the work you could obtain from this source would be less than that available from the original high-temperature source, according to the Carnot efficiency. One hundred units of thermal energy at 1000°C has the potential of doing more work than 100 units of energy at 500°C, both relative to the same low-temperature sink; therefore, the available energy of the higher temperature source is greater.

In an irreversible process, there is a loss in available energy; as a spring with friction slows down, its mechanical energy is transformed into less useful thermal energy, manifested in the increased temperature of the surrounding air molecules and the spring. For an ideal engine operating between two temperatures, the maximum efficiency is achieved by the Carnot cycle, which is one form of a reversible cycle. Any other engine will have a smaller efficiency between the same two temperatures. Less work will be obtained, so there is a loss of available energy that *could* have been obtained if a reversible engine had been employed.

We know from the first law that energy is conserved. However, *it is not the amount of energy in the world that is decreasing, but its ability and availability to do work.* In any fuel-consuming process, there will be a loss of ability to do work. As the high-quality energy of a hot reservoir or the mechanical energy of a compressed spring is converted into thermal energy, there is a degradation of the availability of energy. As heat flows from a hot reservoir to a colder one, we eventually reach the point at which both reservoirs have the same temperature and so useful work cannot be obtained.

Using these concepts, another way to state the second law of thermodynamics is

> It is impossible to convert a given quantity of heat energy completely into work. In an energy conversion process, energy is always degraded in quality, so that its ability to do work is reduced.

Energy is never destroyed, but it can reach a state in which it can no longer serve any useful purpose. Thermal energy can be converted into work only if there is a transfer of heat from a hot source to a colder sink. *It is not the conservation of energy that is important but rather how efficiently energy can be used to produce an end result and minimize total fuel consumption.* For a given environment, the source with the highest potential usefulness is the one at the highest temperature.

It is in the production of low-temperature heat that the most wasteful consumption of energy occurs. Why waste the high-quality energy contained in a hot source only for the production of low-temperature heat? We need to emphasize the effectiveness of a fuel for the performance of a particular task and to minimize the amount of fuel needed for a process. This approach emphasizes end use rather than generation efficiencies, but it is beyond the scope of this book to pursue it further.

G. Summary

The first two laws of thermodynamics are fundamental to understanding energy conversion processes. The first law states that energy is conserved; the heat added to a system is equal to the work done by that system plus the change in its total energy. If we consider a cycle in which the system returns to its original state—so there has been no net change in the total energy—then the first law mandates that the total work done by the system is just equal to the net heat added (heat energy in minus heat energy out).

Many of the energy conversion devices in operation today are heat engines. All such engines make use of a flow of heat from a hot source to a cold sink, with the production of some useful work. The second law limits the amount of work obtained from a heat engine. The heat energy that flows from the hot source cannot be converted entirely into work; some heat has to be discharged into the environment. The maximum efficiency possible for a heat engine operating between a hot source at temperature T_H and a cold sink at temperature T_C is the Carnot efficiency:

$$\text{Maximum efficiency} = \left(1 - \frac{T_c}{T_H}\right) \times 100\% \text{ (where temperature is in K)}$$

The total entropy of a system (the measure of its disorder) increases in a physical process. The direction of entropy change is like an arrow of time. As a heat source cools off to lower temperatures, the work available from that source declines. To conserve energy resources, we must try to match a source to the particular task that must be accomplished; end-use efficiency, as well as energy conversion efficiency, is very important.

Internet Sites

For an up-to-date list of Internet resources related to the material in this chapter, go to the Harcourt College Publishers website at **http://www.harcourtcollege.com**. The links are in the *Energy: Its Use and the Environment* site on the Physics page. General energy related sites and some guidelines for using the World Wide Web in your class are on the inside front cover of this book.

References

Chapter 4

Commoner, B. 1977. *The Poverty of Power: Energy and the Economic Crisis.* New York, Bantam.

Cutnell, J., and K. Johnson. 2001. *Physics.* 5th ed. New York, John Wiley & Sons.

Kreith, F., and R. West. 1997. *Handbook of Energy Efficiency.* Boca Raton, FL, CRC Press.

Rifkin, J. 1980. *Entropy: A New World View.* New York, Bantam.

Rifkin, J. 1989. *Entropy: Into the Greenhouse World.* New York, Bantam.

Schipper, L. 1976. Raising the Productivity of Energy Utilization. *Annual Review of Energy,* 1.

Schipper, L., et.al. 1992. *Energy Efficiency and Human Activity: Past Trends, Future Prospects.* Cambridge, U.K., Cambridge University Press.

QUESTIONS

1. Why can't a body *contain* heat?
2. Give examples of the first law of thermodynamics in which no work is done on the system; give an example in which no heat is added.
3. What is the total energy of a system? What compromises this energy? What is this a function of?
4. Give an example when heat is released from water as it undergoes a change of phase.
5. Is it possible to increase the temperature of gas in a cylinder without any energy transfer as heat? Explain.
6. A Thermos bottle is often used to keep hot liquids hot and cold liquids cold. In terms of heat transfer processes, explain its features.
7. Illustrate all three methods of heat transfer by a wood-burning stove. Sketch these methods.
8. Describe how steam from a tea kettle could be used for a heat engine. How would you define the efficiency of this engine? (See Activity 7 in Chapter 2.)
9. Why is a condenser used on a steam locomotive?
10. Why is a condenser needed in a steam electric plant? Why isn't it better just to recycle the low pressure steam rather than condense it, releasing heat to the environment?
11. In public hearings over a proposed electrical generating plant, opponents demand that a construction permit not be granted until the efficiency of the plant can be doubled. What statements can be made with respect to this demand?
12. Design a heat engine for use in a desert, using the sun as the energy source and water as the working fluid. How does its operation depend on the area of the solar collector?
13. Electrical power plants dissipate about two thirds of their input energy into the environment, primarily in the form of hot water, at about 20°C (35°F) above ambient temperatures. What would be some of the challenges in using this "waste heat" for the heating of buildings near the power plant?
14. How would you explain to a group of citizens why waste heat has to be released in the operation of a steam-generating plant?

15. Why are heat sources at high temperatures more useful than those at low temperatures?
16. A geothermal plant (such as the Geysers in California—see Chapter 18) uses steam produced underground. The steam enters a turbine at a high temperature and pressure, emerging at a pressure below atmospheric pressure. Why is a condenser needed in this plant?
17. What is entropy?
18. If entropy is always produced in real-life processes that occur in isolated systems, how do you explain such phenomena as the increase in order emerging from the construction of an automobile?
19. What illustrations can you give of the second law of thermodynamics in your everyday life?
20. Can you think of any examples in which the entropy of a system decreases? What is the change of entropy of the surroundings?
21. Can you cool a kitchen by leaving the refrigerator door open?

PROBLEMS

1. If 80 Btu of heat energy are added to 2 lb of water at 40°F, what will be the final temperature of the water?
2. How many Btu's of heat energy are needed to raise the temperature of 12 gallons of water from 50°F to 130°F? How much would this cost if electricity for the water heater costs $.08/kWh?
3. How much electrical energy (in kWh) is needed to heat the water in a well-insulated electric hot-water heater of capacity 40 gal from 20°C to 50°C (68°F to 122°F)?
4. How long will it take to heat 40 gal of water from 70°F to 120°F with a 20-kW immersion heater?
5. It takes an electric tea kettle with 20°C water 5 min to reach boiling (100°C). How long will it take for all the water to boil away, assuming the same rate of heat addition?
6. How much energy will it take to melt a block of ice of mass 15 kg? If this same energy was used to lift the ice, to what height (in meters) could it be raised?
7. What wavelength of electromagnetic radiation would you expect from a radio station operating at 120 MHz?
8. A simple heat engine might make use of the warm air around New York City. Energy could be taken as heat from the atmosphere (assume 30°C) and rejected as heat to the Hudson River (10°C). What is the maximum efficiency of such an engine for the conversion of thermal energy into mechanical energy?

9. A coal-fired electrical generating plant has an efficiency of 38%. The temperature of the steam leaving the boiler is 550°C. What percentage of the maximum possible efficiency does this plant obtain? (Assume the temperature of the environment is 20°C.)
10. What is the efficiency of an engine that has 3000 J of heat added during combustion and loses 2200 J of heat in the exhaust?
11. Would you be willing to back financially an inventor who is marketing a device that she claims takes in 25,000 J of heat at 600 K, expels heat at 300 K, and does 12,000 J of work? Explain.

FURTHER ACTIVITIES

1. An alternative to heating water in a metal pot is to use a "paper pot." Take an 8.5″ × 11″ piece of paper, fold it as shown to make a box-shaped container, and staple or tape the corners together. Fill it up about halfway with water and put it above a flame (from a stove or Sterno can). Check that you can raise the temperature of the water to almost 100°C without igniting the paper. Why is this?

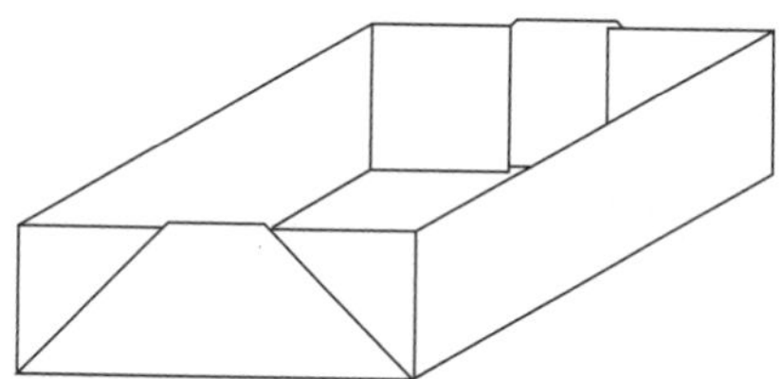

2. Which heats faster, land or water? To study differences in heat capacity, place equal volumes of dry sand (or soil) and water (both at room temperature) into two small jars.
   (a) Place both on a hot plate (or in an oven or under a heat lamp) for 4 to 5 minutes.
   (b) Stir each one a few times and record the final temperatures.
   (c) Which has the greatest heat capacity?
   (d) Bring both jars to the same temperature. (You might have to hold one away from the heat for a while.)
   (e) Insert a thermometer into the center of each sample and remove the jars from the heat.
   (f) Read and record the temperatures every 5 min for about 30 min. (Don't stir the samples.)
   (g) Which sample cooled off the fastest? Was this expected? (Consider the methods of heat transfer.)

3. As a contest or a fun activity, devise a way to bring an ice cube to class with a minimum amount of melting. (No Thermos bottles allowed.)
4. Construct a room temperature map. Measure and record the air temperatures at locations over an entire room at three vertical positions: 6 in. off the floor, at waist height, and over your head. Check near windows and doors especially. Record on three separate room maps, drawing in isothermal lines (lines connecting points at the same temperature) as needed. Describe the presence of heat sources and heat sinks in the room. Is there much convective heat transfer in the room according to your data?
5. Suppose the hot-water faucet in your bathroom has a small drip, say one drop per second. Design an experiment that would measure the water flow rate in gallons per day. How much would this drip cost you (in energy costs for a month) if you used an electric water heater that raised the temperature of the incoming water by 70°F and electricity cost 10¢/kWh?
6. This activity explores relationships between a fluid's density and convection currents. Does hot water rise and cool water fall?
   (a) Make several ice cubes that have been dyed with food coloring.
   (b) Place an ice cube gently in a beaker of warm water.
   (c) Observe the ice-water mixture for several minutes.
   (d) Empty the beaker and refill it with warm water. Add as much salt as will dissolve in the water.
   (e) Repeat Steps b and c.
   (f) What differences did you observe?

5

Home Energy Conservation and Heat-Transfer Control

A. Introduction
B. Building Materials
C. House Insulation and Heating Calculations
D. Site Selection
E. Impact of Energy Conservation Measures
F. Cooling
G. Air Conditioners and Heat Pumps
H. Summary

A. Introduction

We learned in Chapter 4 that heat is transferred from hot to cold objects by the processes of conduction, convection, and radiation. Because the amount of our total energy budget that goes into the heating and cooling of buildings is so large, it is important that we learn how to control the exchange of heat with our surroundings. The effective design of a dwelling and the choice of the materials used can reduce energy consumption by a factor of 2 over conventional designs. **Retrofitting** a structure, that is, making modifications in an existing house without major construction or design changes, can also bring about significant reductions in heating and cooling costs. The use of energy-efficient appliances and the addition of solar energy features to a house are other ways in which energy consumption can be lowered; these latter items will be discussed in Chapters 6 and 10. The previous chapter gave some of the basic principles involved in heat transfer, while this chapter will detail some of the ways in which effective design and choice of materials can reduce energy consumption.

B. Building Materials

Material that is a good conductor of electricity, such as a metal, is usually a good conductor of heat. An insulator, such as fiberglass or Styrofoam, will retard the flow of heat from one object to another. We say that it provides a high "resistance" to heat transfer. Air can be a good insulator, especially if it is motionless (to prevent heat transfer by convection); thus, porous materials such as fiberglass have an excellent resistance to heat flow thanks to the air trapped inside. The choice of different clothing materials to reduce heat loss from the body, or to permit it, depends on this feature; fabric made from wool is more porous than polyester so it is preferred in cold weather as it is a better insulator.

The color of a material also influences heat transfer when radiation is concerned. A black object is both a better emitter and a better absorber of radiant energy than a white or shiny-surfaced object. To be in thermal equilibrium with its surroundings, an object that is a good emitter must also be a good absorber. Conversely, a poor emitter will be a poor absorber. Starting from the same temperature, a hot, black object will radiate energy faster than a similar hot, light-colored object. Ice water will remain cold longer if it is placed in a silver-colored dish rather than a charcoal-colored one. Light-colored clothes are preferable on hot, sunny days, as they are poor absorbers of radiant energy. The absorber plates on solar collectors are painted flat black to better absorb the sun's energy. (Chapter 6 discusses this topic more thoroughly.)

If you enjoy cream with your coffee, have you ever stopped to consider when to add the cream? Should you add it initially or just before drinking so that the coffee will be hottest when you are ready to drink it? The answer is that you should add the cream as soon as possible. The cream makes the black cof-

ACTIVITY 5.1

Study the effect of color on absorption:

1. Paint four identical cans with colors black (flat), green, light gray, and white. Fill the cans with equal quantities of cold water and cover the tops with cardboard through which a thermometer has been inserted. Expose the cans to sunlight for one hour, and measure the final temperatures.
2. Do a similar experiment using three identical glass jars (e.g., baby-food jars). Paint the outside of one jar black and the inside of another black (or use blackened water). Also use a plain, clear jar. Which jar's water reaches the highest temperature? Why?

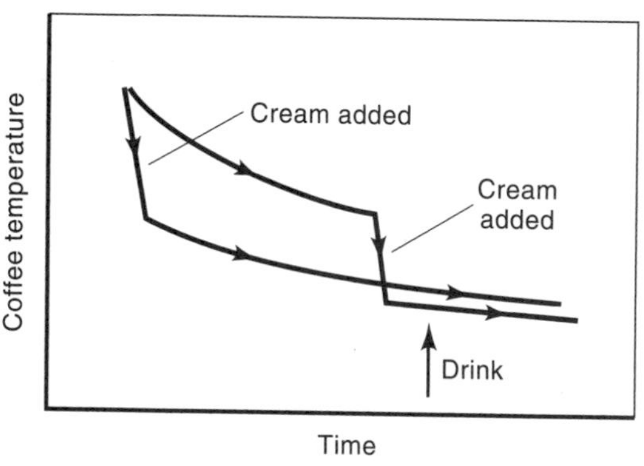

FIGURE 5.1
For coffee to be the hottest when you are ready to drink it at a later time, you should add the cream initially, not just before drinking, as these data for the two cases illustrate.

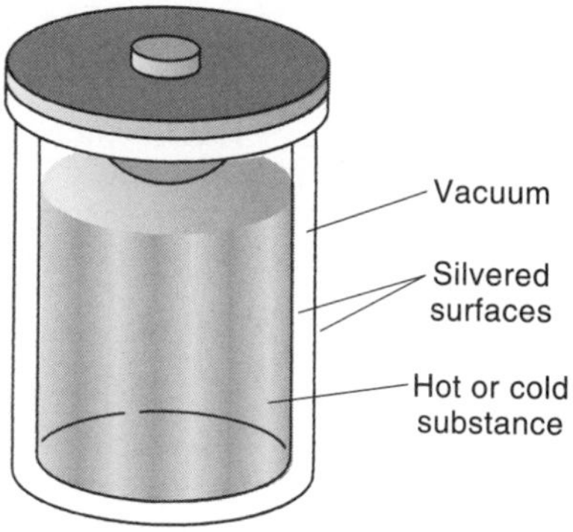

Vacuum
Silvered surfaces
Hot or cold substance

FIGURE 5.2
Thermos bottle (cutaway view).

fee lighter in color, so it radiates more slowly. Also, the temperature will drop when the cream is added, so the rate of heat transfer by radiation and conduction will decrease as a result of the smaller ΔT. Data for this experiment are graphed in Figure 5.1.

One device for reducing heat transfer is the Thermos (Fig. 5.2). The Thermos is a bottle inside a bottle, with the inner vessel separated from the outer one by a vacuum. The evacuated region prevents the transfer of heat from or to the inner vessel by convection or conduction. The inner walls of both vessels are silvered, which cuts down on losses (or gains) from radiation. Glass and cork (or, in recent models, plastic) are poor conductors, so heat transfer by conduction through these materials is negligible.

C. House Insulation and Heating Calculations

Increased insulation of houses is one of the easiest and most cost-effective means of reducing energy consumption. The addition of insulation to a house can save as much as 50% on heating and cooling bills. Nationwide, savings of at least 15% of heating and cooling demands could be made by having all houses meet the federal minimum property standards (Table 5.1).

The rate of heat flow via conduction was given previously as

$$\frac{Q_c}{t} = \frac{k \times A \times \Delta T}{\delta}$$

Table 5.1 FEDERAL HOUSING ADMINISTRATION (FHA) MINIMUM INSULATION RECOMMENDATIONS

	Pre-1978	New
Ceilings	R-19	R-38 (12″ fiberglass)
Walls	R-11	R-19 (6″ fiberglass)
Floors	R-11	R-22 (7″ fiberglass)

where k is the material's thermal conductivity, A its area, δ its thickness, and $\Delta T = (T_2 - T_1)$.

A measure of the resistance of the material to heat flow is the thermal resistance R, often called its **"R-value,"** given by $R = \delta/k$. The higher is the R-value, the better are the insulating properties of the material. Using this notation, the rate of heat flow is given by

$$\frac{Q_c}{t} = \frac{1}{R} \times A \times \Delta T$$

R is a function of both the type of material and its thickness.

The R-values of some common building materials are given in Table 5.2, in conventional units of ft^2-h-°F/Btu. Note that a 6-in. thickness of fiberglass insulation has an R-value of 19 in these units, equivalent to 8 ft of brick (Fig. 5.3). A sheet of polyurethane board (sometimes called "High-R" sheathing), has about six times the insulating strength of a sheet of plywood of the same thickness.

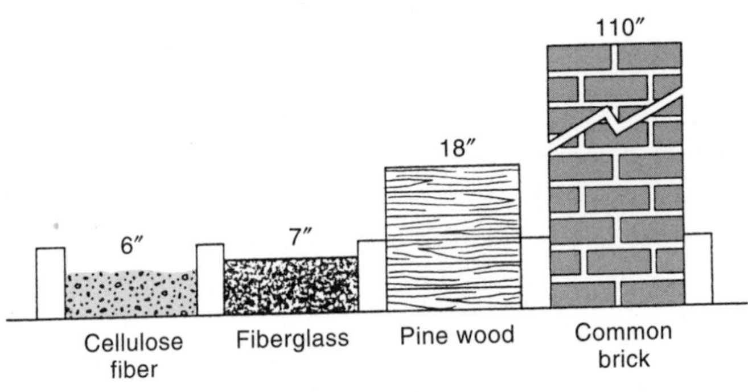

FIGURE 5.3
To obtain an R-value of 22 ft^2-h-°F/Btu, you would have to use the indicated thicknesses of various materials.

Table 5.2 R-VALUES OF COMMON BUILDING MATERIALS*

Material	Thickness	R-Value (ft^2-h-°F/Btu)†
Hardwood	1″	0.91
Softwood	1″	1.25
Plywood	½″	0.62
Concrete block	8″	1.04
Concrete brick	1″	0.20
Sheetrock (gypsum board)	½″	0.45
Fiberglass insulation	3½″	10.9
Fiberglass insulation	6″	19.0
Expanded polystyrene board	1″	4.0
Expanded polyurethane board	1″	6.3
Cellulose insulation	1″	3.7
"Thermax" or "High-R" sheathing	1″	8.0
Flat glass	⅛″	0.88
Insulating glass	¼″ air space	1.54
Insulating glass	½″ air space	1.72
Wood subfloor	25/32″	0.98
Hardwood floor	¾″	0.68
Nylon carpet	1″	2.0
Tile		0.05
Asphalt roofing shingle		0.44
Asbestos shingle		0.21
Steel	1″	0.0032
Copper	1″	0.0004
Wood siding (lapped)	½″	0.81

*Source: American Society of Heating, Refrigeration, and Air-Conditioning Engineers (ASHRAE) "Handbook of Fundamentals", 1991.

†In the metric system, the units for R are m^2-°C/W; R-Value (metric) = R-Value (English) × 0.57.

One pane of window glass is equivalent to about 1 in. of hardwood or 5 in. of brick in thermal resistance.*

EXAMPLE

Calculate the total heat transfer for 12 h through an insulated window (two glass panes with an air gap of ¼ in.) that measures 4 ft by 7 ft, when the outside temperature is 5°F and the inside temperature is 65°F.

Solution

From Table 5.2, the R-value for this type of window is 1.54 ft^2-h-°F/Btu. The area is 28 ft^2, and the temperature difference is $\Delta T = 65°\text{F} - 5°\text{F} = 60°\text{F}$. Therefore,

$$Q_c = t \times \frac{1}{R} \times A \times \Delta T = 12 \text{ h} \frac{1 \text{ Btu}}{1.54 \text{ ft}^2 - \text{h} - °\text{F}} \times 28 \text{ ft}^2 \times 60°\text{F}$$

$$= 13{,}100 \text{ Btu}$$

In the exterior walls of a house, you will find a combination of various materials. From the outside to the inside, there may be exterior siding, plywood, fiberglass, and Sheetrock. To find the *total* thermal resistance of such a composite structure, you simply add the R-values for the individual layers: $R_{\text{total}} = R_1 + R_2 + R_3 + \ldots$.

As an example of a composite structure, consider an exterior wall that consists of ½-in. plasterboard, 3½-in. fiberglass insulation, ¾-in. plywood, and ½-in. lapped wood siding. The overall resistance of this wall (Fig. 5.4) is $R_{\text{total}} = 0.45 + 10.9 + 0.94 + 0.81 = 13.1$ ft^2-h-°F/Btu.

The combined effects of conduction and convection are included in tabulated R-values. The convective resistance depends strongly on the velocity of the air moving along the surface. The R-values in Table 5.2 assume that the air films on the outside of the building are those created by a wind of 10 mph. If the wind velocity increases to 15 mph, the R-value of the air film is decreased by 50%. Planting trees and shrubs near a house to shield it from harsh winter winds will do much to reduce heat transfer by convection.

Heat loss by natural convection on the inside of a window is also increased significantly if air is allowed to move freely past the window. Consequently, drapes that touch the floor—together with cornices or valences—can be very

*The tabulated "resistance" of window glass is the sum of the conductive and convective resistances; that is, an insulating value has been given to the films of air that lie next to the surfaces of the material. This inclusion is especially significant for windows because the glass pane itself (usually about ⅛-in. thick) has a low R-value.

Superinsulated office building in Toronto, Canada, that uses no furnace. Heat from machines, lights, and people is stored and circulated. South-facing windows with insulating glass and excellent insulation are used. (COURTESY OF ONTARIO POWER GENERATION)

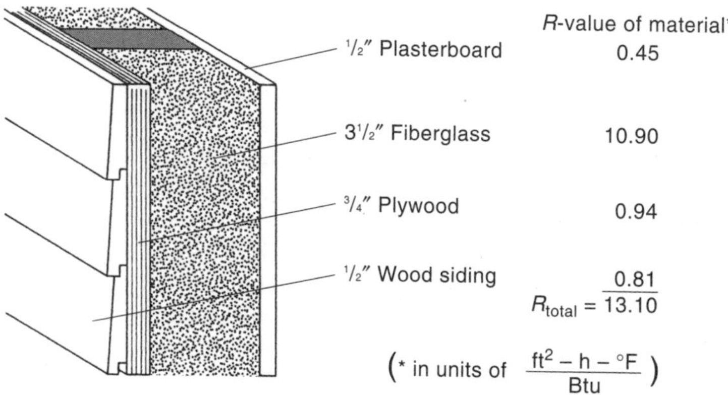

FIGURE 5.4

Example of the calculation of thermal resistance value for a composite wall.

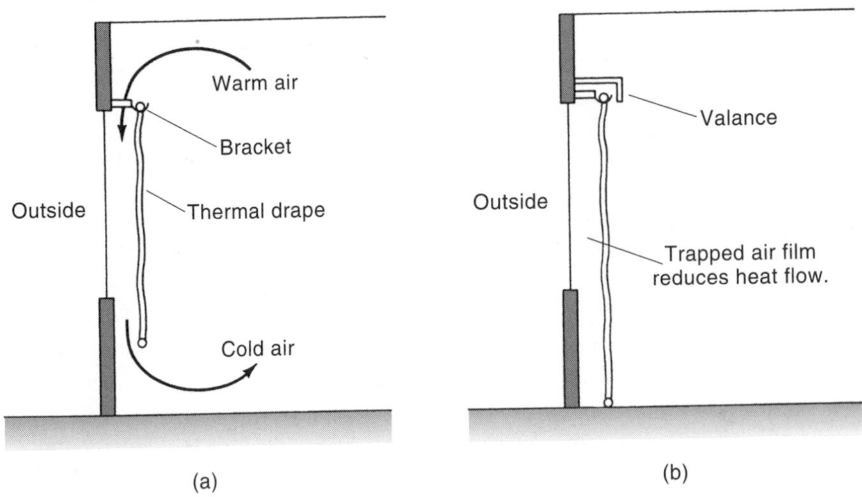

FIGURE 5.5
(*a*) Thermal drapes mounted in this arrangement will contribute to heat losses rather than stop them. (*b*) This arrangement is better for reducing convective losses.

helpful (Fig. 5.5) by maintaining a layer of still air next to the window. From Table 5.2 you can see that wider gaps between two glass panes in an insulated window are not much better. A storm window with a 3½-in. air gap has about the same R-value as one with a ½-in. air gap. Even though conduction will decrease because of a thicker air space, convection losses become more prominent the wider the gap. (Some homeowners used to stuff crumpled newspapers between the studs of old uninsulated houses within the wall cavity. However, this proved a fire hazard and a source of nesting material for mice.)

Windows are important for providing daylight and interaction with the outside world, but they can also be the source of large heat losses and gains. About 35% of the energy requirements for an insulated home are a result of heat loss through the windows. A single-pane window, not exposed to sunlight, can lose 20 times as much heat as an equal area of well-insulated wall. For this reason, double-pane windows are required for newly constructed buildings in most parts of the country. The energy saved by the use of storm windows in an average home in a cold climate is about 2 million Btu per year per window. This is approximately a 1% savings in the fuel used for heating an average home (1500 square feet) *for each* storm window installed. However, even though the energy savings is high, the installation cost might be prohibitive for many families. In most areas it will take from three to ten years for a storm window to pay for itself in fuel cost savings. Since Americans move on the average of once every five years, the economics of storm window installation are not as promising as those of house insulation, unless fuel costs rise even more. A cheaper alterna-

A tight-fitting window shade or quilt can reduce heat loss by forming a layer of still air next to the window. The quilt itself also provides insulation.

tive is the addition of plastic storm windows over a single pane; these are ten times less expensive, although they will need to be replaced every year or two. (See Example at end of this section.)

A feature added to many higher quality windows in recent years to improve their energy performance is "low-e coatings." This "low-emissivity" coating is a very thin, transparent metal coating applied on the inside glass surface of a double-pane window. These coatings are good heat reflectors and poor heat emitters. In the winter, they will reflect heat back into the room. The temperature of the glass will also be warmer, meaning less of a ΔT between a person and the window surface. In the summer, the low-e coating will block the shorter wavelength infrared solar radiation; thus reducing heat gain. These windows look clear because 75% of the visible light is transmitted, compared to 82% for conventional double-pane windows.

Other heating losses come from **air infiltration.** The cold air that leaks into the house from cracks around windows, doors, and the foundation must be heated (Fig. 5.6). Infiltration can be responsible for almost 50% of home heating requirements! The amount of air infiltration depends on the tightness of the house; an average house has about one complete air change per hour. Heat loss (or gain) from infiltration can be reduced by a factor of about 2 by weather-stripping and caulking around windows and doors (Fig. 5.7). This investment can often pay for itself in one-half year.

Infiltration heat losses can be calculated by the "air-exchange" method. Each fresh volume of air that leaks in from the outside and exchanges with the warmer air in the house must be heated to the inside temperature. The heat

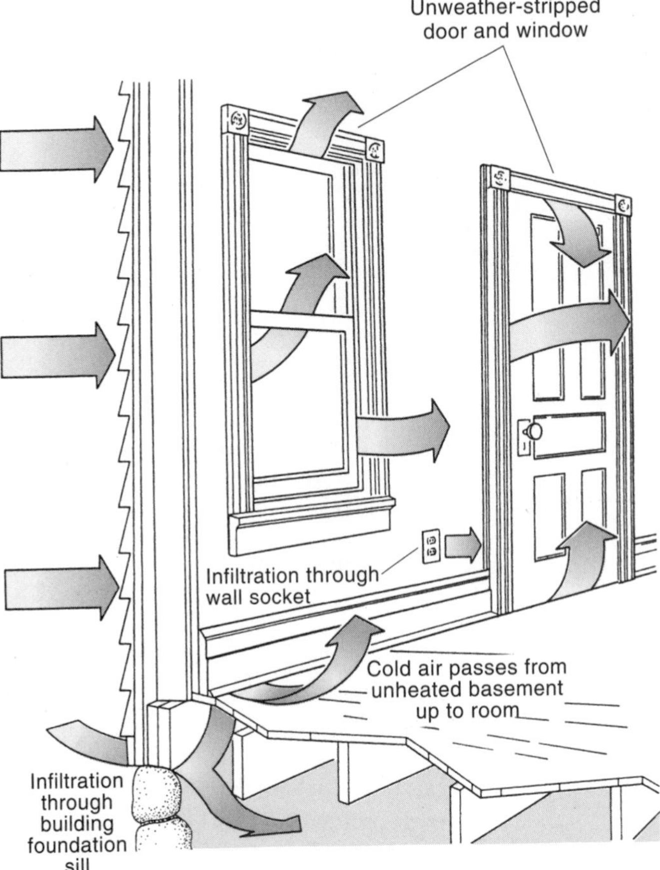

FIGURE 5.6
Cold air infiltration can account for 50% of the energy needs of a house. Wherever there is a way for the air to get in, it will. Winds can significantly increase infiltration rates.

that must be added to the air depends on the volume V (in ft^3) of the house, the temperature difference ΔT between the inside and outside air, the volume heat capacity of air (0.018 Btu/ft^3-°F), and the number K of air changes per hour. Taking these constants into account, the rate of heat loss for infiltration is written as

$$\frac{Q_{\text{infil}}}{t} = 0.018 \times V \times K \times \Delta T \text{ Btu/h}$$

K is usually measured by installing a fan in an exterior door frame. One can then measure the rate of decrease of a tracer gas within the house and calculate

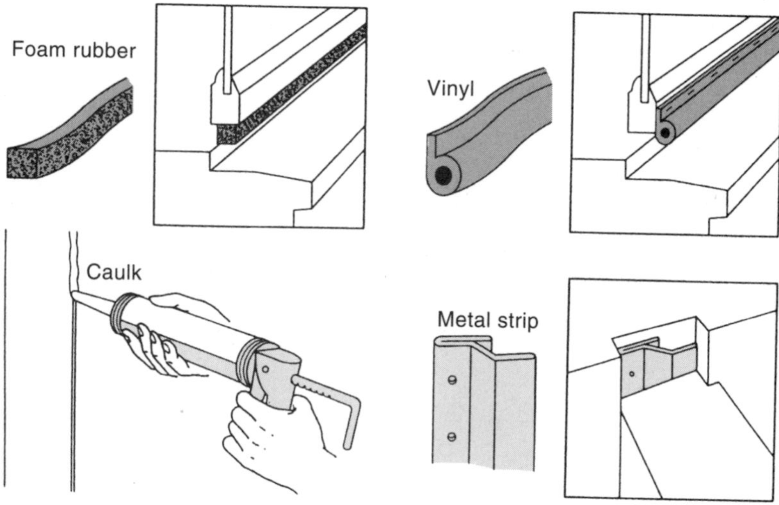

FIGURE 5.7

Types of weather stripping. Caulking all cracks and weather-stripping all windows will reduce air infiltration. Such work is easy to do and very cost-effective (with payback times of one heating season or less).

the number of air changes per hour. Average well-maintained and insulated houses have K values of about 0.5 to 1.5 per hour, while superinsulated houses have $K \approx 0.2$ (which might cause a problem of indoor air pollution; see Chapter 8). Old, uninsulated houses can have K values from 3 to 6, depending on the degree of maintenance.

Let us calculate the hourly heating load for an entire house. We must first calculate the terms $1/R_{\text{total}} \times A$ for each of the exterior surfaces of the house (sides, roof, floor, windows) and add them. This sum is multiplied by the

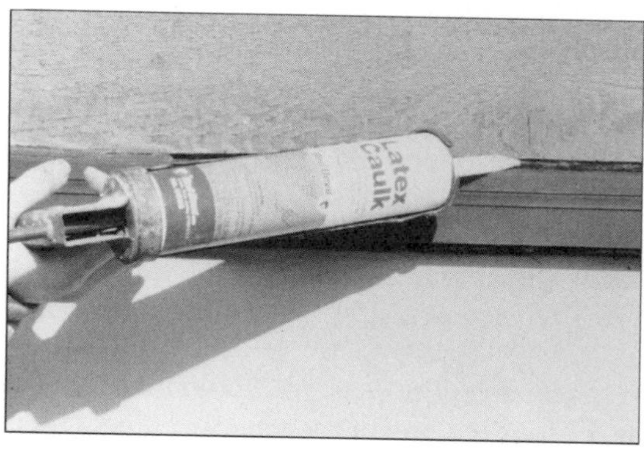

Caulking around the foundation of a home reduces air infiltration.

temperature difference ΔT. Infiltration losses must also be included. The following example illustrates the calculation of the heating load for a house.

EXAMPLE

Calculate the hourly heating load of a house in which all exterior surfaces have a thermal resistance of R = 14 ft²-h-°F/Btu, the inside temperature is 65°F, and the outside temperature is 5°F. The dimensions of the house are 30 ft × 50 ft × 10 ft.

Solution

The conductive heat loss is

$$\frac{Q_c}{t} = \frac{1}{R} \times A \times \Delta T$$

The area of the exterior surface is

$$\begin{aligned}
&\text{Walls: } (30 \times 10) \times 2 + (50 \times 10) \times 2 = 1600 \text{ ft}^2 \\
&\text{Roof, floor: } (30 \times 50) \times 2 = 3000 \text{ ft}^2 \\
&\text{Total surface area: } = 4600 \text{ ft}^2
\end{aligned}$$

So

$$\frac{Q_c}{t} = \frac{1}{14} \text{ Btu/ft}^2\text{-h-°F} \times 4600 \text{ ft}^2 \times (65 - 5)\text{°F} = 19{,}700 \text{ Btu/h}$$

Using the previous equation, infiltration losses (for one air change per hour) are

$$\frac{Q_{\text{infil}}}{t} = 0.018 \text{ Btu/ft}^3\text{-°F} \times (30 \times 50 \times 10) \text{ ft}^3 \times 1/\text{h} \times (65 - 5)\text{°F}$$

$$= 16{,}200 \text{ Btu/h}$$

The total hourly heating load is therefore 35,900 Btu/h. (Note that this house has no windows or doors!)

To size the heating system for a house (i.e., to find the Btu/h capacity needed for your heating unit), use the outside temperature T_o closest to the lowest expected in your locality (called the design temperature): $\Delta T = 65° - T_o$.

The efficiency of the heating unit must be taken into account also. From the previous example, if the design temperature were 5°F, the furnace (assuming a natural-gas furnace with 70% efficiency) should have a rating of 36,000/0.70 = 51,000 Btu/h.

To calculate the heat losses for an entire year, you could multiply $1/R_{\text{total}} \times A$ for the entire house by the ΔT for each day of the year when heating is needed and by the number of hours in the day, and add. However, such a calculation is much easier if you use the concept of a "degree-day." For any one day, the degree-day (DD) is found by averaging the high and low temperatures for that day and subtracting from 65°F.

$$\text{Degree-day (DD)} = 65°\text{F} - T_{avg}$$

A Celsius degree-day is 1.8 times bigger than a Fahrenheit degree-day. Negative degree-day numbers are taken as zero. For a year, the total number of degree-days is the sum of all the individual degree-days. These totals range from 14,000 in Alaska to below 2000 in the southern United States; the median value for U.S. homes is about 5000 Fahrenheit degree-days per year (Fig. 5.8).

EXAMPLE

March 18 had a high temperature of 42°F and a low of 20°F. How many degree-days were there?

Solution

The mean temperature for the day is $(42 + 20)/2 = 31$°F. Using the 65°F base, the number of degree-days is $65 - 31 = 34$ DD.

Using this concept of degree-days, the annual total heating needs for conductive losses can be written as

$$Q_{total} = \sum\left(\frac{1}{R} \times A\right) \times (24\ \text{h/day}) \times (\text{number of annual degree-days})$$

where the sum $\sum$ includes all the exterior surfaces.

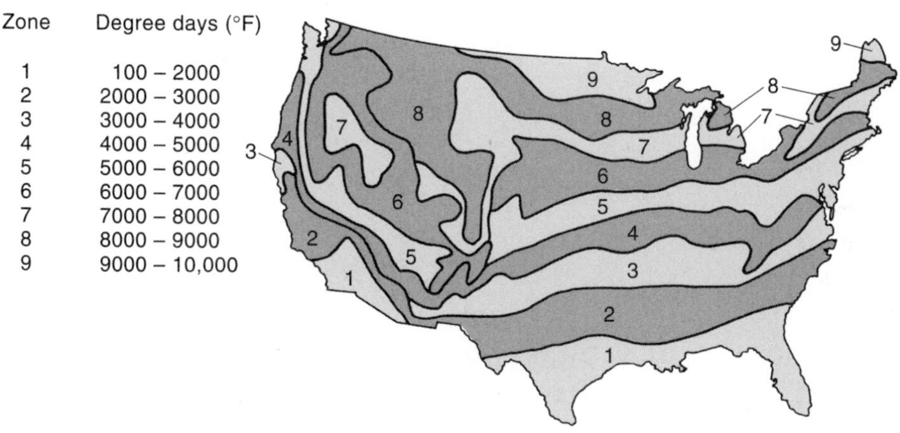

FIGURE 5.8
Annual heating degree-days (DD).

If the insulated house of the previous example was located in Buffalo, New York, with 7000 degree-days per year, the total conductive heating load for the house for a season would be

$$Q_{\text{total}} = \frac{1}{14}\ \text{Btu/ft}^2\text{-h-°F} \times 4600\ \text{ft}^2 \times 24\ \text{h/d} \times 7000\ \text{DD} = 55.3 \times 10^6\ \text{Btu}$$

We must add to this the infiltration losses of 0.018 Btu/ft^3-°F × 15,000 ft^3 × 1/hour × 24 h/d × 7000 DD = 45.4 × 10^6 Btu, for a total heating load of 101 × 10^6 Btu. We abbreviate 10^6 Btu (or one million Btu) as MBtu.

Assume the cost of electricity is $0.07 per kWh; this is equivalent to ($0.07/kWh) × (1 kWh/3413 Btu) × (10^6 Btu/MBtu) = $20.51/MBtu. Then the cost for heating the house in Buffalo with electricity for a season would be $20.51/MBtu × 101 MBtu = $2071. If a furnace is used, the total heating load must be divided by the furnace's efficiency. If we used natural gas at a cost of $0.60 per therm (1 therm = 100,000 Btu's), or $6.00/MBtu, with a furnace efficiency of 70%, the cost would be $6.00 × 101/0.70 = $865. Appendix C presents Home Energy Audit forms that you can use to calculate the heating needs of your house (with windows and doors).

EXAMPLE

Suppose that you lived in an apartment with single-pane windows and paid for your utilities. Your landlord resembles "Scrooge." Let's calculate the savings to you if you installed a plastic storm window. (Assume 5000 annual DD.)

Solution

Single-pane window glass has an R-value of 0.88 ft^2-hr-°F/Btu, while a storm window (even with the plastic sheet installed on the inside of your window) has an R-value of 1.72 (for ½-in. air space).

For a 3 ft × 3 ft window, the annual heat loss for the single-pane window is

$$Q_1 = (1/0.88) \times 9\ \text{ft}^2 \times 24\ \text{h} \times 5000\ \text{DD} = 1{,}227{,}000\ \text{Btu/y}$$

With a storm window, the annual heat loss will be

$$Q_2 = (1/1.72) \times 9 \times 24 \times 5000 = 628{,}000\ \text{Btu/y}$$

Therefore, the savings for one window is

$$Q_1 - Q_2 = 599{,}000\ \text{Btu/year}$$

If your fuel were natural gas at $7.00/MBtu, then your annual cost *savings* would be

$$\frac{599{,}000}{1{,}000{,}000} \times \frac{\$7.00}{\text{MBtu}} = \$4.20\ \text{per year per window}$$

A plastic sheet might cost $2.00, so the installation paid for itself in about one half of the heating season.

D. Site Selection

A very important factor in convective heat loss is the wind. Heat losses through walls and roofs as well as through window, door, and foundation cracks directly depend on the wind velocity. Heat load calculations indicate that an increase in wind velocity from 10 to 15 to 20 mph will increase the building infiltration by 100% and 200%, respectively, for non-weatherstripped windows and doors on the windy side of the building. The R-values given earlier in this chapter included terms for an interior and exterior air film next to a window. The R-value of the outside air film assumed a 10 mph wind. In fact, in the R-value of 1.54 for a double-pane window, about half of the value comes from the thin air film on the inside, which is not moving.

Winds should be blocked during cold periods and used for cooling during warm periods. Windbreaks should be used to reduce wind velocities around the house. A garage situated on the northwest side of a house can be effective in blocking the prevailing winter winds. Trees, bushes, and man-made windbreaks can also be effective. The best place on a hill for a building is midway up the hill (on the leeward-downwind side of the winter winds), as this position is both away from the high winds at the top of the hill and out of the cold air masses that could be trapped at the bottom of a valley (Fig. 5.9).

Site selection should also take into account the path of the sun. By orienting a house with its longest side facing south, solar heating features can be more easily used. South-facing windows can capture some of the sun's radiation during the winter months, while north-facing windows will receive none of the

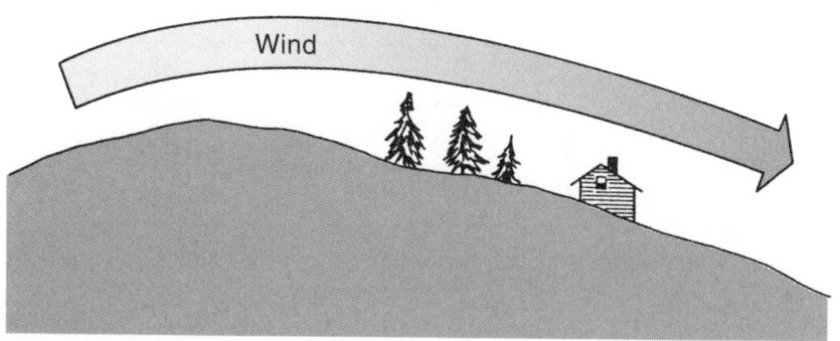

FIGURE 5.9

Planting trees on the leeward side of a hill can substantially reduce the wind velocity over the site. Vegetation or walls can block or deflect natural air-flow patterns and so reduce convective heat loss.

sun's direct light during this time. The north windows will provide a large surface for heat loss, especially due to the prevailing northern winds, while the south-facing ones will offer a net gain in heat energy. The heat addition in one day through a south-facing window as a result of the sun's radiation will typically be *greater* than the heat loss in a 24-hour day through conduction and convection. This will be covered further in Chapter 6.

E. Impact of Energy Conservation Measures

Table 5.3 shows the impact of energy conservation measures in reducing annual heating requirements for new houses in three parts of the country. With increased insulation, infiltration control, and night-time thermostat setback, as outlined in Figure 5.10, the annual heating bill can be reduced by almost 50% in each area. Some of these design modifications would be more difficult to implement: R-19 wall insulation using fiberglass would require a 6-in. wall cavity or the addition of rigid board insulation to the outside of conventional walls, under the siding.

Home Energy Audits are a great place to start in evaluating a building's heat load. They establish a baseline and can pinpoint those areas in a house that should receive attention first. Many utilities are required to make low-interest loans available for energy conservation projects. They will also do a free home energy audit and usually require one as a prerequisite to applying for such a loan. (See Appendix C.)

Table 5.3 IMPORTANCE OF ENERGY CONSERVATION MEASURES*

		Annual Heating Requirements (MBtu/y)				
City	**Degree-Days**	**Base Case**	**With Setback**	**% Savings**	**Increased Insulation**	**Total % Savings**
Boston	5630	54.6	46.6	15%	26	52%
Denver	6280	56.1	47.6	15%	27	52%
Nashville	3580	34.7	28.4	18%	14	60%

	Walls	Ceiling	Basement	Windows	Infiltration
Base Case:	R-11	R-19	Not insulated	Double	0.7 air changes/h
Conservation: (Thermostat set back 8°F at night)	R-19	R-30	R-10	Triple	0.5 air changes/h

*Source: See Figure 5.10.

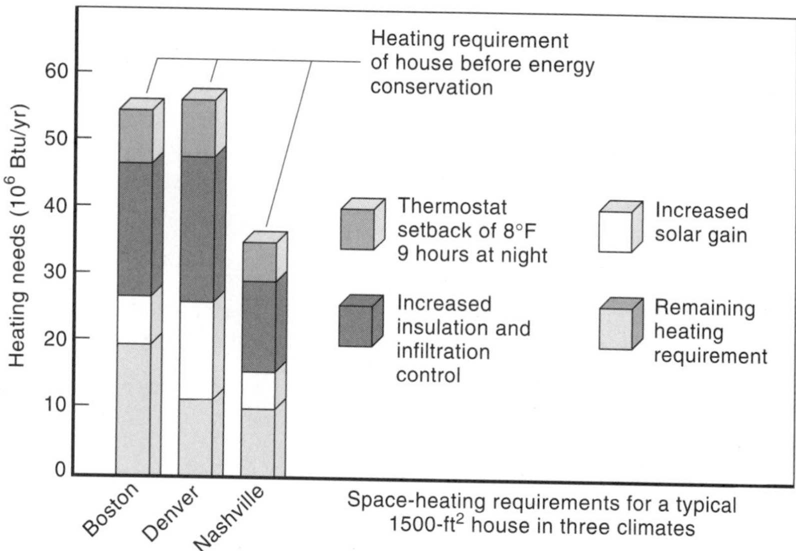

FIGURE 5.10

Summary of the effects of energy conservation measures on space heating requirements for a typical 1500-ft^2 house in three climates.
(LEWIS, D., AND J. KOHLER. 1981. PASSIVE PRINCIPLES: CONSERVATION FIRST. *SOLAR AGE* [SEPTEMBER].)

F. Cooling

Basically the same equations used in the previous section for heating can also be applied to calculating simple cooling loads. The number of cooling degree-days (the average of the day's high and low temperatures minus 75°F) for the year replaces the annual heating degree-days. Insulation and weather stripping are still important to reduce heat gain. Heat gains will also come from sunlit walls and roofs and windows. The intensity of solar radiation as well as the outdoor temperature must be considered. Cooling-load calculations take into account the effect of a time delay in indoor temperature changes caused by thermal storage. Additional cooling is needed later in the day if the sun has been out all day. There are also heat gains inside from lights and appliances and people; a person seated at rest generates about 350 Btu/h of heat and 550 Btu/h if doing light work (such as typing). Appliances and people can make up to 60% of the air conditioning load in an office building.

Air conditioning has completely changed the demography of the southern United States. Places once thought too hot or too humid to live have been transformed into preferred regions for work and retirement through the use of electrically powered air conditioners. The demand for air conditioning has been rising at a rate of almost 10% per year in the United States. Today, two thirds of all households have some form of air conditioning (40% have central air), while 83% of commercial space is air conditioned. Air conditioning accounts for about 5% of the energy used in households in the United States, about one tenth that of the space-heating budget. However, in many regions of the southern United States, the money spent for air conditioning is greater than that for space heating.

Less expensive solutions to summer cooling loads can be achieved through many types of **passive cooling** techniques. The main objective here is to control the heat a building gains from its environment. These principles involve:

- Site selection—location, orientation, vegetation
- Architectural features—surface-to-volume ratio, overhangs, window sizing, shades
- Building skin features—insulation, thermal mass, glazing

Shading is vital in warm climates. Surrounding vegetation, window shades, and overhangs can reduce interior daytime temperatures by 10 to 20°F. Natural ventilation is very important in passive cooling. The moving air can evaporate perspiration from the body and make you feel cooler. Proper attic ventilation is necessary in summer and winter. Adequate ventilation requires vents located at high and low points in the roof to create a natural air flow. Windows on opposite walls of the house are also effective in allowing proper air flow. Since warm air rises, skylights and high windows can act as "thermal chimneys"; the replacement air can be taken from a cooler shaded area on the north side of the house or through an underground tube (a "coolth" tube). These cooling methods are illustrated in Figure 5.11.

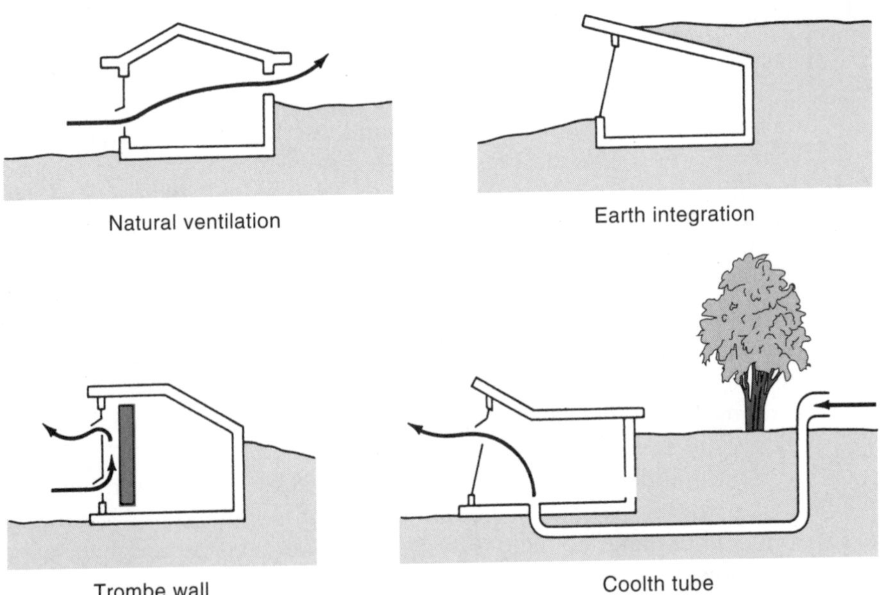

FIGURE 5.11
Passive solar cooling techniques use natural convection and the earth as a heat sink.

Evaporative cooling is another important method, especially in hot, dry climates. Evaporation is a cooling process and single-stage rooftop coolers can be very effective, although the water pump does require additional energy.

In regions of high humidity (such as the southeastern United States), many of these passive techniques won't work. Evaporative cooling is hampered by the high humidity. Radiative cooling at night doesn't work well since the nights are warm and muggy. Natural ventilation must also cope with mold and mildew problems. One technique to provide cooling in such conditions is the use of a desiccant to absorb moisture from the air. During the night, air is circulated through a desiccant in the attic to remove some of the moisture. During the day, the sun dries out the desiccant.

As we know, insulation is important in reducing both heating and cooling loads. A new approach to passive cooling uses "radiant barriers"—materials (such as aluminum foil) that restrict the amount of low wavelength infrared radiation that passes across the attic. Usually the temperature gain by the roof is passed down to the attic floor by radiation and then conducted through the ceiling into the house. Aluminum foil placed against the top of the attic, or over the ceiling insulation (with holes to allow vapor to escape), acts as a good reflector and a poor emitter (Fig. 5.12). Radiant barriers can also be

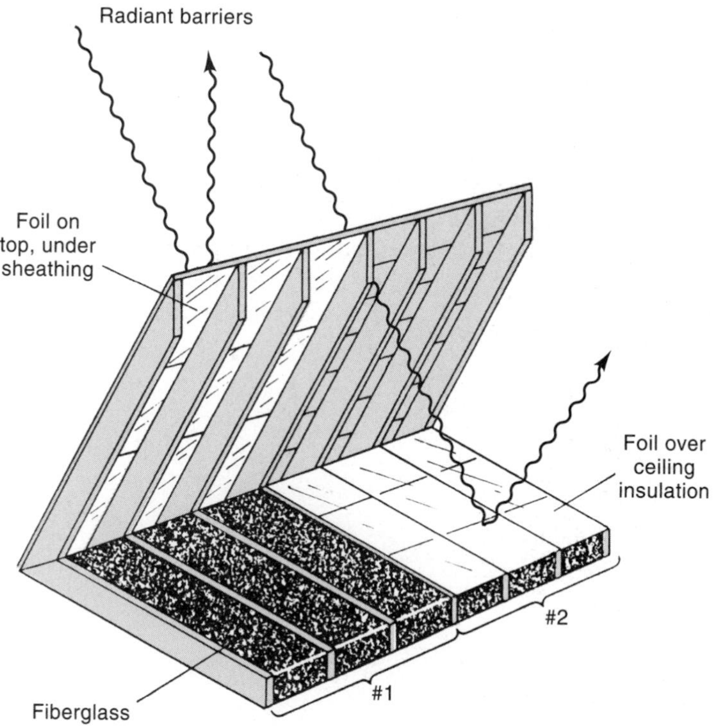

FIGURE 5.12
Radiant barriers can reduce heat gain in the attic of a house.

placed in the walls. An air space is needed next to the barrier to remove the heat that is absorbed by the foil and not emitted.

G. Air Conditioners and Heat Pumps

An air conditioner is similar to a refrigerator in that heat from inside a cooled space is transferred to the warmer outdoor space. This flow of heat from cold to hot is not a violation of the second law of thermodynamics since it does not occur spontaneously. Energy is added to the system via electricity or natural gas combustion or solar energy to accomplish this task.

The four basic components of an air conditioner are shown in Figure 5.13. A working fluid such as Freon-12 (or one of the new, non-CFC refrigerants) is used. In the evaporator (the coils inside the air conditioner or refrigerator), the liquid Freon absorbs heat from the warmer surrounding air and turns into a gas. The electrically driven compressor raises the pressure and temperature of the gas and forces it into the condenser coils outside the house. Because the temperature of the gas is higher than the outside air, the gas condenses, liberating heat. The liquid then flows back through a pressure-reducing valve to the evaporator coils, and is cooled through expansion.

A device that is becoming very popular in certain sections of the United States for space conditioning is the heat pump. A **heat pump** is a heat engine operating in reverse. This device can be used to heat or cool a space. As the name implies, a **heat pump** (Fig. 5.14) can be used to extract or pump heat either into or out of a space. For home heating, a heat pump extracts heat from outside air, even at temperatures below freezing, and transfers it indoors. During hot weather, the cycle is reversed; heat is extracted from the indoor air and expelled to the outside as in an air conditioner or refrigerator. Thus, a heat pump is essentially an air conditioner and a heating unit all in one. Its advantage in heating lies in the fact that when it extracts energy from the outside surroundings, it transfers to the heated space at least twice as much energy as is

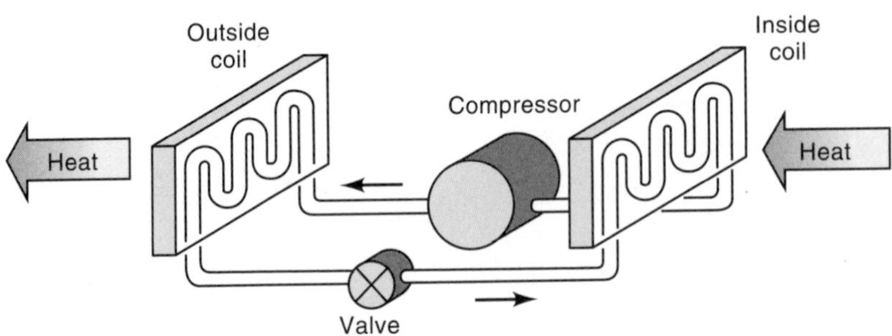

FIGURE 5.13
Air conditioner operation.

A residential heat pump.
(CARRIER CORP.)

used to run the heat pump. Consequently, the cost of heating a home with a heat pump will be about half that of an electric resistance heating system (but still more expensive than a gas furnace).

The heat pump was introduced into the U.S. market in 1932, but developed a bad reputation in its early years because of its lack of reliability. In

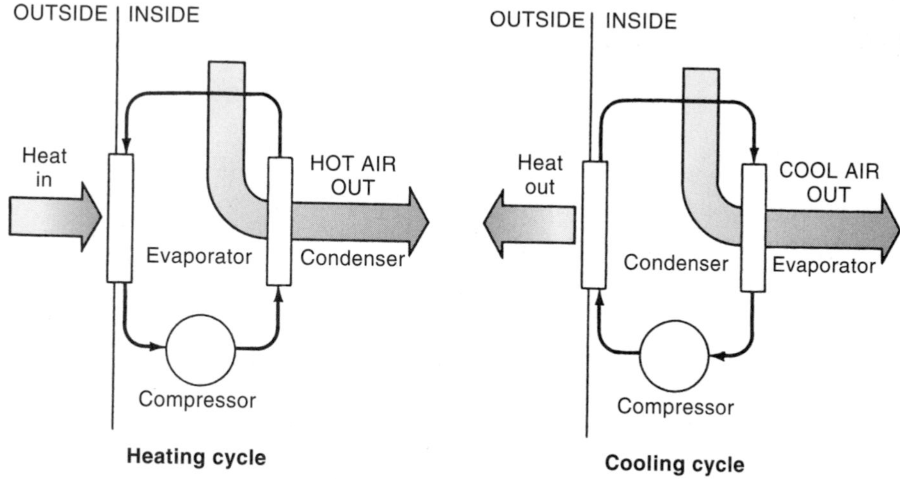

FIGURE 5.14
Basic components of a heat pump system.

the late 1960s, it started to make a comeback with better reliability, and achieved a rate of growth of almost 7% per year in the late 1960s and 1970s. More than 1 million units a year are sold now. Heat pumps are most economical to operate in geographical regions in which both heating and cooling are needed. They are most efficient when the outside temperature is above 10°F. (With low winter temperatures, the efficiency drops to that of an electric resistance heater.)

The basic components of the heat pump are similar to those explained previously for the air conditioner. In the *heating mode,* the outdoor coil is the evaporator and the indoor coil is the condenser. During this mode, (1) the refrigerant (in a liquid state) absorbs heat in the evaporator from the outdoor air, turning into a gas. (2) The compressor raises the pressure of the gas, increasing its temperature. (3) The hot gas then flows through the condenser coils that are inside the heated space. Because the temperature of the gas is higher than the room's, the gas transfers heat to the room and condenses to a liquid. (4) The liquid then flows back through a pressure-reducing valve to the outdoor evaporator coils, being cooled through expansion.

In a *cooling cycle,* the evaporator and condenser coils exchange roles. The direction of flow of the refrigerant is reversed with a special valve. Like an air conditioner or refrigerator, the cold refrigerant absorbs heat from the warmer room, increases its temperature further by passing through the compressor, and then transfers heat through the condenser coils to the outside air.

The performance of a heat pump is expressed not in terms of the ratio of energy output to energy input, but instead in terms of the ratio of energy output to *work* input. The work input is the energy going into the compressor (the quantity we pay for) and is the denominator in the expression for performance. This ratio is called the **coefficient of performance (C.O.P.)** of a heat pump:

$$\text{C.O.P.} = \frac{\text{heat transferred}}{\text{electricity input}}$$

In terms of power, we can also define the C.O.P. as the ratio of the *rate* of energy transfer into or out of the house to the compressor shaft power. For example, a heat pump with a cooling capacity of 30,000 Btu/hr and a compressor power rating of 3.2 kW has a C.O.P. of

$$\text{C.O.P.} = \frac{30{,}000 \text{ Btu/h}}{3.2 \text{ kW} \times (3413 \text{ Btu/h/kW})} = \frac{30{,}000}{10{,}922} = 2.75$$

In direct electrical resistance heating, every Btu put into the heaters in the form of electricity is released as one Btu of heat. In this example, a heat pump is therefore 2.75 times cheaper to operate than heating with an electrical resistance system.

The C.O.P. of a heat pump decreases as the outside temperature drops. The lower the outdoor temperature, the less heat there is available to be moved, and therefore the lower the C.O.P. Also, if heat is to be extracted from 10°F (−12°C) air, the temperature of the working fluid in the evaporator must be below 10°F for heat to be transferred. This puts a greater strain on the compressor, which must raise the temperature of the refrigerant by a larger amount. A performance curve for a standard heat pump is shown in Figure 5.15. Most units are operated at C.O.P.s greater than 1.5.

Because a heat pump's performance varies with the local climate, it is better to speak of a seasonally averaged C.O.P. The **seasonal performance factor (SPF)** is defined as

$$\text{SPF} = \frac{\text{total output}}{\text{total energy consumed}}$$

This number will be lower than the value of the C.O.P. a manufacturer might provide.

To reduce the spread between condenser and evaporator temperatures and increase the C.O.P., the evaporator coil can be placed in a storage tank or reservoir containing well water. It can also be buried in the earth, where the temperatures in the winter are higher than that of the outside air. The storage tank water can also be heated by using simple solar collectors. (Even unglazed [without glass] collectors could be used, since the temperature gain of the tank water does not need to be large.) These "solar-assisted" heat pumps can increase the heating C.O.P. by a factor of 2.

Although a heat pump can provide two units of heat for every one unit of electricity required to run it, the environment is not necessarily getting a bargain. This is because conventional power plants need almost 3 units of heat to produce 1 unit of electrical energy. If the C.O.P. of a heat pump is 2.1, then the overall efficiency (from primary energy to final output) is about 0.7, comparable to the efficiency of an older oil or gas-fired furnace.

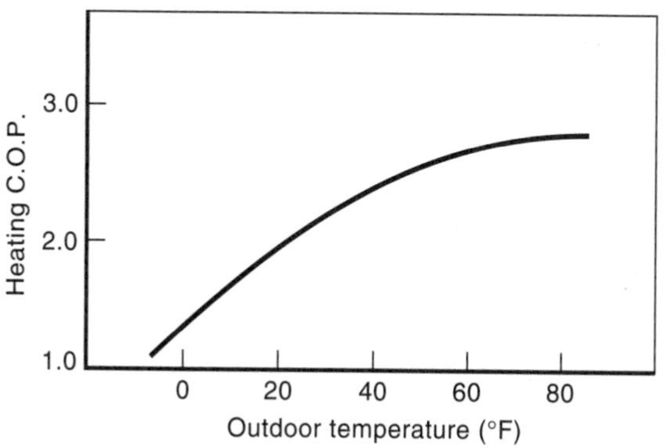

FIGURE 5.15
Heat pump coefficient of performance (C.O.P.) curve.

Whether heat pumps are more economical than other space heating methods depends on the local climate. Figure 5.16 compares the costs of using a heat pump, electric resistance, oil, and natural gas. For the heat pump, several seasonally averaged values of C.O.P. are shown. For example, if a heat pump has a C.O.P. of 2.0 and electricity costs \$0.07/kWh, then the cost of heating with the heat pump will be the same as for an oil-fired furnace at \$0.95 per gallon of oil or for a natural gas-fired furnace at \$0.72 per 100 ft^3 of gas.

Heat pump manufacturers suggest that the heat pump is most economical in locations in which both heating and air conditioning are needed. The capacity of the heat pump should be selected to meet the required cooling capacity, not the heating capacity. Excess cooling capacity of a unit will result in more frequent compressor cycling (and so reduced life expectancy), and provide inadequate humidity control. Therefore, if heating requires a larger load than cooling, supplemental heating will be required (to avoid oversizing the heat-

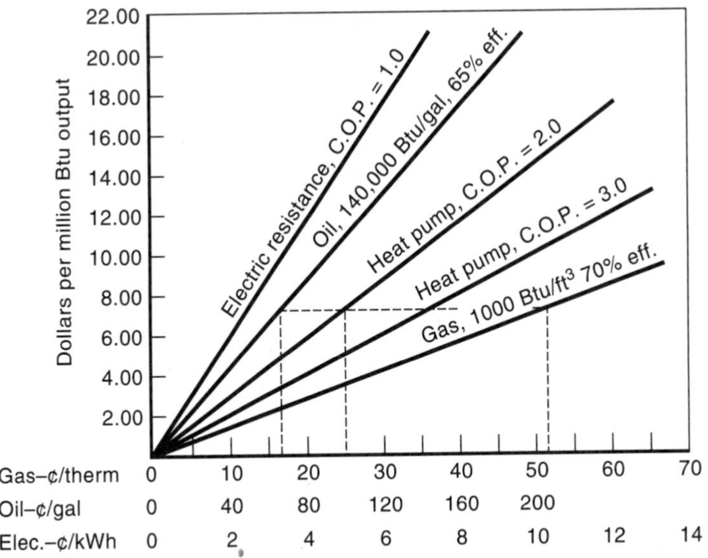

FIGURE 5.16

Relative costs of different types of heating fuels. To determine the cost of using a heat pump, find the unit cost of electricity on the *x*-axis and move vertically to the appropriate heat pump C.O.P. curve. Then move horizontally to the *y*-axis to find the operating cost per million Btu. Find the intersection of this value with the line for gas or oil, and read off its cost on the *x*-axis. For example (as shown with dashed lines), using a heat pump with C.O.P. = 2.0 at an electricity cost of \$0.05/kWh is equivalent to using gas at \$0.52/therm or fuel oil at \$0.67/gal (with the furnace efficiencies shown). (CARRIER CORP.)

pump system). This increases the price of the whole heating system, but even in moderate climates, heating this way can be cheaper than with oil furnaces and certainly cheaper than electrical resistance systems.

H. Summary

Household energy conservation can proceed in many directions and has a substantial impact on your utility bill. This chapter has focused primarily on reducing heat gains or losses in a house by increased insulation and infiltration control.

Table 5.4 lists some things that can be done to reduce energy consumption in a house through retrofitting, that is, changes that can be made with no major design or construction modifications.

Methods for calculating the heating load for a house use the equation for the rate of heat transfer via conduction:

$$\frac{Q_c}{t} = \frac{1}{R_{\text{total}}} \times A \times \Delta T$$

where R_{total} is the total thermal resistance of the material, A is the area of that surface, and ΔT is the temperature difference across the material. To find the

Table 5.4 RETROFITTING—THINGS TO DO TO CONSERVE ENERGY IN THE HOUSE

Caulk and weather-strip around all openings and cracks
Insulate (including floors over unheated spaces)
Insulate hot water pipes; insulate water-heating unit
Add storm windows and doors (consider using plastic)
Install insulated shades, panels, or draperies on all windows
Ventilate attic and crawl space
Conduct general repairs to reduce air infiltration
Install clock thermostat on furnace
Tune-up furnace and water heater annually
Set water-heater thermostat at 120°F
Insulate furnace ducts (hot air system) and heat supply lines
Replace shower heads with water-saving type
Build a vestibule around exterior doors
Replace incandescent lighting with fluorescent lights
Consider task lighting
Redesign landscaping; add foliage to block or divert wind
Paint interior walls with light, warm colors
Provide exterior air sources for furnace and/or fireplace
Install glass doors on fireplace or install air-tight stove in fireplace opening
Install energy-saving appliances

total heat loss rate from a house, contributions from all surfaces (with different A's and R's) must be summed. To find the total heat loss via conductive losses for a season, Q_{total}, we use the degree-day (DD) concept, where a heating DD = $(65 - T_{avg})$. Then

$$Q_{total}\,(\text{Btu}) = \left(\sum \frac{A}{R_{total}}\right) \times 24 \times DD_{annual}$$

Infiltration losses must be added to this to obtain the total heating load for the house.

Internet Sites

For an up-to-date list of Internet resources related to the material in this chapter, go to the Harcourt College Publishers website at **http://www.harcourtcollege.com**. The links are in the *Energy: Its Use and the Environment* site on the Physics page. General energy related sites and some guidelines for using the World Wide Web in your class are on the inside front cover of this book.

References

Chapter 5

Anderson, B. 1977. *Solar Energy: Fundamentals in Building Design.* New York, McGraw-Hill.

Chiras, D. 2000. *The Natural House.* White River Junction, VT, Chelsea Green Publishing.

Energy for Planet Earth [collection]. 1990. *Scientific American,* 263 (September).

Hirst, E. 1986. *Energy Efficiency in Buildings.* Washington, D.C., American Council for an Energy-Efficient Economy.

Kleinbach, M., and C. Salvagin. 1986. *Energy Technologies and Conversion Systems.* Englewood Cliffs, NJ, Prentice-Hall.

Lewis, D., and J. Kohler. 1981. Passive Principles: Conservation First. *Solar Age* (September).

Nisson, N., and D. Gautan. 1995. *Superinsulated Home Book.* New York, John Wiley & Sons.

Robinson, S. 1978. *The Energy Efficient Home.* New York, New American Library.

Rosenfeld, A., and D. Hafemeister. 1988. Energy-Efficient Buildings. *Scientific American,* 258 (April).

Sathaye, J., and S. Tyler. 1991. Transitions in Household Energy Use in Urban China, India, the Philippines, Thailand, and Hong Kong. *Annual Review of Energy,* 16.

Shurcliff, W. 1980. *Thermal Shutters and Shades.* Andover, MA, Brick House.

Swenson, D. 1983. *Heating Technology: Principles, Equipment, and Application.* Belmont, CA, Breton.

Williams, R. H., G. S. Dutt, and H. S. Geiler. 1983. Future Energy Savings in U.S. Housing. *Annual Review of Energy,* 8.

QUESTIONS

1. Sketch the methods of heat flow through a storm window.
2. A hot briquette is taken from the fire and placed on a metal cookie sheet; discuss the types of heat transfer that take place as the briquette cools.
3. What methods can be used to reduce infiltration in a house?
4. What things can be done to reduce the rate of heat transfer through a single-pane window plus storm window?
5. Why is a fishnet-type undershirt good to wear in cold weather?
6. What types of shoes are good to wear indoors in cold weather?
7. To keep an iced drink cold for the longest period of time, what color cup would you use, and why?
8. If you were roofing your house with asphalt shingles, what color would you select? Assume you lived in a moderately sunny but cold winter and hot summer climate. Would there be a difference if you lived in a semitropical climate?
9. What factors affect infiltration into a house?
10. What is the average number of annual heating degree-days or cooling degree-days in your location? (Check the Web.)
11. What produces the cooling in a refrigerator?
12. Why can a heat pump be thought of as a heat engine running in reverse?
13. While air conditioners had a dramatic effect on living conditions in industrialized countries, many people in developing countries live in hot, humid climates with no air conditioning. If you were to live in such a situation for three summer months, what steps would you take to survive the climate during the day and night?
14. If you were paying for the energy used to heat and/or cool your room during the school year, what steps would you take to reduce these costs?

PROBLEMS

1. A wall is made up of four elements, as follows:

   ½″ wood siding (lapped)

   ½″ plywood sheathing

   3½″ fiberglass

   ½″ Sheetrock

   Using the R-values of Table 5.2,how many Btu per hour per square foot will be lost through the wall when the outside temperature is 50°F colder than the inside?

2. Heat loss through windows is substantial. What percentage savings will be gained by covering a double-pane window with 2-in. sheet of rigid polystyrene board?
3. Considering only conductive heating losses, calculate the heating needs in Btu/h of a home constructed of 2-in. hardwood planks plus ½″ of lapped wood siding. Sheet rock (½″) covers the inside of the dwelling. The size of the house is 28 ft × 40 ft × 10 ft. For simplicity, assume the roof is flat, covered with 1-in. plywood plus asbestos shingles, and has 3½″ of fiberglass insulation. There is wood subfloor (25/32″) and a hardwood floor (¾″) over an unheated crawl space. Assume 25% of the wall areas are single-pane windows. The outside is 40°F colder than the inside.
   (a) What will be the percentage savings in heating needs if this house is insulated with a total of 6″ of fiberglass in the attic?
   (b) What will be the percentage savings in heating needs if this house is insulated with 3½″ of fiberglass in the walls (ceiling remains the same 3½″)?
   (c) What percentage savings is achieved by replacing the single-pane windows with double-pane windows (½″ air space)?
4. The following problem can illustrate the economics of insulating the floor above a vented crawl space.
   (a) If the composite structure of the floor is made up of carpet (with an R-value of ½ ft^2-h-°F/Btu), subfloor ($R = 1.0$), and air space between the joists ($R = 0.8$), then find the total R-value.
   (b) If 6″ of fiberglass ($R = 19$) is added between the joists, then find the percentage reduction in the heat loss.
   (c) If there are 6500 degree-days in this area, and the price of fuel is $10 per million Btu, then find the heating cost per square foot per heating season for the uninsulated floor.
   (d) If 6-in. fiberglass costs $0.40 per square foot, what will be the payback time (as a result of energy savings) on this installation?
5. (a) For the house in Problem 3, calculate the infiltration heat losses in Btu/h when $\Delta T = 40$°F. Assume two air changes per hour.
   (b) What percentage of the heat loss is because of infiltration?
6. Sketch the floor plan of the room in which you live, noting windows and doors and exterior surfaces. Estimate (through a calculation) the heat *loss* from your room for 24 hours when the inside temperature is 70°F and the outside is 20°F. Estimate the heat *gain* to your room (excluding solar contributions) when the outside temperature is 90°F and inside is 70°F. Assume all interior surfaces are at 70°F.
7. A room air conditioner has a capacity of 6000 Btu/h. Would this be sufficient to maintain the temperature of a small hut at 70°F when the outside temperature is 95°F? Assume the hut is 10 ft × 10 ft × 6 ft and the exterior surfaces are made of 1-in. softwood.

FURTHER ACTIVITIES

1. An interesting activity with which to study the use of insulation as well as drawing inferences from graphed data uses two identical soda pop cans.
   (a) Wrap each can with a different insulating material, holding it in place with tape or rubber bands (or a bag, in the case of loose fill). Make sure you wrap the can tops and bottoms, but leave a hole above the pop-top opening for a thermometer. Leave one can unwrapped. From Table 5.2, estimate the R-value for each can.
   (b) Ready a thermometer for each can by wrapping a rubber band around its middle, which will support it when inserted into the can. (The thermometers shouldn't touch any part of the cans.)
   (c) Fill each can to within 1 cm of the top with hot water (70°C). (Avoid wetting the insulation.)
   (d) Read and record the temperature of the water in each can for about 20 minutes (taking readings once every minute).
   (e) On a graph, plot the temperature of each can versus time.
   (f) Calculate the relative R-value of each material by comparing the total temperature change of each can over the same time interval. Compare with the estimates of Part (a).
2. Repeating all the steps, add the effect of forced convection by blowing air on the cans with a fan. Graph the temperature of the water versus time and compare the temperature change of each can with that of the same can in the absence of any wind.
3. Check out the rate of energy loss by radiation from hot objects with the following experiment. Pour the same amount of boiling water into two cups—one white and one dark—and cover with cardboard through which a thermometer has been inserted. Record the temperature once per minute for about 20 minutes for both cups and compare the results.
4. A home energy audit enables you to assess ways in which you can reduce your heating bills. Such an audit uses principles covered in this chapter. The outputs of this audit are the total energy consumed during the heating season, its cost, and a breakdown of your heating losses by room and surface (walls, ceiling, windows, floor, plus infiltration). To perform this audit, carefully measure the outside surface areas of each room in your house. The forms for this audit (developed by the author), plus detailed instructions, are found in Appendix C.

5. Make a "draft meter" for measuring infiltration by taping a piece of tissue, thin plastic wrap, or Christmas tree tinsel to a pencil. On a moderately windy day, check around windows, doors, electrical outlets, and similar openings for places through which air is entering. Make a list of these and indicate corrective actions.

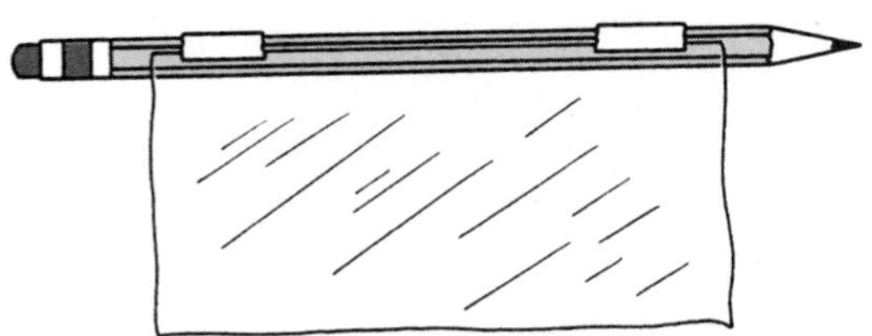

6

Solar Energy: Characteristics and Heating

A. Introduction

Renewable energy today provides about 9% of the world's energy (up to 22% if all uses of biomass are included) and 8 to 10% of the U.S. needs. However, in many parts of the world these percentages are increasing significantly. Wind energy is the fastest growing energy resource in the world today. In the 1990s, it has had a staggering 37% per year growth rate in Europe. This is followed by photovoltaics, with a 24% per year growth worldwide. Hydroelectric energy today provides 19% of the world's electricity. Some recent studies suggest that renewables could rise to a 30–40% share by 2050, assuming global policy efforts are made to address environmental issues, especially climate change.

Renewable energy resources can be categorized into the forms of radiant solar, wind, hydropower, biomass, and geothermal. Each of these forms has

Table 6.1 RENEWABLE ENERGY RESOURCES AND USES

Radiant	Space heating and cooling (active and passive) Domestic hot water, swimming pools Electricity (photovoltaics) Solar furnaces, thermal electric
Wind	Electricity (wind turbines) Mechanical (water pumping, grinding)
Hydro	Electricity, mechanical (waterwheels)
Biomass	Heat (direct combustion), electricity Fuels (gas, liquids)
Geothermal	Electricity, district heating

many uses, as listed in Table 6.1. For the United States, hydroelectric power is the largest contributor to our renewable resources—accounting for 51%, and provides 4% of our total energy production (Fig. 6.1). Biomass (wood and agricultural waste) is also very important, especially worldwide. Radiant solar is used for heating buildings, for domestic hot water, and for the production of electricity with solar cells and heat engines.

Renewable energy resources offer many advantages to an energy-hungry world. They can be used in many ways, offer minimal environmental problems, and can be harnessed with appropriate technology. They particularly offer hope to the developing countries whose economic growth rates are seriously hampered by high energy costs. The potential offered by these resources is immense. Every day, the earth receives thousands of times more energy from the sun than is consumed in all other resources. The states of North Dakota, South Dakota, and Texas have enough wind energy to meet all the

Solar heated home

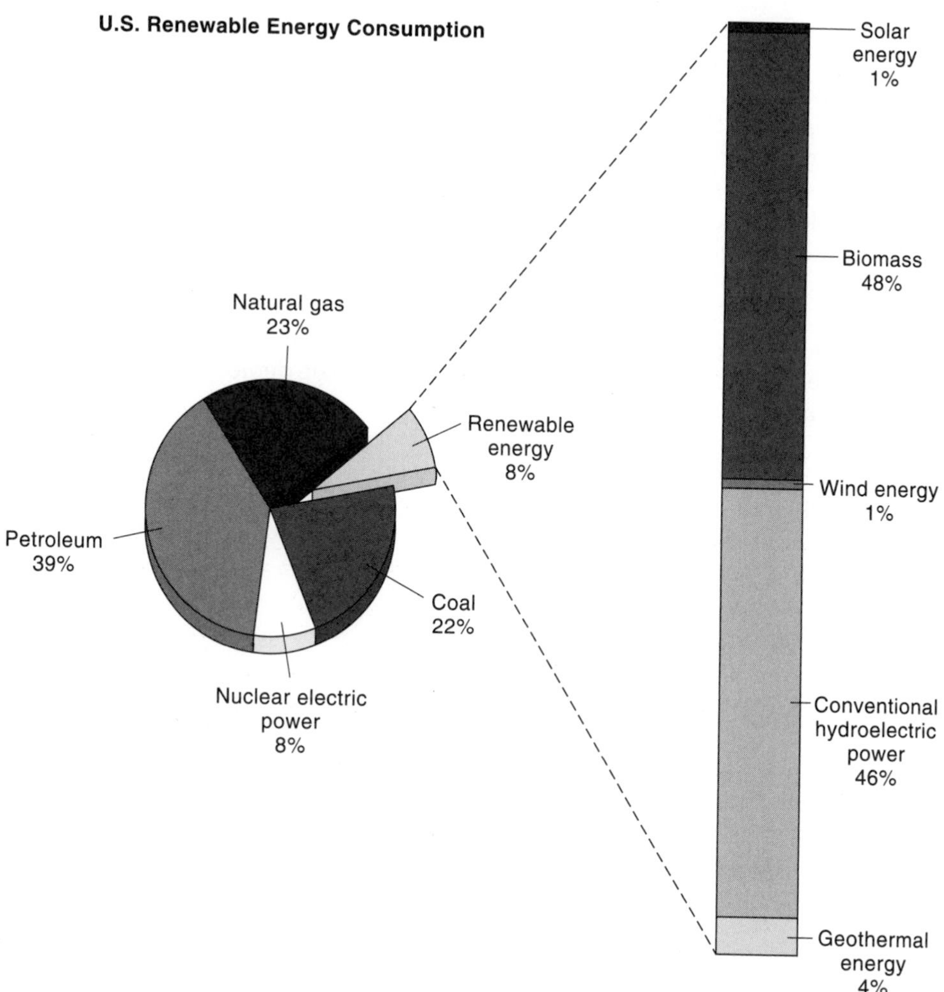

FIGURE 6.1

U.S. renewable energy consumption (by source), as share of total energy, 1999. The generation of electricity accounts for about one half of the renewable resources. (USEIA)

electricity needs of the United States. A 140 × 140 mile parcel of land in Arizona covered with solar cells could also meet the entire electricity needs of the United States.

In view of these facts, one might wonder why renewable energies are relatively underused. The primary reason is economics, especially when the cost of renewable energy generation is viewed in comparison with lower priced commercial fuels. Prices for renewable energies have dropped considerably in the last two decades, but other factors limit their widespread use. There continue to be technical and economic uncertainties, inadequate documentation and evaluation of the many solar projects that have been undertaken worldwide, lack of coherent government strategy, and just plain skepticism on the part of many energy decision makers, all obstacles to the employment of renewable energies. There are also difficulties with energy storage, complicated by the fact that these resources are diffuse, time dependant, and season dependant. Nevertheless, renewable energy sources will be available long after our fossil fuels run out. They are economically and politically less risky than many conventional supplies (especially oil), whose costs are much less predictable. Many utilities seek to decrease their risks by diversifying their supply sources, and so wind energy and photovoltaics are being developed more carefully today. Their flexibility and small size allows them to be deployed much faster to meet increased demands.

The topics that will be considered in this chapter include the nature of solar radiation and solar heating for space and hot water. We have the necessary background from previous chapter discussions. Solar cell principles (photovoltaics), energy from the wind, and the conversion of plant biomass (and industrial and agricultural wastes) into clean fuels, heat, and electricity will be taken up in Chapters 12 and 17.

B. Characteristics of Incident Solar Radiation

The energy from the sun reaching the earth per day (averaged over a year) ranges from 600 Btu/ft^2/d (6800 kJ/m^2/d) for northern Europe to 2000 Btu/ft^2/d (23,000 kJ/m^2/d) for arid regions near the equator. The amount of solar radiation reaching the earth is called the **insolation**—short for "incident solar radiation." For the continental United States, the annual average insolation per day on a horizontal surface ranges between about 1100 and 1900 Btu/ft^2/d (12,000 to 22,000 kJ/m^2/d). (Other units for insolation are discussed in the next section.) In one year the amount of solar energy falling on the entire United States is about 2000 times the energy available from our present annual coal production. A summer day's sun at noon provides a square meter of land with a power input of almost a kilowatt. With current technology, the sunlight falling on a typical house can provide from one third to one half of the heating needs of that house anywhere in the United States, even in areas of persistent cloudiness.

The predominant nuclear reaction in the sun responsible for this energy is the fusion of hydrogen nuclei into helium nuclei. In these reactions a tremendous amount of energy released as matter is converted into energy: about 4 billion kilograms of matter per second! While the interior temperature of the sun is more than 40 million degrees Celsius, the gases at its surface are at about 6000°C. Figure 6.2 shows the relative intensity of the electromagnetic radiation received at the top of the earth's atmosphere as a function of wavelength. About 9% of the radiation is ultraviolet, or very short wavelength, about 40% is

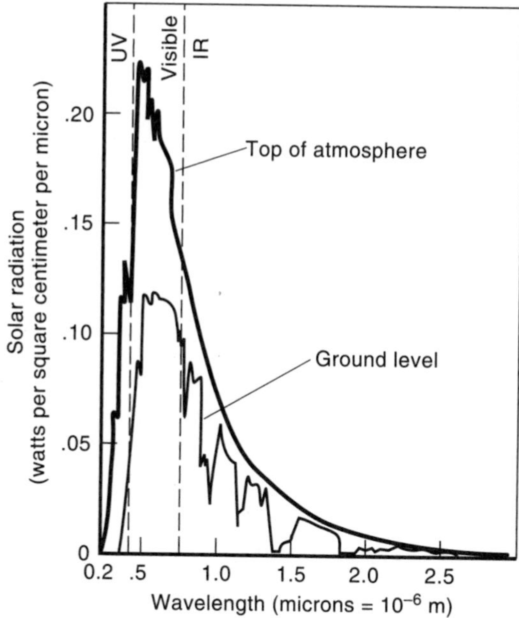

FIGURE 6.2
Spectrum of solar radiation reaching the earth at the top of the atmosphere and at ground level. (The minima in the ground level spectrum are a result of the absorption by water vapor, CO_2, O_2, N_2, and ozone [O_3].) About 40% of the solar radiation is in the visible region.

in the visible region, and about 50% is infrared, or long wavelength. However, only about half of this radiation reaches the surface of the earth. Much of the ultraviolet radiation is absorbed by oxygen, nitrogen, and ozone in the earth's upper atmosphere. Some of the longer wavelength radiation is selectively absorbed by water vapor and carbon dioxide in the lower atmosphere.

About 19% of the radiation received at the earth's atmosphere is absorbed by clouds and other gases and 31% is reflected back into space by the clouds and the atmosphere (Fig. 6.3). The fraction of light reflected from the earth and its atmosphere is called the **albedo.** The remaining 50% of the incident solar energy reaches the surface of the earth and is almost all absorbed (3% is reflected). The relatively constant temperature of the earth is a result of the energy balance between the incoming solar radiation and the energy radiated from the earth. Most of the infrared radiation emitted from the earth is absorbed by CO_2 and H_2O (and other gases) in the atmosphere and then reradiated back to earth or into outer space. This reradiation back to earth is known as the "**greenhouse effect,**" and it maintains the surface temperature of the earth about 40°C higher than it would be if there were no absorption (i.e., the average surface temperature would be about –15°C if we had no atmosphere). This energy balance, and our effect on it through the emission of CO_2 from fossil fuel combustion, is discussed in more detail in Chapter 9.

The amount of insolation reaching the top of the earth's atmosphere is about 1360 W/m^2 or 430 $Btu/ft^2/h$. This number, called the **solar constant,** varies only slightly with time. There are many ways of tabulating insolation,

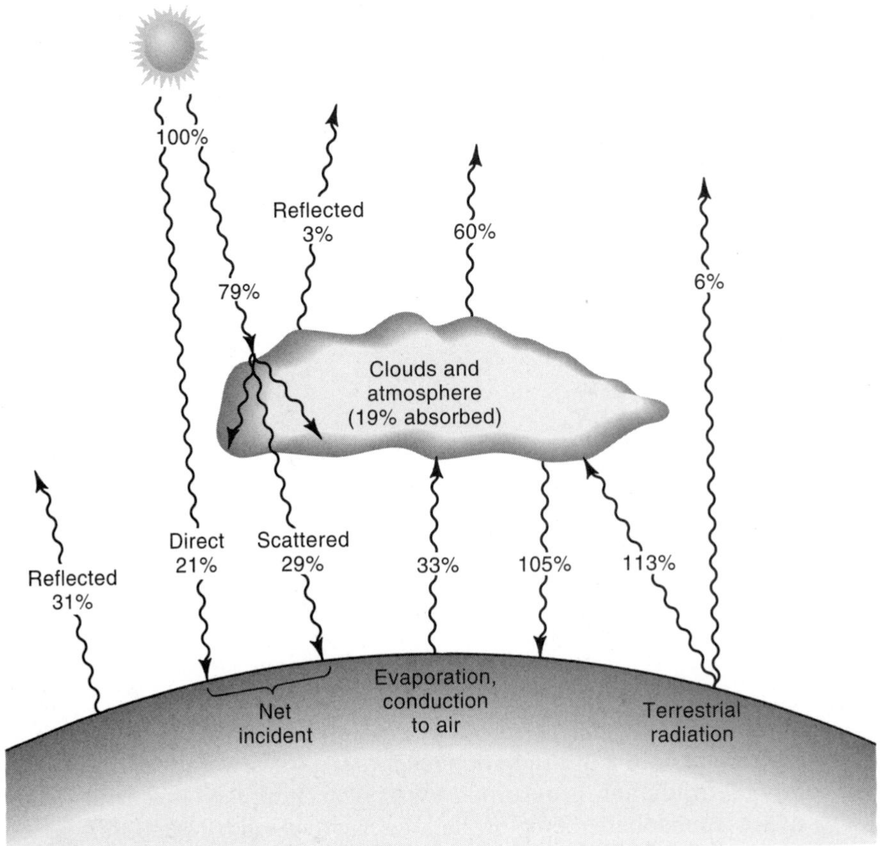

FIGURE 6.3

Energy balance for the earth. The earth receives about 50% of the incident solar radiation: 21% is from direct radiation and 29% is scattered through the clouds. The energy leaving the earth's surface comes from evaporation and conduction to the atmosphere (33%), and infrared radiation (noted here as terrestrial radiation). Most of the infrared radiation (113%) is absorbed by the atmosphere and reradiated back to the surface (the "greenhouse effect"). In order to have temperature equilibrium at the earth's surface, the energy input must equal the energy output. For this figure, 50% (incident radiation) = 3% (reflected) + 33% (evaporation) + 14% (net terrestrial radiation: 113% + 6% − 105%).

depending on the units used and the time interval selected; some conversion coefficients are listed in Table 6.2 for future reference.

The insolation received at any particular location on the earth's surface may vary between 0 and 1050 W/m^2 (330 Btu/ft^2/h), depending on the latitude, the season, the time of day, and the degree of cloudiness. The first two factors are a result of the geometry of the earth's orbit about the sun. The orbit of the earth

Table 6.2 CONVERSION COEFFICIENTS FOR INSOLATION

1 Btu/ft²/hr = 3.16 W/m²
1000 W/m² = 317 Btu/ft²/h
1 Langley = 1 cal/cm² = 3.69 Btu/ft²
1 Btu/ft² = 11.35 kJ/m²

Solar constant:*
1354 W/m²
429 Btu/ft²/h
1.94 Langleys/min
4870 kJ/m²/h
1.52 HP/sq. yd.

*Incident solar radiation at top of earth's atmosphere for unit area perpendicular to the sun's rays. About 50% of this insolation reaches the earth's surface.

around the sun is nearly circular, but the axis about which the earth itself spins is tilted relative to this plane of motion at an angle of 23.5° (Fig. 6.4). Consequently, the North Pole is tilted toward the sun during the Northern Hemisphere's summer and away from the sun in the winter. Therefore, the Northern Hemisphere is exposed to more hours of sunlight in the summer (reaching a maximum on June 22, the summer **solstice**), and the amount of solar radiation striking a horizontal surface is greatest in the summer. In the winter, the insolation is spread over a larger horizontal area due to the tilt angle and the sun's rays must pass through a greater depth of atmosphere, so less radiation reaches the earth because of absorption and scattering from the

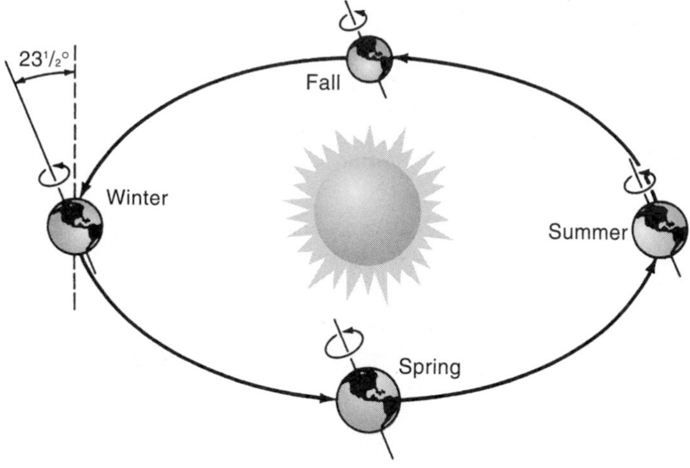

FIGURE 6.4
Motion of the earth around the sun, illustrating the seasons and the tilt of the earth's axis.

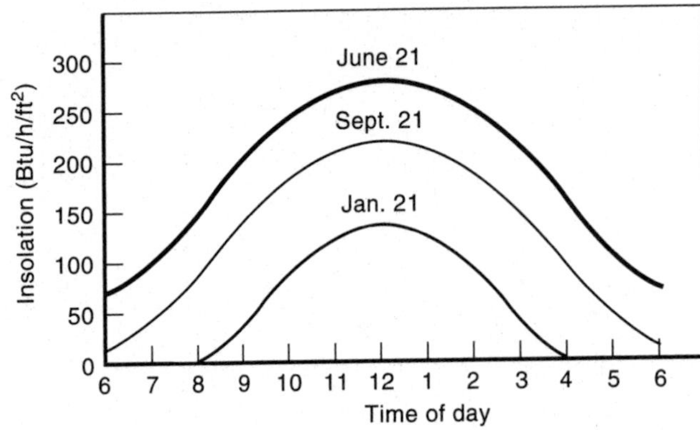

FIGURE 6.5
Insolation values for a clear day on a horizontal surface located at 40°N latitude, as a function of the month and the hour of the day. (AMERICAN SOCIETY OF HEATING, REFRIGERATION, AND AIR-CONDITIONING ENGINEERS (ASHRAE). 1977. *APPLICATIONS OF SOLAR ENERGY FOR HEATING AND COOLING OF BUILDINGS*. NEW YORK, ASHRAE.)

atmosphere. These two effects are shown in Figure 6.5 on the graph of clear-day insolation on a horizontal surface at 40°N latitude as a function of time of day for three different months. Unfortunately, the insolation is at its lowest value in the winter when the need for heat is greatest.

The sun's elevation, or angle above the horizon, is called its **altitude.** The altitude of the sun is a function of your latitude; the farther north you are, the lower in the sky the winter sun will be. At solar noon (the time at which the sun is due south) on December 21 in Boston, the altitude is 24°, while in Miami it is 42°. Figure 6.6 shows the sun's apparent path through the sky for three differ-

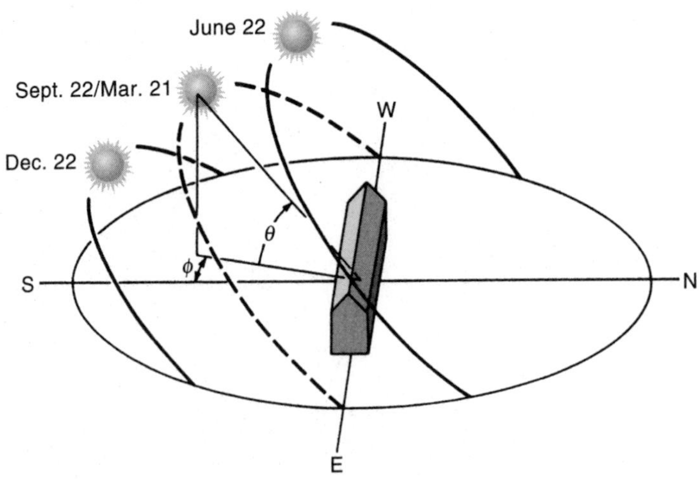

FIGURE 6.6
Yearly and hourly changes in the sun's position in the sky for 40°N. Also shown are the solar altitude θ (angle above the horizon) and the solar azimuth ϕ (angle from true south).

ent times in the year. For 40°N latitude, the altitude at solar noon is 74° on June 21 and 27° on December 21. (For 24°N latitude, these numbers are 89° and 43°.) This figure shows that as fall moves into winter, the sunrise and sunset points of the sun's motion across the sky move gradually southward. The days get shorter and the path gets lower in the sky. The solar **azimuth** is the angle of the sun from true south and represented by the symbol ϕ in Figure 6.6. For 40°N latitude, the azimuth goes from –122° to +122° on June 21; on December 21, the azimuth goes from –53° to +53°. That is, the sun rises quite south of east in December.

The insolation reaching a surface is composed of a *direct* beam from the sun (the radiation that casts a shadow), a *diffuse* component (radiation scattered from clouds and coming from the entire sky), and a *reflected* component (radiation scattered from the ground), as shown in Figure 6.7. Insolation is usually measured on a horizontal surface. For a surface tilted with respect to the horizontal, all three components vary with the tilt angle. On cloudy days, solar collectors can collect about 40% of the diffuse sunlight. Figure 6.8 shows the effect of two collector tilts on the insolation received, as a function of month. The sun is lower in the sky (at a smaller altitude) during the winter, so the insolation on a vertical surface in the winter will be greater than on a horizontal surface. The opposite is true in the summer. This situation can be used for heating a house.

Data collected for the United States as a whole has resulted in the map of Figure 6.9 displaying the average daily insolation striking a horizontal surface. (To apply these data to surfaces that are tilted, such as a solar collector, requires trigonometric calculations.) Because space heating needs are a function of the time of year, economic considerations require a knowledge of both the average insolation and the average outdoor temperature for each month, data that can be obtained from a Climatic Atlas. Such data for some U.S. cities are given in Appendix D. A few

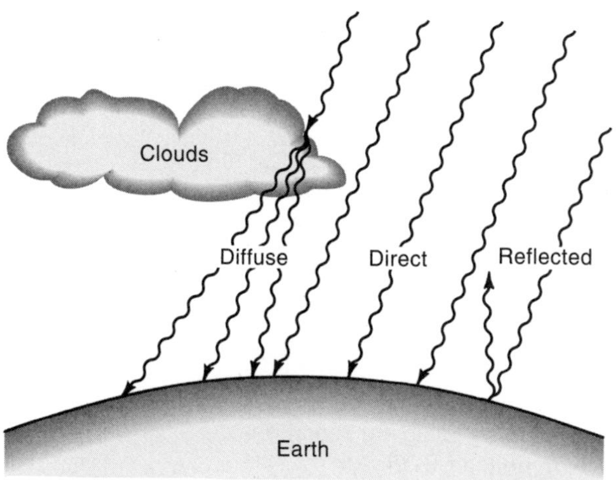

FIGURE 6.7
Components of solar radiation.

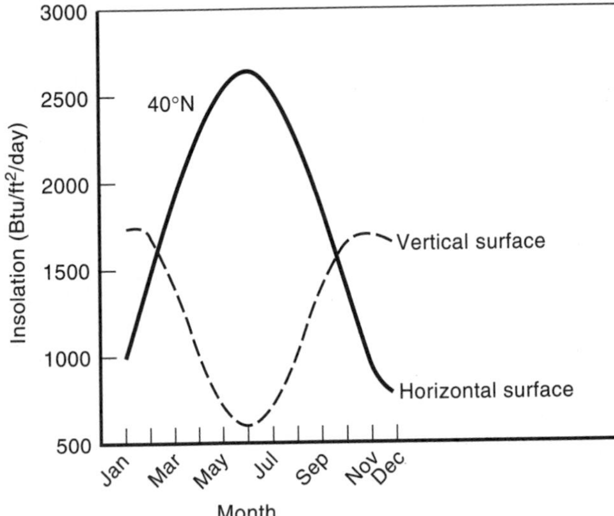

FIGURE 6.8
Daily clear-day insolation as a function of month and collector orientation.

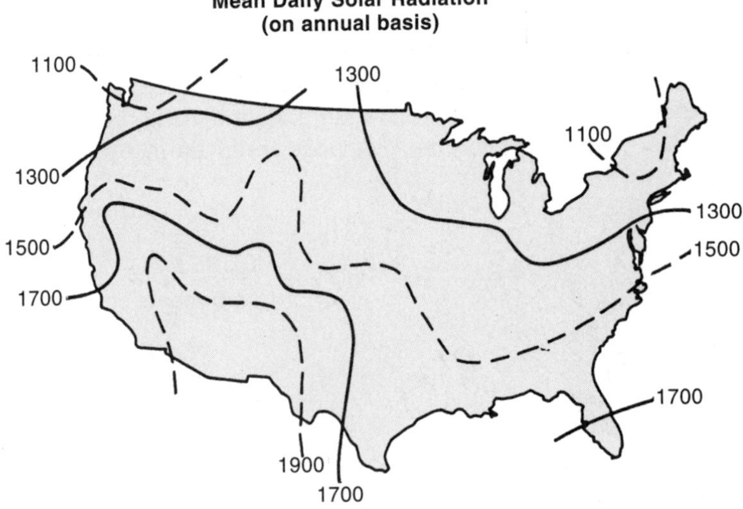

FIGURE 6.9
Mean daily solar radiation (on an annual basis) for radiation incident on a horizontal surface, in units of $Btu/ft^2/d$.

Table 6.3 VARIATIONS IN INSOLATION FOR SELECTED CITIES

		December		March		June		September	
City	**Latitude**	I_H*	I_T**	I_H	I_T	I_H	I_T	I_H	I_T
Miami	26°N	1292	1770	1829	2012	1992	1753	1647	1691
Los Angeles	34°N	912	1496	1641	1936	2259	1920	1892	2114
Washington, D.C.	38°N	632	1068	1255	1493	2081	1790	1446	1605
Dodge City	38°N	874	1652	1566	1942	2400	2040	1842	2106
East Lansing	42°N	380	638	1086	1347	1914	1646	1303	1498
Seattle	47°N	218	403	917	1165	1724	1465	1129	1332

*Insolation on a horizontal surface (in Btu/ft^2/d)

**Insolation on a surface tilted at an angle equal to the latitude (in Btu/ft^2/d)

examples of insolation data are given in Table 6.3, which shows the variation of incoming radiation with season and location. Note that Dodge City, Kansas, and Washington, D.C., both at the same latitude, have quite different insolation.

C. History of Solar Heating

The use of solar energy for heating can be traced back to antiquity. Archimedes reportedly used mirrors to direct the sun's rays on an attacking enemy fleet in 212 BC, igniting their sails at a distance of several hundred feet. More than a thousand years ago, the Anasazi Indians in the American Southwest built homes in the sides of cliffs to make use of the low altitude of the sun for passive solar heating in the winter and the cliff overhang to provide shielding from the rays of the sun in summer. In the 17th and 18th centuries, scientists concentrated the sun's rays with mirrors or lenses to melt metals. Antoine Lavoisier (1743–1794), often called the father of modern chemistry, achieved temperatures close to 1700°C (3100°F) by using the sun, higher than any obtained up to that time. One of the most powerful solar furnaces in use today is in Sandia, New Mexico (Fig. 6.10), where temperatures of 3000°F are obtained.

Solar steam boilers were developed in the late 19th century to produce steam to run engines. The Frenchman August Mouchot ran a printing press in 1878 using steam produced in a device similar to that shown in Figure 6.11. An interesting commercial endeavor in the 1910s in Egypt used a solar boiler to provide steam for the operation of irrigation pumps (see Focus On 6.1: Early 20th-Century Egyptian Solar Power Plant). Another large-scale project using the sun's rays for heating occurred in the 1870s in Chile, where a 50,000 ft^2 solar

FIGURE 6.10

Concentrated sunlight from 1775 mirrors strikes a target at the Solar Test Facility in Sandia, New Mexico. The sun melted a quarter-inch-thick steel plate in 2 minutes. (SANDIA LABORATORIES)

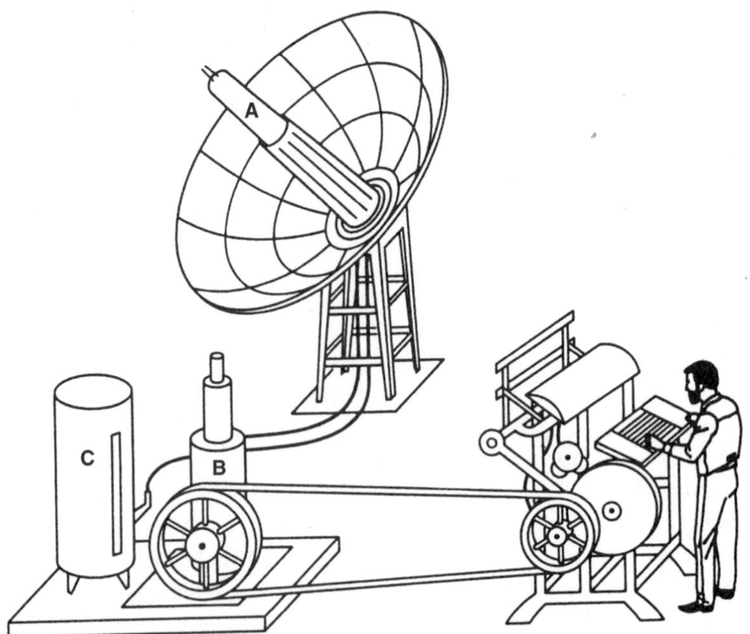

FIGURE 6.11

Solar steam engine, Paris, 1878. Water was heated by the sun at the focus of the concentrating dish (*A*). The steam produced was used to run a steam engine (*B*) whose mechanical output ran a printing press. The water was supplied from tank (*C*).

Focus On 6.1

EARLY 20TH-CENTURY EGYPTIAN SOLAR POWER PLANT

In 1912, the American engineer Frank Shuman put into operation the first large-scale solar power plant in Cairo, Egypt. Its job was to provide irrigation water from the Nile. He used a trough-type parabolic collector to focus the sun's rays onto a black metal pipe to produce steam. The system's peak output was about 50 kW. The total collector area was 13,000 $ft^2 = 1207\ m^2$. Let us see if this output is reasonable by calculating the maximum output from this system. The maximum insolation on the collector surface (in June) is $1207\ m^2 \times 1200\ W/m^2 = 1576\ kW$.

Let us assume that all this solar energy is converted to the thermal energy of the steam, bringing it to a temperature of 100°C. The conversion of this heat energy to useful mechanical energy (to run the irrigation pump) occurs through a heat engine. Assuming that the surrounding environmental temperature is 20°C, the maximum (Carnot) efficiency is given by

$$\text{Max. eff} = \frac{(T_H - T_C)}{T_H} = \frac{(373 - 293)}{373} = \frac{80}{373} = 0.21 \times 100\% = 21\%$$

(See Chapter 4.) This yields a maximum useful work output of $1576\ kW \times 0.21 = 330\ kW$. Therefore, a 50 kW output is certainly possible. (A more modern version of this solar power plant, used to generate electricity, is described in Chapter 12.)

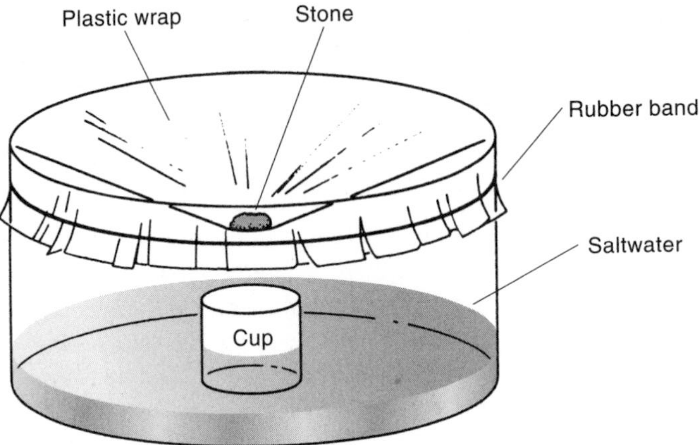

FIGURE 6.12
Solar desalination project using a cup and plastic wrap.

still was constructed to provide 6000 gal per day of fresh water. This project used slanting roofs of glass over trays of saltwater; the water evaporated from the trays, condensed on the glass, and ran down the glass to collection receptacles. An illustration of this technique on a much smaller scale is shown in Figure 6.12. This shows a simple project for desalination using a piece of plastic both to trap the incident solar radiation and to provide a surface for condensation of the water.

The use of solar energy for cooking goes back at least to the latter half of the 18th century. In 1767, the Swiss scientist H. B. DeSaussure obtained temperatures high enough for cooking in an insulated box with several glass covers. Mouchot extended this idea in the 1860s by using a parabolic reflector to focus the sun's radiation on a blackened copper container (holding the food) that was enclosed in a glass vase (Fig. 6.13). He was able to bring 3 L of water to boil in

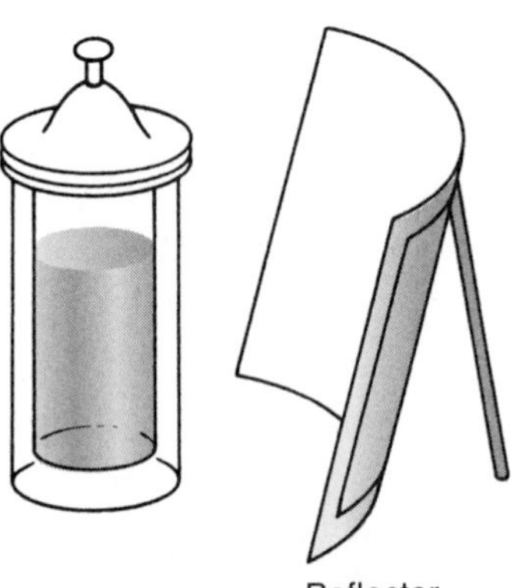

FIGURE 6.13
Mouchot's solar pot was able to bring 3 liters of water to a boil in 1.5 hours.

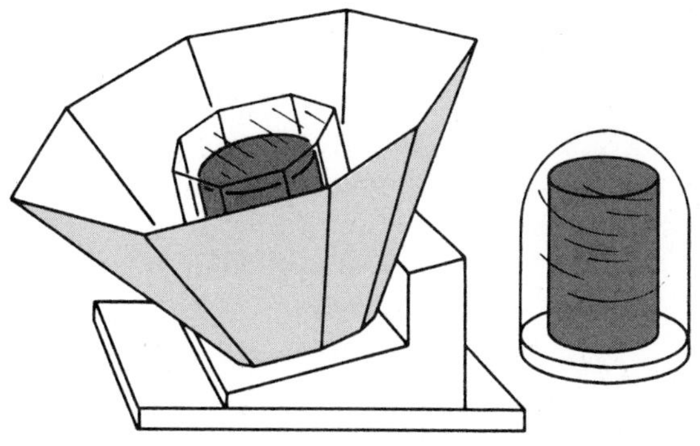

FIGURE 6.14
Adams's solar cooking apparatus, India, 1878. Sunlight is reflected to the blackened metal container, containing the food, as shown in the insert. The metal container is enclosed in a glass jar. (SCIENTIFIC AMERICAN, 1878)

1.5 hours. Other classic experiments with solar cookers were carried out by William Adams in India in the 1870s. He wrote that he was able to cook the rations (meat and potatoes) of seven soldiers in 2 hours in January, the coldest month of the year in Bombay. A drawing of one of his cookers, published in *Scientific American* in 1878, is shown in Figure 6.14. In the United States, much of the work on the development and testing of solar cookers or "hot boxes" was done by Maria Telkes in the 1950s. One of these models is shown in Figure 6.15; auxiliary reflectors were used to reach higher temperatures. Heating of the pot inside the hot box is done by direct absorption and by convection.

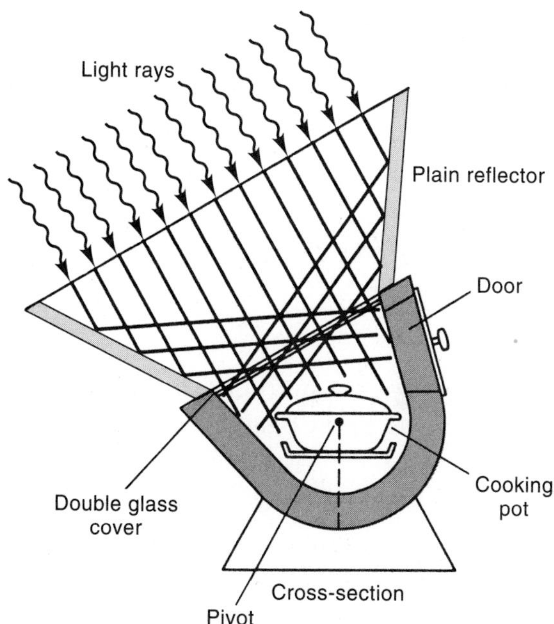

FIGURE 6.15
Telkes's oven. The design features a fixed cooking pot and a moveable reflector.

EXAMPLE

Collector Area for a Solar Oven
If we assume that a useful rate of heat energy needed for cooking is 150 W, then what is the area of the oven (including reflectors) that is needed to intercept this amount of radiation?

Solution

The collector area is calculated by assuming that the solar oven is 20% efficient and that 85% of the insolation is the direct component.

Assuming a noontime insolation rate of 900 W/m^2, the direct component is $900 \times 0.85 = 765$ W/m^2.

The energy needed for the cooker (150 W) is equal to the direct component times the efficiency times the intercepted area:

$$765 \times 0.20 \times \text{area} = 150 \text{ W}$$

Therefore, the area is $150/(765 \times 0.2) = 0.98$ m^2.

Since the early experiments with solar ovens and cookers, many attempts have been made to popularize this concept, especially in developing countries. However, this has not met with a great deal of success. In contrast to the United States, where only about 1% of the total energy consumption is used for cooking, as much as 75% of the total household energy consumption in developing countries is used for cooking. The fuel most often used for cooking, where the majority of people live in rural settings, is firewood and animal dung. The gathering of fuel (Fig. 6.16) requires many hours a day of work; overgathering can accelerate desertification and lead to soil erosion and the removal of soil nutrients. (Energy-efficient cooking stoves are discussed in Chapter 17.)

ACTIVITY 6.1

An interesting experiment to study solar water heating is as follows. Find four plastic (not Styrofoam) cereal bowls. Spray paint the inside of two of them white and the other two black. Find two clean plastic bags that will cover the bowls. Add the same amount of cold water (about 1-cm deep) to each bowl. Cover a white and a black bowl with plastic bags and put all four in the sun. Monitor the temperature T of all four bowls for 1 hour. What bowl's water reached the highest temperature? Graph T versus time (t).

FIGURE 6.16
Women collecting firewood in Burkina Faso. (W. BENNETT)

D. Overview of Solar Heating Today

Today in the residential and commercial sector, solar heating is used primarily for swimming pools and for **domestic hot water (DHW).** The sale of such systems continues to rise slowly but steadily at about 5% per year. These systems are covered in the next section. Solar space heating of homes, though not as widespread because of costs, is covered later in this chapter.

Solar heating saw a large growth in the late 1970s and early 1980s, beginning with the oil embargo of 1973 and subsequent rising oil and electricity prices. A solar tax-credit program provided additional incentive for people to install solar heating systems. This program gave people the chance to deduct directly from their taxes the amount they spent on solar systems, up to a specified maximum. However, the end of tax credits in 1985 brought a sudden halt to the industry's growth. By 1987 the solar DHW market dropped by more than 95% to early 1970 levels. A benefit of the boom in sales in the early to mid-1980s was greater public recognition of solar energy and major advances in the performance and reliability of solar heating systems. Many states have again instituted state tax credits for the installation of solar systems. For example, California offers a tax credit equal to 10% of the cost of a solar system; North Dakota, 15%; and North Carolina, 25% of the cost of an active or passive system (up to $1000). Some utilities offer rebates to their customers who replace electric water heaters with solar DHW. The Sacramento Municipal Utility District is working toward replacing 12,000 electric water heaters with solar ones. This is encouraged by rebates (from $600 to $1400) and low-cost financing. Their experience so far has shown that about 67% of the hot water load can be economically met by solar. The savings in electricity has been about equal to the customer's monthly loan payment. The utility's incentive in this program comes from its avoided costs in the generation of electricity, that is, the investment that would have had to be made to increase their generating capacity.

All solar heating systems have several features in common—namely a collection device, a storage facility, and a distribution system (Fig. 6.17). The solar heating of either domestic water or of homes is accomplished in two different ways: actively or passively. An **active solar system** is one in which the fluid (water or air) that the sun has heated is circulated by a fan or a pump. The solar collector for space heating is similar to that used in a DHW system. A **passive solar system** uses no external power but allows the fluid (usually air) heated by the sun to circulate by natural means. Passive solar has strong economic advantages, especially when it comes to space heating. Passive solar homes being built today can save as much as 50% of heating costs for only a 1% to 5% increase in construction costs. However, passive solar features must be integrated into the design of the building from the outset. Today, 7% of new homes built in the United States are designed with passive solar features.

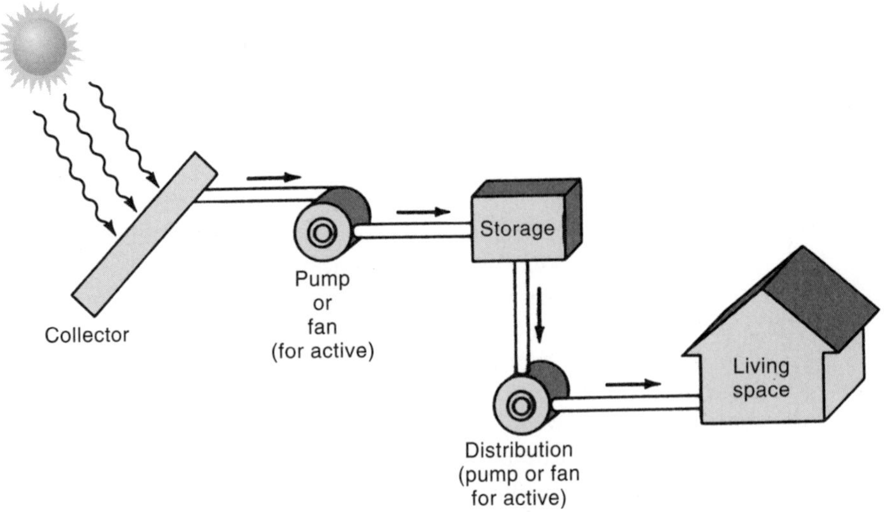

FIGURE 6.17
General features of a solar heating system (active or passive).

E. Solar Domestic Hot Water

The most popular use of solar energy heating systems today is to provide hot water for domestic purposes (DHW) or for swimming pools. Currently, 7% of the collectors sold today are used for DHW and 93% for swimming pools. The swimming pool collectors generally operate at temperatures less than 110°F, while the ones for DHW operate in the range of 140° to 180°F. The number of homes with installed solar DHW systems has increased from about 3500 in 1970 to almost 2 million today. Prices run about $3000 to $5000, but government and utility incentives lower the actual prices to the consumer. (See Focus On Solar Water Heater Performance later in this section.)

DHW systems can be divided into three types: active systems using flat plate collectors, batch water heaters, and passive (or thermosiphoning) systems. The most common systems for DHW or swimming pools uses a "flat plate collector" (FPC). The basic building block of this collector is a thin, flat metal plate to absorb the sun's radiation. Water in tubes is in contact with the absorber plate and circulated by a pump to take away the heat. The plate is painted flat black to increase its absorption, and it is usually covered with one or two sheets of glazing (glass or plastic). For swimming pools, the absorber plate is usually unglazed since lower water temperatures are acceptable. The heating of water in a glass-covered container is similar to the situation in an agricultural greenhouse or in a parked car that has all its windows closed. Even on a cold winter day, when the sun shines, a car can heat up to uncomfortable temperatures. The glass serves two purposes in the collector: It acts as a radiation shield to trap heat emitted from the absorber plate and as a lid to suppress heat loss by simple vertical convection. The glass is opaque to the long wavelength radiation emitted by the heated plate but can transmit 90% (per glazing) of the incident radiation. Insulation behind and on the sides of the absorber

Flat plate collector to preheat water for domestic hot water uses. The house also uses passive solar heating. (C. SALVAGIN)

plate cuts down on conduction losses. The cross-section of a conventional flat plate collector is illustrated in Figure 6.18, with heat losses shown. Water temperatures of 160°F to 180°F can be achieved, although collectors are usually operated at lower temperatures where they are more efficient.

There are many different designs for the absorber plate, a few of which are shown in Figure 6.19. The tubes carrying the water are soldered onto the absorber plate or sandwiched between two metal plates. The water can also flow over the metal plate without tubes, called a "trickle-type" collector. It is extremely important that the tubes carrying the water make good thermal contact with the absorber plate. A poor press-fit union between tube and plate might be 100 times less efficient than good solder connections. Another type of absorber plate uses a synthetic black rubberized mat consisting of closely spaced tubes and fins (webbing). This is used primarily for low temperature swimming pool heating.

The solar collector should be placed facing south (in the Northern Hemisphere); its angle of inclination will depend on its intended use. The maximum insolation on the collector will occur when the collector is perpendicular to the sun's rays. For domestic hot water needs, insolation all year long is important, so the collector's optimum tilt from the horizontal should be at an angle about equal to the latitude.

Figure 6.20 is a diagram of a solar DHW system that could be used in most northern climates. The FPC is mounted on the roof, with the other components

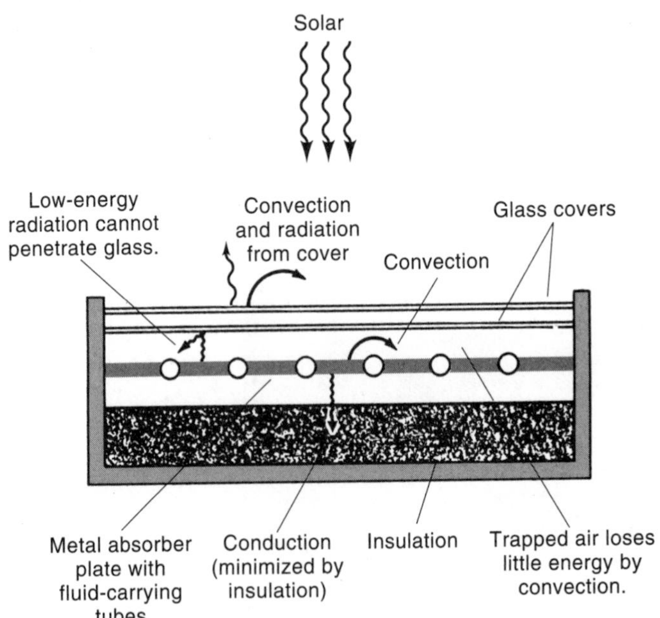

FIGURE 6.18
Cross-section of a flat plate collector (FPC) showing heat losses and gains.

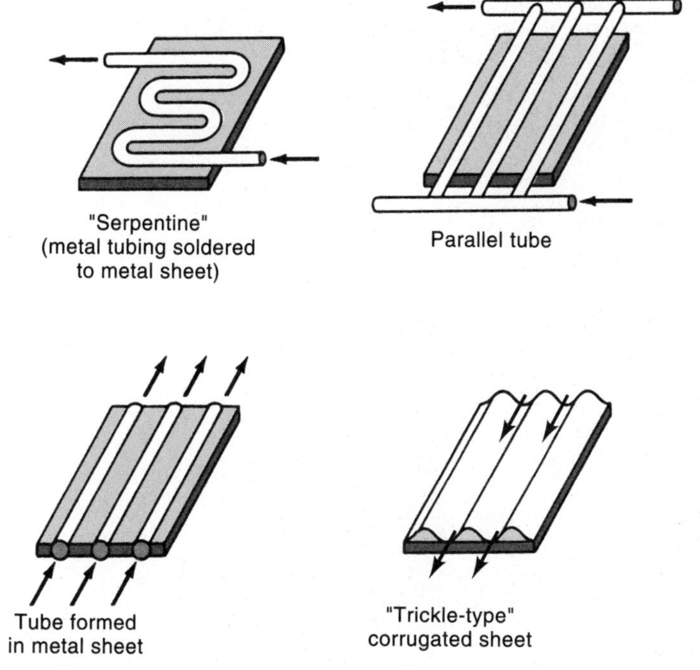

FIGURE 6.19
Flat plate collector absorber plates.

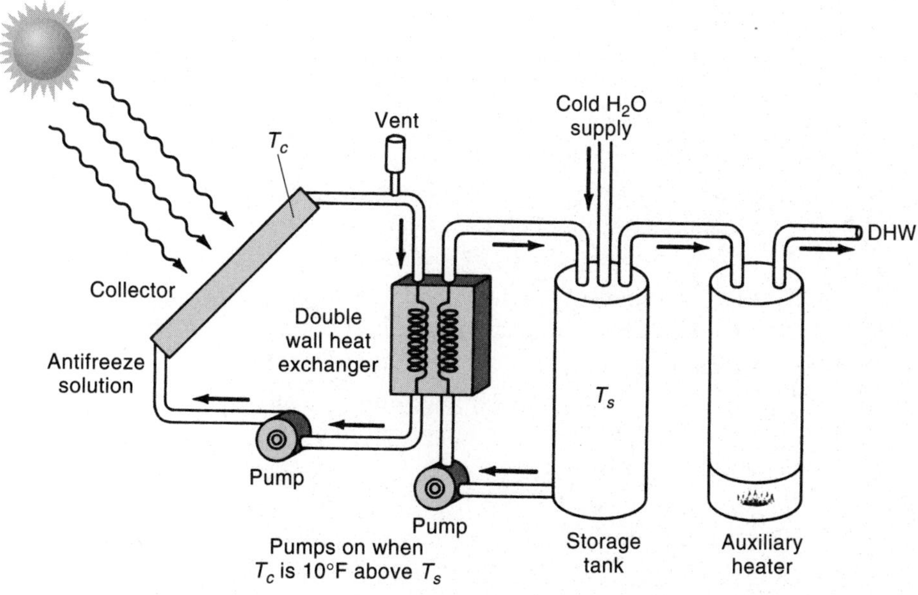

FIGURE 6.20
Solar domestic hot water system.

ACTIVITY 6.2

Experiment with different glazing materials for a solar collector by doing the following "solar energy in a can" activity.

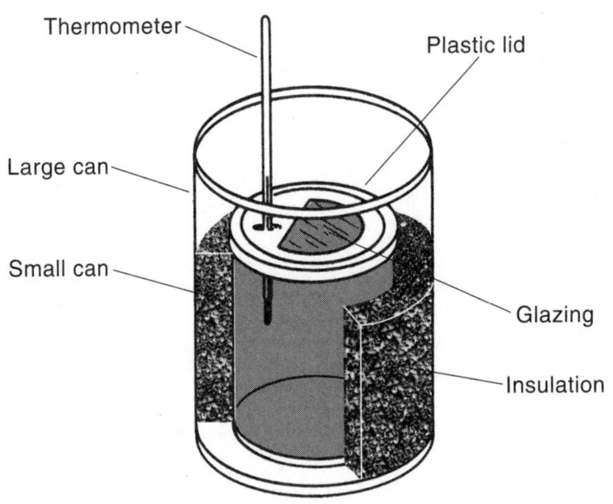

1. Begin by gathering a 3-lb coffee can and a 1-lb coffee can, both with plastic lids.
2. Cut out the center of the smaller can's lid, leaving a 1-cm rim plus a small area through which a slit can be cut to hold a thermometer (see figure).
3. Paint the inside of the smaller can black.
4. Choose a glazing material (plastic wrap, waxed paper, polyethylene, glass) and stretch it across the small can, holding it in place with the plastic lid.
5. Place the smaller can inside the larger one, fill the gap between them with insulation (fiberglass, cellulose, shredded newspaper, etc.) and insert the thermometer.
6. Place the solar collector in the sunlight so that the tops of the cans face the sun. Take a temperature reading each minute for 30 minutes.
7. Repeat steps 4 through 6 for various glazing materials, including once with no glazing and once with glass.
8. Plot temperature versus time for each trial. According to your graphs, which glazing worked best?

within the house. The use of antifreeze and a heat exchanger provides freeze protection. The system makes use of a backup tank to heat the water (by electricity or natural gas) to desired temperatures if the solar system does not supply hot enough water. The area of the FPC depends upon the insolation and the hot water requirements. See the following Example for a sizing illustration.

EXAMPLE

Calculate the size of collector needed to heat 80 gal of water per day from 50° to 130°F in March in Los Angeles. Assume the insolation is 1700 Btu/d/ft^2 and the FPC efficiency is 50%.

Solution

Recall from Chapter 4 that $Q = mc\Delta T$, where m is the mass and c the specific heat of the substance. Then the heat Q needed is

$$Q = 80 \text{ gal} \times 8.3 \text{ lb/gal} \times 1 \text{ Btu/lb-°F} \times (130 - 50)\text{°F} = 53{,}100 \text{ Btu/d}.$$

The heat available from the FPC will be Q = insolation × area × efficiency or

$$53{,}100 \text{ Btu/d} = 1700 \text{ Btu/d/ft}^2 \times \text{area} \times 0.5$$

Therefore, the collector area = 62 ft^2.

Batch water heaters or "bread-box" heaters are popular, inexpensive systems for preheating water using the sun. They have been in use for over 100 years. The design is quite simple: A black tank inside an insulated box with a glass cover absorbs solar energy to heat the domestic water. Cold city water replenishes the water in the tank whenever a hot water faucet is opened. The output from the black tank usually flows into a conventional water heater where it can be further heated, as needed. Figure 6.21 shows a two-tank system that can supply the needs of a four-person family. Insulated coverings are put over the glass at night. A $500 system of this type can achieve a payback in 5 to 10 years.

In addition to the batch water heater, another type of passive solar water heater uses the **thermosiphon** approach in which water flows from the collector to the tank under natural circulation. In this method, the storage tank is placed *above* the collector. The water heated in the collector is less dense than the incoming colder water and will rise into the tank (Fig. 6.22). Under full sun, the temperature of the water can rise by 15° to 20°F in a single pass through the FPC. These systems are very popular in the Middle East and are usually mounted on the roof. There are 800,000 such DHW systems in Israel, serving 70% of the population.

It is important at this point to emphasize **energy conservation** when discussing domestic hot water usage. Energy costs for DHW can be reduced by

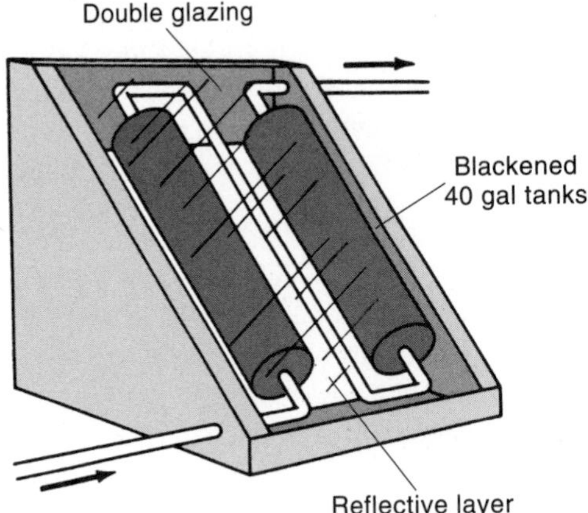

FIGURE 6.21
"Bread-box" or batch water heater for domestic hot water.

both minimizing the amount of hot water used and by shifting to a renewable energy source such as solar energy. Several things can be done:

- Reduce water heater thermostat to 120°F.
- Insulate the water heater with R-19 fiberglass; this can reduce standby energy losses by about 15%. These water heater "blankets" are widely available for about $10 and are easy to install.

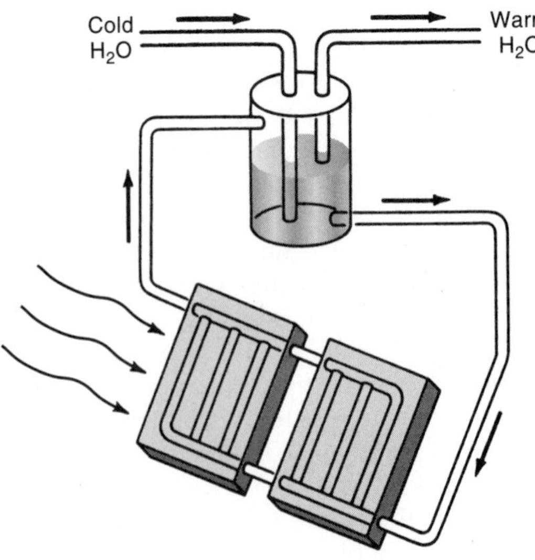

FIGURE 6.22
Thermosiphoning domestic hot water system. The can must be above the collectors.

Focus On 6.2

SOLAR WATER HEATER PERFORMANCE

The Florida Solar Energy Center, the state of Florida, and the Department of Energy (DOE) have launched a program to install solar water heaters in houses of low income people. Over 800 such units have been installed and monitored for performance. Twenty-four-square-foot flat plate collectors were installed with existing 50 gallon electric water heaters. This smaller system provided about half of the hot water needed per residence, but cost about 40% less than the larger 40 ft^2 system. Post-installation inspections were a critical part of the success of this program.

- Use flow-restriction showerheads or low-flow aerators for the sink. These reduce hot water use by about 50%.
- Install a "graywater" heat recovery system. This system preheats incoming water by using some of the heat in the waste water from the shower and laundry that would normally be disposed of down the drain.
- Install an automatic timer on an electric water heater to turn it on only during times of use.
- Consider an "on-demand" or tankless water heater to heat only on-demand, as water passes through it. These units can boost the water temperature for dishwashers, and so allow a further thermostat setback in the conventional water heater.

F. Passive Solar Space Heating Systems

For passive solar space heating, the house itself acts as the solar collector and the storage facility (Fig. 6.23). Heat energy flow is by natural means: No mechanical devices such as pumps and fans are used. The object is to let sunlight in through south-facing windows and store this energy: One must keep it out during the summer, usually by using roof overhangs to protect the windows from the sun that is high in the sky. Passive systems make use of the fact that the amount of solar energy transmitted through south-facing glass during a clear day is *greater* than the heat lost through those same windows over a

FIGURE 6.23
The Brookhaven house: An energy conservation house at the Brookhaven National Laboratory in New York State uses a greenhouse as a major passive solar feature. Fuel consumption is about one fourth the normal usage of a house of similar size in the same climate. (BROOKHAVEN NATIONAL LABORATORY)

24-hour period. To reduce overheating effects and store the incoming solar energy, passive systems make use of the material of the house itself. Some objects have the capacity to absorb large amounts of heat energy. These objects are made from materials such as concrete, water, and stone and are called **thermal mass.**

The essential elements of a passive solar system are:

1. excellent insulation,
2. solar collection (with south-facing windows), and
3. thermal storage facilities.

Passive systems can be categorized into three types:

1. direct gain,
2. indirect gain, and
3. attached solar greenhouse.

In **direct-gain** systems, large south-facing windows are used to admit the sunlight. Thermal storage material—concrete, slate, water, or bricks—is placed in the house to absorb the solar radiation. (It must be placed so it is exposed to the direct radiation.) The concrete floor of Figure 6.24 is one such example. Massive masonry floors and walls absorb the radiant energy during the day and radiate it back into the room during the night. Without thermal storage, indoor temperatures can rise to more than 90°F in the spring and fall. Figure 6.25 shows the temperature performance of a building using solar direct gain coupled with substantial thermal mass; a pleasant temperature can be maintained and temperature extremes avoided. The adobe houses of the Southwest are good illustrations of buildings with appropriate solar gain and sufficient thermal mass spread throughout the structure to provide warmth during the winter and coolness during the summer.

An **indirect-gain** system collects and stores the solar energy in one part of the house and uses natural heat transfer (convection and conduction) to distribute

Slate is used as thermal mass in this greenhouse. (BROOKHAVEN NATIONAL LABORATORY)

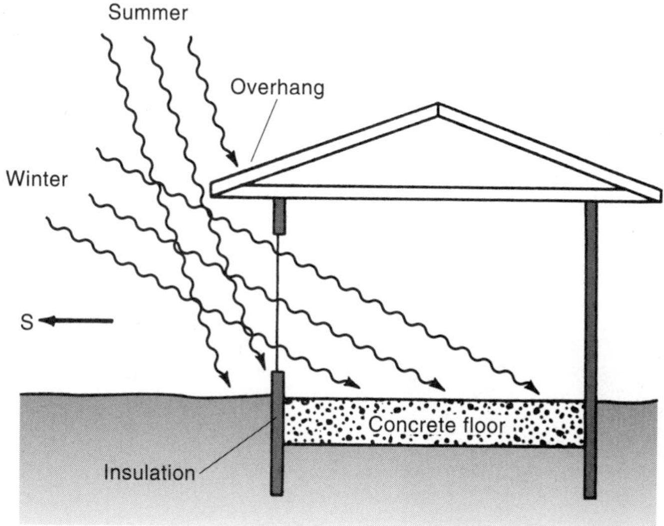

FIGURE 6.24

Passive solar system—direct gain. South-facing windows act as solar collectors. Moveable insulation is used to cover the windows at night to reduce heat loss. A massive concrete floor acts as a storage device and prevents overheating. The overhang blocks the summer sun.

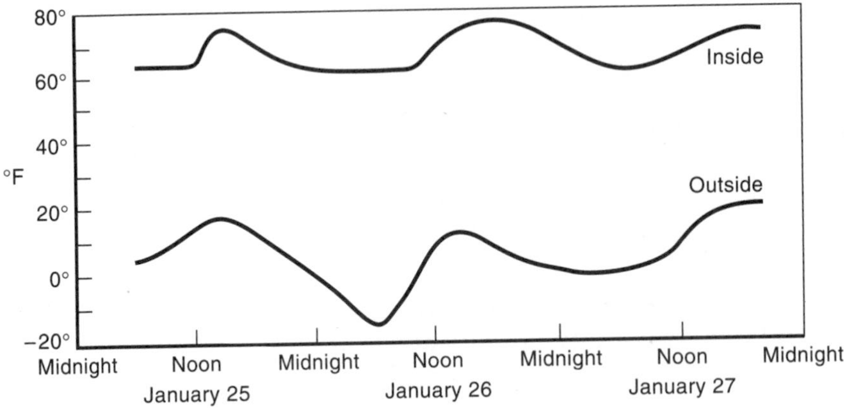

FIGURE 6.25

The performance of a passive solar commercial building (the Conservation Center, Concord, New Hampshire) during three sunny but cold winter days. Heating was with direct gain (large double-glazed, south-facing windows, with no night insulation). Thermal storage consists of a dark slate floor over a 4-inch concrete slab and phase change materials in the walls. Even though the ambient temperature ranged from 20°F down to −15°F, no auxiliary heat was used.

this heat to the rest of the house. A good example of this system uses the **Trombe wall** (Fig. 6.26). Here a massive black masonry or water wall is placed about 10 cm behind a south-facing glass area. The solar radiation passing through the glass is absorbed by the wall, heating its surface to temperatures as high as 150°F. This heat is transferred to the air trapped between the black wall and the window. The warm air rises through vents in the top of the thermal storage wall into the living

An attached greenhouse can be added to a house after construction and can provide additional space heat.

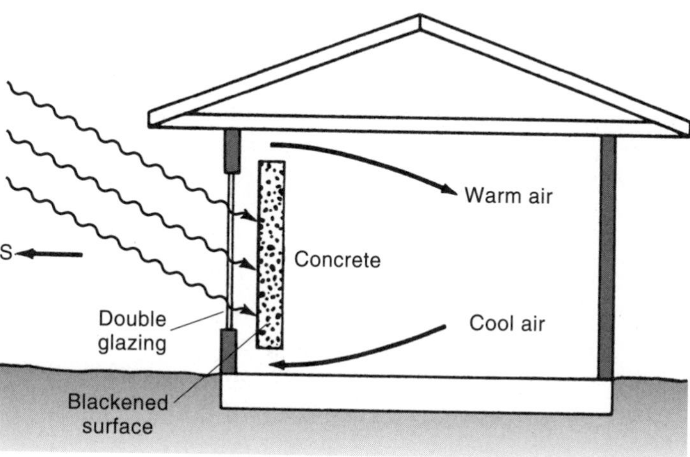

FIGURE 6.26

Trombe wall. The concrete wall acts as a solar collector and a heat storage medium.

space, and is replaced by colder air through vents in the bottom. At night the vents are closed to prevent the reverse process from occurring. At night, heat conducted through the wall is distributed to the living space by radiation and convection from the wall's inner face.

Another variation of solar passive design uses an **attached greenhouse** on the south side of the house (Fig. 6.27). This acts as an expanded thermal

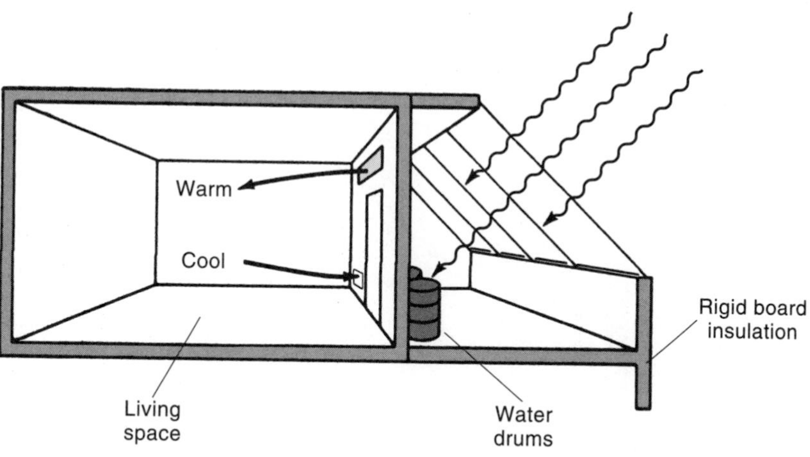

FIGURE 6.27

Indirect gain, using an attached greenhouse. As a combination of direct and indirect-gain systems, the water drums and masonry floor of the attached greenhouse provide needed heat storage.

storage wall. This can serve for both food and heat production, sharing some of its heat with the adjoining house. As with all such systems, heat storage and excellent insulation (especially over the windows at night) are necessary features. Concrete floors and water-filled drums are common energy storage devices.

Another example of a passive solar system is a **thermosiphoning** air panel collector (Fig. 6.28). This air panel is powered by the difference in pressures between the solar heated air and the cooler room air that enters from the bottom. (As seen in the detail of the figure, the air flows *behind* a corrugated metal absorber to reduce heat loss by convection.) Because such collectors can be easily retrofitted on the south side of a building, they are popular additions. These air heaters usually do not incorporate thermal storage in the system, so they are best used in buildings with large daytime heating loads, such as schools and office buildings.

While solar energy is not generally recognized as an energy conservation technique, it is used to reduce the consumption of fossil fuels. Passive features can be included in new house construction at little additional cost and can provide a substantial part of the space heating needs. Table 6.4 examines such energy savings for houses in three cities that we have examined before. (See Table 5.3 for the impact of other energy conservation measures.) In all three cities the south-facing window area was increased from 50 ft^2 to 250 ft^2 to take advantage of solar gain. Calculated savings of about 65% to 80% in commercial fuel usage resulted.

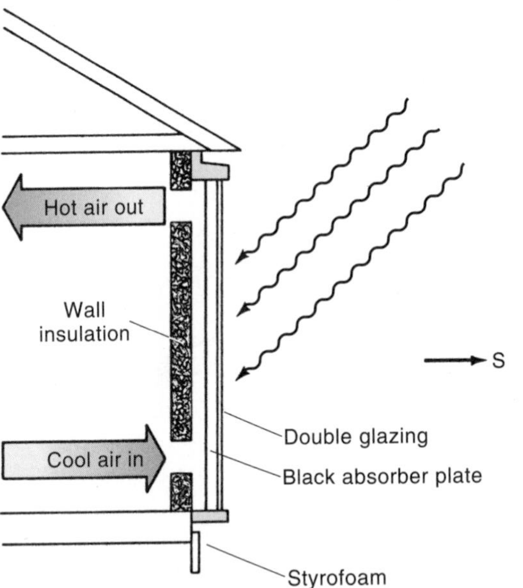

FIGURE 6.28
Thermosiphoning air panel collector.

Table 6.4 EFFECT OF INCREASED SOLAR GAIN

	Annual Heating Requirements (MBtu/y)			Total % Energy Savings		
City	**Base Case**	**Increased Conservation***	**Increased Direct Gain****	**From Conservation**	**From Solar**	**Total**
Boston	54.6	26	19	52	+ 13	= 65%
Denver	56.1	27	11	52	+ 28	= 80%
Nashville	34.7	14	9	60	+ 14	= 74%

*See Table 5.3 for details.

**Increase south-facing windows from 50 ft^2 to 250 ft^2 and add R-4 nighttime insulation.

(Lewis, D., and J. Kohler. 1981. Passive Principles: Conservation First. *Solar Age* [September].)

G. Active Solar Space Heating Systems

Active solar space heating systems for homes have been around for some time, but their popularity has not been translated into large numbers of installations, primarily because of economics. Pioneer work on solar heated homes was carried out in the United States at MIT under H. Hottel, in Colorado under G. Lof, and in the Washington, D.C. area under H. Thomason. More than 50% of the heating needs of these test homes were supplied via solar energy. However, because of the availability and low cost of other fuels, maintenance and corrosion difficulties, and problems of energy storage, the use of solar energy for heating increased slowly and was limited primarily to demonstration units until the mid-1970s. The number of homes supplemented with solar energy heating systems increased from several dozen in the early 1970s to thousands in the mid-1980s. Government programs during this time provided a chance for solar manufacturers to test new concepts and equipment. Today, very few active solar energy space heating systems are being added to individual residences. In Europe, active systems for district heating needs (using a central heating plant) are being constructed. There are some interesting physics concepts (especially in the area of thermal storage) that can be covered in a brief overview of this technology.

Solar Collectors

Active solar space heating systems must make use of a FPC of the type described in the previous section, some kind of thermal storage, and a mechanical means to transfer the heat from storage into the living space. The fluid used to

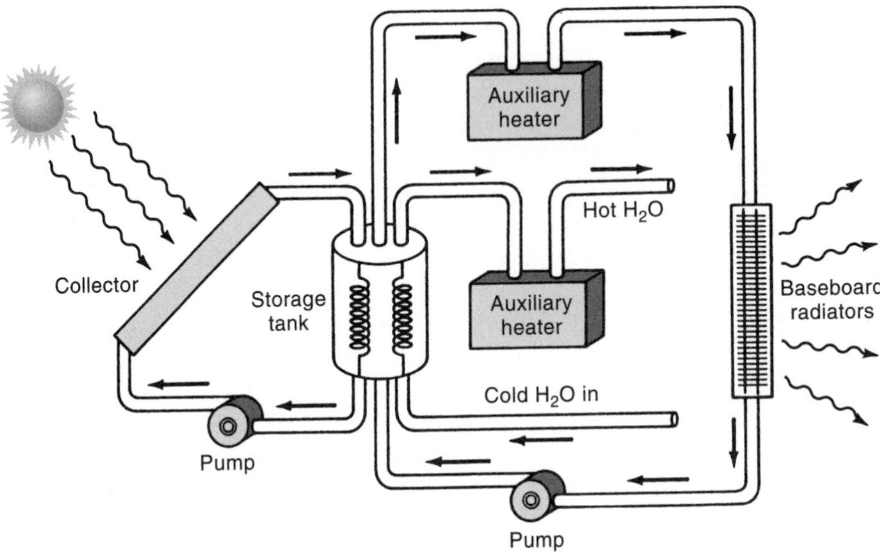

FIGURE 6.29
Basic space heating and domestic hot water system.

transfer heat from the FPC into storage and into the house is usually water or air. A simple system for space heating and DHW is shown in Fig. 6.29.

Flat plate collectors are usually roof mounted, and a well-insulated thermal storage tank is typically located in the basement. Heat is delivered to the house from the storage tank by baseboard radiators. Heat exchangers are used to transfer heat from the collectors to the storage tank; ethylene glycol or the safer propylene glycol (antifreeze) is added to the water passing through the collector to prevent freezing. For both space heating and hot water requirements, auxiliary heaters are employed to compensate for days of poor insolation. The pump in the solar collector loop of this system is controlled by a *differential thermostat:* The pump is turned on only if the temperature of the water in the collector is several degrees above that of the water in the storage tank.

Air can also be used as the working fluid in the collector. An air system with rock storage might look like that of Figure 6.30 and Figure 6.31. The hot air distributed to the house can come directly from the collectors or from the storage facility. Air costs less than a water system, won't freeze, and won't cause problems if it leaks. However, it is not as efficient a heat transfer medium as water, needs a larger storage facility, costs more to run because the fans require more electricity to operate, and is more difficult to retrofit because of the size of the ducts used. The air used for space heating is usually at a temperature of 90° to 120°F, while the water in baseboard radiators is from 140° to 160°F. Since the sun is lower in the sky during the winter months, the collector position should be placed at a large angle (measured from the horizontal) in order to maximize

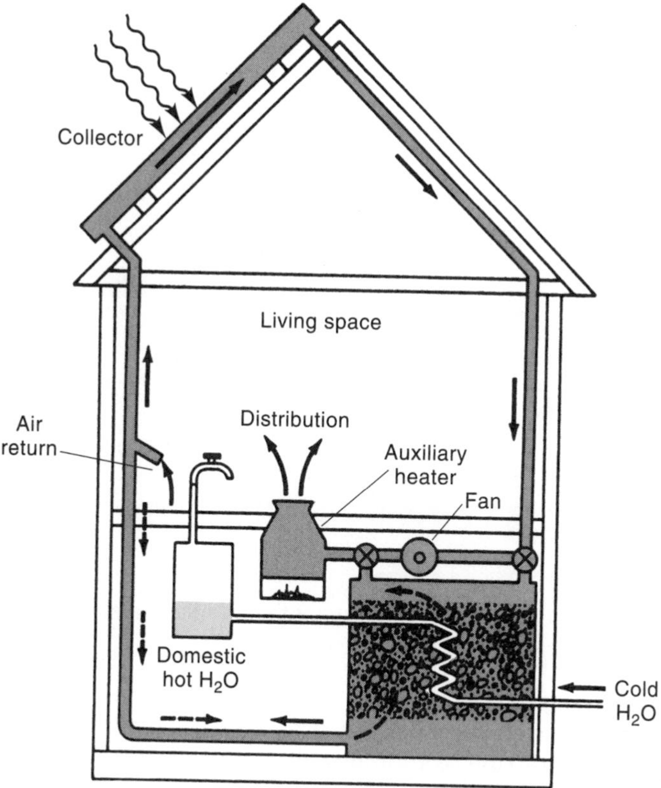

FIGURE 6.30

Hot-air flat plate system. Air transfers heat from the collector either directly into the rooms or into the rock storage bin (*solid line*). When heat is being removed from storage (*dashed line*), the air flow is in the opposite direction so that as much heat as possible can be picked up from storage. Water for domestic use is preheated in the storage bin.

FIGURE 6.31

This active air system provides about 30% of the heat needs of this house in upstate New York. (C. SALVAGIN)

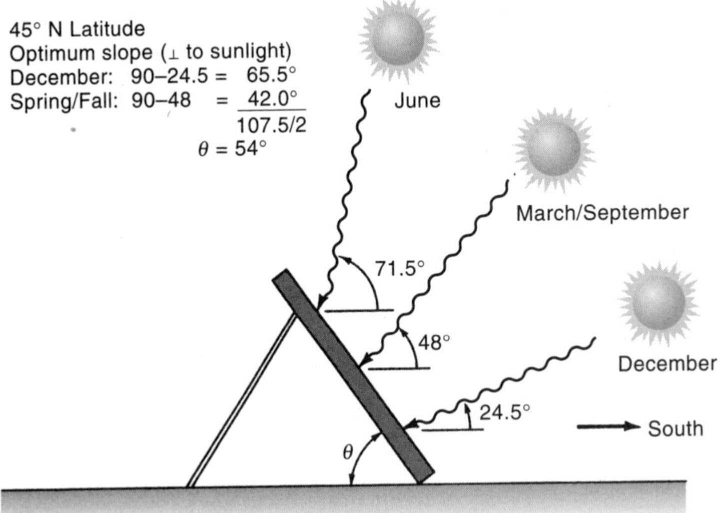

FIGURE 6.32
Calculating collector tilt angle from the horizontal for space heating.

the total insolation on it. (Figure 6.32 illustrates the calculation of tilt angle for 45°N latitude.) A useful rule of thumb is that the collector for space heating should be inclined at an angle equal to the local latitude plus 10°.

To calculate how many square feet A of FPC are needed to provide a house with a quantity of heat Q, one needs to know the average insolation I, and the efficiency of the collector ε (for converting the sun's energy into useful energy that will be delivered by the collector fluid into the house).

$$Q = I \times \varepsilon \times A$$

A numerical example of a sizing calculation follows.

EXAMPLE

Sizing Collectors for Space Heating
How many square feet of collector are required to provide all the thermal energy needed to heat a house for one day when the heat load is 20,000 Btu/h? Take the mean daily insolation on the collector surface as 1800 Btu/ft^2/d and the collector's efficiency as 50%.

Solution

Recall that $Q = I \times \varepsilon \times A$. The thermal energy Q needed for a day will be 20,000 Btu/h × 24 h/d = 480,000 Btu/d. The solar energy collected in one day will be 1800 Btu/ft^2 × 0.50 × collector area A. Therefore,

$$A = 480{,}000 \text{ Btu/d}/900 \text{ Btu/ft}^2/\text{d} = 533 \text{ ft}^2$$

This is about one half the roof area of a typical two-story house! At a price of about $55 per square foot for a solar heating system, you can see why not many active solar heating systems are being built.

H. Thermal Energy Storage

One important requirement of a solar energy heating system is that it is able to store energy for nighttime use and for cloudy days. An important criterion in selecting a storage medium is the *specific heat* of the substance. Different materials absorb different amounts of heat in undergoing the same temperature increase. As you might remember from Chapter 4, the relationship between a temperature change ΔT and the amount of heat Q added (or subtracted) is given by

$$Q = mc\Delta T$$

where c is the specific heat and m the mass of the substance. Table 6.5 lists some specific heats for different materials per pound and per cubic foot.

Table 6.5 THERMAL ENERGY STORAGE MATERIALS

	Specific Heat	Density		Heat Capacity*	
Material	**Btu/lb-°F**	**(lb/ft³)**	**kg/m³**	**(Btu/ft³-°F)**	**(kJ/m³-°C)**
Water	1.00	62	1000	62	4186
Iron	0.12	490	7860	59	3521
Copper	0.09	555	8920	50	3420
Aluminum	0.22	170	2700	37	2430
Concrete	0.23	140	2250	32	2160
Stone	0.21	170	2700	36	2270
White pine	0.67	27	435	18	1220
Sand	0.19	95	1530	18	1540
Air	0.24	0.075	1.29	0.02	1.3

*Heat capacity = specific heat x density

Using an active water system for space heating, an average-sized house located in a mild winter climate might have a 1500-gallon water storage tank, which can store enough thermal energy to continue heating the house for 4 to 6 days. For air heating systems, a rock bin is used as the storage media. For our average-sized house, a rock bin might occupy a volume of 280 ft^3; these rocks would average about 1 inch in diameter and have a total weight of 7 tons. A numerical example of thermal storage follows.

EXAMPLE

Thermal Storage

Using the house and collector of the previous example, how many gallons of water would be necessary to store enough thermal energy for 3 days of space heating? Assume the water in the storage tank begins at 150°F and has a useful lower limit of 90°F (i.e., the temperature change the water will undergo will be $\Delta T = 60°F$).

Solution

The heat that must be provided by the storage system is

$$Q = 3 \text{ d} \times 480{,}000 \text{ Btu/d} = 1{,}440{,}000 \text{ Btu}$$

We can use the relationship $Q = mc\Delta T$, where in this case Q is the heat taken from the storage container as its temperature is lowered by $\Delta T = 60°F$. For water, the specific heat is 1.0 Btu/lb-°F. Therefore, 1,440,000 Btu = mass × 1.0 × 60°F. We find that the mass = 25,000 lb; since there are 8.3 lb/gal, the amount of water required is ~3000 gal.

Other media for thermal storage are *phase-change materials.* At relatively low temperatures, these materials undergo a change of phase from a solid to a liquid, as ice does at 0°C. The amount of energy absorbed per unit mass without a temperature change is called the "heat of fusion." When the reaction proceeds in the opposite direction at night as the material cools, the material solidifies and heat is released into the room. One common group of substances being used in this respect are "eutectic salts." These are salts, such as sodium sulfate decahydrate (also known as "Glauber's Salt"), that are combined with water. At 91°F, this salt melts with the addition of 108 Btu/lb. Conversely, when the temperature drops below 91°F, 108 Btu/lb of heat energy is released as the salt solidifies. The properties of a few commonly used phase-change materials are listed in Table 6.6.

Table 6.6 PROPERTIES OF PHASE-CHANGE MATERIALS

Material	Density (lb/ft^3)	Heat of Fusion (Btu/lb)	Melting Temperature (°F)
Glauber's salt (sodium sulfate decahydrate, $Na_2SO_4 \cdot 10H_2O$)	91	108	88–90
Hypo (sodium thiosulfate pentahydrate, $Na_2S_2O_3 \cdot 5H_2O$)	104	90	118–120
Paraffin	51	75	112
Ice	57	144	32
Calcium chloride hexahydrate, $CaCl_2 \cdot 6H_2O$	102	75	84–102

I. Summary

Most systems for heating using solar energy have a collector, thermal energy storage, and a distribution system. Active systems (including domestic hot water) usually make use of a flat plate collector through which water or air moves to transfer the collected energy. A pump or a fan is used to move the fluid between the collector and a storage tank. A passive solar system uses the south-facing windows of a house as the collector and natural means of heat transfer. Thermal mass (water or rock) within the house is used to store the energy and reduce temperature fluctuations during the day and night.

To determine the required size of the solar collector, one must know the incident solar radiation (insolation) on the collector (measured in units of W/m^2 or $Btu/h/ft^2$), the amount of heat that must be delivered for domestic hot water (DHW) or space heating, and the collector efficiency. For DHW applications, the collector should be tilted from the horizontal at an angle equal to the latitude.

Internet Sites

For an up-to-date list of Internet resources related to the material in this chapter, go to the Harcourt College Publishers website at **http://www.harcourtcollege.com**. The links are in the *Energy: Its Use and the Environment* site on the Physics page. General energy related sites and some guidelines for using the World Wide Web in your class are on the inside front cover of this book.

References

Chapter 6

American Society of Heating, Refrigeration, and Air-Conditioning Engineers (ASHRAE). 1977. *Applications of Solar Energy for Heating and Cooling of Buildings*. New York, ASHRAE.

Anderson, B., and M. Riorden. 1996. *The New Solar Home Book*. 2nd ed. Andover, MA, Brick House.

Anderson, B., and M. Wells. 1996. *Passive Solar Energy: The Homeowner's Guide to Natural Heating and Cooling*. 2nd ed. Andover, MA, Brick House.

Brown, N. L. 1980. Renewable Energy Resources for Developing Countries. *Annual Review of Energy*, 5.

Darrow, K., and M. Saxenian. 1993. *Appropriate Technology Sourcebook: A Guide to Practical Books for Village and Small Community Technology*. Stanford, CA, Appropriate Technology Institute.

Halacy, B., and D. Halacy. 1992. *Cooking with the Sun: How to Make and Use Solar Cookers*. San Mateo, CA, Morning Sun Publishing.

Halacy, D. S. 1973. *The Coming Age of Solar Energy*. New York, Avon.

Holdren, J. P., G. Morris, and I. Mintzer. 1980. Environmental Aspects of Renewable Energy Sources. *Annual Review of Energy*, 5.

Lovins, A. B. 1978. Soft Energy Technologies. *Annual Review of Energy*, 3.

Mazria, E. 1979. *The Passive Solar Energy Book*. Emmaus, PA, Rodale Press.

McDaniels, D. 1991. *The Sun*. Melbourne, FL, Kreiger Publishing.

McIntyre, M., ed. 1997. *Solar Energy: Today's Technologies for a Sustainable Future*. Boulder, CO, American Solar Energy Society.

Reif, D. 1981. *Solar Retrofit: Adding Solar to Your Home*. Andover, MA, Brick House.

Robinson, S. 1978. *The Energy-Efficient Home*. New York, New American Library.

Shurcliff, W. 1979. *New Inventions in Low-Cost Solar Heating*. Andover, MA, Brick House.

Weider, S. 1982. *An Introduction to Solar Energy for Scientists and Engineers*. New York, John Wiley & Sons.

Weinberg, C., and R. Williams. 1990. Energy from the Sun. *Scientific American*, 263 (September).

QUESTIONS

1. Give several ways that you could increase the amount of solar energy delivered to a square foot of material.
2. What is the best angle at which to place a solar collector if the radiation is primarily diffuse?
3. During what season will more insolation be delivered to a vertical south-facing window on a clear day?
4. What would you propose to give a homeowner "solar access"—that is, regulations prohibiting owners of adjacent property from erecting structures that would block their sunshine?

5. What are three fundamentals of a passive solar home? What are the three types of passive solar homes?
6. Sketch a floor plan for a house designed to be heated by direct-gain passive solar. Make the house two-stories tall, with three bedrooms. Note the orientation.
7. Sketch the design for an attached greenhouse that one could add to a house to provide supplementary heat. Note its different features.
8. Discuss the operation of a batch water heater for domestic hot water. Identify its components on a sketch.
9. Design a solar heater for a swimming pool, using material one can find around the house or easily purchase at a local hardware store.
10. The design of a solar collector must take into account the reduction of heat losses. Sketch a cross-sectional view of a flat plate collector and show the ways in which heat is transferred from the absorber plate.
11. What are the disadvantages and advantages of locating solar collectors on a vertical, south-facing wall?
12. The efficiency of a solar collector decreases as the temperature difference between the collector fluid and the outside air temperature increases. Why? How does this influence the desired water flow rate through the collector?
13. What types of lifestyle changes would be brought about by living in a house whose space heating and hot water needs are supplied almost entirely by the sun? Think of a daily living schedule. What differences would there be in your answer for active versus passive solar space heating?
14. What is the optimum angle of tilt (from the horizontal) for a solar collector located at 30°N latitude placed for the purpose of domestic water heating? For space heating?
15. What are the advantages to using thermal mass on a direct-gain passive solar system?
16. Solar energy can be stored by raising the temperature of water or rocks. A storage system will lose its stored energy by heat losses from the holding tank to the outside air. Even though water is a poor conductor of heat, the circulation of the warmer water to the top and colder water to the bottom will transfer heat within the tank. With this background, why are rock piles (or small pebbles) good energy storage materials for the heat energy removed from hot air even though their heat capacity is small?
17. One scheme for the storage of solar energy uses eutectic salts. These salts undergo a change in state from solid to liquid at temperatures around 100°F. When they melt, they will absorb energy with no change in temperature. On return to the solid state, energy is liberated. Compare this situation to the use of ice cubes in a drink.
18. Sketch a block diagram for an active solar domestic hot water system.
19. What institutional barriers exist to the expansion of solar energy technologies?

20. There are many descriptions of solar ovens available on the Web. Sketch two such models, and describe the differences in their operation, noting physics principles.
21. In your community, identify some passive solar heated homes. Make observations based upon the principles discussed in this chapter. Refer to the Web home page of a local/regional solar energy association and identify several public facilities using solar energy.

PROBLEMS

1. Using the data of Table 6.3, graph the mean daily solar radiation on a horizontal and a tilted surface for both Dodge City, Kansas, and Washington, D.C.
2. Solar energy can be attractive not only environmentally but also economically. If an electric clothes dryer has a power rating of 5000 watts and is used for 1 h per day, how much money can be saved in a month by using the experimental backyard clothes dryer of Figure 6.33, assuming electricity costs 9 cents per kWh?

FIGURE 6.33
Backyard solar clothes dryer.

3. Suppose the solar radiation is 850 W/m^2 and you can collect 20% of the energy that falls on the reflecting surface of a solar hot dog cooker. If you need 240 W for the cooker, what is the minimum collector area required?

4. What size flat plate collector is needed to supply a family's domestic water needs in March in Denver, Colorado? Assume 80 gal per day are needed (1 gal = 8.3 lb), ΔT = 70°F for the water, and that the exchanger system has an average efficiency of 40%. The collector tilt angle is equal to the latitude. (See Appendix D.)
5. If the insolation on a flat plate collector is 800 Btu/ft^2/d, how large must the collector be to provide 30,000 Btu/h of heat for 1 d? The collector efficiency is 40%.
6. Water has the highest specific heat of any ordinary material, which means that it can hold a good deal of thermal energy. A cubic foot of water stores about 62 Btu/°F, while 10 ft^3 of water will store 620 Btu/°F. Rocks have a much smaller specific heat but a much greater density than water. If the specific heat of rock is 0.2 Btu/lb-°F and the density is 170 lb/ft^3, how many cubic feet of rock are necessary to store 620 Btu/°F?
7. What maximum percentage of the 40,000 Btu/h heating needs of a house in Minnesota in January can be met with a flat plate collector of area 700 ft^2? Assume that the collector is tilted at an angle equal to the latitude, and the system efficiency is 50%. Use tables in Appendix D.
8. Find several prices on the Web of FPCs to be used for providing DHW. Assume that you will need 3 units (usually 3 ft × 8 ft). Add in a storage tank and pump. Using local electricity or natural gas prices, how long would it take this system to pay for itself if it provided 50% of your annual DHW needs? Assume 80 gal/d is needed for DHW, with a water temperature rise of 70°F required.

FURTHER ACTIVITIES

1. To use solar energy as an additional heating source, the siting of your house is very important.
   (a) Beginning with a "bird's eye" view, sketch your house from the top, use a compass to determine its orientation to the sun, and indicate directions. (Check the magnetic declination* for your area.)
   (b) Locate the prevailing winds and mark them on your sketch.
   (c) Determine the noontime angle (altitude) of the sun for December and June.
   (d) What shading (natural or from other buildings) exists to block sunlight from reaching your house?
   (e) What potential exists for using solar energy for your home?

*Magnetic declination is the difference between true North (geographic North Pole) and north as measured by a compass. This value varies from point to point. In New York City, the magnetic declination is 12° W, while in Los Angeles it is 16° E.

2. What color absorbs the sun's heat best? Check this out by covering the end of a thermometer with construction paper of different colors and recording the temperatures after 10 minutes in the sun or under a heat lamp. (Make sure the conditions are the same for each color. Place the thermometer on a newspaper to reduce heat loss to the ground.) As an alternative, place an ice cube on top of each colored sheet of construction paper. Which melts first? Would you get the same result if the construction paper was on top of the ice cube?

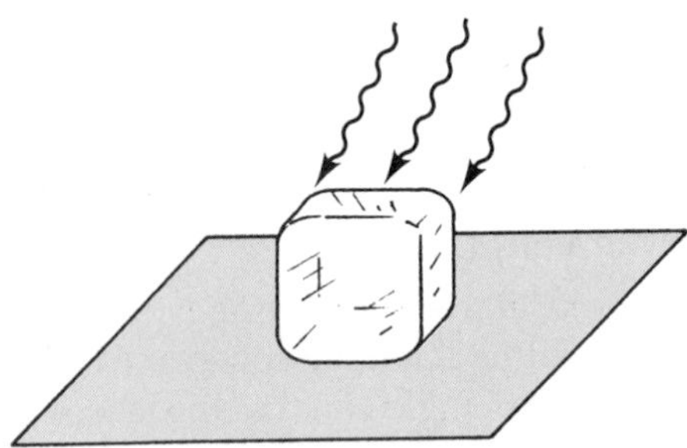

Would a house with a white roof be cooler than a house with a dark roof?

3. Construct a model of a passive solar house, using a Styrofoam ice chest or a shoe box. Cut a large "south-facing" window in one side of the box, and use plastic as the glazing. Insulate the structure (especially the shoe box). Use rocks or cans of water as the thermal mass. (Be realistic and don't fill up the entire living area.) Stick a thermometer into the side of the house. Set the house in the sunlight and measure the inside temperature every 5 minutes for 1 hour. Now put the house in the shade and monitor the inside temperature as a function of time for 1 hour. See Figure 6.34.

FIGURE 6.34
The passive solar house model can be arranged as a class contest, with appropriate prizes for design and for highest inside temperature. The model can be removed from the sun and its temperature measured in the shade for a period of time.

4. You can measure insolation in your locality by the following activity.

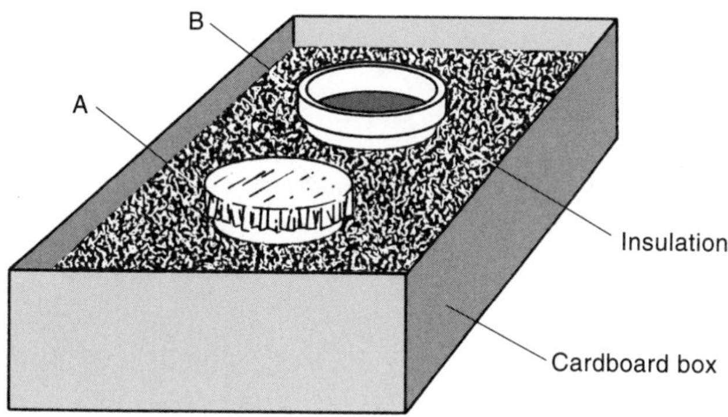

(a) Fill two Styrofoam cups with the same measured amount of water.
(b) Add food color or ink to one cup to make the water as black as possible to absorb sunlight. Cover it with clear plastic or glass.
(c) Cover the top of the other cup with aluminum foil to reflect sunlight.
(d) Put both cups into a cardboard box and surround them with fiberglass or cellulose insulation.
(e) Put the box in the sun for 20 minutes (preferably near noontime).
(f) Stir the water in the cups and record the temperatures. Call the temperature of the black water T_B and the other one T_A.
(g) Measure the area of the water surface in each cup (in cm^2). Note that area = $\pi \times (\text{diameter})^2/4$.
(h) The horizontal insolation is

$$I = \frac{\text{volume of water in one cup (ml)} \times (T_B - T_A)}{\text{area (cm}^2) \times (20 \text{ min})}$$

(This number should be about 1 cal/cm^2/min.)

5. To determine local incident solar radiation, a solarimeter is a useful instrument. A simple solarimeter can be constructed for less than $10. You need a solar cell (one square inch is good), a current-measuring meter (a milliammeter, with a full-scale reading of 200 mA), and 3 ft of coated or insulated magnet wire (no. 30 gauge). Use a small cardboard box for a mounting, and arrange the parts as shown in the following illustration. Twist or solder the wire connections together. (The magnet wire is used to "shunt" some of the current from the solar cell away from the meter.)

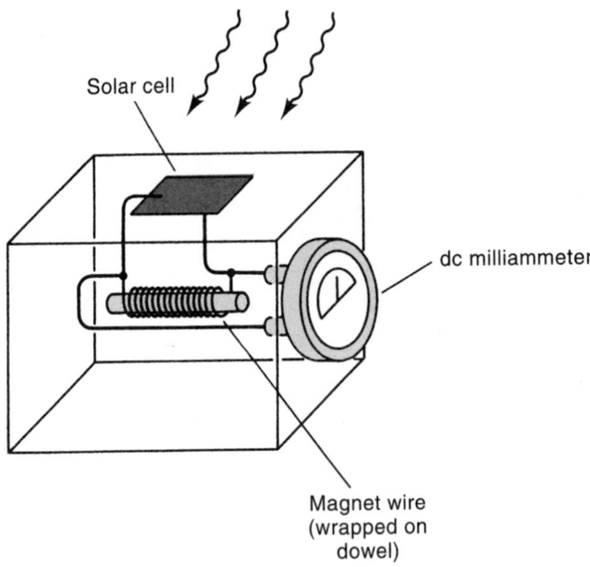

To calibrate the solarimeter, it is best to put your instrument alongside an instrument that has already been calibrated. You can then put a scale on your meter, marked in units of $\text{Btu/ft}^2\text{/h}$ or W/m^2. In lieu of this procedure, you can make an approximate calibration scale by observing the meter reading for the radiation received on the horizontal solar cell on a clear, sunny day around noon. Noting your latitude and the day of the year, you can find what this value should be by looking in Appendix D. Say the value from the Appendix is 600 W/m^2. If your meter reading later in the afternoon is half of the maximum reading, then your incident radiation at that time is 300 W/m^2.

6. Construct a modular solar collector that uses air as the transfer fluid, as shown in the figure. Because this collector is designed for experimentation, one side is hinged so that different configurations for glazing, insulation, and absorber plate can be tried. The air can pass in front or in back of the absorber plate, depending on which holes are plugged with corks. The collector can be set at a wide range of angles.

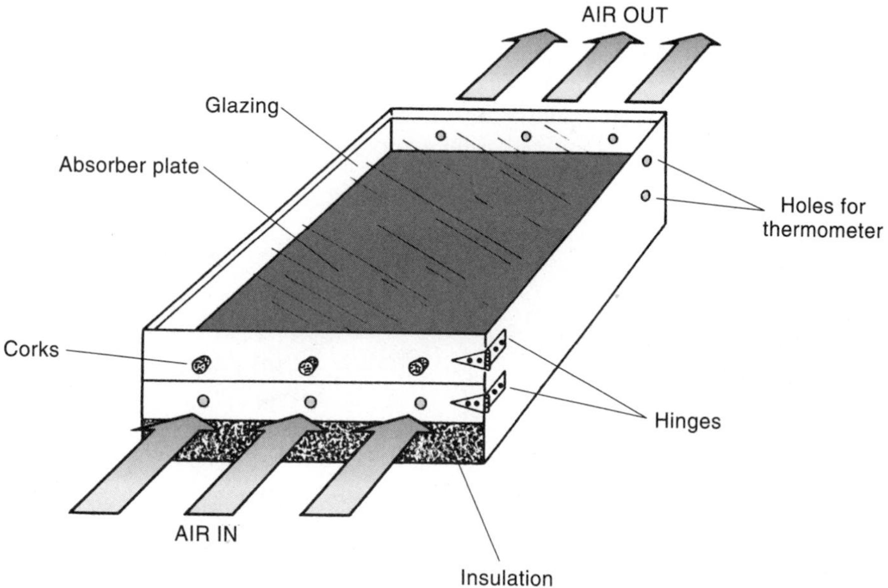

7. Construct a device that will boil a cup (240 ml) of water, using only radiant solar energy. This activity can be done as a contest, in which each team starts with room-temperature water and uses their heater to try to reach the highest temperature in the shortest time.
8. A simple, inexpensive solar panel for heating water can be built as follows. Construct a shallow tray of wood, using a plywood sheet for the bottom and 1-by-2-inch studs for the sides. Line the tray with a thick sheet of black plastic. Insert an inlet tube near the bottom of the tray and an outlet tube near the top. Connect them with a plastic tube painted black that snakes back and forth in the tray—be careful of kinks in the tubing. (No pump is necessary because hot water rises.) Cover the top of the tray with one sheet of clear plastic (such as 4-mil polyethylene) or with two sheets separated by ½ inch. (These sheets could be on a separate frame, hinged to the tray to allow access to the collector.) Use a 5-gal can for the water reservoir, which must be placed *above* the tray. The inlet tube comes from the bottom of the can and the outlet tube from the top. Place the collector facing the sun. Fiberglass insulation behind and on the sides of the tray will make the system more efficient, yielding water temperatures above 100°F.

9. Another group competition similar to that in Further Activity 7 is to cook an egg for each team member using only radiant solar energy. The egg(s) can be cooked using any method, and can be in any form (fried, boiled, poached, etc.). The solar cooking device can be moved during the contest. Application of the concepts from Chapter 4 and 6 should be used in the design.

This parabolic dish allows eggs to be cooked in a few minutes using direct sunlight.

7

Energy from Fossil Fuels

A. Introduction

Today, approximately 90% of the commercial energy resources used in the world come from the fossil fuels—coal, oil, and natural gas. However, except for coal, the supply of these fuels at their present rate of use might not last much longer than the lifetimes of the people now living. The fuel most commonly used, oil, appears to be in the shortest supply worldwide, but continues to be that commodity on which we rely most heavily. Indeed, dependence on oil, especially for transportation, leaves many countries quite vulnerable to a major economic disaster in the event of an oil shortage or interruption in supply. Japan imports almost all of its oil (principally from the Middle East), while some countries whose economies are growing rapidly today (such as Korea, Thailand, Singapore) depend greatly on imported oil.

In the United States, coal accounts for the largest share of domestic energy production, followed by natural gas and crude oil. In the 1990s, growth in the economy and low prices for crude oil led to growth in the consumption of coal, natural gas, and oil. The increase in coal consumption occurred primarily in the electric utilities, whose output grew at 2.2% per year. In the 1990s, natural gas demand rose at about 2.3% per year in almost all sectors (residential, industrial, and electric utilities). Increased petroleum* consumption occurred primarily as a result of increases in transportation; consumption of motor gasoline rose 15% in the 1990s; fuel used per car increased less (12%) due to better fuel economy (mpg). Passenger vehicles consumed 85% of our imports!

A side effect of low crude oil prices is the decrease in exploration for oil and gas. When the price of crude oil rose in 1981, exploration was at record levels. There were 681 crews engaged in exploration, rotary rigs in operation were 3970, and exploratory wells drilled were 17,500 per year. As oil prices collapsed in the mid-1980s, there were drastic cutbacks in exploration. By 1993, these numbers had decreased to 79 for crews, 754 for rotary rigs in operation, and 3100 for exploratory wells drilled. Through most of the 1990s, low prices continued to dampen exploration; about 3000 exploratory wells per year continued to be drilled. Spectacular increases in oil prices in 1999 and afterward (Fig. 1.14) reversed this trend.

B. Resource Terminology

The terminology used to describe the status of fossil-fuel resources seems ambiguous at best. Words such as reserves, known reserves, and undiscovered resources are used frequently and many times incorrectly. The differences are important and will be discussed in this section. How much of a particular resource remains unused or undiscovered in the ground is hard to predict. Since our predictions are always based on incomplete exploration, there will always be some left to discover. And even though a certain amount of a resource is thought to exist, economic and technical factors often affect how much of it can be extracted.

We will use a method developed by the U.S. Geological Survey for categorizing the different types of oil resources, called the McKelvey diagram (Fig. 7.1). We start with a rectangle that shows all the resources (oil in this example) that exist in a particular area (such as the United States). The vertical axis displays the increasing cost of the finished product, beginning with the upper-left corner as the least expensive to recover. The horizontal axis displays the increasing uncertainty of discovery. **Reserves** occupy the upper-left part of this rectangle and are defined as those resources that are well known through geologic exploration and are recoverable at current prices and with current technology. Undiscovered resources occupy the right side of the dia-

---

*Oil is the liquid portion of petroleum.

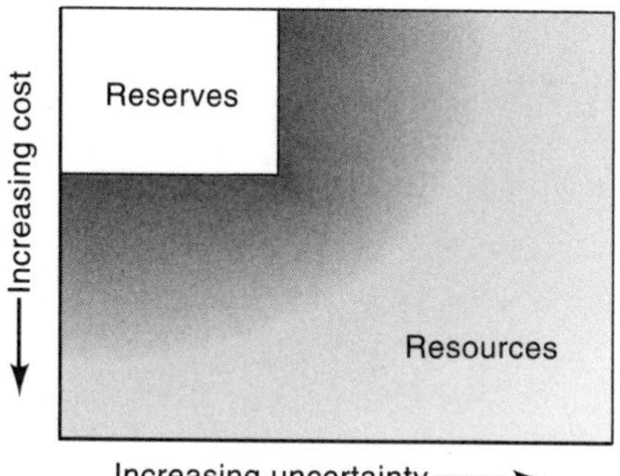

FIGURE 7.1
The McKelvey diagram for categorizing reserves and resources relates the variables of geological certainty and economic feasibility.

gram, while the lower-left corner displays those resources that are known to exist but are too expensive to extract today. The lines separating economic from uneconomic or known from undiscovered change with time, so the magnitude of reserves changes. Reserves are sometimes divided into *proven* reserves, *indicated* reserves, and *inferred* reserves. Proven reserves are those that are reasonably certain of being produced from known reservoirs under existing economic and technological conditions. Indicated reserves are those amounts believed recoverable from known fields using improved recovery techniques. Inferred reserves are those deposits expected in identified fields but not yet measured.

The size of the reserves box is established by the energy industries themselves. Unfortunately, some of the data may be underestimated, since some states tax known reserves. Still, they are the best numbers available. As an industry exhausts old reserves, it strives to keep up its inventory by adding new reserves. However, the addition of new oil reserves in the United States historically has not been able to keep up with consumption (Fig. 7.2). From 1962 to present, we have not been able to replenish the domestic oil consumed in any particular year with the discovery of new reserves.

Just how do we know how large is the box of total oil resources? One method estimates this size by assuming that the amount of oil contained in a unit volume of a particular type of geological structure is a fixed amount, and then estimates the total volume of that type of structure that exists in the United States or in the world. The amount present is then the estimated number of barrels per cubic meter times the estimated volume of that type of geologic structure. Another method is the "behavioristic approach," in which the history of oil production is extrapolated into the future to estimate the amount left in the ground. A famous example of this approach is the forecast by M. King Hubbert in 1969, introduced in Chapter 1. The principle behind these bell-shaped production curves (see Fig. 1.11) is that the historical use of a fuel begins with a time of continuous increase. Eventually the inexpensive or technically easy deposits are exhausted; the use of the fuel reaches a maximum and then decreases as prices rise and fuel substitutions

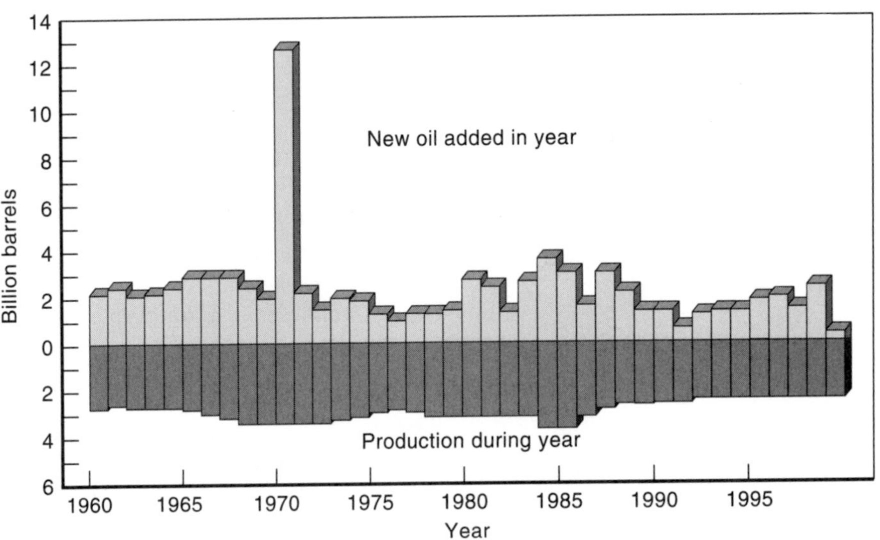

FIGURE 7.2
U.S. crude oil reserves. Above the line are annual additions to reserves (the 1970 peak is Alaskan oil), and below the line is annual production. Note that the addition of reserves is less than our production. (USDOE)

grow. Figure 7.3 shows one such curve for world oil production. (Figure 1.12 shows U.S. oil production.) This curve suggests that by the year 2050 the production of oil from world resources will be at only 10% of present levels.

Many resource analysts do not feel that the behaviorist approach can or should be applied to make predictions about total resources available. The argument is that this is a geological question, which should be answered by using geological data. The use of past history, they say, is not relevant to making pre-

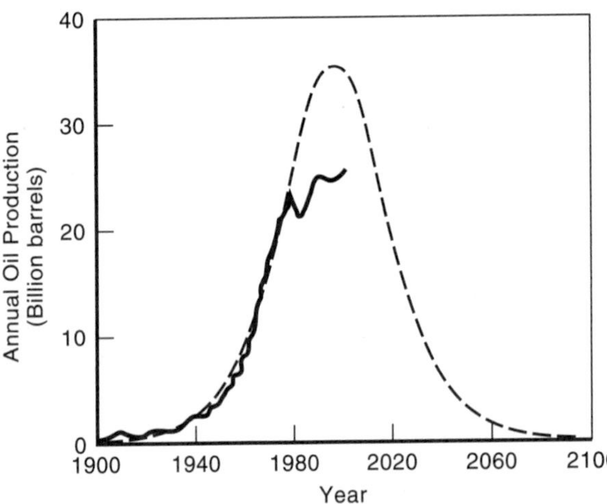

FIGURE 7.3
Hubbert oil-depletion curve for world oil production. The solid line shows actual production, using updated Department of Energy data. (M. KING HUBBERT, *AMERICAN JOURNAL OF PHYSICS*, VOL. 49, NO. 11, 1981)

Table 7.1 WORLD AND U.S. PROVEN RESERVES: 1998

	World	United States	Lifetime*
Oil	1020×10^9 bbl 5.9×10^{18} Btu	22×10^9 bbl 0.12×10^{18} Btu	8 y
Natural gas	5090×10^{12} ft^3 5×10^{18} Btu	165×10^{12} ft^3 0.17×10^{18} Btu	9 y
Coal	1.09×10^{12} tons 27×10^{18} Btu	0.58×10^{12} tons 14×10^{18} Btu	500 y
Tar sands	300×10^9 bbl 1.7×10^{18} Btu	22×10^9 bbl 0.12×10^{18} Btu	8 y
Shale oil	150×10^9 bbl 0.87×10^{18} Btu	20×10^9 bbl 0.11×10^{18} Btu	6 y

*Ratio of U.S. reserves to 1998 U.S. production rate. (U.S. Energy Information Administration)

dictions. There are no substitutes for geological exploration. The behaviorist approach does not take into account new technological advances for discovery and for extraction of the resource. Annual production depends on technology as well as cost and demand.

Table 7.1 lists U.S. and world reserves, as of 1998, for oil, natural gas, and coal. Also shown are U.S. and world consumption rates for each of these fuels. (As noted in previous paragraphs, there is a great deal of uncertainty in estimates of resources.) The ratio of reserves-to-consumption gives some estimates of the longevity of that fuel under current consumption rates. But it must be remembered that reserves will be augmented substantially by resources as prices rise or new technology becomes available.

C. Oil

In the year 1922:

- The first issue of *Reader's Digest* is published
- Gandhi is imprisoned for civil disobedience
- Jimmy Doolittle flies coast to coast with one stop in 22.5 hours
- New York Appellate Court rules that the station wagon is not a commercial truck but a passenger car
- Geological survey indicates that the United States oil supply will last for only 20 years

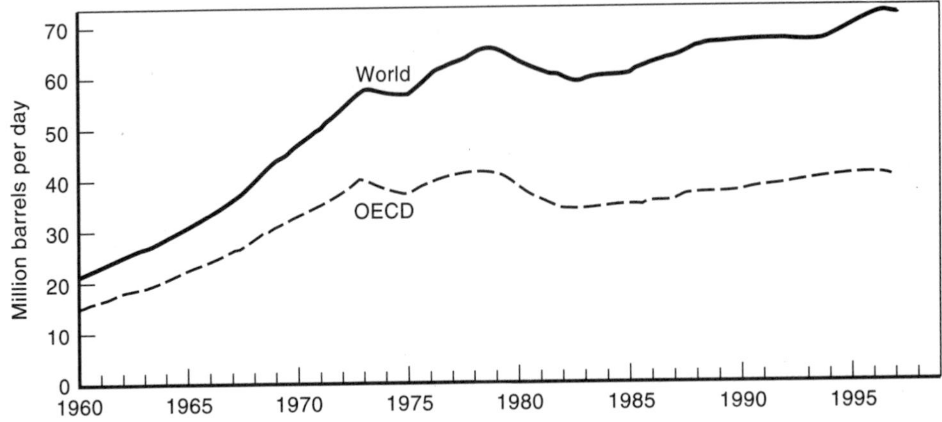

FIGURE 7.4
World petroleum consumption, 1960–1997. The Organization for Economic Cooperation and Development (OECD) countries include the United States, Japan, Western Europe, and Canada. (UNITED STATES ENERGY INFORMATION ADMINISTRATION)

The world's appetite for oil continues to grow (Fig. 7.4). In 1998, the world's total daily consumption was 74 million barrels per day (MBPD), of which 19 MBPD, or about 25%, were consumed in the United States. (About 51% of this oil was imported.) Besides consuming more oil than any other country, the United States produces more than any country except Saudi Arabia. However, the United States has only about 2% of the world's known oil reserves, only enough to last about seven years if used at current rates. Even though such figures might seem deceiving, because more oil will certainly be discovered domestically, new discoveries are not expected to alter this lifetime appreciably.

Figure 7.5 shows world reserves of oil by region and world oil use by country. Note the nonuniformity of oil's distribution. The oil fields in the Middle East contain 60% of the world's oil in 0.5% of its land area. Table 7.2 lists world oil reserves by country for 1998.

While the first commercial oil well was drilled in Titusville, Pennsylvania, in 1859 (Fig. 7.6), petroleum's uses for medicine, lamps, and other purposes date back to biblical times. The pitch from asphalt was used in Egypt and Babylonia around 2500 BC for waterproofing and roads; the Greek and Roman armies used it for their weapons. The Chinese drilled for oil and gas before 1000 BC and used them for heating and lighting. The use of oil grew rapidly after 1859 because it could be used as a substitute for whale oil. Prices for whale oil (used for lighting) were increasing as a result of a scarcity of whales.

As we have discussed throughout this book, oil is crucial to the world's economy, especially in applications in which substitution is difficult—such as

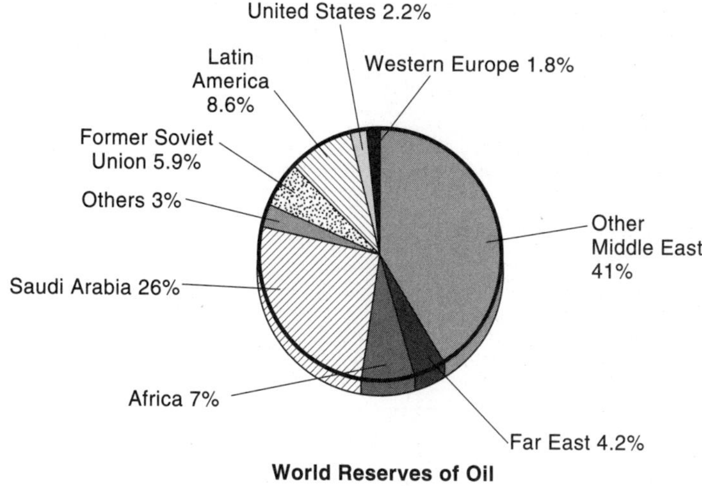

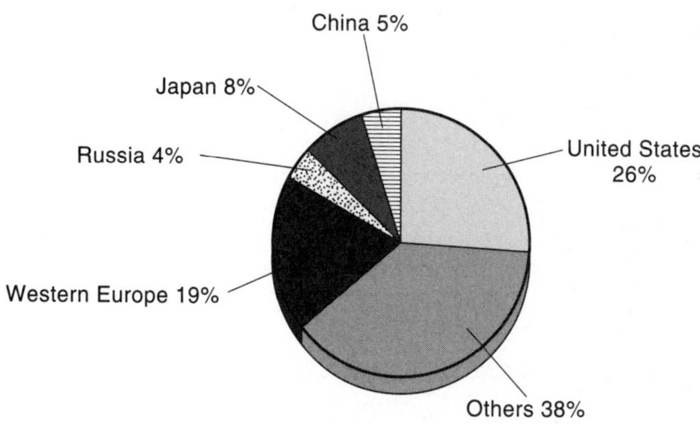

FIGURE 7.5

World oil reserves and oil use by country, 1998. (UNITED STATES DEPARTMENT OF ENERGY)

transportation, agriculture, and petrochemicals. Two thirds of the oil consumed in the United States goes for transportation (Fig. 7.7).

Petroleum is a mixture of crude oil, natural gas in solution, and heavy, thick, asphaltic semisolids. All petroleum deposits contain natural gas, but not all natural gas deposits contain oil. Petroleum is a complex mixture of

Table 7.2 WORLD OIL RESERVES: 1998

Country	Reserves (Billions of Barrels)
Saudi Arabia*	262
Iraq*	112
United Arab Emirates*	98
Kuwait*	96
Iran*	93
Venezuela*	72
Russia	49
Mexico	40
Libya*	30
China	24
United States	21
Nigeria*	17
Norway	10
Algeria*	9
United Kingdom	5
Indonesia*	5
Canada	5
India	4
Egypt	4
Qatar*	4

*Organization of Petroleum Exporting Countries (OPEC) member
(American Petroleum Institute)

hydrocarbons (compounds of hydrogen and carbon), with an average ratio of H to C of about 1 to 7 by mass. Some compounds have only one carbon atom, while a few have as many as 100. No two oils ever contain the same mix of compounds. Petroleum also contains small percentages of compounds of vanadium, nickel, and sulfur.

FIGURE 7.6

The world's first commercial oil well, Titusville, Pennsylvania, 1859. On the right is Colonel Edwin Drake, who conceived the idea of drilling for oil and used the primitive rig pictured in the background. (DRAKE WELL MUSEUM)

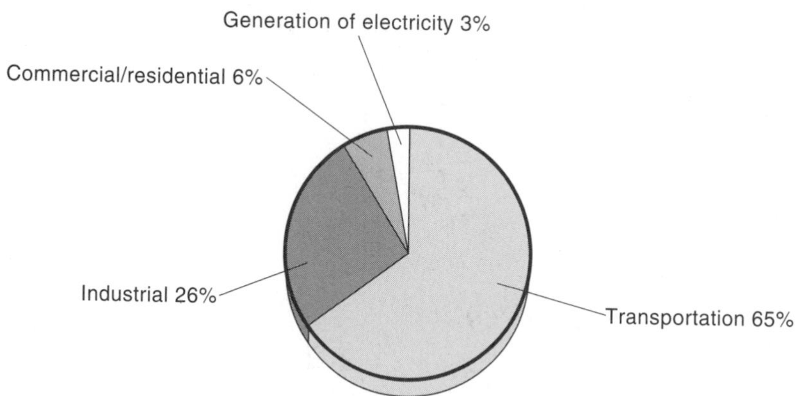

FIGURE 7.7

U.S. oil consumption by end use, 1999. (UNITED STATES ENERGY INFORMATION ADMINISTRATION)

Focus On 7.1

PIPELINE POLITICS IN THE FORMER SOVIET UNION

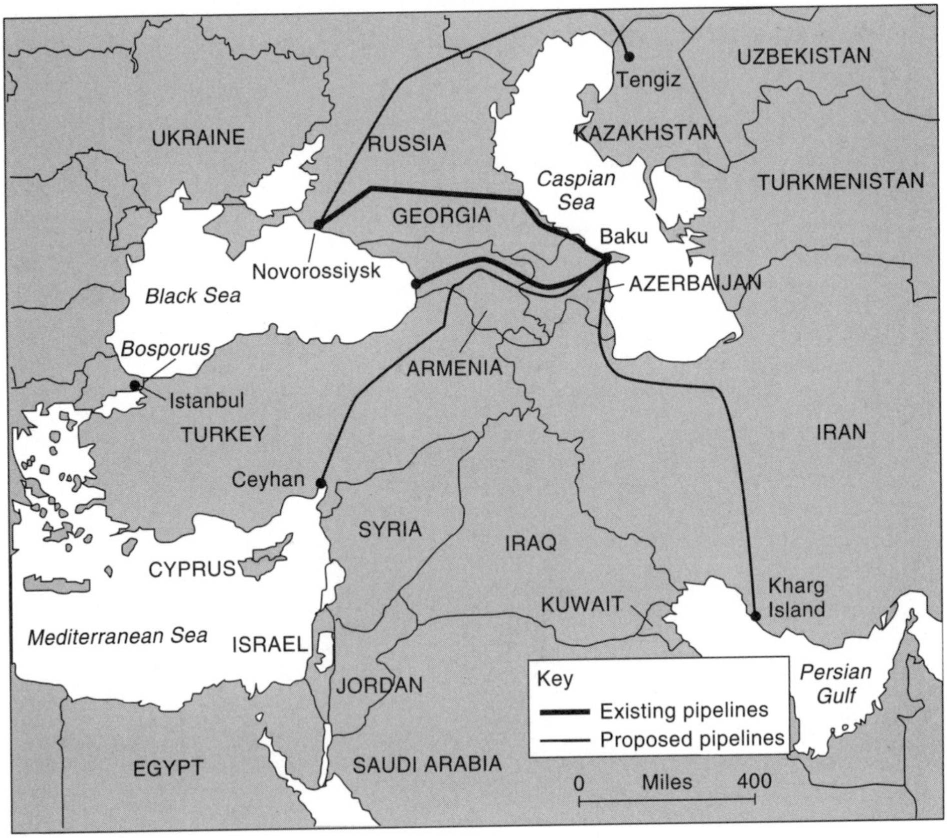

Prior to 1989 and the collapse of Communism, the former U.S.S.R. had the third-largest oil reserves in the world. Economic and political problems in Russia have left the oil industry in shambles. Facilities have been allowed to deteriorate and have not been repaired. An increase in oil production is unlikely and may actually decline. Now, the new republics have a share in the oil resources and are hoping to use them to better their own economic positions. A prime example of this is Kazakhstan and its 5 billion barrel oil production project at Tengiz (see map). This field is the largest discovered in the world in the last 25 years. Western oil companies are involved in this project but are facing difficulties because of ethnic tensions. Other problems arise because of Russia's reassertion of claims to the oil wealth of these republics and the difficulty of siting a pipeline route for the oil exports. The Caspian Sea is landlocked and new pipelines are needed for the vast oil and gas reserves of this region. Routes through Russia to the Black Sea bring Turkey into the picture, which wants the oil revenues but is raising environmental concerns about too many oil tankers going through the Bosporus Strait. At the beginning of 2000, prospects were good for building a multibillion dollar pipeline across Turkey to the Mediterranean (from Baku, Azerbaijan, to Ceyhan, Turkey). Another possible route for a pipeline to the Persian Gulf is through Iran, but it is politically risky.

Crude petroleum must pass through a series of stages in the refining process to convert it into useful products. In **refining**, the first step is distillation, which separates the different parts of petroleum through their different boiling points. The petroleum is initially heated and then passed into a fractionating tower (about 40 m high). The various petroleum products condense at different temperature levels in this tower and are collected. The heaviest fractions (or products) collect at the bottom of the tower, while gasoline condenses near the top. (Some gases do not condense and so are taken off at the top and added to the natural gas.) Figure 7.8 shows a simplified picture of a refinery.

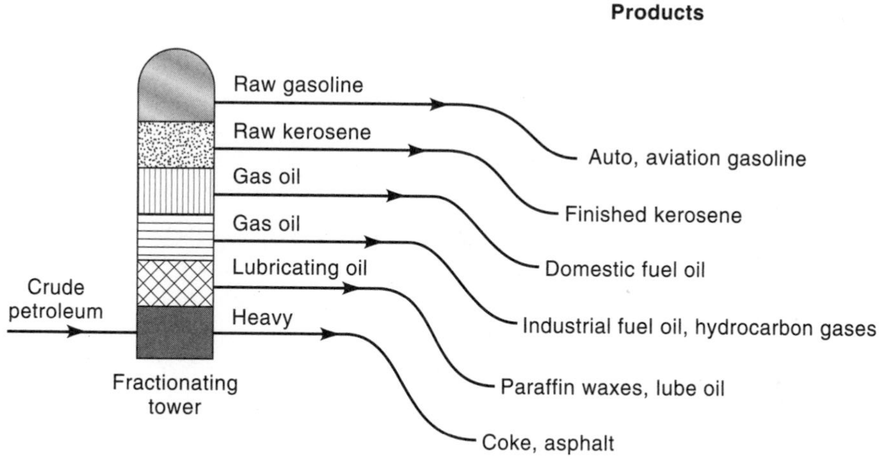

FIGURE 7.8

At the refinery, the petroleum fractions are separated thermally in the fractionating tower, then individually treated to produce the products listed.

Most of the products from the tower are treated further chemically and/or thermally to produce such products as gasoline, heating oil, jet fuel, diesel oil, paraffins, and asphalt. Gasoline is the most important petroleum product, accounting for about 45% of the refinery output. The refinery output can be adjusted according to the season to produce more or less gasoline and heating oil. About 10% of all petroleum used in the United States provides the chemical industry with raw materials such as methane, ethane, benzene, and toluene. These are used in the production of fertilizers, plastics, solvents, nylon, synthetic rubber, and so on.

Petroleum originates from the decay of organic material, usually marine life, that has been converted to petroleum over millions of years under the high pressures and temperatures that are associated with deep burial. The oil (and natural gas) formed under these conditions can migrate through the surrounding rocks and accumulate in deposits, from which it can be extracted. These deposits are found within reservoir rocks such as sandstone, shale, and limestone, which hold oil the way a sponge holds water. These rocks are quite porous (having voids or openings) and permeable, allowing the liquids or gases to move. For the oil to accumulate and not escape, it is essential that the reservoir rock be covered with impermeable or nonporous rock that will act as a trap to prevent the upward migration of the oil and gas. The most common type of structural trap or barrier is the "anticline," caused by the upfolding of the earth's strata. As the oil moves up in the anticline, it is trapped by an arched-over caprock, such as impervious shale (Fig. 7.9). Another trap is the "salt dome," resulting from the upward intrusion of a salt plug from great depths into the surrounding strata.

The search for oil has concentrated on the search for salt domes or anticlines. Early geologists concentrated on observing surface features and recording the inclination of sedimentary layers or beds on a map. Such surface studies have been expanded more recently by the use of seismology to "see" the structure of the

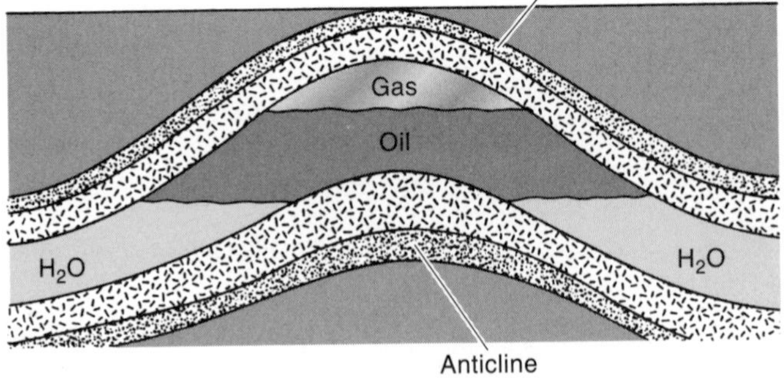

FIGURE 7.9

Typical petroleum trap. The upper boundary of the trap is an impermeable layer of rock called a roof or caprock.

Focus On 7.2

OIL SPILLS

The largest oil spill in the United States occurred on March 24, 1989, when the tanker *Exxon Valdez* went off course in Prince William Sound near the southern coast of Alaska and hit submerged rocks. About 11 million gallons of oil were released into the water. The oil slick spread over thousands of square kilometers, killed hundreds of thousands of birds, and damaged the area's gigantic fishing industry. Hundreds of kilometers of shoreline were covered by a thick, black goo. Forty-five million dollars was spent on wildlife rescue and rehabilitation, and more than 10,000 workers used steam and rags and buckets to clean up the coastline. In 1992, the U.S. Coast Guard and state of Alaska declared the cleanup complete. There have been five record harvests of salmon in Prince William Sound since 1989. Signs are that the eagles have recovered.

The most massive oil spill in world history occurred in 1991 during the Persian Gulf War when about 250 million gallons of crude oil were dumped deliberately into the Persian Gulf, equivalent to a fleet of supertankers running aground. The fires set at the 600 oil wells in Kuwait consumed about 4.6 MBPD. It took a year to put out the fires that blackened the sky and contributed 20,000 tons of SO_2 per day to the atmosphere—about one half of what all U.S. utilities contribute.

(ANCHORAGE DAILY NEWS/GAMMA LIAISON)

earth beneath the surface. (This is discussed further at the end of this chapter in a Special Topics section on *Physics of Oil and Gas Exploration.*)

As the search for new oil deposits continues, it becomes more and more difficult to achieve success with conventional drilling techniques. Since oil and natural gas are often found together, their recovery methods are similar. They are usually trapped within porous rock that acts as a reservoir. The reservoir has a cap or roof of impermeable rock, which prevents the upward migration of the fuels. Water is denser than oil, so it lies at the bottom of the trap, below the oil and natural gas. A well drilled into a reservoir initially will produce results by natural flow, as underground pressures will force the oil up—yielding "gushers" in some cases. To increase and sustain removal, a pumping unit is commonly employed, providing a partial vacuum at the top of the well, as with a straw in a soda. Additional (*secondary*) recovery is achieved by injecting water into the reservoir to provide increased pressure. Even with secondary recovery, less than a third of the oil in the reservoir is taken out (Fig. 7.10). To further increase petroleum extraction, *enhanced* or *tertiary* recovery methods are used. The most common such method is to inject steam into the ground. The steam's heat makes the oil less viscous and allows it to flow more easily out of the porous rock and into the well, so that it can be pumped to the surface. Another method involves injecting CO_2 or nitrogen directly into the oil to increase the pressure in the reservoir and free the oil from the rock. A third method uses chemicals such as polymers in place of gases to force molecules of oil out of the reservoir rock. It is estimated that the use of enhanced recovery methods can increase oil production from a given reservoir by 10 to 20%.

The United States has more drilled wells per square mile than any other country in the world. As it becomes more and more difficult to extract oil, the oil discovered per drilled foot has been decreasing in almost a linear fashion since the 1940s. The average well depth today in the United States is almost 8000 ft, and only 30% of drilled wells are successful. As the technology needed to extract the oil advances, the energy expended per drilled foot increases. Some resource analysts suggest that we eventually will reach a point at which the energy obtainable from a well will equal the energy put into that well.

Most increases to U.S. reserves are expected to come from offshore drilling and very deep wells. About 10% of our crude oil and 21% of our natural gas comes from offshore waters today. Except for a 3-mile-wide band of state waters, the federal government owns the subsurface mineral rights out to international waters, and it leases tracts for oil exploration and production. Some estimates suggest that half of U.S. oil reserves lie in such waters. However, exploration in the middle 1980s off the East Coast (in the Baltimore Canyon—off of New Jersey and Maryland) proved quite disappointing. Also, environmental concerns about oil spills have prompted the withdrawal of millions of acres of offshore land from leasing.

Offshore drilling is very expensive (about ten times as costly as an onshore well per drilled foot), and it is a monument to 20th-century engineering. Drilling rigs can operate in water more than one mile deep, although the major-

OIL RECOVERY
(How to pump more from the ground)

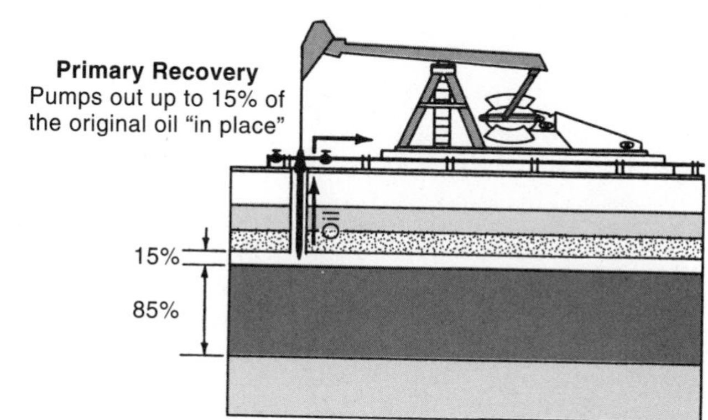

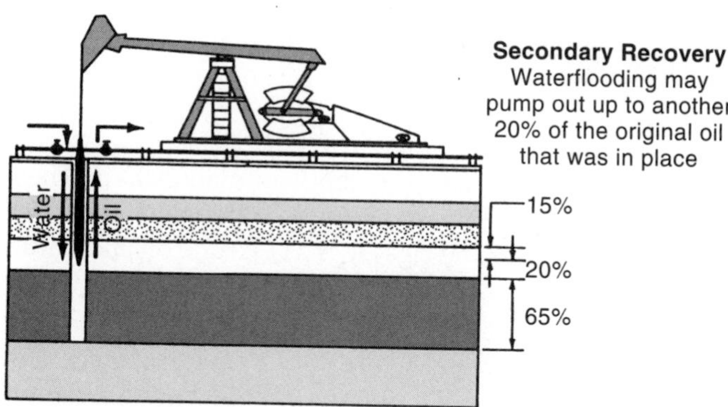

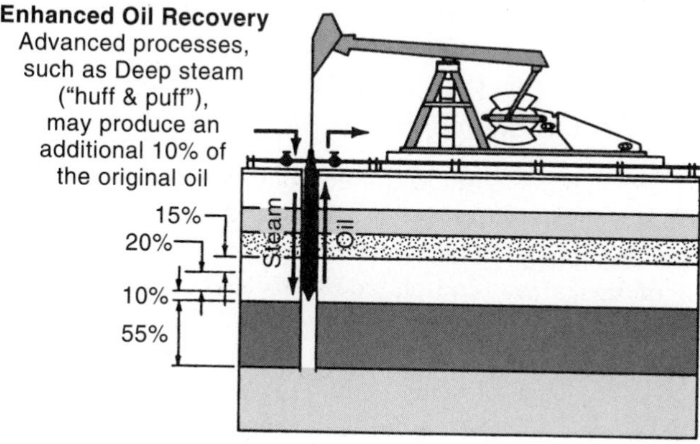

FIGURE 7.10

Enhanced oil recovery methods. (UNITED STATES DEPARTMENT OF ENERGY)

FIGURE 7.11
Offshore drilling rig operating in the Baltimore Canyon area.
(AMERICAN PETROLEUM INSTITUTE)

ity in the United States have been in waters less than 2000 ft. These rigs are moored to the seabed by many lines or held in position over the well by motor-driven thrusters aboard the drilling rig (Fig. 7.11). Crude oil and natural gas are usually brought to shore through pipelines.

Focus On 7.3

OIL FROM ALASKA: PAST, PRESENT, AND FUTURE DILEMMAS

In 1968, the largest oil field in the United States was discovered on the North Slope of Alaska—with reserves of about 10 to 20 billion barrels. The 800 mile long pipeline that runs between Prudhoe Bay and Valdez (on the Gulf of Alaska) was completed in 1977 and carries about 1.2 MBPD—about 20% of U.S. domestic production (see map).

As production from the North Slope decreases, attention has been turned to the Arctic National Wildlife Refuge (ANWR). Oil companies would like to drill for oil on the north coast of the Refuge, where they have access to the existing Trans-Alaska pipeline. Oil reserves there are estimated at 3 to 5 billion barrels, one of the best remaining prospects for oil in the United States.

There is strong opposition to this plan. The Refuge is a fragile, ecologically valuable area and home to many animal species, including caribou and snow geese. Opponents feel that the potential degradation of this wilderness area in exchange for a small (10%) increase in our oil reserves is not worth it. They argue that simple energy conservation measures would save more oil, faster, and at a lower cost. Various attempts over the years made by members of Congress to open about 10% of the Refuge for drilling have been defeated, but high world oil prices provide reason for annual debate.

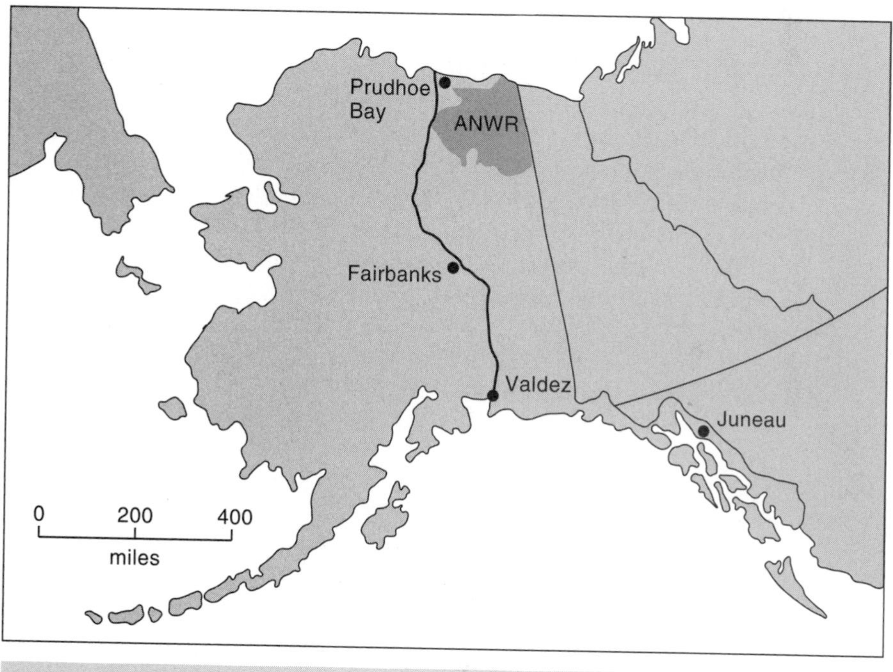

D. Natural Gas

Natural gas is a mixture of light hydrocarbons, primarily methane (CH_4). As with crude oil, it is formed from decayed organic material. It may be mixed with oil (at pressures found in the reservoirs) or trapped in regions in which crude oil is not abundant. Natural gas found alone in reservoirs is called *nonassociated gas*, and when it is found in the same reservoir as crude oil it is called *associated gas*. Some theories suggest a nonbiological origin of natural gas, coming from deep within the earth. (The ramifications of such theories would lead to different methods of prospecting and different locations.) The first well dug in the United States for natural gas was near the shores of Lake Erie in Fredonia, New York, in 1821. However, before the gas industry could expand, it was necessary to develop a pipeline system to deliver the fuel to the customer. After World War II, a high-pressure pipeline network was constructed to serve the entire continental United States. Vermont was the last state to be included in the network in 1966. Today, the United States has more than 1 million miles of pipeline.

From the end of World War II until the 1970s, U.S. natural gas use had a phenomenal rate of growth. Consumption quadrupled between 1950 and 1970, twice the growth rate of oil. Gas consumption dropped in the 1970s and 1980s as there was a perceived scarcity of natural gas, but now it is growing at about 2% per year. One of the constraints to gas production in the 1970s was the imposition of price controls on interstate sales, many times below the price of an equivalent amount of energy in other forms. Deregulation occurred in 1985 and brought about an increase in U.S. gas production.

Natural gas is inexpensive, clean burning, and available. It is a very good substitute for oil and helps us reduce our dependence on imported oil. Natural gas has many uses—space heating, water heating, as fuel for boilers (industrial and utility), in transportation, and as chemical feedstock (for ammonia, fertilizers, plastics, synthetic rubber, etc.) Figure 7.12 gives a breakdown of the uses for

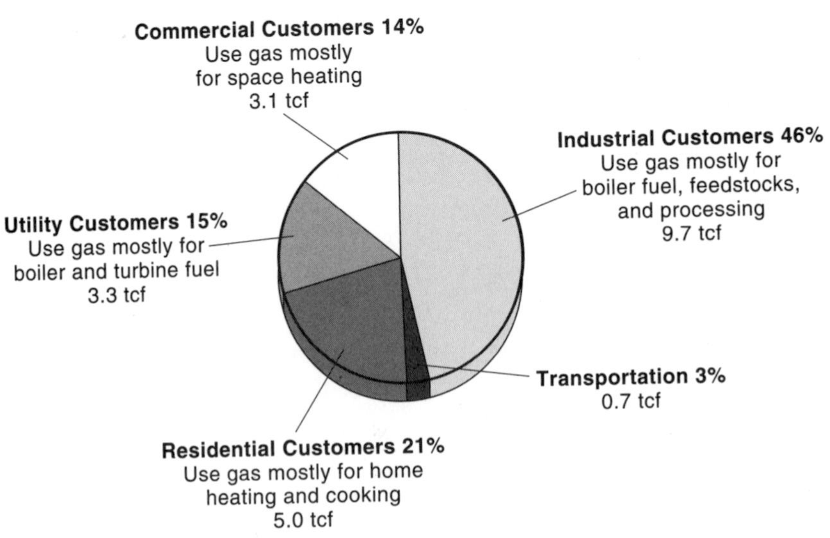

FIGURE 7.12

U.S. natural gas consumption by end-use sector: 1999. (UNITED STATES ENERGY INFORMATION ADMINISTRATION)

gas in 1998 for different sectors. Natural gas accounts for more than 50% of the direct fossil-fuel inputs to the residential, commercial, and industrial sectors.

World reserves of natural gas are estimated at more than 5000 trillion cubic feet (tft^3), enough to last 61 years at the current consumption rate of 82 tft^3/yr. The greatest reserves by far are found in Russia, estimated to be about 1700 tft^3. In the United States, 50% of the gas is found in the Gulf Coast region. United States annual gas production peaked in 1973 at 24 tft^3, leveling off in the late 1980s at less than 20 tft^3. In recent years, additions to reserves and production have been about the same, leading to larger estimates of supply and increased use of natural gas in many areas, including as fuel for electrical-generating facilities (Fig. 7.13).

Additional unexploited resources of natural gas in the United States can be found, albeit at substantially higher prices. Such sources are from (1) gases trapped within coal beds, (2) Devonian shale that underlies much land in the eastern United States, and (3) gas in so-called "tight sands" in the Rocky Mountains. As opposed to conventional gas deposits, in which gas is obtained by drilling a well into the reservoir, the gas in these areas is found trapped within material that has very low permeability (small chance of penetration) to natural gas. Thus, if there are no natural channels through which the gas can flow to the well, the shale or tight-sand strata must be fractured to make channels in order to stimulate gas flow. This can be done by explosives or by water injected into the well under high pressure. Technology for these schemes is still being developed.

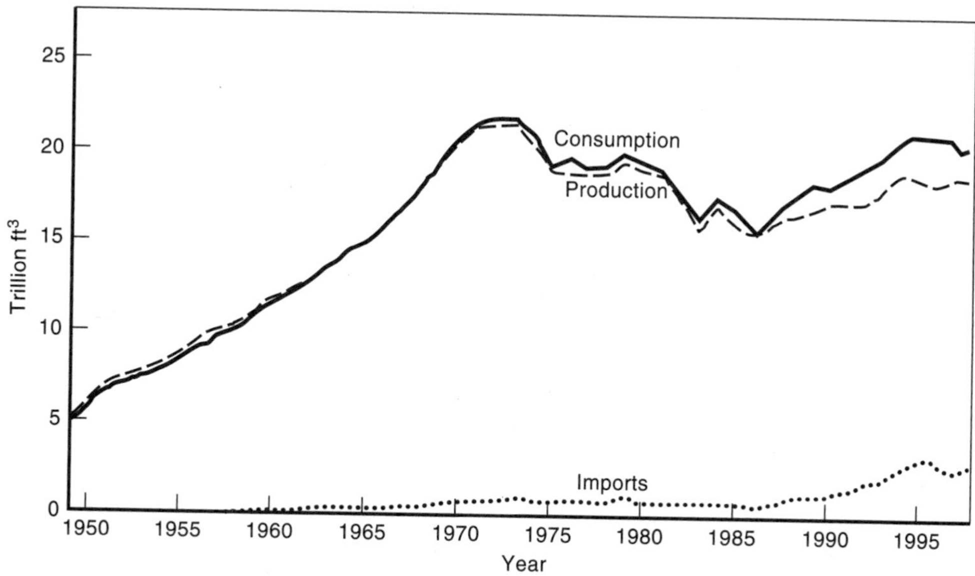

FIGURE 7.13

U.S. natural gas consumption by year: 1950–1999. (UNITED STATES DEPARTMENT OF ENERGY)

Natural gas used to generate electricity in the 1990s grew by about 22%, and is expected to continue to increase in the future. Gas-fired electrical-generating units are less expensive, are more environmentally benign (they produce almost no SO_2 and only a third of the CO_2 of a coal-fired plant of the same size), and have shorter construction times. Although the United States now has excess electrical-generating capacity, this excess is diminishing, maybe disappearing by the early 2000s. Gas-fired units will be smaller (100–200 MW) and allow utilities to invest in new capacity almost as they need it, minimizing the costs of producing electricity. Many of the new gas-fired units will not be owned by utilities, but instead will be part of "co-generation" efforts by independent power producers. (*Co-generation* involves both the generation of electricity from a turbine-generator and the productive utilization of thermal energy from the exhaust gases. The steam or process heat is sold to a nearby industry. See Chapters 10 and 11.)

E. Coal: An Expanding Role

Coal is America's most abundant fuel. The United States has been called the "Saudi Arabia of coal." Figure 7.14 shows world recoverable reserves of coal; the United States has about one quarter of the total reserves. The U.S. Geological Survey estimates our coal resources at more than 3 trillion tons, with approximately 300 billion tons of this being recoverable economically with present technology. Of recoverable U.S. fossil fuel reserves, 80% are coal, compared to less than 3% for oil and 4% for natural gas. Yet today coal supplies only 23% of our energy needs, down from 70% in 1925.

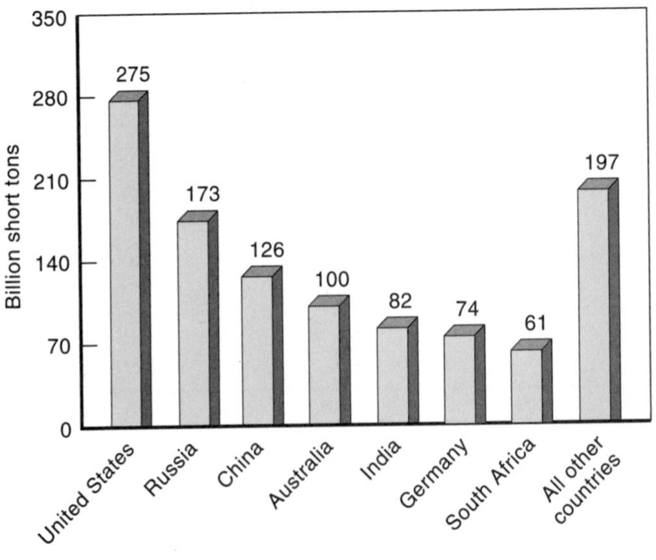

FIGURE 7.14
World recoverable reserves of coal. (UNITED STATES ENERGY INFORMATION ADMINISTRATION)

EXAMPLE

If all of our energy needs were to be supplied by coal, then at our present level of energy consumption, how long would our coal reserves last?

Solution

Current U.S. energy use is about 97×10^{15} Btu/y. One pound of bituminous coal provides about 13,000 Btu, so one ton releases 26×10^{6} Btu. In 1 year, the number of tons consumed would be

$$\frac{97 \times 10^{15}\ \text{Btu/y}}{26 \times 10^{6}\ \text{Btu/ton}} = 3.6 \times 10^{9}\ \text{tons/y}$$

Therefore our reserves would last

$$\frac{300 \times 10^{9}\ \text{tons}}{3.7 \times 10^{9}\ \text{tons/y}} = 81\ \text{y}$$

This is a misleading figure, because (1) coal could not supply all of our needs (some applications require liquid fuels), (2) we have not included any losses demanded by the second law of thermodynamics, such as the approximate 65% loss in the generation of electricity, and (3) we have assumed a zero–growth rate in energy consumption. A growth rate of energy consumption of 3% per year would reduce this number to 50 years, while 5% per year growth yields 35 years to expiration.

Twice as much of our energy consumption comes from oil as from coal. Broader substitution of coal for oil is not a simple matter. Much of our petroleum goes into the transportation sector. Liquids made from coal do offer some promise, but research and development is slow and the economics are currently unfavorable. Ninety percent of coal used today is burned by utilities and independent power producers to generate electricity. In industrial boilers, environmental restrictions have hampered conversion to coal. It seems increased coal use is constrained by demand and not supply.

Types of Coal

Coal was formed from plant material that accumulated in swamps millions of years ago. This vegetation decomposed into peat; as the land subsided, the peat was covered by mud and sands, which formed the mudstones and sandstones found on top of coal seams today. Over thousands of years, the peat was compacted by geological pressures and gradually transformed into the present coal seams. It is estimated that 20 ft of plant material were required to form 1 ft of

Table 7.3 RANKS OF COAL*

Rank	Carbon (%)	Energy Content (Btu/lb)
Lignite	30	5000–7000
Subbituminous	40	8000–10,000
Bituminous	50–70	11,000–15,000
Anthracite	90	14,000

*(P. Averitt, U.S. Geological Survey Bulletin 1412, 1975)

coal. Coal comes in four main classifications or **ranks,** according to the amount of carbon it contains (Table 7.3). The youngest coals are called **lignites.** The geological pressures from the ground above and temperatures have been lower for these, and so they have high water content and lower heating values. Under increased heat and pressure, **subbituminous** coal is formed. Although their water content is high, these coals are of current interest because of their low-sulfur content and low mining cost. They are found primarily in the surface mines of the Great Plains. With additional pressure and heat, the next step in the formation of coal yields **bituminous** coal, the most plentiful type of coal. For this type, the heating value is high. Large deposits occur in the eastern and midwestern United States. However, its sulfur content tends to be high—more than 2% by weight. Finally, we have **anthracite** coal, a very hard coal with a high heating value. It was popular for home heating because it lacks dust and soot and burns longer than other types of coal. However, the supplies of anthracite are very limited and are now found mainly in Pennsylvania. In each stage of development, the percentage of carbon in the coal increases.

Coal Production and Consumption Patterns

Consumption of coal in the United States remained relatively constant during the 30 years after the end of World War II, at about 600 million tons per year. However, during this time its percentage contribution to total energy consumption declined dramatically, from about 40% to 18% (Fig. 7.15). This decline was primarily a result of the availability of clean-burning and inexpensive oil and natural gas. The markets drastically changed as well. Railroads were the largest single users of coal in the 1940s, consuming 125 million tons annually, but today their use of coal is negligible as diesel and electric engines have replaced the coal-fired steam locomotive. Industrial and residential uses also dropped appreciably during those 30 years. The only sector whose use of coal increased was the utilities, going from 50 million tons in 1940 to 1037 million tons in 1998. Today, electric utilities consume about 90% of the total coal produced. Coal's contribution to domestic energy consumption has been rising since the middle

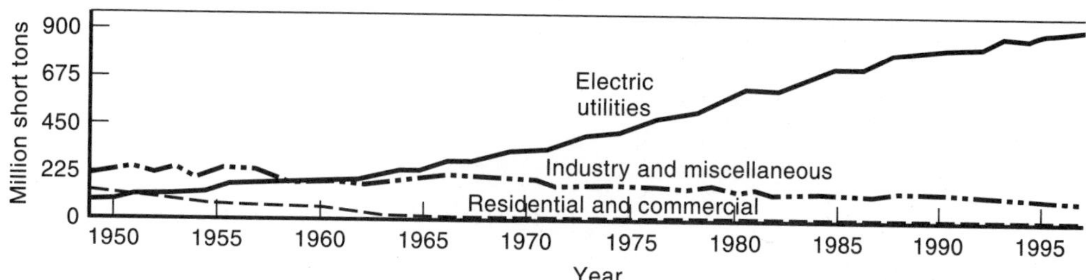

FIGURE 7.15

U.S. coal consumption by sector: 1950–1998. (UNITED STATES ENERGY INFORMATION ADMINISTRATION)

1970s, growing at about 4% per year. United States coal production reached 1000 million tons in 1996 and is expected to continue at this level or higher.

Coal production has undergone significant geographical shifts since the 1940s. In general, coal production has shifted to the western United States, and from underground to surface mines. Figure 7.16 shows the U.S. coal supply

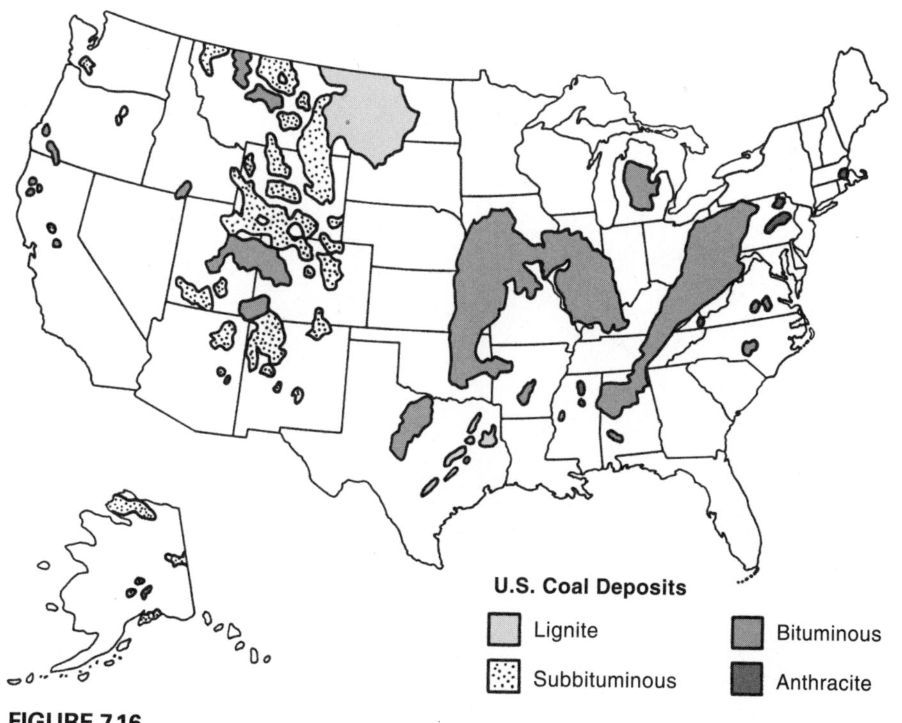

FIGURE 7.16

U.S. coal deposits. (NATIONAL MINING ASSOCIATION)

Underground coal mining equipment. (AMERICAN PETROLEUM INSTITUTE)

regions by type of coal. Although more than half of our coal production today comes from the Appalachian region, the low-sulfur content coal of the Midwest and West will show most of the growth in the future because of the sulfur-emission limitations of the Clean Air Acts. (See Chapter 8.)

Strip Mining

Coal occurs in seams or beds, varying in thickness from inches to more than 100 ft. It is mined underground and in surface or strip mines. About 60% of the coal produced today is by strip mining, from seams lying fairly close to the earth's surface (Fig. 7.17). In strip mining, the earth and rock above the seam—called

FIGURE 7.17
Aerial view of a Montana strip mine. (U.S. DEPARTMENT OF THE INTERIOR)

the "overburden"—is removed and placed aside. The exposed coal is broken up, loaded into trucks, and hauled away. Ideally, the overburden is then replaced and the surface restored for future use by planting. The machines used to remove the overburden are gigantic (Fig. 7.18). Their large draglines can reach more than the length of a football field and pick up 180 yd^3 (about half the size of a house) in a single bite.

The strip mining of coal poses some important environmental problems, the resolutions of which are critical to expanded coal production. Strip mining in the eastern United States has wrought some catastrophic effects in the past. Large areas of land have been left with no use except as mudholes. The removal of the topsoil and its vegetation, without replacement, leads to erosion. The exposed land is not capable of supporting vegetation or, consequently, wildlife. Today, federal legislation (Surface Mining Control and Reclamation Act of 1977) requires reclamation of strip-mined areas, returning the land to its original productivity. Reclamation is done by carefully replacing the topsoil and revegetating it. However, it is still not clear that the disturbances to the land can be totally rectified. One problem is the amount of water available for revegetation. In dry areas (with a rainfall of less than 10 in. per year), revegetation might be impossible. Mining may also upset local water supply and drainage systems. "Acid mine drainage" is another difficult problem. In this situation, the sulfur in the exposed coal combines with oxygen and water vapor to form sulfuric acid, H_2SO_4. The acidic water is hazardous to surrounding vegetation and aquatic life.

Strip mining in the western states will rise more sharply in the years ahead. Western surface mines have coal seams that are 50 to 100 ft thick (compared to 5- to 10-ft seams in the east), the overburden is only 30- to 40-ft deep, the sulfur

FIGURE 7.18

Dragline used in strip mine to remove coal. (U.S. DEPARTMENT OF THE INTERIOR)

content is low, and surface mining is far safer and less labor intensive than underground mining. The output from a surface mine in tons per hour is approximately eight times the output of an underground mine.

Even with some of the environmental difficulties associated with the strip mining of western coal, it is clear that the role of coal in the nation's energy picture will be increasingly important in the years ahead. What other constraints are there on its expanding role?

Constraints to Coal Supply and Demand

Deterrents to increased use of coal can come from the supply side, the demand side, or both. We will mention only briefly some of these constraints to demonstrate the complexities involved in moving toward increased use of coal.

Constraints on coal *supply* come from several areas. One has already been discussed: the resolution of the environmental impacts associated with surface mining. The speed of the development of western coal depends on the success of rehabilitating these lands. Social problems also will arise as a result of rapid coal development in the West. Boom towns will develop as a result of a large influx of miners and construction workers. In many instances this will lead to overloading of school systems and public services. A detriment to underground mining is the availability of coal miners. Underground coal mining is the most dangerous major occupation today, and so it might be difficult to attract the new workers needed for the projected increase in the work force.

As with all large endeavors, things cannot change overnight. It takes three to seven years to open a large coal mine, and even this lead time is being lengthened by a several-year backlog of orders for new mining equipment. Once the coal is mined, there must then be sufficient means for transporting the coal to the population centers of the East. Currently, 65% of all coal mined is shipped by rail. Consequently, thousands of new hopper cars and locomotives are needed.

Increased *demand* for coal in the short term will come from its increased use for the generation of electricity. Today, 80% of the electric power generated from fossil fuels comes from coal. However, major constraints to increased coal use are posed by existing air quality standards, in particular SO_2 emission standards. *Acid rain* (from chemical reactions between sulfur and nitrogen oxides and water vapor in the atmosphere) is a major issue today, especially in the eastern United States, Canada, and western Europe. Harmful effects to fish life and forests have been substantiated. However, sulfur dioxide emissions have declined over the past 15 years, as a result of the new pollution-control devices and the use of cleaner burning coal and coal combustion techniques (such as fluidized bed combustion) as well as increased use of gas-fired units. Reliable SO_2 control devices called scrubbers or flue gas desulfurization units can remove as much as 90% of the sulfur dioxide. (See Chapter 8.) Other concerns with the use of coal (and oil and gas) are climatic effects resulting from CO_2 emissions.

F. Future Sources of Oil

Synthetic Oil and Gas from Coal

Even with the vast reserves of coal available to the United States, there are many cases in which coal cannot be used directly as a source of energy, since it is not a fluid and it is not as clean burning as oil or gas. In large installations such as power plants, scrubber-system technology to reduce SO_2 emissions can make coal a clean substitute for oil or gas; but in such sectors as transportation, solid coal is not a viable choice. However, coal can be converted into forms of oil and natural gas that are called synthetic fuels or "synfuels." The **gasification** of coal is a straightforward process that entails adding hydrogen to the coal. It is an old process, first used in the early 1800s in various "gas works." This "coal" gas was produced by heating coal in the absence of air, yielding a gas whose heating value was about 500 Btu/ft^3, about half that of natural gas. In the United States, several demonstration coal gasification units have been constructed that can convert 200 to 400 tons per day of high-sulfur coal into gas.

The conversion of coal into oil, or syncrude, is called **liquefaction.** As in gasification, the basic process is the heating of coal in the presence of hydrogen, with the separation of the gases and liquids produced. Several programs are in the demonstration stages, but the cost of producing liquid fuels from coal is still quite a bit higher than the market price of oil. Even though the technology is not currently economically attractive, one must realize that it takes a good deal of time to develop such technologies and synfuel production cannot just be turned on like an oil spigot to meet demand for liquid or gaseous fuels.

Oil Shale and Tar Sands

More than 100 years ago, Native Americans in the western United States demonstrated an interesting phenomenon native to their region: the burning of rocks! Medicine salespeople in the eastern states in the past century marketed certain rocks as providing both good lamp oil and ideal medicine. Both groups were making use of rock that contained a substantive amount of oil: **oil shale.** Millions of years earlier, nature had trapped decaying vegetation in sedimentary rock, some deposits of which are so rich in shale oil—an organic compound called "kerogen"—that they actually can burn. Normally shale oil is produced by crushing and heating the rock, producing 15 to 30 gallons of oil per ton. As early as 1855, the Mormons in Utah produced oil from shale, but interest waned with the discovery of oil in Pennsylvania that was easier and cheaper to produce.

Deposits of shale in the United States constitute a huge resource of oil. In the rich deposits alone (those with greater than 30 gallons per ton of shale), there are about 110 to 130 billion barrels of oil—several times the current estimates of U.S. recoverable oil reserves. Most of the shale that is rich in oil is found in one geological formation: the Green River area of the Rockies (located in adjoining sections of Colorado, Utah, and Wyoming). Resource estimates of 2000 billion barrels of shale oil have been made for this region, with economically recoverable amounts placed at about 600 billion barrels. World resources are placed at about three times this figure.

The process for extracting oil from shale is relatively simple. The shale is mined, crushed, and heated to 330° to 480°C (800° to 900°F)—called *retorting*—to release the oil. The process is straightforward since there are no high pressures or difficult catalytic procedures involved. Most of the shale is mined underground, as the overburden is sometimes 300 m thick, and the reprocessing is done on the surface.

Another important source of crude oil is the *tar sands*. These are similar to oil shale in that their oil cannot be recovered by drilling in conventional methods. Tar sands contain a highly viscous asphalt-like oil, called "bitumen." To recover the oil, the sands are mined and then mixed with hot water or steam to extract the bitumen. The largest resources of tar sands in the world are found in Alberta, Canada. These deposits are estimated to contain about 900 billion barrels of oil. The United States' tar sands contain about 25 billion barrels, 1% of the world's resources.

G. Summary

Fossil fuels are projected to constitute the majority of our energy supplies well into the 21st century. Estimates of the amount of a resource thought to exist are always tenuous, especially for petroleum and natural gas. Estimates are usually given for *reserves,* which are those resources that are well known through geologic exploration and are recoverable at current prices and with current technology. Forecasts of future resource production are often made by using past history; the Hubbert bell-shaped curves are such examples. Of special interest is petroleum, which supplies 40% of the world's energy needs today. The Hubbert curves suggest that by the middle of the 21st century, oil production will be only 10% of present levels.

Oil and natural gas accumulate beneath the earth's surface in reservoir rocks such as sandstone and are capped with impermeable rock that prevents their upward movement. To increase petroleum production from a reservoir, enhanced recovery techniques are used. "Secondary" recovery uses water to increase the natural pressures within a reservoir, while "tertiary" recovery uses steam, gas, or chemicals to increase the pressure and free the oil from the rock.

The United States, with more reserves of coal than any other country, is called the "King of Coal." Coal was formed from decaying plant material and is classified according to the amount of carbon it contains. Bituminous coal has a relatively high-heating value and is our most plentiful type. Coal appears in seams or beds. When the seams lie close to the earth's surface, they can be extracted by strip mining. More than half of the coal mined in the United States today is recovered by strip mining, primarily in the western states. Such coal has a relatively low-sulfur content. The demand for coal will be constrained primarily by environmental considerations because of its SO_2 and CO_2 emissions.

Coal can be converted into oil and natural gas by liquefaction and gasification processes. Expansion to commercial production is unlikely unless the prices of these "synthetic fuels" are reduced. Similar arguments can be made about oil that is found in shale rock in the western United States. Enormous deposits of shale oil exist, but environmental and economic factors will inhibit their development in the near future.

Internet Sites

For an up-to-date list of Internet resources related to the material in this chapter, go to the Harcourt College Publishers website at **http://www.harcourtcollege.com.** The links are in the *Energy: Its Use and the Environment* site on the Physics page. General energy related sites and some guidelines for using the World Wide Web in your class are on the inside front cover of this book.

References

Chapter 7

Atwood, G. 1975. The Strip Mining of Western Coal. *Scientific American,* 234 (December).
Borowitz, S. 1999. *Farewell Fossil Fuels: Renewing America's Energy Policy.* New York, Perseus Books.
Cochran, N. 1976. Oil and Gas from Coal. *Scientific American,* 235 (May).
Cuff, D., and W. Young. 1986. *The United States Energy Atlas.* 2nd ed. New York, Macmillan.
Dargay, J., and D. Gately. 1995. The Response of World Energy and Oil Demand to Income Growth and Changes in Oil Prices. *Annual Review of Energy,* 20.
Darmstadter, J., and R. W. Fri. 1992. Interconnections Between Energy and the Environment. *Annual Review of Energy,* 17.
Energy for Planet Earth. 1990. *Scientific American,* 263 (September).
Gever, J., R. Kauffman, D. Skole, and C. Vorosmarty. 1986. *Beyond Oil: The Threat to Food and Fuel in the Coming Decades.* Cambridge, MA, Ballinger.
Gordon, D. 1991. *Steering a New Course.* Cambridge, MA, Union of Concerned Scientists.
Helm, J., ed. 1990. *Energy: Production, Consumption, and Consequences.* Washington, D.C., National Academy Press.
Hodgson, B. 1990. Alaska's Big Spill. *National Geographic* (January).
Hubbert, M. K. 1969. Energy Resources. *Resources and Man.* San Francisco, W. H. Freeman.
Hubbert, M. K. 1971. Energy Sources of the Earth. *Scientific American,* 224 (September).
Kleinbach, M., and C. Salvagin. 1986. *Energy Technologies and Conversion Systems.* Englewood Cliffs, NJ, Prentice-Hall.
Linden, H. R., W. W. Bodie, B. S. Lee, and K. C. Vyas. 1976. Production of High-Btu Gas from Coal. *Annual Review of Energy,* 1.

Menard, W. 1981. Towards a Rational Strategy for Oil Exploration. *Scientific American*, 244 (January).
Miller, T. 1999. *Living in the Environment*. 11th ed. Belmont, CA, Brooks/Cole.
Prindle, D. F. 1984. Shale Oil and the Politics of Ambiguity and Complexity. *Annual Review of Energy*, 9.
Raven, P., L. Berg, and G. Johnson. 1998. *Environment*. 2nd ed. Philadelphia, Saunders College Publishing.
Rogner, H. H. 1989. Natural Gas as the Fuel for the Future. *Annual Review of Energy*, 14.
Rose, A., W. Labys, and T. Torries. 1991. Clean Coal Technologies and Future Prospects for Coal. *Annual Review of Energy*, 16.
Scientific American. 1990. *Managing Planet Earth*. New York, W. H. Freeman.
Yergin, D. 1991. *The Prize*. New York, Simon & Schuster.

QUESTIONS

1. What are the reasons for the decrease in U.S. crude oil production since 1970?
2. The routes for oil and gas pipelines, as discussed in the section on "Pipeline Politics," show both one that goes to the Mediterranean and one that goes to the Persian Gulf. Considering world economic growth, why would this latter route have some advantages?
3. Use the World Wide Web to determine the status of drilling for oil in the Arctic National Wildlife Reserve. What percentage of U.S. oil reserves are estimated to be found there?
4. The 1989 Alaska oil spill caused a good deal of environmental damage to the wildlife and fish of that area. Cleanup efforts have been massive. From the World Wide Web, determine what is currently known about the success of such efforts.
5. With deregulation of the electric utility industry, many new electrical-generating projects are being planned by a host of companies. Identify what electrical power plants in your state have been proposed that will use natural gas. Note completion dates and power output.
6. What are the advantages of strip mining? What problems will there be in restoration of the mined lands?
7. What social problems might be expected in the boom towns of the West that could arise because of strip mining or oil-shale expansion?
8. What constraints might be expected to decrease demand for coal production?
9. Discuss the analogy of sipping soda with a straw and the enhanced recovery of oil.
10. What are the differences between secondary and tertiary enhanced recovery methods?

PROBLEMS

1. Suppose the cost of oil is $30 per barrel. If the price increases by 10%, how much would you expect the cost of a gallon of gasoline to increase by?
2. If all of our needs for oil were to be provided by that available from the estimated reserves located in the Arctic National Wildlife Reserve, how long would that supply last, assuming no growth in demand?
3. A gas-fired turbine generator for producing electricity has an efficiency of about 50%. How many cubic feet would be needed to produce 1000 MWe for 1 year? (Note conversion factors in Appendix B.)
4. If the demand for electricity continues to grow at about 2% per year, how many more MWe would be needed to meet that demand in 10 years? If all of that need was met with gas-fired turbine generators, how much more gas (in ft^3/yr) would be required? What fraction of our present use of natural gas is this?
5. For a gasification project being planned, 26,000 tons of coal per day will be used to produce 250,000,000 ft^3/d of natural gas. If the coal is rated at 8700 Btu/lb and the gas at 950 Btu/ft^3, what will be the efficiency of this plant?

SPECIAL TOPIC

Physics of Oil and Gas Exploration

Exploration for oil and natural gas trapped beneath the earth's surface has progressed over the past many decades from the hunch of the wildcatter to the scientific methods of the geophysicist. Two methods used to infer locations of oil and natural gas will be discussed in this section, gravimetric and seismic, with an emphasis on the principles of the latter.

Gravimetric Exploration

The gravitational force between two objects is proportional to the mass of each body and inversely proportional to the square of the distance r between them:

$$F = \frac{Gm_1m_2}{r^2}$$

where G is the universal gravitation constant.

If the earth were a perfect homogeneous sphere, the gravitational force on an object (its weight) would be the same everywhere on the earth for the same altitude. However, the earth is not a perfect sphere (it looks like a pear, with a bulging at the equator and a flattening at both poles) and so the distance from the center of the earth to sea level varies. The density of the earth also is not the same from place to place, because of local deposits of such matter as iron ore or oil. This latter fact is used in petroleum exploration. In gravity surveys, local variations in the gravitational force at the same altitude are observed with the use of a **gravity meter,** which might be a mass suspended from a spring (Fig. 7.19).

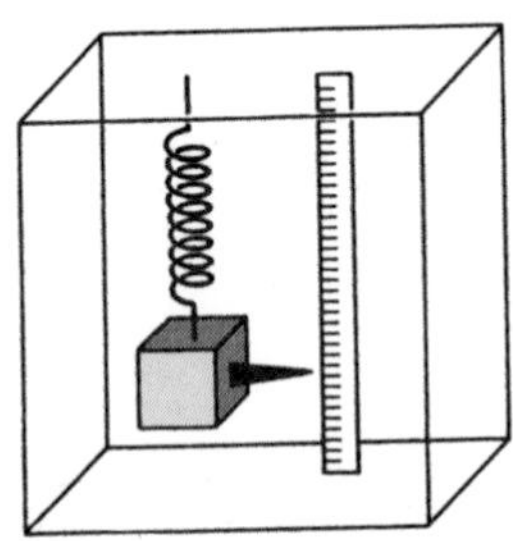

FIGURE 7.19
A gravity meter. The force acting downward on the weight extends the spring. Variation in the earth's local density can be measured by observing changes in the position of the pointer.

As the mass is moved from place to place, its weight changes due to irregularities in the earth's density, and so the stretch of the spring changes. From these variations, petroleum deposits can be located. For example, a buried salt dome in shale rock would produce a small decrease in the attraction of gravity. Salt domes are particularly interesting in that petroleum and natural gas deposits have been discovered around such structures.

Seismic Exploration and the Nature of Waves

A more popular and modern technique in petroleum exploration is the use of **seismic waves** which provide geophysicists with a powerful tool for imaging rock structures inside of the earth. This method uses shock waves that are created in the ground by an explosion or ground thumping. The traveling shock waves strike discontinuities in the physical properties of the underlying rock strata and are partially reflected back to the surface and detected. The time required for the reflected wave to return to the surface indicates the depth of the discontinuity. From measurements over a large area, a profile of the different structures beneath the earth's surface for that locality can be put together, from which possible sites for entrapped oil and gas can be inferred. Before examining this "echo-sounding" procedure in more detail, we digress for a brief discussion of sound waves.

Transverse and Compressional Waves

The subject of waves was introduced in Chapter 4 in a discussion of electromagnetic radiation. Waves are a very common phenomenon in everyday experience, and they tie together several areas of physics. A common example is water waves, which can be produced by dropping a rock into a pool of still water. Ripples spread out from the source of the disturbance. It is not the water that travels along with the ripples, but the disturbance itself. The water molecules move up and down as the disturbance passes; a cork on the water moves up and down as the ripples pass by it. The disturbance carries energy transmitted from the source through the medium (water in this case).

Water ripples on a pond.

Another example of wave motion is the waves on a rope that are produced when your hand gives a sudden jerk on one end (Fig. 4.14). Here the disturbance is in the form of a pulse or hump that is transmitted down the rope, away from the source. Again, the rope particles themselves do not move along the rope, but only move up and down as the disturbance passes. These waves, as well as those for the water ripples, are called **transverse waves** (Fig. 7.20a). This means that the particles of the medium move perpendicular or transverse to the direction of travel of the wave.

A shock wave is a different type of wave. The disturbance, which might be energy emanating from an explosion, compresses the surrounding medium (e.g., air), which compresses the adjoining medium, and so forth. The disturbance propagates out from the initial vibration in the form of a **compressional** or **longitudinal wave** (Fig. 7.20b). As in the case of transverse waves, the particles of the medium do not carry the energy transfer by their own motion but simply vibrate back and forth along the direction of the wave as the disturbance passes.

Velocity of Sound Waves

Common characteristics of all waves are their wavelength, frequency, and velocity. These three quantities are related through the formula

$$\text{Velocity} = \text{wavelength} \times \text{frequency}$$

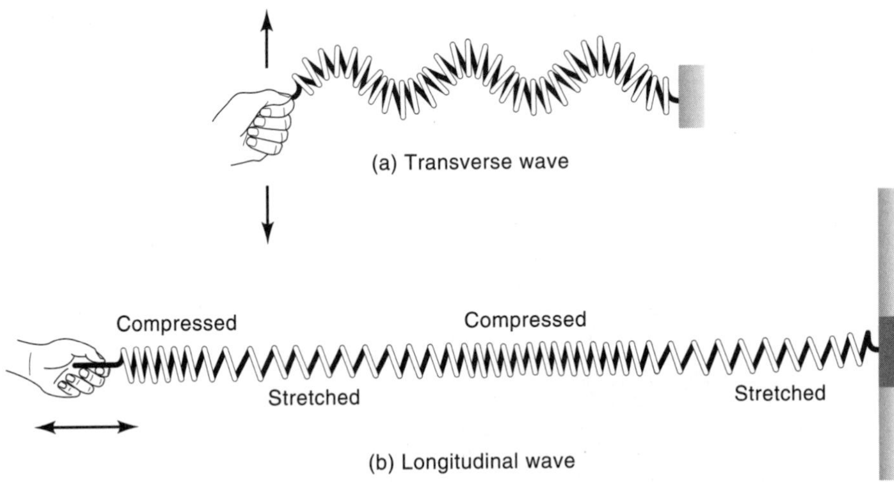

FIGURE 7.20

(*a*) Transverse wave: The motion of the medium is perpendicular to the direction in which the wave travels. (*b*) Longitudinal wave: The motion of the medium is in the same direction as that in which the wave travels.

Table 7.4 SPEED OF SOUND IN VARIOUS MATERIALS

Material	Speed (m/s)
Air	330
Water	1460
Clay rock	3480
Copper	3560
Brick	3650
Steel	5000
Aluminum	5100
Granite rock	6000

(See Chapter 4.) A fundamental difference between light waves and sound waves is that sound needs a medium to travel in, but light can travel in a vacuum. Sound is related to the mechanical vibration of matter while light is related to the vibration of electric and magnetic fields. If you put a ringing alarm clock into a bell jar and pumped out the air, you would not be able to hear the ring anymore. But you would still be able to see the clock by the light it reflects.

The speed of a wave depends on the nature of the medium in which it is traveling. The speed of sound in air is 330 meters per second (1100 ft/s or about 750 mph), about a million times smaller than the speed of light. In general, the speed of sound is higher in metals and solid material than in air and liquids (Table 7.4). To find out if a train is coming, a useful trick is to listen for its approach by putting your ear to the track, since the metal is a better conductor of sound than air.

Reflection and Refraction of Waves

Seismic exploration for oil and natural gas makes use of another general property of waves—reflection and refraction. The bouncing of a tennis ball off a wall is like the **reflection** of a particle. For a ball or for waves, the law of reflection from a surface is followed: The angle of incidence is equal to the angle of reflection. (Each of these angles is the angle between the waves and the "normal" or a perpendicular line to the surface.) Figure 7.21 illustrates this example for light being reflected from a flat mirror.

When a wave in one medium encounters another medium, such as light going from air into water or a sound wave going from material of one density

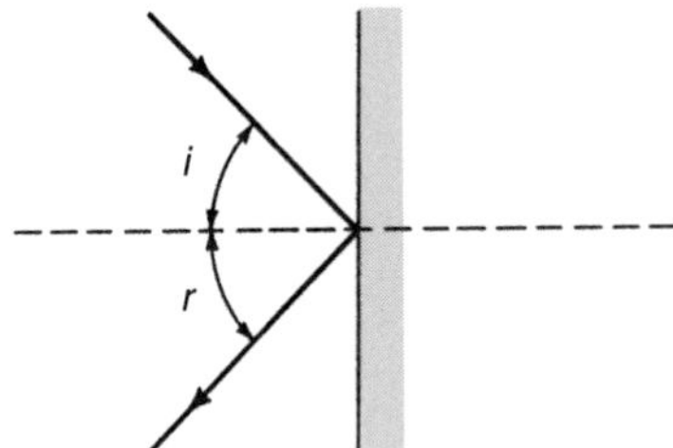

FIGURE 7.21
The law of reflection for waves: The angle of incidence i = the angle of reflection r.

into another, part of the wave is reflected and part is transmitted into the other medium. The wave transmitted into the other medium is deflected, or **refracted,** because of the difference in speed of the waves in the two media. If you place a pencil in a glass of water, the pencil appears to bend at the surface for this reason. When light enters a denser medium, the light is bent toward a line normal to the surface. Glass lenses focus light by the property of refraction. In looking for fish underwater from above the water's surface, you should know that the apparent position of the fish is not where it really is: The light is bent away from the normal as it enters the less dense air, making the fish appear shallower than it really is (Fig. 7.22).

Seismic Exploration Methods

With this background, let us return to our topic of seismic exploration. Discontinuities or interfaces between different types of rock are important in locating oil deposits. (Recall that rock structures can trap oil.) A seismic wave produced by an explosion or thump on the surface will propagate

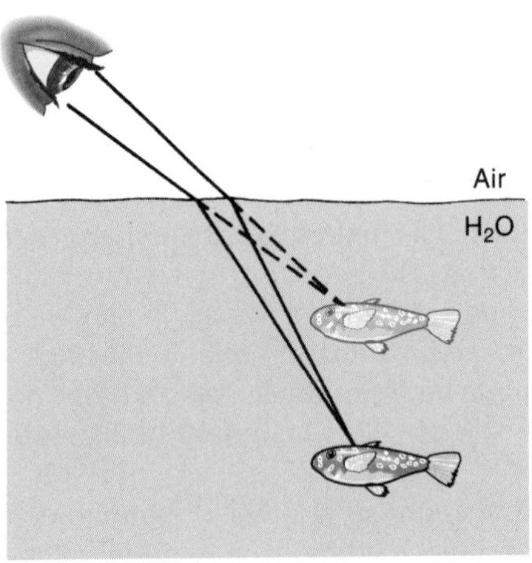

FIGURE 7.22
Submerged bodies seem closer to the surface than they actually are because of refraction of light at the air–water boundary.

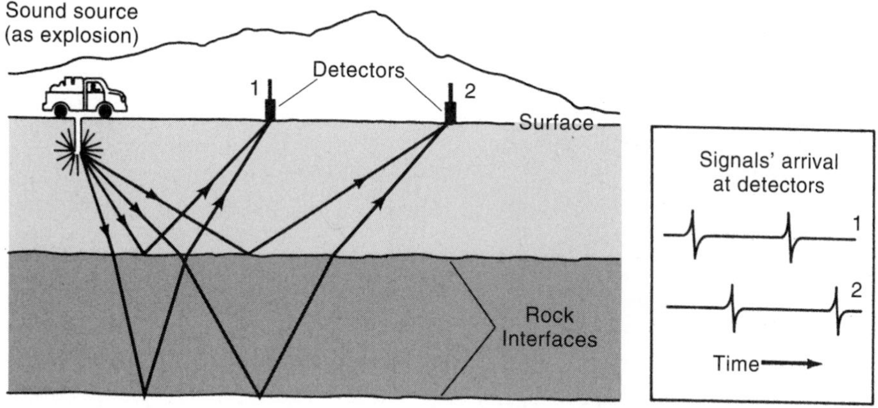

FIGURE 7.23

Seismic exploration for petroleum. The arrival of sound waves at the detectors after their reflection and refraction from underground rock boundaries will yield information about the surrounding geologic structure; if the explorers are lucky, it will reveal the presence of oil reservoirs. (AMERICAN PETROLEUM INSTITUTE)

through the ground. The wave will be partially reflected back to the surface when it encounters a physical discontinuity in the rock structure or a liquid reservoir (Fig. 7.23). Part of the wave will be refracted into the next layer and then back later into the top layer. At the surface, geophysicists collect data on the reflected and refracted shock waves with an array of special microphones (seismometers) located at different distances from the explosion that convert the mechanical vibrations into electrical signals to be recorded by a computer. The time required for the reflected wave to return to the surface indicates the depth of the discontinuity, because distance = speed × time. The times required for the refracted waves to reach seismometers gives information about the speed of the waves in that medium. The intensity or strength of the returning signal also provides information on the structural features of the crust. Porous rock in which natural gas is trapped reflects a much stronger signal than denser rock or rock filled with water. Detecting oil directly is very difficult due to the waves being insensitive to water and oil proportions.

The result of "shooting" seismic data is a series of profiles that provide a two-dimensional cross-sectional image, or slice, of the earth's crust. A seismogram (Fig. 7.24) is the recording of a series of pulses for each detector indicating reflected and refracted waves that follow different paths from source to detector. The seismogram is helpful in identifying subsurface geological features from which information on geologic formations favoring the occurrence of oil can be inferred. Seismic surveys are also done at sea.

FIGURE 7.24
Seismic operations print-out map for oil location. Each horizontal line represents a detector at a different distance that records pulse strength (y-axis) versus time of travel of the shock wave (x-axis). (AMERICAN PETROLEUM INSTITUTE)

Airguns, the source of the shock waves, are towed behind a ship. There are some ambiguities in these techniques but such seismic mapping can provide geological information for an underwater area that can result in a 50%-successful drilling rate.

QUESTIONS

1. In conducting a gravity survey over a wide area, how would you distinguish between petroleum and ore?
2. Using the equation for the gravitational force, how does your weight vary as a function of your position on the earth? Would your mass vary in the same way?
3. What are the similarities and the differences between light waves and sound waves?
4. If the frequency of a seismic wave is 100 Hz, what will be its wavelength in granite rock? (See Table 7.4.)
5. If you wished to shoot an arrow at a fish, where do you aim? Why?
6. In using seismic waves for the location of petroleum, how does the geophysicist distinguish between the different rock interfaces?

8

Air Pollution and Energy Use

A. Introduction

Energy resources provide some of the crucial building blocks of a modern society and make possible many of the conveniences that we enjoy today. However, the quality of life is measured by more than material possessions; human health and well-being and the nature of our social systems have important, perhaps fundamental, roles. We realize that we are part of a larger picture and our well-being cannot be separated from the well-being of the planet. In the 1960s "ecology," or the study of the interrelationships between organisms and their environment, was popularized as the public began to appreciate the delicate balances that exist in nature and, in some cases, became aware of the stewardship role that they had a responsibility to exercise.

As introduced in Chapter 1, power production and energy use can bring about significant adverse environmental effects. What harm to our environment, to our health, and standard of living will come from the expanding

Focus On 8.1

EASTERN EUROPE AND THE ENVIRONMENT

"She's so small," Jeffrey whispered.

Dr. Sova's smile was tinged with sadness. "This is a very big baby. Very healthy. She's almost ready to go home. We have many premature births with weights at seven hundred grams, or about one and one-half pounds. Those are the ones who cause us the greatest worry."

The baby was more than tiny. She was so small as to appear incapable of life. A rib cage smaller than his fist. A head that could have fit within the palm of his hand. Incredibly fragile arms and legs and hands and feet, like the limbs of a tiny china doll.

"The problem is not just one of survival," Dr. Sova went on. "The problem is the quality of survival. Cerebral palsy. Blindness. Mental ability. All of these are unanswered questions with premature babies at this stage of their development."

"This region has forty times what is considered to be the maximum safe level of dust in the air," Katya told him, "and sixty times the level of lead in both air and the water. Half of all rivers in Poland are so polluted that they are not even fit for industrial use; their water will corrode the intake pipes. Almost two thirds of Cracow is without any sewage treatment at all; everything is simply dumped into the Vistula River. New studies show that the level of chemicals in the air has reached critical levels."

Air pollution in Copsa Mica, Romania, dubbed the most polluted city in the world. (ANTHONY SUAU/GAMMA LIAISON)

"Sulfur dioxide," Dr. Sova recited. "Carbon dioxide, carbon monoxide, heavy metals, iron, and just plain soot. This region has one ton per square meter of dirt fall from the sky each year, the highest on earth."

"The effect on people's health must be devastating," Jeffrey said.

"Especially the children," Dr. Sova agreed. "Within this region, ninety percent of all children under the age of five suffer from some pulmonary disorder at one time or another. One half of all four-year olds suffer from some chronic disease, two thirds of all six-year olds, and three quarters of all ten-year olds. Again, these figures have only in the past six months become collected. Under the Communists, all records of our children's health were classified top secret, and no such collation of data was permitted. All we could tell you was that too many of our children were ill for too long. Far too long." (Bunn, T. Davis. *The Amber Room.* Minneapolis: Bethany House Publishers, 1992.)

The fall of Communism in 1989 in Eastern Europe brought new-found freedom for millions of people, but also revealed to the world an environmental nightmare of poisoned air, water, and land. In the name of progress and industrial production, environmental concerns had been ignored for decades with long-term effects on the land and its people. Many power plants burned brown coal (lignite) with high sulfur and ash content. Some of the coal contained high contents of cadmium and arsenic. The lack of air pollution control devices, and the age of many of the plants, produced health problems in the young and the old; acid rain damaged millions of acres.

exploitation of our energy resources? One of the most obvious consequences of increased energy consumption is the topic of this chapter—the pollution of the air around us. Air pollution is a concern to almost all people today, especially those who live in metropolitan areas.

Air pollution does not respect boundaries—state or national. It affects people and plants far from its source. Some of the worst cases of air pollution, however, lie much closer to home—in the large urban areas or megacities of the world. Mexico City has the world's worst overall air pollution, while serious levels of SO_2 and particulates exist in Beijing, Rio de Janeiro, Seoul, and Shanghai. Los Angeles has a serious problem with (ground-level) ozone, the primary constituent of smog.

In this chapter we will study the nature and origin of pollutants in the air, the effects of air pollution on human health and the environment, and the steps necessary to reduce those emissions. Air pollution standards and the trade-offs made between energy and consumption, economics, and environmental quality will be considered also.

B. Properties and Motion of the Atmosphere

The air about us consists of a mixture of gases. The concentrations of gases in normal (clean) dry air are given in Table 8.1. The main components of air are the molecules of nitrogen (78%) and oxygen (21%). (There is no such thing as an air molecule.) The atmosphere also contains other substances that are added by natural sources such as growing or decaying vegetation, dust from the soil, and fumes from volcanic eruptions. We will define air "pollutants" as those substances added by humans that are toxic or irritant to animals, vegetation, or property.

Pressure

The earth is immersed in a "sea of air," with 99% of the mass of the atmosphere lying below an altitude of 33 km (20 miles). Compared to the size of the earth, this is a fairly thin covering: If the earth were a circle drawn to the size of a silver dollar, the atmosphere would be the thickness of a pencil line.

Table 8.1 CONCENTRATION OF GASES IN NORMAL DRY AIR

Gas	Concentration (ppM)*
Nitrogen	780,900
Oxygen	209,400
Argon	9300
Carbon dioxide	315
Neon	18
Helium	5.2
Methane	1.0–1.2
Krypton	1
Nitrous oxide	0.5
Hydrogen	0.5
Xenon	0.08
Nitrogen dioxide	0.02
Ozone	0.01–0.04

*ppM = parts per million of gas by volume, which refers to the number of molecules per million molecules of air.

Because we live at the bottom of this sea of air, we experience a pressure on us as a result of the weight of the air above. **Pressure** is defined as force exerted per unit area:

$$\text{Pressure} = \frac{\text{force}}{\text{area}}$$

The units of pressure are pounds per square inch (psi) in the English system and newtons per square meter, called pascals (Pa), in SI.

As you can see from this equation, the pressure is increased if you reduce the area over which a force, such as your weight, is applied. For example, if a woman put all of her weight on the back of a pair of high-heeled shoes, the pressure she will exert on the floor is greater than that exerted by an elephant standing on all four feet! Conversely, pressure can be reduced by increasing the area over which the force is exerted. You have a better chance of crossing thin ice (without falling through) if you crawl because you have reduced the pressure you exert on the ice by increasing the area over which your weight is distributed.

EXAMPLE

A 5-year-old child weighs 48 lb and wears size 1 shoes, each with a surface area of 16 in.^{2}. Compare the pressure exerted by the child to that exerted by a 160-lb man who wears size 10 shoes, each with a surface area of 42 in.^{2}.

Solution

The pressure exerted by the child is 48 lb/(2 × 16 in.^{2}) = 1.5 psi. For the adult, the pressure is 160 lb/(2 × 42 in.^{2}) = 1.9 psi, or 25% more.

In the sea of air that surrounds us, the force on us is the weight of all the air above us. At sea level, atmospheric pressure is 14.7 psi, or about 100,000 Pa. (At the top of Mount Everest the air pressure is about 5 psi.) It would take a column of water 34 ft (or 10 m) high to yield a pressure at its bottom equal to the pressure exerted by air at sea level (a column 20 mi high).

Figure 8.1 shows some examples using air pressure. Atmospheric pressure enables us to use rubber suction cups. A glass of water can be held upside down with a piece of paper on its top because the air pressure acting on the paper is equal to or greater than the water pressure. Drinking with a straw makes use of pressure differences to force the liquid into your mouth: When you suck on a straw, the pressure in your mouth is decreased, so the liquid is

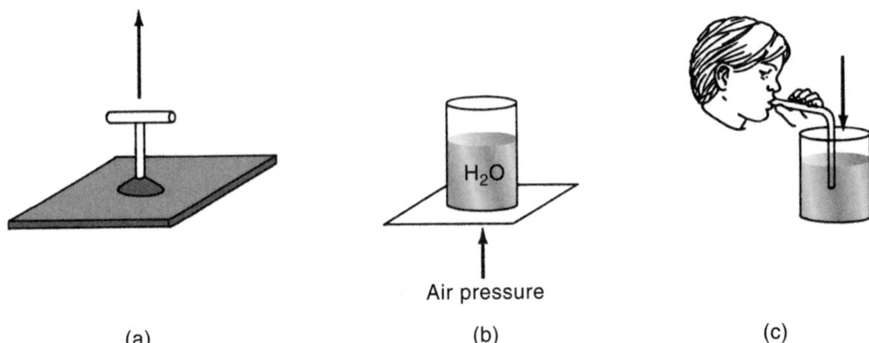

FIGURE 8.1

Examples of atmospheric pressure. (*a*) The maximum weight you can lift with a 4-in.-diameter suction cup is area × pressure $= \pi(2)^2\text{in.}^2 \times 14.7$ lb/in.^{2} = 184 lb. (*b*) Turning a glass of water (covered with a piece of paper) upside down forces some of the air out of the glass. The pressure of the remaining air plus the weight of the water is less than atmospheric pressure, so the water remains in the glass. (*c*) Sipping soda: Atmospheric pressure forces the drink up the straw to the region of lower pressure in your mouth.

pushed up the straw by the larger atmospheric pressure on the liquid at the bottom of the straw. It is not pulled up by the partial vacuum.

Atmospheric pressure is important to meteorologists, as regional differences in pressure cause the weather to move from high to low pressure regions. "The barometer is rising" is usually a sign of approaching good weather. Air pressure is a function of temperature. Variations in temperature on the earth's surface, caused by uneven heating from the sun and by physical features (mountains, oceans, etc.), give rise to pressure differences that produce the winds or horizontal air circulation. The barometer shown in Figure 8.2 is a standard method of measuring atmospheric pressure. (See Further Activities num-

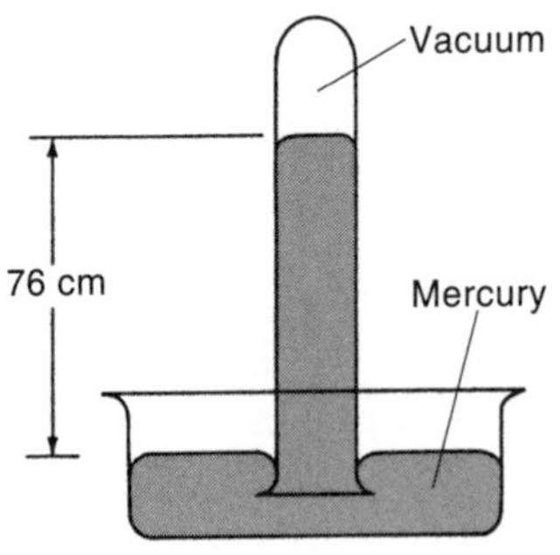

FIGURE 8.2

A barometer. Atmospheric pressure acting on the mercury causes the mercury to rise 76 cm in the tube. (If the fluid was water, this height would be 1000 cm.)

ber 1 at the end of the chapter.) The tube is evacuated at the top so there is no pressure exerted by any air. The fluid in the tube exerts a pressure equal to its density times its height. At equilibrium, this is equal to the pressure of the atmosphere on the fluid in the dish. You can see why mercury rather than water is used as the fluid; mercury is about 13 times denser than water, so a shorter column can be used.

Buoyant Force and Air Temperature Profiles

Vertical motion of the air is a result of the density variations in the atmosphere. Air that is warmed by the ground is less dense than cold air, and so rises. It rises because of the "buoyant force" acting on a volume of warm air (called a "parcel"). The **buoyant force** on an object is the upward force exerted on it by the fluid in which it is immersed. Because of this force, an object immersed in water appears to weigh less than when in air. For example, a block of aluminum that weighs 10 lb in air and appears to weigh 7 lb in water has a 3-lb buoyant force acting on it.

The buoyant force is a result of the difference in pressures on the top and bottom of the object immersed in a fluid (Fig. 8.3). There is a greater pressure pushing up on the bottom of an object than pushing down on its top, because the bottom part is at a greater depth under the fluid. **Archimedes' principle** states that *the buoyant force on an object is equal to the weight of the fluid displaced by that object.* An object will float if the buoyant force on it is equal to its weight. Since weight = density × volume of object, another way of stating this is

> A solid object will float if its density is equal to or less than the density of the medium it is in.

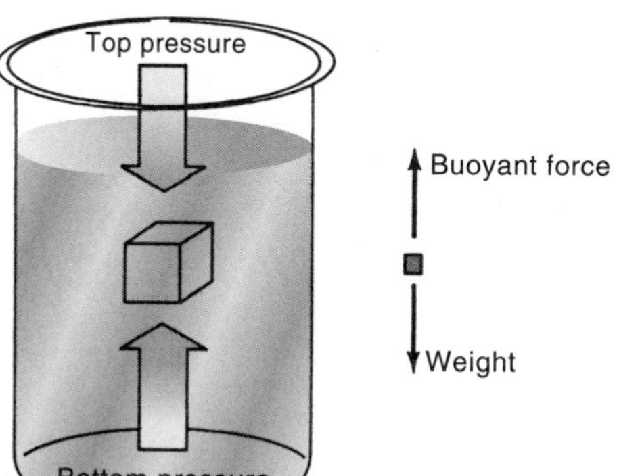

FIGURE 8.3
The buoyant force on a submerged object is equal to the weight of the displaced fluid. This is a result of the difference in pressure between the top and the bottom of the object. An object will float if the buoyant force is equal to or greater than the object's weight.

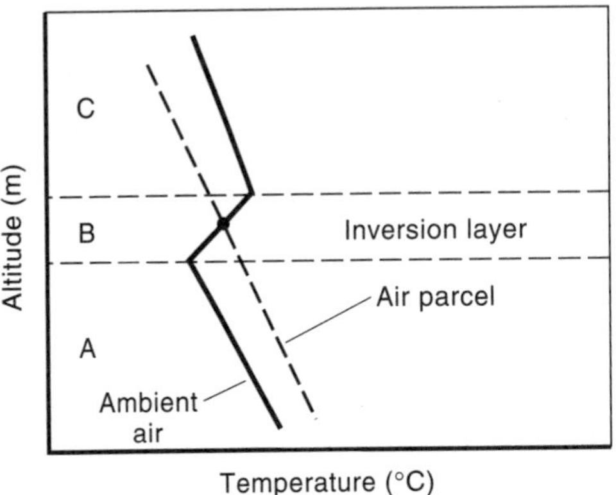

FIGURE 8.4
The temperature of ambient air normally decreases with increasing altitude. A parcel of air will rise until its temperature is equal to that of the surrounding air. An elevated inversion layer (a region in which the temperature of the ambient air increases with altitude) will put a lid on the rising air parcels.

We're accustomed to seeing ice cubes float in water or a beach ball float in a lake; these objects have a density less than that of water. However, we also know that steel and concrete boats will float, even though the densities of steel and concrete are much greater than that of water. This is because the boat is not a solid object and has a large amount of air contained in its hull, so that the density of the *entire boat* is less than that of the water.

If an object's density is less than that of the liquid in which it is immersed, it will settle in a position, partly submerged, such that the weight of the displaced liquid—the buoyant force—is equal to the object's own weight. If the density of the object is equal to the density of the surrounding fluid, then there will be no net force on the object and it will remain at rest at the position in which it is placed. In the context of this chapter, a parcel of warm air will rise until its density equals the density of the surrounding air.

As warm air rises, it expands (becomes less dense) and cools down. The energy used for expansion is taken from the thermal energy of the air. Feel the air released from an inflated bicycle tire tube: It can be quite cold. The normal temperature profile of the atmosphere is shown in Figure 8.4. The air temperature decreases at a rate of about 7°C per kilometer (about 3.5°F per 1000 ft) above the earth's surface. An air parcel will rise because of the buoyant force on it *until* its temperature equals that of the surrounding air, that is, until its density is equal to that of the surrounding air. If the warm-air parcel cools *less rapidly* than the surrounding air, its temperature will always be *above* the temperature of the surrounding atmosphere, so the equilibrium point will not be reached and vigorous mixing of the parcel and surrounding air will occur.

Natural Dispersion of Air Pollutants; Temperature Inversions

The vertical motion of the air, plus the winds, are very important mechanisms for dispersing air pollutants. Vertical mixing will be restricted if a **temperature inversion** exists. This occurs when a warm layer of air lies *above* a colder part, in effect putting a lid over the region and therefore stopping atmospheric dispersion of air pollutants. As the smoke from, say, a power plant ascends into the atmosphere, it expands and cools down. As long as it is warmer than the surrounding air, it will be less dense and so will be buoyed up even higher. However, if it gets to a warmer (less dense) region, it will stop rising. This low-lying temperature inversion layer will trap most of the pollutants below it (Fig. 8.5).

A temperature inversion occurs most commonly because of radiation cooling. During the night the earth continues to radiate and so cools down; the air in contact with the earth will cool to a lower temperature than the air up higher. This temperature inversion is usually dissipated in the morning by the heat of the sun. However, a heavy morning fog will inhibit this dissipation, as the sun's rays will not penetrate sufficiently well to warm up the bottom layer. You might have seen the effects of such an inversion early in the morning as smoke from chimneys stays close to the ground.

Temperature differences are also responsible for general global air circulation. The sun does not heat all regions of the earth uniformly. At the equator the sun's rays strike the ground almost perpendicularly, but they strike at progressively larger angles for larger latitudes. As air is heated near the equator, it rises and moves toward the poles, subsiding as it cools. It is replaced by cooler air, which flows back toward the equator. (This convection is similar to the air circulation patterns in a room.) Patterns are more complicated than this, however, because of the rotation of the earth. The air circulation patterns break down into several "cells," as seen in Figure 8.6, giving us our prevailing westerly winds.

FIGURE 8.5
The sun sets through a layer of heavy smog over downtown Los Angeles.
(CORBIS/BETTMAN)

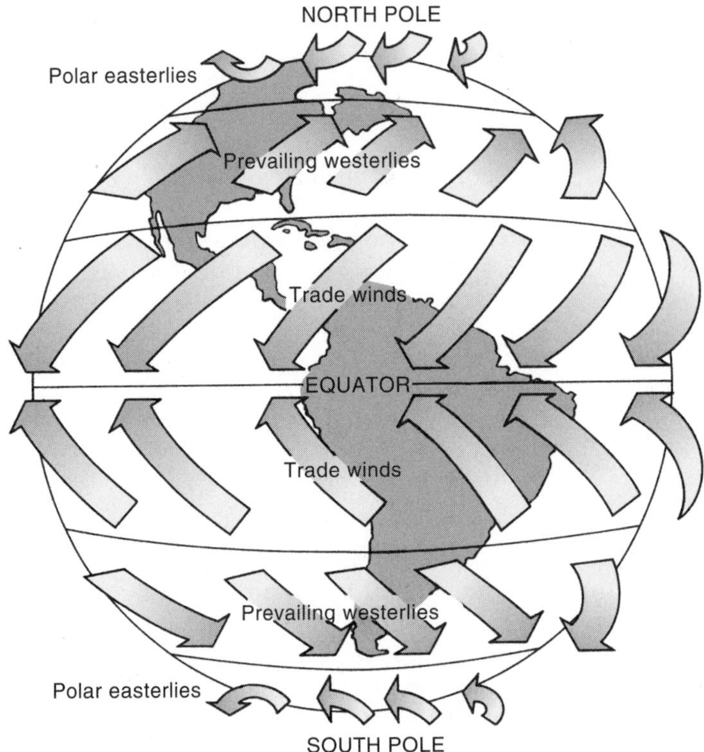

FIGURE 8.6

Earth's wind engine. The rotation of the earth produces a complex wind pattern, as the cold polar area sinks toward the equator and the warm tropical air rises and moves toward the poles.

The general air circulation patterns are important because pollutants can travel large distances before settling out or being absorbed. International complications can arise when airborne emissions from one country affect the environment of another. The overall movement of air in the United States is from west to east. Strong disagreements exist between the United States and Canada over emissions from large coal-fired power plants in the midwest. Evidence indicates that such emissions can lead to acid rain.

C. Air Pollutants and Their Sources

Air pollutants are generally considered to be those substances added to the air by human activities that adversely affect the environment. These pollutants are in the form of gases, small particles of solids (**particulates**), or small droplets of

Stack monitoring to characterize the emissions from a modern coal-fired power plant. (UNITED STATES DEPARTMENT OF ENERGY)

liquid suspended in a gas (called **aerosols**). The pollutants emitted in the greatest amounts by human activities are carbon monoxide, sulfur oxides, particulate matter, hydrocarbons, and nitrogen oxides. Every year more than 150 million tons of these pollutants are emitted by our activities into the air over the United States. This is comparable to 3.3 lb (1.5 kg) per day per person, which is also about equal to the amount of garbage generated per person per day. The levels of air pollution in a particular area depend on the type and amount of pollutant emitted by a source, the ways in which the pollutants are released (for example, high stacks), and meteorological conditions leading to the dispersal of the pollutants.

In general, pollutants come from either stationary sources such as power plants and industries or moving sources such as motor vehicles. We are affected not only by these primary pollutants themselves but also by the products of chemical reactions that these pollutants undergo in the atmosphere, such as photochemical smog. The U.S. Environmental Protection Agency (EPA) gathers and analyzes emission and air quality concentration data for five principal pollutants: CO, sulfur oxide, particulates, volatile organic compounds, and nitrogen oxides. Figure 8.7 shows the quantity of air pollutants emitted in the United States in 1997 by type and source. The largest amount emitted is carbon monoxide. The total adverse effects from this gas, however, are less than those from particulates or sulfur dioxide, so "tons emitted" is not an adequate representation of their impact. As discussed in the section "Air Quality Standards," later in this chapter, air quality in the United States has improved considerably over the past several decades. Between 1970 and 1997, while U.S. population increased by 31%, and vehicle miles traveled increased by 127%, total emission of these five principal pollutants decreased by 30%. However, almost half of our population reside in areas in which at least one air quality standard has been exceeded in recent years.

Our major air pollutants originate from other sources besides power production. Industrial processes such as smelting and solid waste incineration, and the exhausts of automobiles, contribute heavily in certain areas. Natural sources can also be significant. The annual emissions from volcanoes, biological

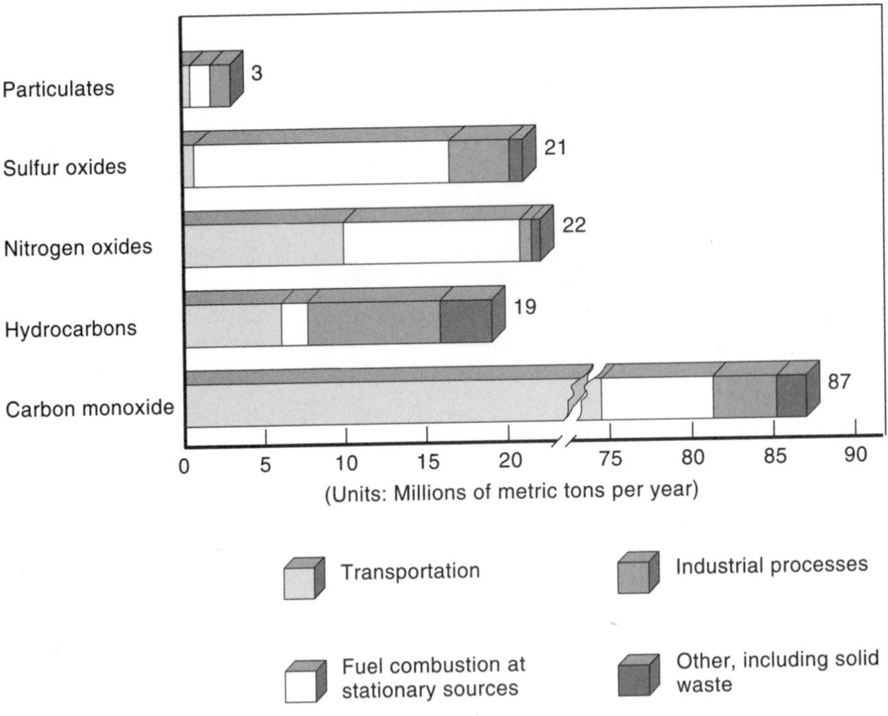

FIGURE 8.7
Emissions of major air pollutants in the United States by source: 1997. Particulates refer to those of size less than 10 microns (PM-10). (UNITED STATES ENVIRONMENTAL PROTECTION AGENCY)

decay, forest fires, and spray from oceans contribute more air pollutants on a worldwide basis than artificial sources. The adverse effects of these emissions are obvious if you have driven through a dust storm, battled a forest fire, or worried about corrosion to property located near the ocean. However, their contributions in industrialized areas are normally small compared to human-made ones. Since there is not much that can be done to control these emissions, we do not usually treat them as "pollutants."

Just as nature provides sources of contaminants, it also provides sinks or scavenging mechanisms to remove pollutants from the air. Absorption by vegetation, the ground, and water, plus oxidation and conversion to precipitates, prevent buildup of pollutants in the atmosphere. Most pollutants remain in the atmosphere only for days. (An important exception is carbon dioxide, as treated in Chapter 9.)

One way of measuring the concentration of a pollutant in the air is to determine the number of molecules of that type found in 1 million molecules of air (nitrogen and oxygen). This is expressed as parts per million, or ppM. Thus,

10 ppM means that a sample of air containing 1 million molecules has 10 molecules of a pollutant. This might sound insignificant, but toxic effects of most pollutants occur when concentrations are only a few parts per million. Another way of rating air pollution concentration is by the mass of pollutant found in a given volume of air. This is expressed as micrograms per cubic meter of air, or $\mu g/m^3$.

Carbon Monoxide

Carbon monoxide (CO) is a colorless, odorless, poisonous gas produced primarily in automobile engines by the incomplete combustion of carbon in the gasoline fuel. If not enough air is provided, CO is produced in the oxidation process $C + \frac{1}{2}O_2 \rightarrow CO$. Carbon monoxide is toxic to humans because it will pass through the lungs into the blood stream and bind to the hemoglobin, thus preventing the hemoglobin from carrying O_2 from the lungs to the body cells. The symptoms of CO poisoning are those associated with oxygen deprivation, such as dizziness, headaches, and visual aberrations. Both the duration of exposure and the CO concentration are important. Average exposure to CO is at concentration levels between 10 and 30 ppM. However, drivers in heavy traffic may be subject to 50 to 100 ppM of CO, and it has been suggested that such pollution may be a cause of increased traffic accidents through a decrease in driver perception. Leaky exhaust systems are quite hazardous, especially in the winter when car windows are closed.

The air quality standard set for CO levels is a maximum of 9 ppM for an eight-hour interval (or 10,000 $\mu g/m^3$). No physiological effects are known to appear at this concentration. Although the total weight of carbon monoxide emitted is more than that of all the other major pollutants put together, it is generally a health hazard only when emissions are highly concentrated, such as in heavy urban traffic.

Sulfur Oxides

The oxides of sulfur, namely SO_2 and SO_3, have long been recognized as important contributors to air pollution. They arise primarily from the burning of fossil fuels and the oxidation of sulfur: $S + O_2 \rightarrow SO_2$. Coal contains as much as 6% sulfur by weight, and its burning accounts for most of the emissions of sulfur oxides, about 20 million tons per year. Sulfur dioxide is a colorless gas with a suffocating odor, especially at concentrations greater than 3 ppM. It makes up about 98% by weight of the emitted sulfur oxides.

Sulfur is added to the environment from natural sources as well as from human activities. The emission of hydrogen sulfide (H_2S, having a rotten egg smell) from decaying organic matter and the release of sulfates (SO_4) from sea spray add about twice as much sulfur to the environment as do human sources of "pollution."

There is a continuous cycling of sulfur through the environment: Much of the SO_2 is precipitated by rain to the ground and oceans, and a lesser amount is taken up by vegetation and gaseous absorption in oceans and lakes. On the average, the amount of sulfur oxides naturally remaining in the biosphere is about 0.2 ppB (parts per billion), certainly minuscule compared to the several ppM (equivalent to several thousand ppB) that have been reported during air pollution alerts.

Sulfur dioxide in the atmosphere has many harmful effects on human health, vegetation, and materials. Epidemiological studies (studies of large populations) and other research indicate that SO_2 concentrations are associated with increased morbidity (illness rate) and mortality (death rate). Sulfur dioxide inhalation can result in damage to the upper respiratory tract, damage to lung tissue, and aggravation of lung disease. Such adverse effects are most pronounced in the very young, the old, and the 3 to 5% of the population with existing respiratory ailments such as bronchitis and emphysema. Increased breathing difficulty might not bother the average citizen, but can be fatal to someone with a respiratory ailment.

Several catastrophes have occurred when populations were exposed to greatly increased air pollution concentrations. One occurred in Donora, Pennsylvania, in 1948, in which 19 people died, and another in London, England, in 1952, in which 4000 people died. In both cases meteorologic conditions led to increased SO_2 and smoke concentrations in those localities, primarily from the burning of fossil fuels. In London, the concentration of SO_2 soared to about 7 times its normal level, and the death rate increased by a factor of more than 3 during this period. The burning of 650 oil wells in Kuwait, at the end of the 1991 Persian Gulf War, produced a variety of health problems, possi-

Oil wells on fire, Kuwait, 1991. (PETER JORDAN/GAMMA LIAISON)

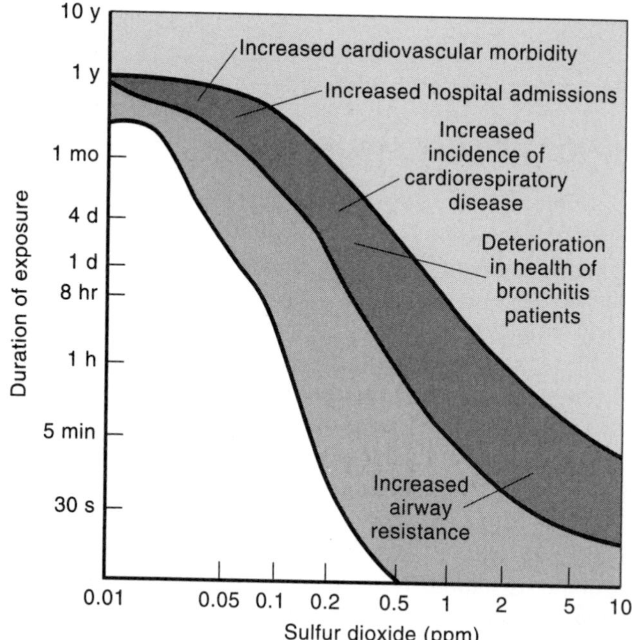

FIGURE 8.8

Effects of sulfur dioxide pollution on health. The figure shows ranges of concentrations and exposure times in which (a) the number of deaths reported was above expectation ▒, (b) significant health effects have occurred ■, and (c) health effects are suspected ■. (NATIONAL CENTER FOR AIR POLLUTION CONTROL, U.S. PUBLIC HEALTH SERVICE, 1967)

bly affecting U.S. soldiers even years later. The effects of SO_2 on health depend on many factors, including exposure time. Figure 8.8 illustrates the range of SO_2 concentrations coupled with exposure times during which significant health effects have been observed.

Another crucial problem associated with the emission of SO_2 is its further oxidation into SO_3 and the formation of sulfuric acid, H_2SO_4, when it reacts with water vapor:

$$2\,SO_2 + O_2 \rightarrow 2\,SO_3$$
$$SO_3 + H_2O \rightarrow H_2SO_4$$

This is the source of the **acid rain** phenomenon. In fact, all SO_2 is eventually converted into either sulfuric acid or what are called **particulate sulfates,** unless it is precipitated or absorbed out of the atmosphere. These particulate sulfates are chemical combinations of a sulfate ion, SO_4, and a metal ion such as iron—yielding, for example, iron sulfate ($FeSO_4$). The mechanism for the formation of these sulfates is not entirely understood, but we know that it is influenced by many factors, including temperature, humidity, ozone concentration, ultraviolet light intensity, and particulate and nitrogen oxide concentrations. Where fossil-fueled power plants emitting SO_2 and particulates coincide with high concentrations of nitrogen oxides from automobile exhausts, we see *synergism* (the working together of two things to produce an effect greater than the sum of their individual effects).

Focus On 8.2

AIR POLLUTION AND THE THIRD WORLD

For most developing countries, industrial development is a top priority. Pollution and its damage to the environment are far from being of prime concern. However, environmental problems are becoming too big to ignore. Acid pollution may prove to be deadlier in the developing nations than it is in the industrialized nations.

China is the world's third largest producer of SO_2. Daily means of SO_2 concentrations in China's large urban areas are between 100 and 400 $\mu g/m^3$, compared to large U.S. cities that are between 20 and 100 $\mu g/m^3$. These findings are not surprising when one considers that coal, burned largely with minimal or no air pollution controls, supplies 75% of all primary energy in China. Sulfur dioxide emissions in China are causing great concern in Japan, where it is believed that rice crops and the fishing industry are being damaged. Indian studies estimate that its own SO_2 emissions will reduce agricultural production across that nation. Damage to fruit and grain crops has already been observed in the eastern Indian states.

The conversion of SO_2 into sulfates occurs at a relatively slow rate, from hours to a few days. Thus the local air pollution problem can become a regional or national problem as well, because these substances can be transported in the atmosphere over hundreds of miles. More details on acid rain will be given later in this section.

Particulates

The presence of very small particles (0.01 to 50 microns* in diameter), or **particulates,** in the atmosphere can lead to pollution problems. Dust from dry soil, volcanic ash, pollen, sea salt spray, and fly ash from combustion processes are some sources of the particulates carried by the atmosphere. They are called "aerosols" because they are solid or liquid matter suspended in air. Particles with diame-

*A micron (μm) is one millionth of a meter. A human hair has a diameter of about 100 microns, we can easily see particles in the air that are about 50 microns across.

ters greater than one micron are usually associated with natural sources, such as dust, while those less than 1 micron in size usually originate from combustion processes. Dust particles can be from 1 micron to 1000 microns in diameter. Particulates can affect breathing, aggravate existing cardiovascular disease, and possibly damage the body's immune system. Small particles with diameters less than 10 microns, denoted as PM-10, are of special concern since these are the ones that can reach the lower regions of the respiratory tract. The symbol for all sizes of suspended particulates is TSP or total suspended particulates.

Seven million tons of TSP are emitted each year into the atmosphere over the United States, principally from industrial smelters and electric power plants. A list of trace elements in a standard mixture of coal is given in Table 8.2. A power plant burning this coal would emit all of these metals from its stack, although not necessarily in the amounts listed. Submicron-sized particulates can escape from the pollution control equipment and enter the atmosphere, where they might reside for days before eventually settling to earth.

EXAMPLE

A particulate with a diameter of 1 micron has a settling velocity of about 0.05 cm/s, so it will drop very slowly to the ground. If it is emitted from a stack that is 200 m tall, how long will it take to reach the ground?

Solution

Recall the discussion of distance and velocity in Chapter 2. Since distance = velocity × time, the time needed to drop to the ground will be

$$t = \frac{d}{v} = \frac{200 \text{ m}}{0.0005 \text{ m/s}} = 400{,}000 \text{ s}$$

or 4.6 days. Therefore, winds can take these particulates quite a distance.

ACTIVITY 8.1

You can study atmospheric particulates (concentration, size, etc.) as follows. Coat a piece of glass (such as a microscope slide) with a thin layer of petroleum jelly. Attach the glass to the radio antenna of your car with wire and drive around town. Examine the slide with a microscope or magnifying glass. How can you be quantitative about the particulates collected?

Table 8.2 TRACE ELEMENTS IN COAL

Element	Concentration (ppM by weight)
Iron	8700
Titanium	800
Manganese	40
Zinc	37
Vanadium	35
Lead	30
Chromium	20
Copper	18
Nickel	15
Cobalt	6
Arsenic	6
Selenium	3
Silicon	3
Thorium	3
Beryllium	1.5
Uranium	1
Thallium	0.6
Cadmium	0.2
Mercury	0.1
Silver	<0.1
Tellurium	<0.1

(National Bureau of Standards, Standard Reference Material)

Hydrocarbons or Volatile Organic Compounds

As their name implies, hydrocarbons are compounds consisting of carbon and hydrogen atoms. They are also known as volatile organic compounds (VOCs). They occur in many forms and are produced both naturally and by human activities such as petroleum use, trash incineration, and evaporation of industrial solvents. Gasoline contains almost 100 different hydrocarbons, including isooctane, heptane, and ethane. The evaporation of gasoline from your car's tank is a source of hydrocarbon pollution. In nature, the biological decay of vegetation releases methane (CH_4, the primary component of natural gas).The aromas in forests are caused by the emission of hydrocarbons called terpenes from the trees. Visibility can be reduced by the blue haze formed by photochemical reactions involving the terpenes.

Although natural sources account for about 85% of the hydrocarbons in the air, the more reactive ones are those associated with human activities. In cities the contribution from natural sources is less than that from artificial sources. Some hydrocarbons become very reactive on oxidation in the atmosphere, and they play an important role in the formation of ozone and photochemical smog, as discussed in the next section. Plant damage from smog can be severe.

Nitrogen Oxides, Photochemical Smog, and Ozone

There are many gaseous nitrogen-oxygen compounds, denoted by NO_x, that exist in various states of oxidation, from N_2O (nitrous oxide, also known as laughing gas) to nitrogen dioxide (NO_2) and dinitrogen pentoxide (N_2O_5). Unlike the other pollutants we have mentioned, the element oxidized in this case (nitrogen) comes not from the fuel but from the air that the fuel combustion processes use. (Air contains 78% nitrogen.) As we saw in Figure 8.7, the oxides of nitrogen are generated mainly by automobiles and electric power plants.

The direct effect of NO_x on humans is not serious at current levels. It is the secondary effects, notably the role of NO_2 in the formation of photochemical smog, that are of concern. **Photochemical smog** has been a serious problem for several decades, starting perhaps with the Los Angeles basin in the 1940s. Eye irritation, reduced visibility, and respiratory ailments are a few of the adverse health effects from smog. Photochemical smog is a combination of different gaseous and particulate pollutants, and its production occurs in a series of reactions.

Solar radiation begins the process by dissociating the pollutant NO_2 into NO and free oxygen.

$$\text{Light energy} + NO_2 \rightarrow NO + O$$

This free oxygen can combine with molecular oxygen in the air to yield ozone, O_3. Normally ozone will react with NO to re-form NO_2,

$$O + O_2 \rightarrow O_3$$
$$O_3 + NO \rightarrow NO_2 + O_2$$

In the absence of hydrocarbons, O_3, NO_2, NO, and O_2 are in rough equilibrium. However, the presence of hydrocarbons (from automobile emissions and industrial plants) changes this cycle. The hydrocarbons react with the free oxygen and NO and NO_2 to form very reactive **organic radicals** that are strong oxidizing agents and can lead to eye irritation. Some of these organic radicals are known as PANs (peroxyacyl nitrates). These radicals interfere with the nitrogen-oxygen cycle by producing more NO_2 and preventing the destruction of ozone, thereby allowing it to accumulate in higher than normal concentrations. NO gas is colorless, but NO_2 absorbs sunlight, giving it the brownish color characteristic of smog.

Ozone occurs in two separate regions of the atmosphere and creates a different concern in each place. While ozone in the upper atmosphere is beneficial to life by shielding the earth from harmful ultraviolet light, high concentrations of ozone at ground level are a major health and environmental concern. In the lower atmosphere, ozone is responsible for smog's odor, can cause damage to plants, and can irritate the lungs. Repeated exposures to ozone can make people more susceptible to respiratory infection, resulting in lung inflammation, and can aggravate asthma. Ozone production is stimulated by sunlight and high temperatures, so that peak ozone levels occur typically during the warmer times of the year. Children active outdoors during the summer are at most risk in experiencing chest pain and cough. The summer of 1988, the third hottest since 1931, produced some peak ozone levels. While emissions of air pollutants over the past 20 years have decreased, more than 60 major U.S. urban areas do not now meet federal air quality standards limiting 8-hour ozone concentrations to 0.080 ppM (parts per million).

The largest contributor to photochemical smog is the automobile. In certain metropolitan areas, especially those with only light industry, such as Los Angeles, the internal combustion engine accounts for as much as 90% of these air pollutants. Because of the importance of radiant solar energy in the production of smog, pollution concentration will vary with the time of day (Fig. 8.9). The reactants necessary for smog production—nitrogen oxides and hydrocarbons—are

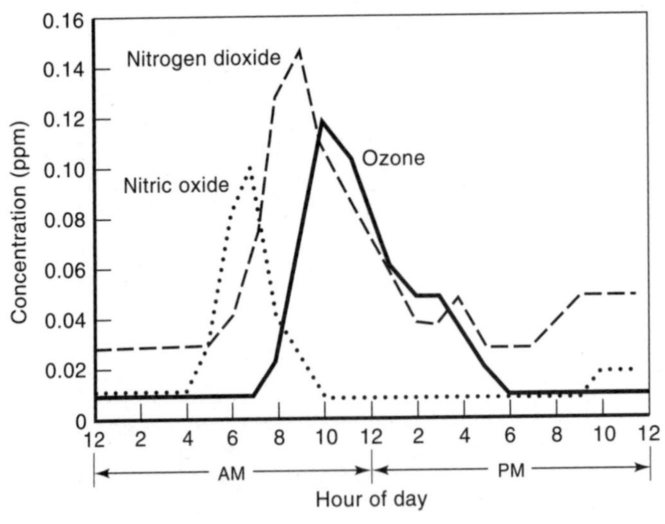

FIGURE 8.9
Variation of NO_x and ozone concentrations with time of day in the Los Angeles basin. (UNITED STATES ENVIRONMENTAL PROTECTION AGENCY)

produced by automobiles during the morning rush hour. If these reactants are trapped over a city by a temperature inversion, then smog production proceeds with the help of sunlight. The production drops off in the afternoon as the wavelengths of sunlight necessary for the reaction are filtered out by the atmosphere.

Acid Rain

In many parts of the United States the rains of summer and snows of winter bring not just hope and vitality but a subtle threat to the environment. The rain and snow must fall through an atmosphere polluted by sulfur oxides and nitrogen oxides, reacting to produce acid precipitation—sulfuric acid (H_2SO_4) and nitric acid (HNO_3). Acid rain can have many harmful effects, including the acidification of lakes—with significant declines in fish population—and damage to vegetation and forests, building corrosion, and possibly damage to human health.

The acidity of any solution is measured on a scale known as the pH scale, which is a measure of hydrogen ion concentration. Hydrogen ions have a positive electrical charge (they have lost their electron). A solution with an equal number of positive and negative electrical charges is neutral and has a pH of 7. Each decrease in pH by one unit represents a factor of *ten* increase in the solution's acidity. A solution with a pH of 6 has 10 times as many hydrogen ions as one with pH = 7. Figure 8.10 shows the pH of some common substances. Unpolluted rain is slightly acidic (pH = 5.6) because of its interaction with atmospheric carbon dioxide to produce carbonic acid (H_2CO_3).

Acid rain has primarily affected the eastern United States and Canada and northern Europe. Particularly strong evidence of increased acidification has been observed in lakes in New York's Adirondack Park. More than 200 high mountain lakes there have lost their native fish populations. Rainwater with pH values between 4.0 and 4.5 (10 to 40 times more acidic than pure rain) is common in the northeastern and southeastern states, and values between 3.0

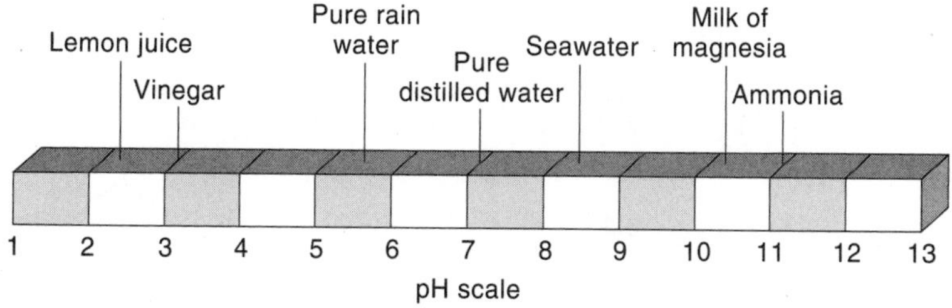

FIGURE 8.10

Acidity is expressed by pH, which is a logarithmic number. A 1-unit change in pH represents a change in acidity by a factor of 10. A solution with a pH of 5 is 100 times more acidic than pure water, which is neutral at a pH of 7.

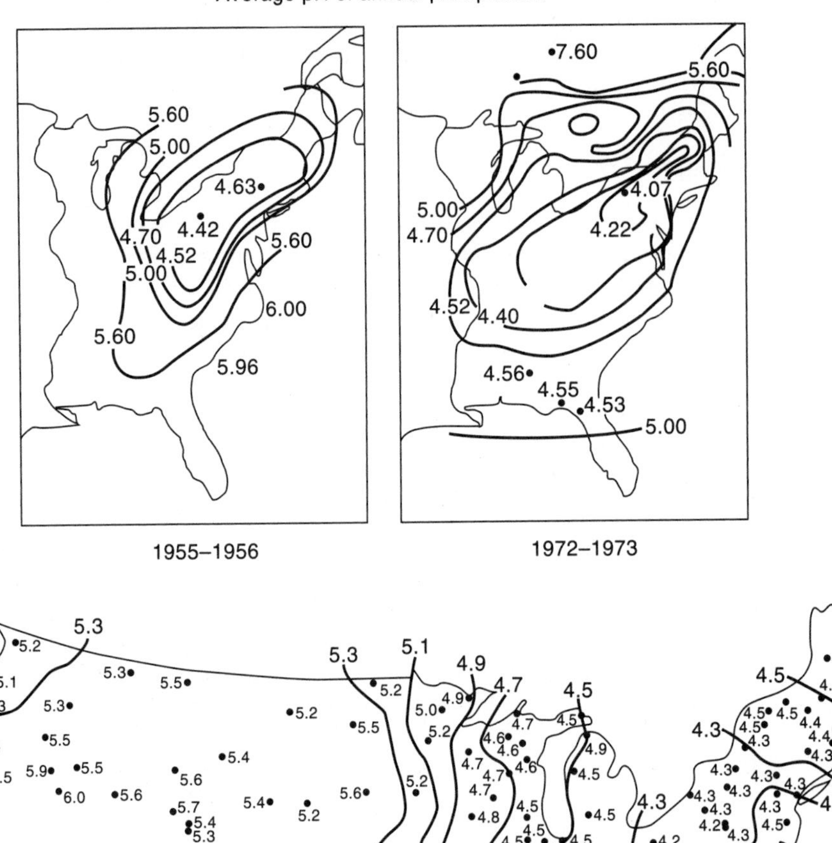

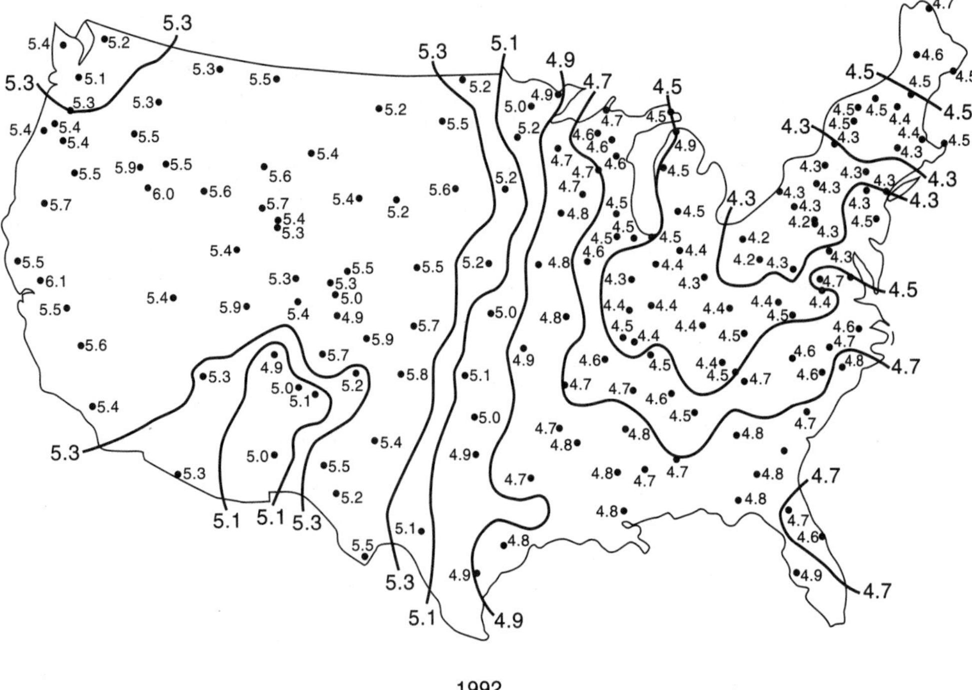

FIGURE 8.11

Change in annual average pH of precipitation in the eastern United States between 1955 and 1992. (UNITED STATES ENVIRONMENTAL PROTECTION AGENCY)

Effects of acid rain on a forest in Europe. (PATRICK PIEL/GAMMA LIAISON)

and 4.0 have been detected in some individual storms. Annual average pH levels less than 4.0 are observed in many northern European countries. Figure 8.11 shows the change in the annual average pH levels for precipitation in the eastern United States between 1955 and 1992.

Figure 8.12 shows the change in the distribution of pH for Adirondack lakes between the 1930s and 1970s. The fish population will be eliminated if the pH drops below 5.0. The lakes most susceptible to acidification effects are those

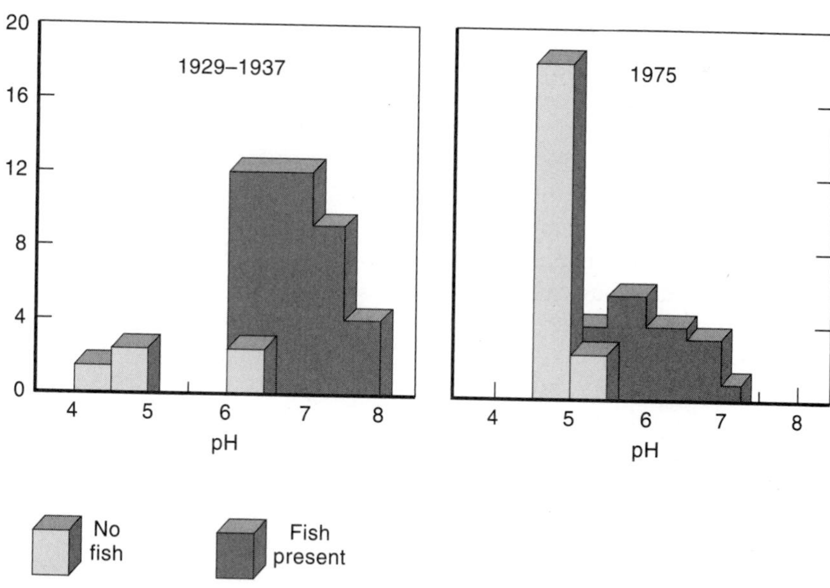

FIGURE 8.12

Change in pH and fish population for 200 Adirondack lakes (in New York) above 600 meters altitude between the 1930s and 1970s. (UNITED STATES ENVIRONMENTAL PROTECTION AGENCY)

that are surrounded by hard, insoluble bedrock with thin surface soils and that have low buffering capacity to neutralize the acids. The effects of acidification are blunted if the surrounding soil is alkaline (usually rich in limestone) or if the water contains basic ions derived from the weathering of rock.

Acid rain also affects crops by suppressing the bacterial decomposition of organic matter and by leaching (removing) nutrients such as calcium and magnesium from soil. In more serious cases acid rain causes lesions on leaves; this reduces the area for photosynthesis and so stunts plant growth. Since 1980, there has been a clear loss in the vitality of U.S. and European forests, but the role of acid rain in this situation is not clear. Forest damage in Europe has been especially severe; in 1995, about 20% of Europe's trees were moderately or severely defoliated. It has been suggested that the acid precipitation adds stress to trees and so weakens them to attack by disease or insects.

We still don't have a complete understanding of the relationship between emissions and deposition of acid rain. While annual SO_2 emissions have decreased by 12% since the 1980s, some lakes still are becoming more acidic. Indeed, the explanation of the formation of (sulfuric) acid rain, as given by the equations $2\ SO_2 + O_2 \rightarrow 2\ SO_3$ and $SO_3 + H_2O \rightarrow H_2SO_4$, is too simple and probably is only part of the process. The term "acid rain" is also a misnomer, as there are other means by which acid is deposited on the earth. There are other forms of *wet deposition,* such as acid snow and acid fog. There is also *dry deposition,* in which acid-forming substances in dry form (such as sulfates) fall to earth by gravity and are transformed to acids on the earth by precipitation. Wet and dry forms of acid deposition seem to be of about equal importance; collectively they will continue to be referred to as acid rain throughout this book.

There are many atmospheric processes and factors that influence the transformation of SO_2 and NO_x into acid rain. It is known that such processes can take a long time. It might take four days before the pollutants either fall to earth as dry deposition or react with moisture in the air to form acids. Consequently, acid rain can be transported many, many miles from the source to the deposition point, crossing state and national boundaries. Canada claims that SO_2 emissions from the United States are causing serious deterioration of its eastern lakes and forests.

Taller smokestacks built in recent years have put pollutants higher into the atmosphere, allowing them to remain aloft longer. The prevailing westerly winds place a large burden on eastern Canada and the northeastern United States from industries in the midwest. Some possible solutions to the acid rain problem involve burning lower sulfur fuel or the renovation of older plants with flue-gas desulfurization units or fluidized beds for cleaner combustion of coal. These methods are discussed in following sections.

Key questions about acid rain remain to be answered. There are disagreements within the scientific community as to the severity of the environmental effects of acid rain. A few examples:

- A U.S. Geological Survey study, determining precipitation chemistry from nine New York State sites from 1965 to 1978, found no significant long-term trends in acidity.

- A panel of the American Chemical Society concluded (1982) that pH values as low as 4.5 could be expected in places because of natural sources of acidity, but the "normal" pH of precipitation remains "debatable."
- The U.S. Interagency Task Force on Acid Precipitation (1982) noted that "there is no indication to date that the soils have become acid because of [acid precipitation] or that forest production is being affected."

Indoor Air Pollution

All the attention so far in the chapter, and in the actions of the public and the government until the past few years, has been in the area of outdoor air pollution. However, studies in the early 1980s found that *indoor* air pollutant concentrations can be many times greater than outside levels, even exceeding Environmental Protection Agency (EPA) standards. Since 80% of our time is spent indoors, some environmental scientists feel that breathing indoor air may cause or aggravate half of all illnesses in the United States and contribute to thousands of deaths a year.

The home or office is a place where oil, gas, kerosene, wood, and tobacco are burned and where furniture and building supplies emit volatile chemicals (Table 8.3). More than 200 different chemicals have been detected in indoor air. A common example is formaldehyde, which is found in plywood, particle

Table 8.3 INDOOR AIR POLLUTANTS

Pollutant	Source	Effect
Radon gas	Uranium in soil	May cause lung cancer
Tobacco smoke	Smokers	Respiratory ailments
Asbestos	Pipe insulation, ceiling and floor tiles	Lung disease, cancer
Fungi, bacteria	Humidifiers, air conditioning systems	Allergies, asthma, Legionnaire's disease
Carbon monoxide, nitrogen oxides	Stoves and heating systems	Headaches, drowsiness, nausea; in high concentrations, carbon monoxide is fatal
Formaldehyde	Plywood, particle board, foam insulation	Irritates eyes, skin, and lungs; causes cancer in animals
Benzene	Solvent cleaners	Suspected of causing leukemia
Styrene	Carpets, plastic products	Damage to kidney, liver

(*Newsweek*, January 7, 1985, p. 58)

board, and foam insulation. Formaldehyde can cause irritation to the eyes, skin, and respiratory system and has driven many people out of a home or building. Radon gas is another important pollutant, accounting for more than 50% of our radiation dose and thousands of cases of lung cancer per year (see Chapter 15). The link between indoor pollutants and disease is difficult to prove, because of the low concentrations, much like low levels of radiation discussed in Chapter 15. However, headaches, nausea, rashes, eye and skin irritation, and abdominal and chest pains might be symptomatic of a "sick building" syndrome.

One of the reasons for high indoor air pollution levels in newer homes is increased "tightening" of the home (weather stripping, caulking, vapor barriers) to reduce infiltration and so reduce home heating and cooling costs. Air exchange rates have decreased by a factor of 2 to 4 in most energy-efficient houses. Increased ventilation (through air to air–heat exchanges) can often reduce pollutant concentrations to safe levels.

D. Air Quality Standards

Today in the United States, more than 150 million tons of pollutants per year are emitted into our atmosphere. Even in small concentrations, these gases and particulates can cause harmful effects to human health, vegetation, and material property. The smog in Los Angeles, the black smoke in our industrial cities, can no longer be considered as signs of "progress" for a modern society. The Pennsylvania and English accidents of the late 1940s and 1950s have made us even more conscious of the severity of air pollution. In evaluating the effects of air pollution, death is not the sole factor to be considered, as people die anyway. Rather, the changes in life expectancy and the quality of life resulting from changes in the concentration of pollutants in the air are the key factors.

An organized attack on air pollution in the United States was instituted in 1970 with the passage of the Clean Air Act Amendments. This Act established a set of national ambient air quality standards (NAAQS) that had to be met beginning in 1975 for six pollutants linked to human health effects: sulfur oxides, nitrogen oxides, particulates, ozone, carbon monoxide, and lead.

The Clean Air Act Amendments of 1970 were added to the 1967 Air Quality Act to provide for national ambient air quality standards, to be set by the EPA. This act was preceded by the (weaker) Clean Air Act of 1963 and the Motor Vehicle Pollution Act of 1965. For enforcement purposes, the United States was divided into 274 air-quality control regions. Each region had to meet limits imposed by the EPA. Areas in which ambient pollutant concentrations are above these limits are required to devise plans to meet these limits; new and modified pollution control devices are required to achieve the lowest emissions, regardless of cost.

In 1990, major revisions were made to the Clean Air Act, reflecting an increased public concern for the environment. The plan concentrated on acid rain, urban air pollution (smog), and toxic emissions. In the first area, proposals were made to cut sulfur oxide emissions by 10 million tons a year from 1980 levels (about 23 million tons) by the year 2010. This will be done by the installation of scrubbers on power plants and the use of lower sulfur fuels. In the area of airborne toxic chemicals, the goal is a reduction of 75% to 90% in those suspected of causing cancer and other serious health problems. This will be done primarily by the installation of additional emission-control equipment, in facilities ranging from chemical plants to local dry-cleaning shops.

To reduce urban smog, an emphasis will be placed on using fuels that minimize CO and ozone emissions. Alternatives include methanol, ethanol, compressed natural gas, and electricity (Table 8.4). Some states (notably California and New York) have decided to deal with urban smog by attacking the most important source—the automobile. By 2003, 10% of all new vehicles (about 22,000) sold annually in California must be zero-emitting vehicles (ZEV)—probably electric vehicles or ones with fuel cells. Originally scheduled for 1996, then 1998, this mandate is opposed by the auto industry, who argue that the projected increased vehicle cost ($7000 to $20,000), mainly due to the batteries, will make mandated sales impossible. Maybe the new hybrid cars (see Chapter 10), which use both batteries and an internal combustion engine, will be the bridge to a pure ZEV.

There are two types of NAAQS, primary and secondary. Primary standards are designed to protect human health, while secondary standards protect human welfare, including effects of air pollution on vegetation, materials, and visibility. Setting air quality standards is a complex task. There is a wide variation in the susceptibility of different persons to air pollutants. There are also synergistic effects to be considered, as air pollution acts in addition to the effects of other substances. There is no disease that is caused only by air pollution. Consequently, it is difficult to agree on safe levels. The standards assume that below a certain level there will be minimal effects and that the benefits gained by further reductions cannot justify the costs of the additional efforts.

Table 8.4 GASOLINE ALTERNATIVES

Fuel	Source	Benefits	Drawbacks
Methanol	Coal, wood, gas	Fewer hydrocarbons and CO_2 High octane	Less energy content
Ethanol	Corn, sugar	Fewer pollutants High octane	Less energy content High fuel cost
Compressed gas	Natural gas	Inexpensive Fewer hydrocarbons and CO_2	Expensive vehicle conversion
Electricity	Fossil, nuclear, solar	No car emissions Power plant emissions easy to control	Limited range Battery cost

Table 8.5 lists the NAAQS in effect for the six pollutants that have standards. The standard concentrations are expressed as ppM or micrograms per cubic meter. The original particulate matter standard included all suspended particles up to 45 μm in diameter. The standards were revised in 1987 to include only those particles with diameters smaller than 10 µm (noted as PM-10). In 1997, the EPA again revised the PM standards to include an indicator for even smaller particulates, those less than 2.5 µm (PM-2.5). Important new studies suggested that significant health effects such as premature mortality and respiratory illness occur from the fine particulates that penetrate deeply into the lungs.

Overall, emissions from all stationary sources have been declining over the past 30 years, a 36% decrease since 1970. Particulate emissions have been reduced by 75% mainly by the installation of control equipment on industrial and utility plants, the use of less coal by industrial concerns, and a decrease in the burning of solid wastes. Carbon monoxide emissions have decreased significantly (down 32%) as a result of federal automobile emission standards, despite an increase in vehicle miles traveled during this period. Hydrocarbon emissions also have decreased 40% because of automobile emission standards, but this has been offset somewhat because of increased industrial process emissions. Nitrogen oxide emissions, however, have remained about the same during this period. This is a result of increased electricity generation and increased automobile emissions; the effect of tougher federal emission standards for NO_x have been balanced out by the increase in total miles traveled by cars. Sulfur oxide emissions have dropped about 35% since 1970, as flue-gas desulfuriza-

Table 8.5 U.S. NATIONAL AMBIENT AIR QUALITY STANDARDS (1997)

Pollutant	Averaging Time	Primary Standard	Secondary Standard
Particulate matter (PM-10)	Annual 24 hours	50 μg/m^3 150 μg/m^3	Same as primary
Particulate matter (PM-2.5)	Annual 24 hours	15 μg/m^3 65 μg/m^3	Same as primary
Sulfur oxides	Annual 24 hours	80 μg/m^3 (0.03 ppM) 365 μg/m^3 (0.14 ppM)	For 3 hours 1300 μg/m^3 (0.50 ppm)
Carbon monoxide	8 hours 1 hour	10 mg/m^3 (9 ppM) 40 mg/m^3 (35 ppM)	None
Nitrogen oxides	Annual	100 μg/m^3 (0.05 ppM)	Same as primary
Lead	Quarterly	1.5 μg/m^3	Same as primary
Ozone	1 hour 8 hours	235 μg/m^3 (0.12 ppM) 155 μg/m^3 (0.08 ppM)	Same as primary

(U.S. Environmental Protection Agency)

tion units and lower sulfur coal have been used. Improvements in ambient air quality (as measured by mean concentrations) over the ten-year period from 1988 to 1997 also have been achieved for all six pollutants that have NAAQS. The improvement for this period is from 15 to 60%, depending on the pollutant. Most areas of the United States meet the NAAQS for SO_2 and lead. However, approximately 100 million people in the United States live in counties that failed to meet at least one air quality standard as measured in 1997. Figure 8.13 shows annual average concentrations of SO_2, TSP, and ozone in selected cities worldwide.

Conflicts certainly have arisen with the Clean Air Act revisions. Everybody wants cleaner air, but no one wants to pay for it. The price tag runs into the tens of billions of dollars annually. The question frequently asked about any such

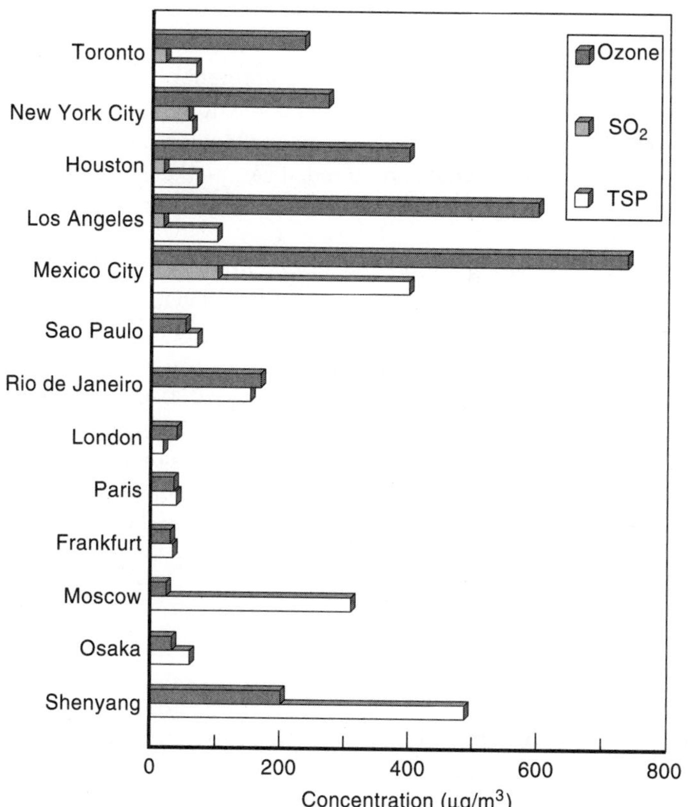

FIGURE 8.13

Comparison of levels of average annual (daily 1-hour maximum) ozone, annual total suspended particulate (TSP) matter, and sulfur dioxide concentrations among selected cities, expressed in $\mu g/m^3$. (UNITED STATES ENVIRONMENTAL PROTECTION AGENCY)

Focus On 8.3

1990 CLEAN AIR ACT AMENDMENTS

The 1990 Clean Air Act Amendments represented the first tightening of federal air standards in 13 years. In a complex set of guidelines, the Act aims to reduce acid rain, urban smog, and airborne toxic substances. An improved enforcement program will help ensure better compliance with the Act.

The Act requires utilities, by the year 2000, to halve their 1980 SO_2 emissions and cut their NO_2 emissions by 30% from the projected level without controls. The reduction in SO_2 emissions reflects a level that is 10 million tons per year below 1980 emission levels. Utilities will have some flexibility in meeting these requirements. For SO_2, retrofitting plants with flue-gas scrubbers or switching to low-sulfur coal most likely will be done to meet the emissions requirement of 1.2 lb of SO_2 per million Btu of output. Utilities also will be able to purchase or trade SO_2 and CO_2 emissions allowances. This means that if a utility can reduce emissions beyond what is required, it can market its savings to another utility, which might find it cheaper to buy allowances rather than to retrofit.

Another part of the Clean Air Act deals with urban air pollution. New standards on car emissions will reduce CO and VOC (volatile organic compounds) emissions to 3.4 and 0.66 g/mi by the year 2000. All cities currently not meeting health standards for ozone and carbon monoxide must attain these standards by the year 2000. Clean fuels will be encouraged. The 189 airborne toxic substances considered to pose a risk to human health or the environment will be reduced from 75 to 90%. These are chemicals in both particulate and gaseous form that originate primarily from the petrochemical and metals industries.

legislation is, "Do we save our planet, or do we save our jobs?" Are these the best means to achieve this goal? Clean coal technology (see Chapter 7) might do a better job of reducing sulfur oxide emissions than the control equipment discussed later in this chapter in the section "Stationary Source Air Pollution Control Systems." Electric cars and mass transit might do a better job of reducing urban smog than alternate fuels. Economically, costs added to cars, steel, and electricity by the installation of pollution control devices will make U.S. products less competitive in the global marketplace. The decisions are tough.

E. Automobile Emission Control Devices

The primary pollutants in automobile emissions are carbon monoxide (CO), hydrocarbons, and nitrogen oxides (NO_x). CO is formed by the incomplete combustion of fuel in the engine, when not enough oxygen is available. Carbon monoxide formation can be reduced in the combustion chamber by using leaner mixtures of gasoline, that is, an excess of air. In this way the combustion is complete and the reaction product is CO_2.

Hydrocarbon emissions result from unburned gasoline. The vapors that come from the evaporation of gasoline are hydrocarbons; thus a car can pollute the air even while it is not running. In fact, in cars made before the late 1960s, 20% of the hydrocarbon emissions were from fuel evaporation. In the combustion chamber, combustion takes place so rapidly that there is a thin layer of gasoline around the edges of the chamber that doesn't get burned. These gases are forced past the piston rings during the normal operation of the engine and accumulate in the crankcase. These "blow-by" gases used to be vented into the atmosphere, using air brought in through the oil filter cap. In 1961, the venting method was eliminated and the crankcase vapors were drawn through the engine back into the combustion chamber. This crankcase ventilation system, which uses a positive crankcase ventilation (PCV) valve to make it a "closed" system, is shown in Figure 8.14a. Evaporation of hydrocarbons from the fuel tank and carburetor is eliminated by capturing the vapors in a storage canister and feeding them to the carburetor when the engine is running. This canister is a small can, containing a charcoal filter.

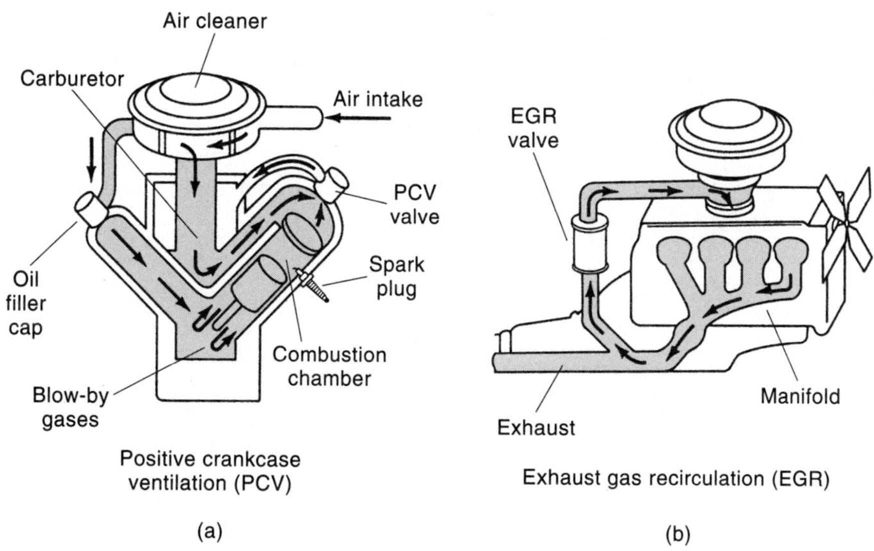

FIGURE 8.14
Automobile emission controls.

Initial pollution controls through engine modifications concentrated on minimizing CO and hydrocarbon emissions by improving combustion. This was done by using leaner fuels and increasing combustion temperatures. The engine's air-to-fuel ratio is the most important factor in determining automobile emissions. As this ratio is increased (leaner fuels), CO and hydrocarbon emissions are reduced. Fuel consumption also declines. However, these measures also increased the amount of NO_x formed as a result of the higher temperatures and increased oxygen required for more complete combustion. Therefore, to reduce NO_x emissions, lower pressures and temperatures were necessary. This was achieved by detuning the engine by retarding the spark timing (so combustion would not occur at maximum air-fuel pressure) and reducing the compression ratio (the ratio of the cylinder volumes before and after piston compression). However, fuel economy declined! Retuning the engine for increased efficiency with no increase in emissions was made possible with the use of the catalytic converter, beginning in 1975. The converter controls CO and hydrocarbon emissions after combustion through their reaction with oxygen on a catalyst surface. Fuel economy of new cars increased 13% from 1974 to 1975, to 13.75 mpg (urban test cycle). To preserve the expensive catalyst from ruin, unleaded gasoline had to be used, with its lower octane rating. Thanks mainly to reduced weight, fuel economy of new cars today has risen to 27.5 mpg.

To further reduce oxides of nitrogen, exhaust gases are circulated to dilute the incoming fuel and air mixture. This dilution of the incoming mixture lowers the peak flame temperature during combustion and thus limits the formation of NO_x. This is called the exhaust gas recirculation system (EGR, the status of which is indicated by a light on many dashboards), and it consists of a series of valves that control how much exhaust gas is put back into the system (Fig. 8.14b).

Emission standards for autos are required by the Clean Air Act Amendments. These are given in Table 8.6 in grams per mile. Another route to increased automobile efficiency and decreased emissions is to change to alternate power plants for the car, such as diesel, gas turbine, and steam engines. Reduced total gasoline consumption also can be achieved by decreasing demand or by decreasing the energy expended per passenger per mile, as with mass transit schemes. (See Focus On 18.4, Mass Transit.)

Table 8.6 AUTOMOBILE EMISSION STANDARDS (grams per mile)

Pollutant	1975	1990	1998
Hydrocarbons	3.4	0.41	0.25
NO_x	3.1	1.0	0.4
CO	34	3.4	3.4

(U.S. Environmental Protection Agency)

Focus On 8.4

MASS TRANSIT

Mass transit has been a subject of considerable debate over the past four or five decades in the United States. Mass transit consists of trains, subways, trolleys, and buses. In recent years, a number of U.S. and world cities have built rail systems at huge costs: Washington, D.C., Atlanta, San Diego, Toronto, Mexico City, Hong Kong—to name a few. However, the number of U.S. riders on rail systems has decreased by about a factor of three in the past 50 years. Part of this reason is the shift to the suburbs, a love of the automobile with its sense of independence, and lower auto costs. With all the talk about mass transit, only 10% of federal gasoline taxes goes to these systems (90% going to highway construction), while federal and state subsidies for automobile use (including tax breaks) continue to be large.

One alternative to the auto for longer distance travel is the magnetic-levitation train (MAGLEV), which can travel at speeds of 500 km/h (300 mph) on a cushion of air through the use of conventional or (eventually) superconducting electromagnets (see Chapter 10). Presently, Japan and Europe cover the distances between cities with new, fast supertrains that travel at speeds of as much as 300 km/h (180 mph), (Fig. 8.15).

Buses offer the cheapest and most energy-efficient approach to mass transit. They are flexible in routing and can couple with park-and-ride systems. Separate express lanes for buses eliminate traffic congestion. Emissions also can be curtailed sharply by using natural gas, fuel cells, or ethanol (See Chapter 17).

FIGURE 8.15
Bullet train and Mount Fuji, Japan. (TRAVELPIX/FPG INTERNATIONAL, LLC.)

F. Stationary Source Air Pollution Control Systems

The primary pollutants from fossil-fueled power plants are sulfur dioxide and particulates. For these plants and other stationary sources, there are several general methods or philosophies for meeting air quality and/or emission standards. They are:

1. Use of fuels with low sulfur and/or low ash content,
2. Removal of sulfur from the fuel prior to combustion,
3. Removal of particulates and sulfur oxides from the flue gas after combustion,
4. Shifting of fuels or power output in response to air quality demands, and
5. Dilution of effluent gases through the use of tall stacks and the natural atmospheric dispersion processes.

In this section we shall describe the control devices used in a power plant to reduce pollutant emissions or pollution concentrations (items 2, 3, and 5 in the preceding list).

Several types of pollution control devices are employed by fossil-fueled power plants after combustion. The types and sizes of pollutant they seek to control are different, and so combinations of various devices are often used within the same plant. Particulate emissions were the first pollutants to receive technical attention because of the visible nature of the pollutants and the ease with which they could be controlled. Ash particles associated with the burning of coal and oil can be anywhere between 0.01 and 100 microns in size. **Particulates** are removed in such devices as settling chambers, cyclone or inertial collectors, electrostatic precipitators, filters, and/or scrubbers.

Settling chambers or **gravitational collectors** are usually the first devices used to treat effluent gases. The flue gas spends a long enough period of time in this device so that the larger particles (>50 microns) drop to the bottom under the force of gravity and are not buoyed out by the gas. After leaving this device, the gases might proceed into an **inertial** or **cyclone collector** (Fig. 8.16) in which they are forced to undergo circular motion. The heavier particles, having more inertia, collide with the collector walls and fall to the bottom. This is analogous to the trajectory followed by objects sitting on a spinning CD: The heavier objects are thrown off first. The cyclone collector can remove more than 99% of the particles larger than 50 microns, but is very inefficient for those less than 5 microns.

The **electrostatic precipitator** has been very successful in reducing particulate emissions from power plants. Almost all new units use such systems, in combination with some of the devices already described, producing results shown in Figure 8.17. The effluent gases pass through a set of metal wires and plates located before the exhaust stack, arranged as shown in the model of Figure 8.18. The wires are given a large (~50,000 V) negative voltage, with the plates at ground potential. The large electric field between the wire and the

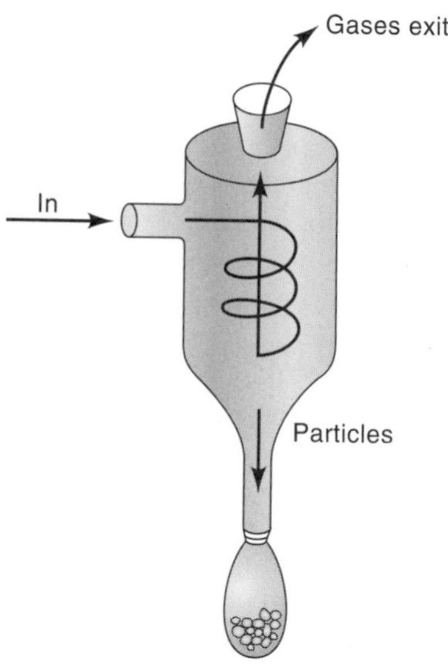

FIGURE 8.16

Inertial or cyclone collector. As the gas undergoes circular motion, the heavier particles collide with the collector's walls and fall to the bottom, where they are collected for disposal.

FIGURE 8.17

A before-and-after sequence showing the effect of an electrostatic precipitator on stack gas emissions from a coal-fired power plant.

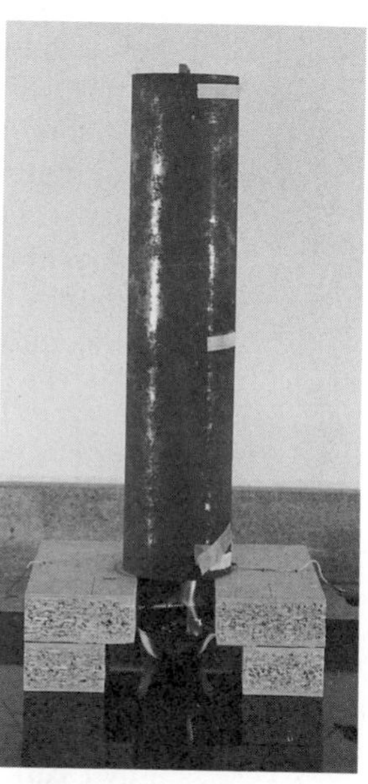

FIGURE 8.18

Working model of an electrostatic precipitator. The sheet-metal cylinder has a taut wire along its center. The wire is given a high voltage by a Tesla coil, and the cylinder is grounded. (A tennis ball can could be substituted for the sheet metal.)

plate ionizes some of the gas near the wire, providing electrons. These electrons are captured by either a gas molecule (such as N_2 or O_2) or a fly ash particle, forming a negative ion that will move toward the positive plate. There the particles are collected and mechanically removed at periodic intervals. The fly ash collected is either taken off site and dumped or sold for the metals that can be extracted. The electrostatic precipitator can remove as much as 99% by weight of the particulates present in the fly ash, but it is inefficient for particles smaller than 1 micron in size. The efficiency for these particles might only be 50%. Unfortunately, the smaller particulates are those that are the most damaging to the respiratory tract.

Another technology for the control of particulates is the **fabric filter.** Acting like a water filter or vacuum cleaner filter, these fabric filters consist of large bags of cotton cloth or glass fiber material that intercept the particles, trapping them on the inside of the bags. The particulates are removed by shaking the bags. Fabric filters are cost competitive with electrostatic precipitators for power plants and have the advantage of 99.9% removal efficiencies. They are especially good for the submicron size particulates, down to about 0.1 micron.

Because sulfur dioxide is a gas, the devices described previously will not work to remove this pollutant. Generally, **scrubbers** are used for SO_2 control in which the combustion gases pass through a water solution spray. The scrubber is used in conjunction with chemicals to remove SO_2 from the exhaust gases. In this system, called **flue-gas desulfurization,** slurries of lime or limestone (calcium carbonate) or dolomite (magnesium carbonate) react with SO_2 in the scrubbers to form solids of calcium or magnesium sulfate, which can subsequently be removed with the other particulate matter:

$$\underset{\text{(gas)}}{2\,SO_2} + 2\,CaCO_3 + O_2 \rightarrow \underset{\text{(solid)}}{2\,CaSO_4} + 2\,CO_2$$

The clean gases leaving the scrubber are reheated to reestablish buoyancy, and then exit through the conventional stack. Sulfur dioxide removal efficiencies as high as 98% have been reported for scrubber techniques. Consequently, most new power plants install scrubbers.

An environmental problem associated with the use of flue-gas desulfurization technology is the disposal of the byproducts of the process. The byproducts either can be discarded as waste or be recovered and sold, generally as sulfuric acid or elementary sulfur. The waste sludge generated in the scrubber process is enormous: Each ton of coal generates approximately 0.4 ton of concentrated aqueous waste. For a 1000-MWe power plant burning 10,000 tons of coal per day, that amounts to 4000 tons per day. This can constitute a major problem in large cities where transportation and landfill costs and disposal-site availability can be major factors. However, the potential adverse health effects from such selected waste disposal sites under adequate environmental controls seem more desirable than random damage to the public through air pollution.

Flue-gas desulfurization systems cost about \$50 to \$80 per installed kilowatt, or about 10 to 15% of the construction cost of a coal-fired plant. The energy required to operate such a unit is about 3 to 7% of the power produced by the plant. This energy is used primarily to reheat the gases from the scrubber (to increase their buoyancy). Scrubber systems represent permanent solutions to SO_2 pollution. However, there is an ongoing controversy between the EPA and the utilities regarding the status of scrubbers, especially with respect to their reliability and their addition to older plants. Many utilities argue that intermittent controls (such as tall stacks for SO_2 dispersal and flexible operating practices) can meet the ambient air standards and should be allowed until scrubber systems have been shown to be technically feasible. Proponents have argued that permanent controls on SO_2 emissions were required under the Clean Air Act to protect public health, and that scrubbers are a demonstrated and economically feasible control technology. Europe and Japan successfully use scrubbers (which function about 90% of the time) with an 85 to 95% SO_2-removal rate.

Another widely used technique to reduce air pollution levels in an area is to release the effluent gases through very tall chimneys or stacks. Since the wind velocity is greater at higher elevations, these tall stacks use the natural atmospheric dispersion processes to dilute the air pollutants. To use a phrase, "the solution to pollution is dilution." Some of these stacks are as tall as any human-made structure (Fig. 8.19) and might be able to stick up above any low-level inversion layer. However, some pollutants (such as acid rain) will just cause trouble somewhere else downwind.

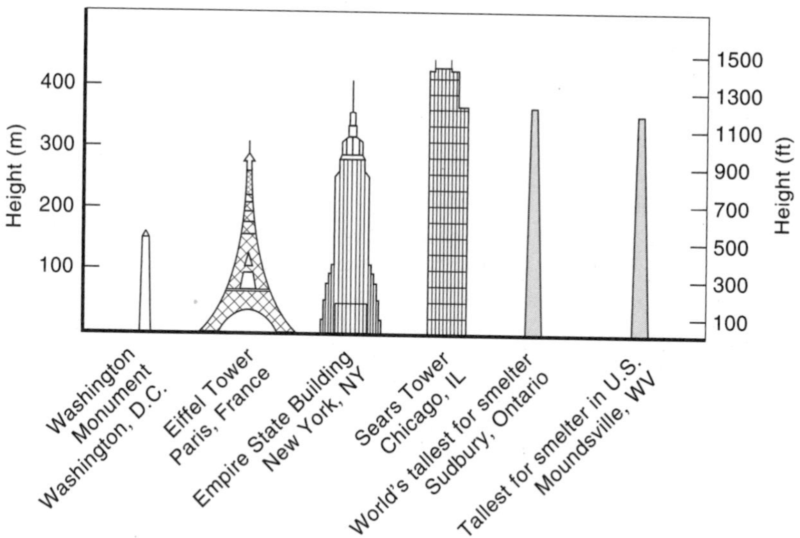

FIGURE 8.19
Chimneys are some of the tallest structures built by humans.

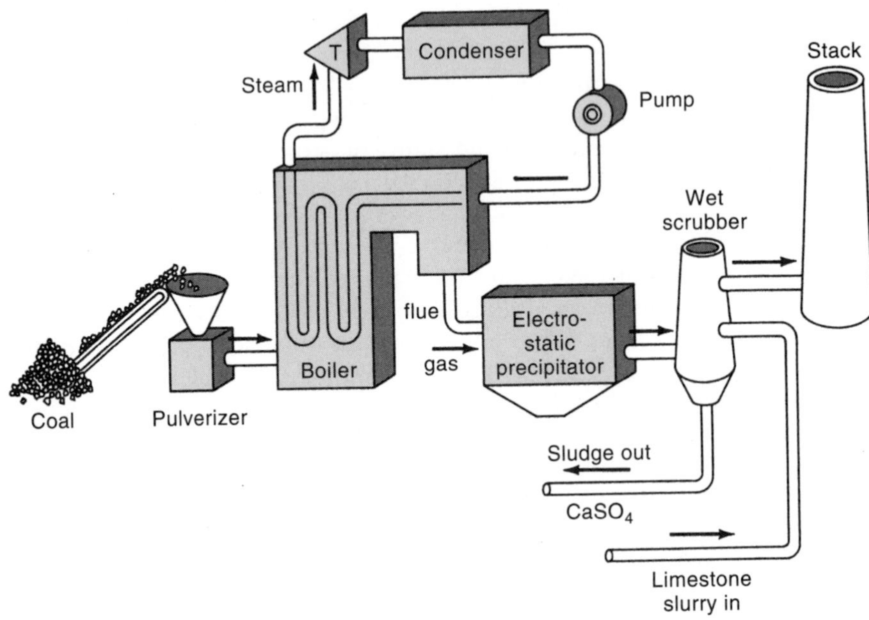

FIGURE 8.20
Schematic of a typical fossil-fuel plant, showing equipment used to remove pollutants from the flue gas or boiler exhaust.

Figure 8.20 shows air pollution control devices as they might appear in a conventional power plant burning pulverized coal. Not all these components would necessarily be present in an older plant.

Another way to meet SO_2 emission standards is to reduce the amount of SO_2 formed before combustion. This can be done either by using low-sulfur coal or by removing SO_2 directly in the combustion process. The refining of high-sulfur coal to a fuel that is clean to burn can be accomplished in several ways. One method is the conversion of coal to synthetic oil or natural gas; another is coal beneficiation-the removal of sulfur from the coal before it is burned. The conversion of coal to synthetic fuels is in the research and development stage in the United States, with several pilot plants operating; however, costs are high. (See Chapter 7.)

Fluidized bed combustion (FBC) removes the SO_2 as soon as it is formed by burning crushed coal on a moving bed of air and sand to which is added limestone (Fig. 8.21). The reaction $CaO + \frac{1}{2}O_2 + SO_2 \rightarrow CaSO_4$ takes place in the boiler, and the calcium sulfate solid can be removed after combustion as particulates in the exhaust gases. Fluidized beds have been used in the chemical industry and in incineration processes since the 1920s. For electric utilities, fluidized bed combustion is a new technology and involves a different firing

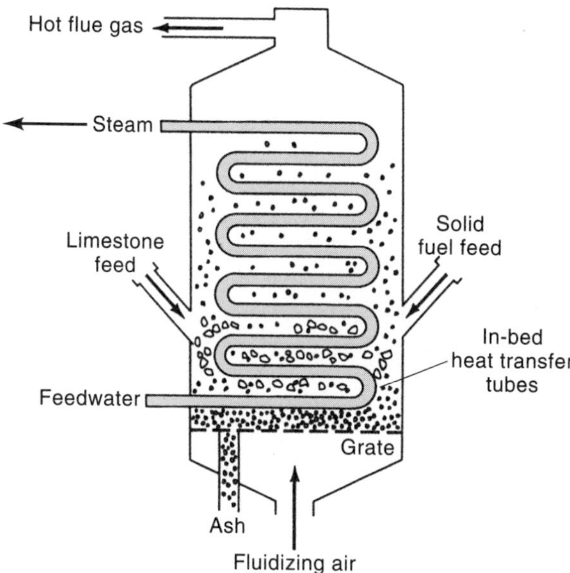

FIGURE 8.21
A fluidized bed combustion (FBC) unit. When air is forced up from below, the bed of ash becomes fluidized (that is, the solids "flow" like a liquid). Crushed solid fuel burning in the bed heats the ash. Limestone added to the fluidized mixture reduces the amount of SO_x emitted in the flue gases.

system than that used in conventional pulverized coal boilers. The advantages of this technology are the high heat transfer rates that result from the turbulent mixing of the fine fuel particles in the boiler, possibilities of using multiple fuels, and the control of air pollution at its source. Fluidized bed combustion also reduces the amount of NO_x formed by lowering the combustion temperature. Economically, fluidized bed combustion competes favorably with scrubbers. Worldwide, many facilities burning coal or wood waste are using FBC. In China there are 2000 small fluidized bed boilers using peat, wood chips, and municipal waste to provide electricity and heat to villages. In the United States a handful of utility boilers in the 100- to 200-MW range are in commercial operation using FBC.

G. Summary

The major air pollutants from stationary and mobile sources are SO_2, particulates, nitrogen oxides, VOCs (hydrocarbons), ozone, and carbon monoxide. Figure 8.7 summarized the emissions for each of the pollutants. Some pollutants are secondary in origin, such as photochemical smog, which is a result of a reaction among primary pollutants (hydrocarbons and NO_x) in the atmosphere in the presence of sunlight. Pollution-control devices for moving and stationary sources have had a positive effect in recent years in reducing the total amount of pollutants emitted, down about 30% from 1970 emission

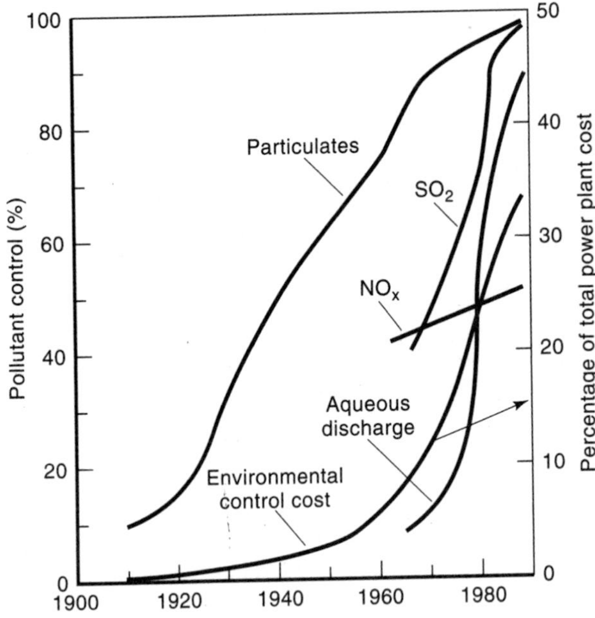

FIGURE 8.22
The rising cost of environmental controls. Pollution controls represent at least one third of the total cost of recently built coal-fired power plants. (*EPRI JOURNAL*)

levels. However, debate continues between lawmakers and industry over the reliability of some of the control devices, the economics of their installation, and the necessity for stricter air quality standards. Many people feel that the most effective and economic way to combat acid rain is by renovating older power plants by adding fluidized bed combustion and flue-gas desulfurization units. About half of currently operating coal-fired power plants were built before 1975 and have no pollution control devices except those for particulate removal.

The units previously described for controlling power plant emissions do not come cheap. Figure 8.22 shows the rising cost of environmental controls from 1900 forward. Pollution controls represent about one third of the total capital cost of recently built coal-fired power plants!

Internet Sites

For an up-to-date list of Internet resources related to the material in this chapter, go to the Harcourt College Publishers website at **http://www.harcourtcollege.com**. The links are in the *Energy: Its Use and the Environment* site on the Physics page. General energy related sites and some guidelines for using the World Wide Web in your class are on the inside front cover of this book.

References

Chapter 8

Ausubel, J. and H. Sladovich. 1989. *Technology and Environment.* Washington, D.C., National Academy Press.

Bates, R. W. 1993. The Impact of Economic Policy on Energy and the Environment in Developing Countries. *Annual Review of Energy,* 18.

Lent, J., and W. Kelly. 1993. Cleaning the Air in Los Angeles. *Scientific American,* 269 (October).

Likens, G. E., R. F. Wright, J. N. Galoway, and T. J. Butler. 1979. Acid Rain. *Scientific American,* 241 (October).

Mehr, C. 1989. Are the Swiss Forests in Peril?" *National Geographic* (May).

Miller, T. 1999. *Living in the Environment.* 11th ed. Belmont, CA, Brooks/Cole.

Mohner, V. 1988. The Challenge of Acid Rain. *Scientific American,* 259 (August).

Murota, Y., and Y. Yano. 1993. Japan's Policy on Energy and the Environment. *Annual Review of Energy,* 18.

Raven, P., L. Berg, and G. Johnson. 1998. *Environment.* 2nd ed. Philadelphia, Saunders College Publishing.

Rose, A., W. Labys, and T. Torries. 1991. Clean Coal Technologies and Future Prospects for Coal. *Annual Review of Energy,* 16.

Smith, K. R. 1993. Fuel Combustion, Air Pollution, and Health: The Situation in Developing Countries. *Annual Review of Energy,* 18.

Stigliania, W. M., and R. W. Shaw. 1990. Energy Use and Acid Deposition: The View from Europe. *Annual Review of Energy,* 15.

Thompson, J. 1991. East Europe's Dark Dawn. *National Geographic* (June).

Torren, I. M., J. E. Cichanowicz, and J. B. Platt. 1992. The 1990 Clean Air Act Amendments: Overview. *Annual Review of Energy,* 17.

Walsh, M. 1990. Global Trends in Motor Vehicle Use and Emissions. *Annual Review of Energy,* 15.

World Resources Institute. 2000. *World Resources.* New York, Oxford University Press.

QUESTIONS

1. How can you increase the pressure exerted by this book on a tabletop without increasing its weight?
2. The device pictured in Figure 8.23 has three differently shaped containers, all of which are connected to the same base. Using the ideas of pressure, in which of these containers will the water level be the highest?

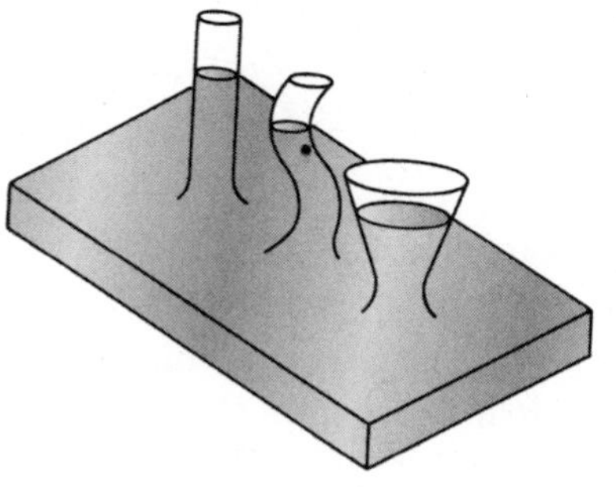

FIGURE 8.23

3. The deepest well from which water can be pumped, by creating a vacuum on the top end of the pipe, is 10 m (34 ft). In drilling for oil, wells of a mile or more in depth have to be made. From this depth, how do you think the oil is recovered?
4. What is the buoyant force on a 2-ton ship that is floating in water? In alcohol? If this ship gets a hole in its side and sinks in the water, how and why does the buoyant force change?
5. A helium-filled blimp (of fixed volume) will rise just so high into the atmosphere and then stop. Why? Will a smaller blimp rise higher than a larger one, if both are filled to the same pressure?
6. The average molecular mass of air is 29, and that for water is 18 (in atomic mass units). Do you expect moist air to be more or less dense than a volume of dry air under the same atmospheric conditions? What about the atmospheric pressure of moist air regions (compared to dry air)? How is the weather forecast based on barometric pressure changes?
7. Why is a dam constructed with a very broad base, as shown in Figure 8.24?

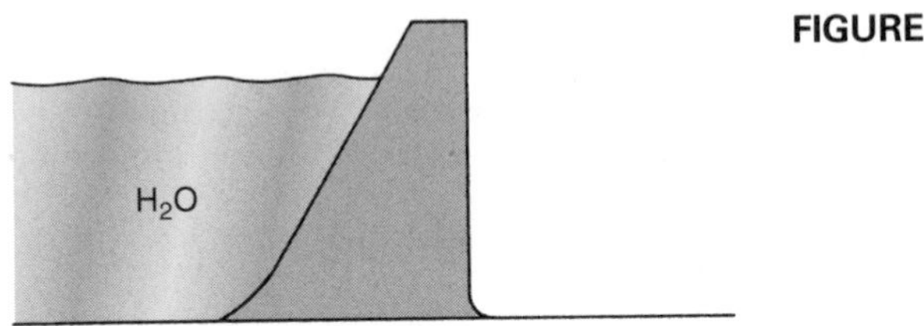

FIGURE 8.24

8. If the temperature of the air in a certain region is increased by heating from the sun, does the air pressure at the ground increase or decrease?
9. What atmospheric conditions promote vertical mixing of the air? Why is this important?
10. What are the difficulties in ascertaining the effects of various air pollutants on public health?
11. What are the distinctions between air standards related to emissions and those related to local atmospheric concentrations of pollutants?
12. What are the effects of the Clean Air Act on the use of coal for electric generating plants? Check the Web.
13. Investigate reasons behind postponements of tighter emission standards for automobiles. Check the Web.
14. Make a table of air pollution control devices and the types of pollutants they can remove from stack gases.
15. Formulate a definition of an air "pollutant." How do the contributions of artificial substances to the air compare to those from natural sources?

16. Why do you expect photochemical smog to be a greater problem in a city like Los Angeles than in a heavy industrialized city like Pittsburgh? What air pollution problems are prevalent in the latter?
17. What are the principles behind the operation of a cyclone collector? For what type of pollutants can this be used?
18. What are some of the limitations in using an electrostatic precipitator to control air pollution emissions?
19. What can be done to reduce "acid rain"?
20. Give several reasons for the use of tall stacks. (Consider wind speeds and temperature inversions.)
21. Why do the gases that exit a scrubber system have to be reheated before being vented to the atmosphere? What does this do to power-plant efficiency?
22. Why might you expect indoor air pollution to be more of a factor for a person living in a house built in 1985 than for one built in 1885?
23. Considering the fact that oil powers the tractors required to harvest the corn that goes into ethanol, discuss the overall consequences of using cleaner burning gasolines.

PROBLEMS

1. Calculate the pressure exerted on your finger if you hold up a 15-lb box of apples.
2. What is the pressure exerted by your finger if you hold back water leaking through a hole in a dike, as shown in Figure 8.25?

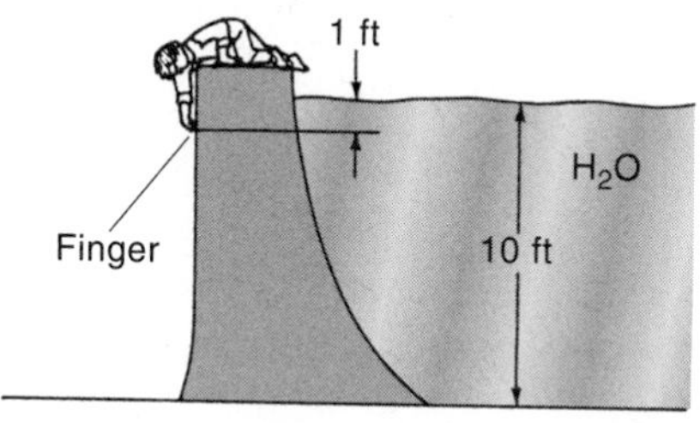

FIGURE 8.25

3. What pressure is exerted at the bottom of a tube of mercury 15 cm high? (Mercury's density is 13,600 kg/m^3.)
4. A swimming pool measures 10 m $\times$ 7 m $\times$ 2 m. Calculate the total force exerted by the water against the bottom.

5. A 1000-MWe plant uses coal with 3% sulfur content. How many tons of SO_2 will be emitted into the air during the operation of the plant for a day? (Use the list of conversions and equivalences located on the inside back cover of this book.)
6. What is the rate of emission (in kg/h) of particulates from a power plant with a thermal output of 3000 MW that burns coal with an ash content of 2%? How much is emitted if an electrostatic precipitator of 95% efficiency is used? If the ash contains 1 ppM of mercury, how many tons of mercury will be emitted by this plant per year?
7. How many kilograms per year of carbon monoxide would your car emit to the atmosphere if it just conformed to the emission standards for automobiles?

FURTHER ACTIVITIES

1. A very simple barometer can be constructed by using a jar with a rubber balloon stretched across the top and held in place with a rubber band. A straw or wooden splint attached to the balloon can monitor changes in atmospheric pressure. Using a marker and a ruler, record the position of the straw several times a day. Compare your data with the times of highs and lows published by the weather forecaster in your area. The air in the jar should be maintained at a constant temperature (Fig. 8.26). (Keep the jar away from windows.)

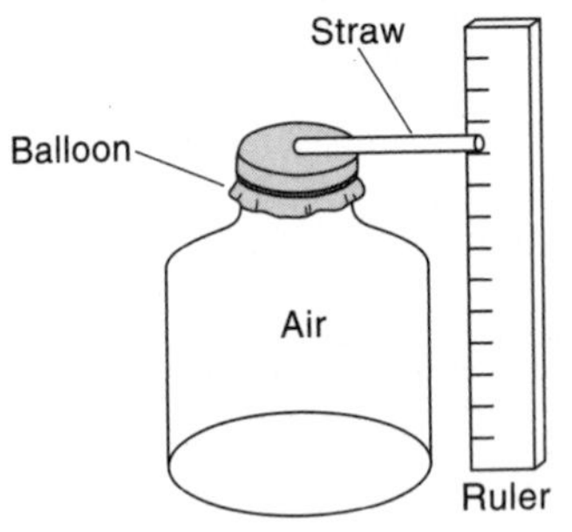

FIGURE 8.26

2. A fun illustration of atmospheric pressure uses a (peeled) hard-boiled egg. Place the egg in the opening of a glass milk bottle in which a fire was just begun using paper or wood shavings. The egg eventually will be pushed into the bottle. Why? (A small-sized egg will work best.) How might you get the egg out of the bottle?

3. Make a model of an electrostatic precipitator (see Fig. 8.18) by using a tennis ball can (to be grounded) with a copper wire in the middle. You can test it out using cigarette or incense smoke if you can get a high-voltage (dc) source such as Tesla coil to attach to the wire (Fig. 8.27).

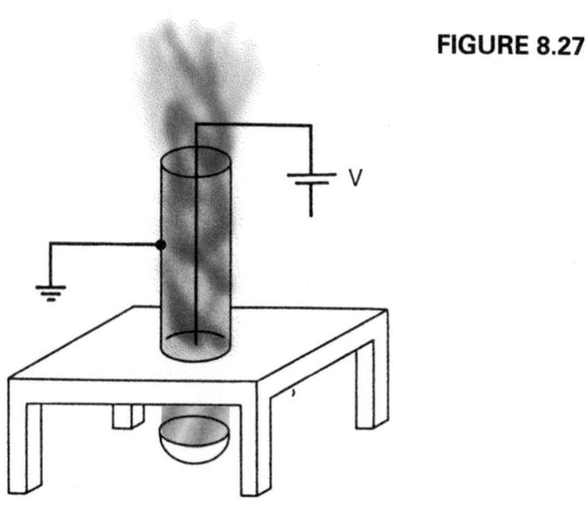

FIGURE 8.27

4. Another good illustration of atmospheric pressure makes use of a soft drink can. Put a little water into the can and heat it to boiling. After heating the can, invert it quickly and put it into a shallow pan of cold water. Watch what happens to the can as it cools down.

9

Global Warming, Ozone Depletion, and Waste Heat

A. Introduction

As discussed in previous chapters, the conversion of fuel into useful energy also produces waste heat and releases many types of pollutants into the air and water. One crucial concern today is the possibility that the earth's climate could be changed permanently by carbon dioxide emissions from the burning of fossil fuels. This "global warming" could bring about climatic effects we have never before experienced. Another atmospheric problem is the depletion of the ozone layer in the upper atmosphere, resulting in increases in skin cancer as a result of increased ultraviolet exposure. This chapter will examine the long-term problems of global warming due to emissions of CO_2 and other gases as well as other problems in our environment caused by the depletion of ozone in the upper atmosphere. We will also study the effects of waste heat on our environment from energy conversion processes, and the recovery of useful energy from some of this waste heat.

B. Global Warming and the Greenhouse Effect

A scientist's laboratory can be a remarkable place. Beakers and scales, voltmeters and microscopes, computers and instruments large and small are but a few of the pieces of equipment found there. One of a scientist's goals is to explore a particular component of nature-property of matter or the interaction of one thing with another (living or inanimate), draw some conclusions, and then extend these conclusions to the world outside the lab. By studying the effects of certain drugs on small animals, for example, we can hypothesize their effect on people; by studying different materials at low temperatures, we can discover characteristics of matter that might lead to the development of new superconductors. These experiments are not always reversible (the animals could be affected forever), but they are carried out on a scale small enough to cause little effect on things outside of the laboratory. However, we are also part of a much larger laboratory—the entire earth itself—and our experiments there are not necessarily reversible or benign. With the combustion of fossil fuels, our atmosphere has become one large experimental laboratory, leading to consequences that might cause disastrous alterations in our climate.

Evidence is mounting that as we continue to burn fossil fuels, we are releasing gases into our large laboratory that can raise the earth's temperature by at least several degrees and potentially change the climate of the entire earth. Once lush farmlands might become dust bowls while some desert regions might bloom. Coastal areas might see rising sea levels, causing the migration of millions of people from such areas as southern Florida and Bangladesh. One of the problems with this massive experiment is that we might not be able to stop it, although it might be slowed down. Scientists have many different viewpoints on this issue and no consensus exists on what actions, if any, should be taken. However, the results of recent studies have brought the United Nations' Intergovernmental Panel on Climate Change (IPCC) to the conclusion that "there is a discernable human influence on global climate from the buildup of greenhouse gases." Justification for this statement comes in part from a good match between computer predictions (for a greenhouse gas-warmed atmosphere) and the earth's actual temperature profile.

The **greenhouse effect** is caused by gases in the atmosphere that absorb certain wavelengths of infrared radiation emitted from the earth that would otherwise radiate out to space. Recall from Chapter 6 on Solar Energy (Fig. 6.3) that the earth's temperature depends on the balance between energy coming in from the sun and the energy that is radiated back into space. About half the

ACTIVITY 9.1

To understand how the greenhouse effect operates, try the following activity. Gather three large glass jars and three thermometers. Place a thermometer inside each jar. Cover one jar with a glass plate, leave the second jar uncovered, and put a potted plant inside the third, covered with a glass plate. Expose to the sunshine, or a light bulb, and record the temperatures every 10 minutes for 1 hour. Graph the results and compare them.

energy entering the atmosphere is absorbed by clouds and particles or reflected back to space. The remainder is absorbed at the earth's surface, warming the land and the oceans. The surfaces reradiate this energy in the form of infrared, or thermal, radiation. Naturally occurring water vapor and CO_2 in the atmosphere absorb certain wavelengths of this radiation. Some of this absorbed heat is then reradiated *back* to the earth. This process keeps the earth's surface about 30°C (54°F) warmer than it would be without an atmosphere. Increased CO_2 concentrations will cause more heat to be trapped within the earth's atmosphere.

Figure 9.1 shows the correlation between atmospheric CO_2 concentrations and temperature changes during two time periods. These data were taken from an analysis of air bubbles trapped within ice core samples collected in Antarctica and Greenland. Over the past 160,000 years, changes in the earth's temperature and CO_2 concentrations seem to be well correlated (see Fig. 9.1a). The cessation of one ice age about 130,000 years ago and another about 10,000 years ago seems to be reflected in strong changes in both temperature and CO_2 (in parts per million). Over the past century (Fig. 9.1b), the data show a slight global warming of about ½°C as well as a 20% increase in atmospheric CO_2 concentrations. Since the beginning of the industrial age, atmospheric CO_2 concentrations have risen 30%. Recent studies have shown that the 20th century was the warmest in the last 1000 years, by far.

Other gases that are accumulating in the atmosphere also play an important role in warming the earth. While carbon dioxide and water vapor only weakly absorb infrared radiation with wavelengths between 7 and 12 µm, other "greenhouse" gases are powerful absorbers of the heat emitted in this range of wavelengths. These gases are methane, nitrogen oxides, and chlorofluorocarbons or CFCs (such as Freon). Although present in very small quantities, these gases remain in the atmosphere for many years and have a greater ability to absorb heat. One molecule of CFC can have the same effect as 10,000 CO_2 molecules. CFCs come only from human activity, while nitrous oxide (N_2O) originates in agricultural and industrial processes. Methane can be traced in part to growing populations of cattle, the decay of organic matter in rice paddies and landfills, and the production of fossil

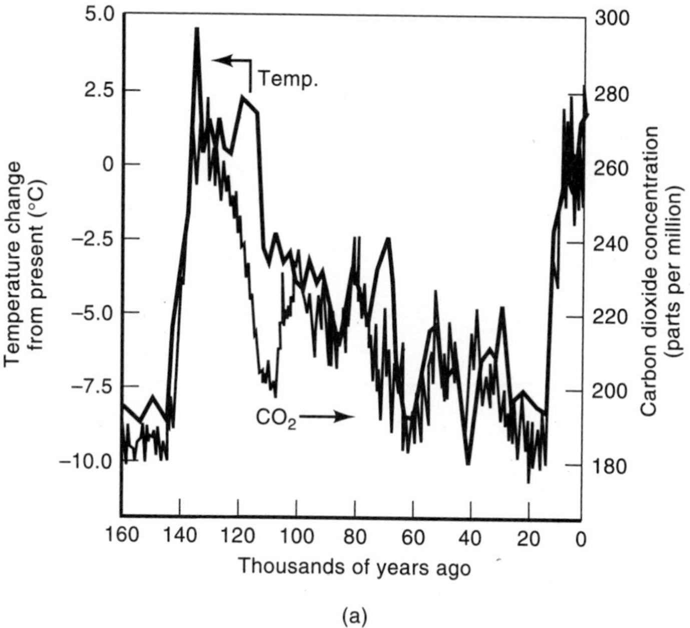

(a)

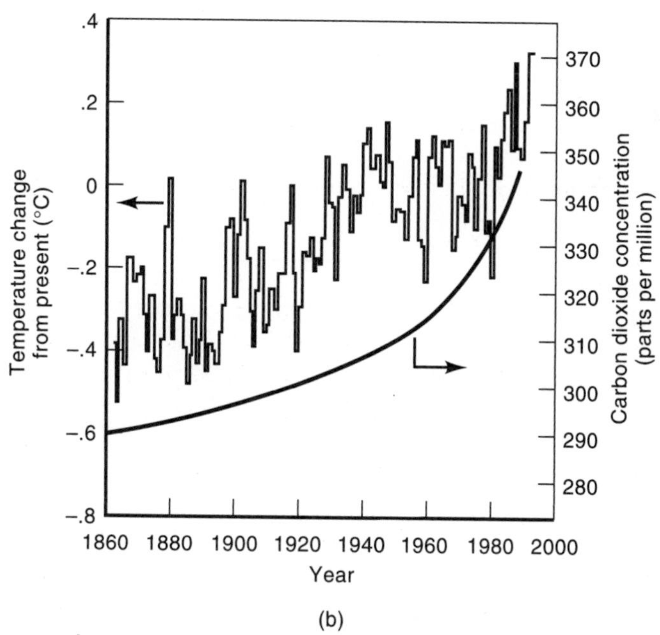

(b)

FIGURE 9.1

The correlation between carbon dioxide concentrations and the earth's temperature over the past 160,000 years is reflected in the top graph (*a*). The correlation is not as obvious over the past 100 years, as shown in the bottom graph (*b*). (See also Figure 1.1.) (H. IKEN. *SCIENTIFIC AMERICAN*, 1989; 261: 74 [SEPTEMBER]).

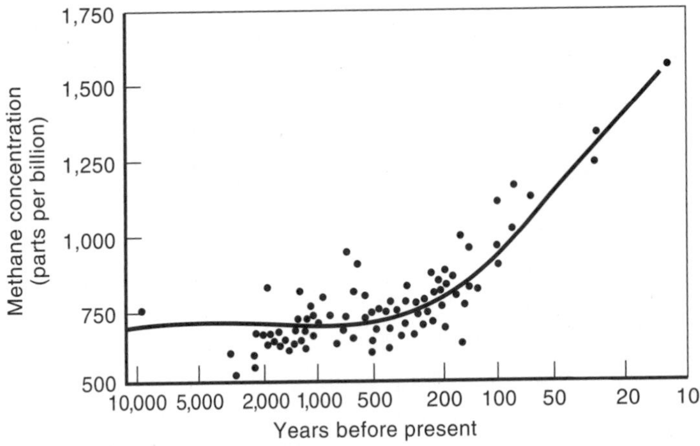

FIGURE 9.2
Global concentration of methane gas over the past 10,000 years indicates a dramatic increase beginning about 100 years ago. The data was obtained from air bubbles trapped in ice in Greenland. (H. IKEN. *SCIENTIFIC AMERICAN*, 1989: 261; 63 [SEPTEMBER]).

fuels. Figure 9.2 shows methane concentrations over the past 10,000 years. Methane concentrations held steady since the end of the last ice age about 10,000 years ago, but began to climb about 100 years ago at a rate of 1% per year.

Table 9.1 provides information about greenhouse gases: their sources, present concentrations, atmospheric lifetimes, and **global warming potential (GWP)—**

Table 9.1 GREENHOUSE GASES

Gas	Sources	U.S. Emissions (MT/y)	GWP*	Atmospheric Lifetime (y)	1995 Concentration (ppM)
CO_2	Fossil fuels, deforestation	5500	1	100	360
Methane	Rice fields, cattle, landfills	300–400	21	10	1.7
Nitrogen oxides	Fertilizers, deforestation	15	310	170	0.31
CFCs	Aerosol sprays, refrigerants	1	1300–12,000	70–100	0.003 (Cl atoms)

*GWP = Global Warming Potential, which is related to a molecule's ability to absorb thermal radiation relative to that of CO_2.

the ability of each greenhouse gas to trap heat. Greenhouse warming is clearly an international issue. Figure 9.3 shows the global distribution of CO_2 emissions, as well as the source of such U.S. emissions. About 60% of the greenhouse gases are CO_2. Energy-related activities account for about 80% of the CO_2 put into the atmosphere each year. Figure 9.4 shows per capita CO_2 emissions for those countries with the highest industrial emissions in 1995. The United States emitted about 21 metric tons/person/year. Comparable numbers from other countries

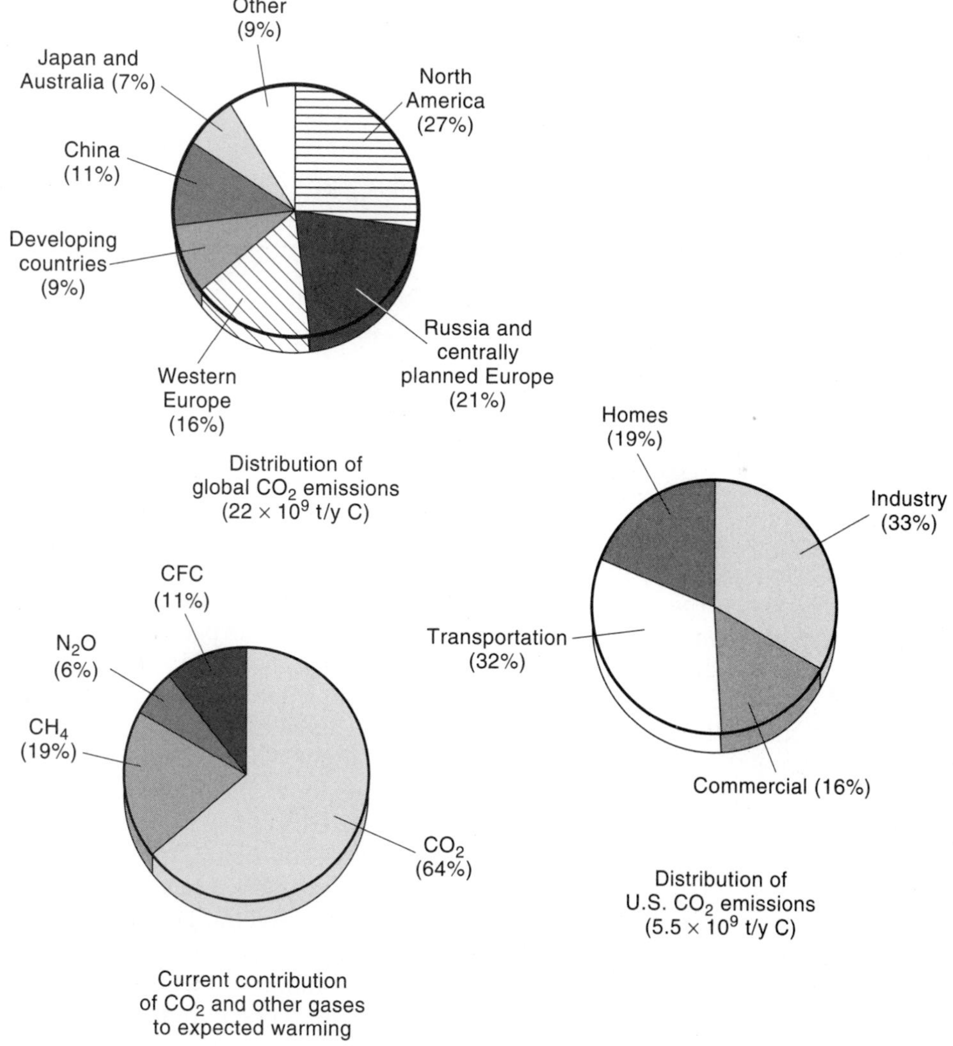

FIGURE 9.3

Distribution of carbon dioxide emissions from fossil fuels, 1996. (ELECTRIC POWER RESEARCH INSTITUTE [EPIR] JOURNAL)

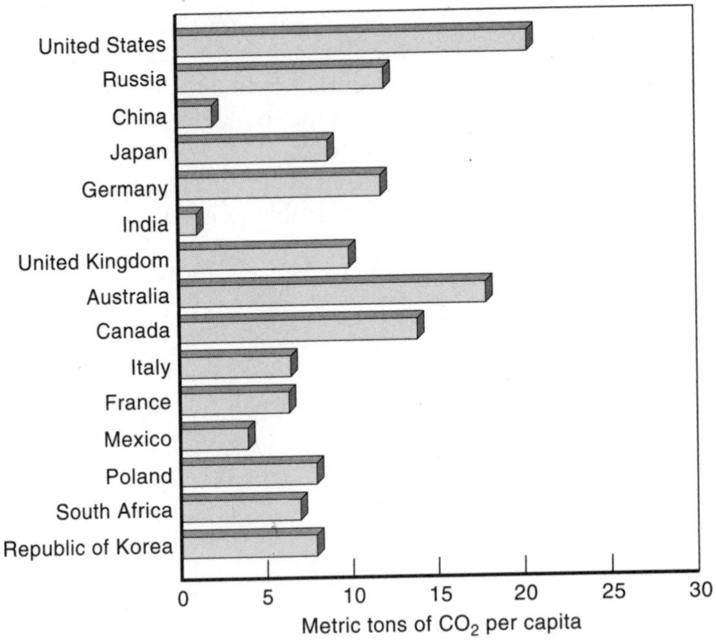

FIGURE 9.4
Annual per capita carbon dioxide (CO_2) releases for the 15 countries with the highest total emissions, 1995. (WORLD RESOURCES INSTITUTE, WASHINGTON D. C.)

are Germany (11), Japan (9.3), China (2.8), and India (1.1). The highest per capita emissions of carbon dioxide are from the United Arab Emirates at about 30 metric tons per year. As developing countries continue to expand, these numbers will grow. Within five years, some experts predict that China will replace the United States with the greatest total CO_2 emissions. Table 9.2 lists annual carbon emissions for the top countries.

One of the first observations of our ongoing "experiment" with greenhouse gases came in the 1970s from measurements of carbon dioxide levels in the atmosphere. Data from the Mauna Loa Observatory, Hawaii, showed that atmospheric CO_2 concentrations were steadily rising, from 315 ppM in 1958 to about 360 ppM today (Fig. 1.1). From such data it has been estimated that if worldwide fossil fuel use continues to grow at the present rate of 1 to 2% per year, a doubling of CO_2 concentration can be expected by the middle of this century. Because CO_2 acts as a thermal blanket in the atmosphere, this doubling of CO_2 concentration could mean an average global temperature increase of from 1.5°C to 4.5°C; in the extreme, the earth's climate would be warmer than it has been in 2 million years. Although the earth has had dramatic changes in its climate (in the last billion years it has experienced four

Table 9.2 EMISSIONS OF CARBON DIOXIDE, BY COUNTRY, 1996

Country	Emissions (billion tons/y)	Per capita emissions (tons/y)
United States	5310	21.2
China	3370	2.78
Russia	1582	10.7
Japan	1170	9.3
India	998	1.06
Germany	862	10.5
United Kingdom	558	9.5
Canada	411	13.8
Republic of Korea	407	9.0
Italy	404	7.1
Ukraine	396	7.7
France	363	6.2
Poland	356	9.2
Mexico	349	3.7
Australia	308	17

(Oak Ridge National Laboratory)

major ice ages), these projected temperature increases are substantial. In the past 125,000 years, temperature variations have been only about ±5°C. In the past 100 years, the earth's mean temperature has risen between 0.3°C and 0.6°C. A temperature increase of 2°C by the year 2050 (as forecast by some climatic models) corresponds to a 0.3°C-per-decade change—seven times the historical rate.

Additional evidence for global warming is found in the rapid recession of glaciers outside the polar areas and a decrease in snow cover in some areas, such as Europe. The glacier on Mount Kenya shrunk by 40% between 1963 and 1987. In terms of temperatures, 10 of the 15 warmest years ever recorded have occurred within the last 16 years. Other observations include earlier springs and later winters over higher latitudes in the Northern Hemisphere and a shift in the geographic range of some animal species toward the poles. A recent British study showed that birds are laying eggs earlier in the spring.

Iceberg and Portage Glacier, Alaska. (GARY BUSS/FPG INTERNATIONAL LLC.)

Determining the impacts of global warming in the 21st century is very difficult and fraught with many uncertainties. This is due largely to the fact that the large computer models used to do the simulations and predict future climates are ill-equipped to simulate how things may change in local regions. Climate models are numerical representations of complex physical processes that depend upon exchange of heat and water between the atmosphere and the ocean, atmospheric compositions, solar radiation, cloud cover, and other environmental conditions. The potential implications of this warming trend are numerous and awesome.

1. Increased global temperatures would not be geologically uniform and would be larger at the poles and could lead to a melting of the polar ice caps, raising the levels of the oceans anywhere from 0.3 to 7 m (1 to 23 ft). Almost all the world's ports could be put under water, as well as coastal farmland in many countries. In this last century, sea levels have risen 10 to 25 cm.
2. Changes in precipitation and altered storm patterns could have serious effects on agriculture through a shift in productive areas and a change in the growing season. One prediction envisions the U.S. grain belt slowly moving north into Canada during this time.
3. A few degrees warmer might seem nicer in the middle of winter, but summer days could become unbearable. In Washington, D.C., it is estimated that 87 days could have temperatures of more than 90°F; there is an average of 36 now.
4. Because the ocean currents are driven by temperature differences between the poles and equator, some areas (such as Europe) could become colder as a result of changes in circulation patterns of the oceans. Some of the most noticeable effects could be increased intensity of tropical storms.

Although most (but not all) scientists agree that increasing carbon dioxide concentrations will affect the earth's atmosphere, the question is how much

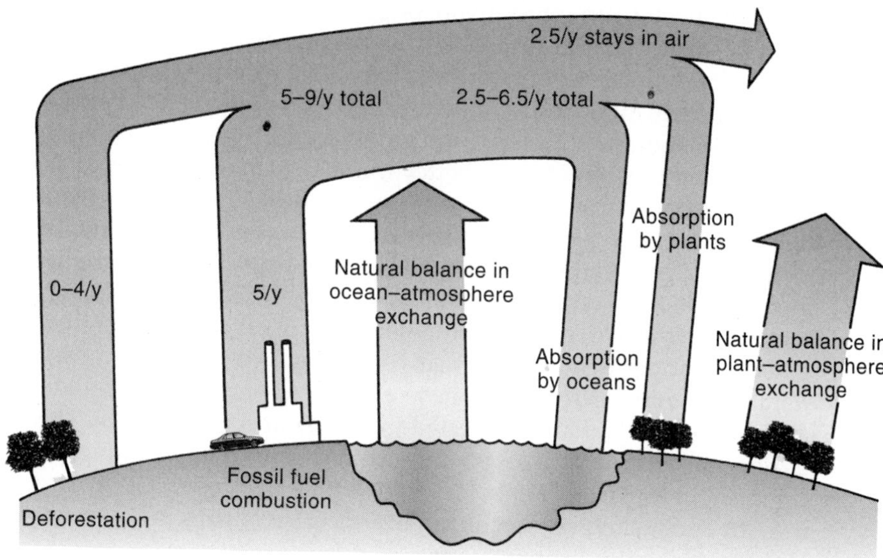

FIGURE 9.5

The carbon cycle of the earth, showing flows of carbon among the air, plants, oceans, fossil fuel combustion, and deforestation. (Emissions in billions of tons per year.) Without human influence, the flows of carbon among the air, plants, and oceans would be roughly balanced. (*EPRI* JOURNAL)

and how fast the temperatures will rise. A key element in predicting that rise is an understanding of the carbon cycle (Fig. 9.5). Without human influence, the flows of carbon between the air, plants, and oceans would be roughly balanced. Fossil fuel combustion adds about 5 billion tons of carbon per year to the atmosphere.* About half of this is absorbed by the oceans and by plants, while the other half remains in the atmosphere. While plants provide a sink for carbon dioxide, deforestation may cause plants to be a net source. The carbon added to our atmosphere by the results of deforestation—burning and decay—is estimated to be about 1 to 2 billion tons per year. Since the oceans contain about 55 times as much carbon as the atmosphere and 20 times as much as land plants, changes in the oceans' ability to absorb and store carbon are crucial in understanding the carbon cycle.

Much of the uncertainty associated with predictions of warming trends depends on an understanding of the sizes of various **feedbacks** that may occur in response to the changes we are making in the climate. By saying that a system has "feedback," we mean that a signal from the output is returned to the

*To convert to CO_2 emissions, multiply by $^{44}/_{12} = 3.67$, which is the ratio of the mass of CO_2 to C.

input to affect it either negatively or positively. Negative feedback will have a cooling effect that will reverse or dampen the warming trend, while positive feedback will increase the warming trend. For example, warmer temperatures will evaporate more seawater and, since water vapor is a better absorber of infrared radiation than CO_2, it might exert a positive feedback on the temperature change by providing a better thermal blanket. However, it could also produce a thicker cloud cover, which will reflect sunlight; this would be an example of negative feedback that would reduce the warming trend. In another example, higher carbon dioxide levels have been shown in the laboratory to enhance crop growth, which will lead to enhanced sinks for CO_2, a negative feedback. However, warmer temperatures will also provide higher rates of organic matter decay in soils, a positive feedback. An increase in the earth's temperature could decrease its snow and ice cover, which would reduce the amount of solar radiation reflected from the earth (the albedo). This would increase the absorption of solar energy and thus further encourage the warming trend. Figure 9.6 examines some of these potential feedbacks to climatic warming. An understanding of which feedback mechanisms are dominant is important in making predictions.

The climatic system of the earth is very complex, and knowledge about climate change is developing rapidly. New findings on the roles of local air pollution and ozone depletion in moderating the greenhouse effect have lowered computer projections on the rate of global warming. There also have been some recent doubts raised on the perceived increase of CO_2 in the atmosphere. Remember that air trapped in ice cores from the Antarctica showed CO_2 concentrations were lower in the ice ages than at present. However, questions have been raised about sampling techniques. Some of the CO_2 in the air bubbles might remain trapped when the ice is crushed and never show up in laboratory measurements.

Even though there is a good deal of uncertainty about the impacts of increasing concentrations of greenhouse gases in our atmosphere, if we wait until it's obvious that warming trends are occurring, it might be too late to do anything about them and their consequences. Debates in both scientific and political circles will occur in the years ahead over the appropriate course of action. One view holds that we do not yet know enough about what is occurring to be able to take appropriate actions. More research is required. This view argues that large uncertainties in climate projections make it unwise to spend large sums of money trying to avert outcomes that may never materialize. A second view states that we should accept the fact that climate change is inevitable and begin now to adapt to warmer temperatures, higher ocean levels, shifts in farmland, and so forth. Another view is that we must begin now to make changes in our lifestyles and technologies to reduce the severity of these potential climatic changes.

What, if anything, can be done to prevent these concentrations from rising to unacceptable levels? Energy policy is certainly one area in which changes can be made to reduce CO_2 emissions. Greater emphasis on energy conservation, economic incentives, renewable energy technologies, and nuclear power

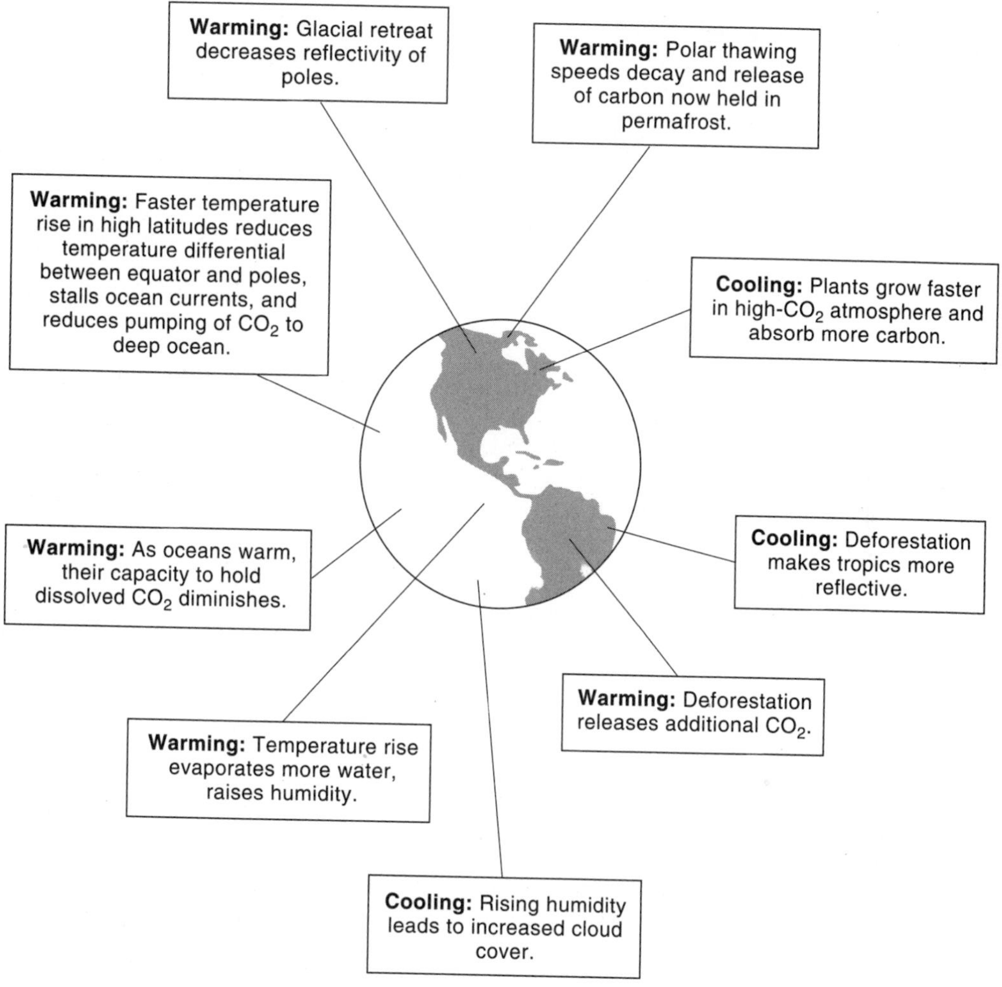

FIGURE 9.6

Potential feedbacks to global warming. Positive feedbacks are expected to increase the warming, while negative feedbacks will probably have a cooling effect. (*EPRI* JOURNAL)

are a few of the options. Natural gas-fired electricity produces about 60% less CO_2 per kWh generated than coal, and the use of gas is growing. We must also be aware of how much our own activities produce CO_2. An American emits on the average of 21 metric tons of CO_2 per year—six times more than a person in a developing country. Table 9.3 depicts CO_2 production for some common activities. Increased energy efficiency will also cut down on emissions while allowing us to maintain the same level of activity.

Table 9.3 INDIVIDUAL CARBON DIOXIDE PRODUCTION

Use	Basis	CO_2 Emissions (lb)
Automobile	Per gallon	20
Electricity	Per kWh (using coal)	2
Natural gas	Per therm (for heating)	12

A recent suggestion to reduce the amount of CO_2 being added to our atmosphere is **carbon sequestration.** The goal here is to capture carbon at its source and direct it to nonatmospheric sinks. This can be done to the CO_2 by (1) absorbing it from the gas stream by contact with a solvent, (2) injection into old oil wells, (3) pumping it into the oceans for trapping in sediments, (4) directing it more efficiently to terrestrial ecosystems, such as forests, vegetation, soils. A Norwegian company is currently sequestering CO_2 in deep salt reservoirs under the North Sea. Much research needs to be done on this approach.

For these approaches to be effective, international strategies will have to be developed and followed. Most governments have a built-in bias against solving long-term problems. Steps to slow the rate of global warming will be costly

Focus On 9.1

CARBON TAXES

One method to reduce greenhouse gas emissions is the imposition of a carbon tax that would be levied on fuels according to their carbon emission intensity. This would encourage consumers to switch to alternative fuels, including electricity produced by wind, photovoltaics, and solar thermal sources. The carbon tax would impact most strongly on coal, which produces 21% more CO_2 than oil per unit of thermal output and 76% more than natural gas. Forcing transitions away from the fossil fuels will be very expensive. The timing is important; a change too early or too late in the development of alternative technologies probably will have a much higher price tag.

and controversial. Regulations on fossil fuel use certainly will be burdensome on some groups, and developing nations will be hard-pressed to agree to controls on fossil-fuel combustion and increased energy conservation.

In June 1992, at the Earth Summit in Rio de Janeiro, 167 nations ratified the U.N. Framework Convention on Climate Change. One of the documents (the "Rio Declaration") stressed that one nation can no longer do anything it wants with its environment without regard to the impact on other countries. The "Climate Convention" treaty established a goal to stabilize (at 1990 levels) carbon dioxide and other greenhouse gases emissions by the year 2000. However, this agreement was voluntary and developed nations failed to attain the emission restrictions. In fact, U.S. CO_2 emissions have grown more quickly than anticipated due to robust economic growth, lower energy prices, and slower gains in energy efficiency and renewal energy technology penetration into the marketplace.

In December 1997, another international conference on the environment was held in Kyoto, Japan. (See Focus On 1.3, The Kyoto Protocol on Climate Change.) The nations developed a protocol that aimed to cut emissions of greenhouse gases by developed countries by 5% from their 1990 levels by the year 2010. (The U.S. target is a 7% decrease from 1990 emissions levels.) However, the Kyoto Protocol does not set any binding limits on emissions from developing countries. Partly due to this, the U.S. Senate has not yet ratified this treaty.

The effects of global warming will not be felt equally around the world. Differences from region to region could be both in the magnitude and in the rate of climate change. Some nations (and regions) will likely experience more adverse effects than others, while some nations may benefit more than others. Poorer nations are generally more vulnerable to the consequences of global warming. These nations tend to be more dependent on climate-sensitive sectors, such as subsistence agriculture, and lack the resources to protect themselves against the changes that global warming may bring. The IPCC (Intergovernmental Panel on Climate Change) has identified Africa as the continent most vulnerable to the impacts of projected changes because widespread poverty limits adaptation capabilities. (See Focus On 9.2, Africa and Global Warming.)

The political consequences of a global temperature rise of several degrees and the accompanying changes in regional climates and agricultural productivity can be quite unsettling. The atmosphere appears to be sensitive to even small chemical changes, and so there is also the possibility of unwanted surprises

ACTIVITY 9.2

Calculate the amount of carbon dioxide emitted annually by your household. You will have to estimate the amount of electricity, natural gas, and gasoline used. Use the approximate conversions of Table 9.3.

Focus On 9.2

AFRICA AND GLOBAL WARMING

Because of widespread poverty and rapid population growth, Africa is the continent most vulnerable to the potential impacts of global warming. Agriculture is the economic mainstay in most African countries, contributing up to 55% of the total value of African exports. In most African countries, farming depends entirely on the quality of the rainy season—a situation that makes Africa particularly vulnerable to climate change. As lands become less productive under new climate conditions, people living there may be forced to migrate to urban areas, where infrastructure already is approaching its limits as a result of population pressure and lack of resources. A warmer climate could also open up new areas for diseases, which will stress the already weak economies.

occurring in shorter times than forecasters might envision. Governments are held accountable on matters involving national security and the basic freedoms of its citizens. Should they also be aware of the environmental costs to a society that has the "freedom" to experiment with its own future?

C. Ozone Depletion

Ozone (O_3) in the earth's atmosphere occurs in two separate locations and presents a different concern for each region. Ozone depletion is a separate issue and not connected to the global warming problem; it is another illustration of the impact of human activity on our climate and ecosystem. The previous chapter discussed ozone in the surrounding air (up to about 10 km above the earth's surface—the troposphere) as an air pollutant and a key component of urban smog. Organic compounds from industry and transportation react with nitrogen oxides to produce ozone. The increase of ozone near ground level (especially in urban areas) is a significant problem. On the other hand, there is also a concern for the decrease in ozone in our upper atmosphere—the stratosphere—10 to 50 km above the earth. Here a layer of relatively high ozone concentration (about 300 parts per billion) is responsible for protecting life on our planet by absorbing much of the sun's dangerous ultraviolet radiation. Today

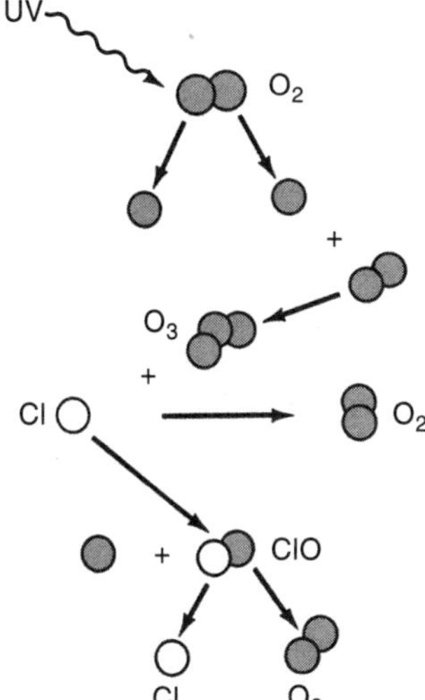

FIGURE 9.7
Ozone in the stratosphere reacts with a free chlorine atom to create chlorine oxide (ClO) and an oxygen molecule. A free oxygen atom breaks up the ClO, allowing the Cl to go on and destroy another ozone molecule.

we are observing a depletion in ozone concentrations in the stratosphere. A NASA study in 1988 concluded that ozone levels over most of the United States fell 2.3% between 1967 and 1987. There was three times as much ozone depletion in the 1980s as there was in the 1970s. Studies in the 1990s showed that ozone concentrations were 9% below normal in both the Northern and Southern Hemispheres, especially at higher latitudes. Such a reduction can cause an increase in skin cancer. It can also cause damage to crops and can destroy the beginnings of the marine food chain.

In 1974, University of California at Irvine chemists S. Roland and M. Molina* proposed that emissions of chlorinated compounds called chlorofluorocarbons (CFCs) could contribute to the destruction of stratospheric ozone (Fig. 9.7). Although inert in the lower atmosphere, CFCs can rise to the stratosphere, where they are chemically decomposed by ultraviolet radiation. This produces a free chlorine atom that vigorously attacks ozone, creating chlorine oxide and an oxygen molecule. Since it doesn't become part of a molecule to be taken out of the reaction, one chlorine atom can destroy as many as 100,000 ozone molecules! Because CFC compounds are inert until they reach the upper atmosphere, they remain potentially dangerous for about 100 years.

CFCs have many commercial applications: aerosol propellants, refrigerants, blowing agents for foams (including those used in fast-food packaging), and solvents (Fig. 9.8). On the basis of the warning of Roland and Molina, as well as other scientists, and at the urging of the National Resource Defense Council, CFC propellants in aerosol spray cans were banned in the United States in 1978.

*Roland and Molina won the 1995 Nobel Prize in Chemistry for this work.

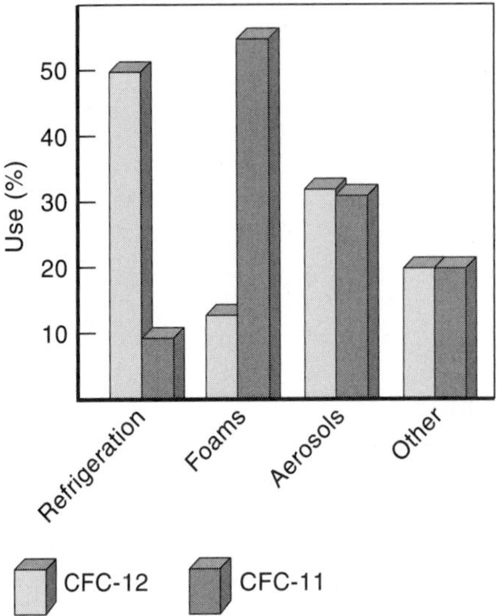

FIGURE 9.8
Uses of CFC-11 and CFC-12. (UNITED STATES ENVIRONMENTAL PROTECTION AGENCY)

(This was not without considerable pressure from the chemical industry to wait until the evidence and facts were known better.) While public concern was appeased by the aerosol banning, CFCs continued to be emitted to the atmosphere through other avenues. The amount of chlorine in the atmosphere is now twice what it was when the alarm was sounded in 1974 and five times what it was in 1950! Table 9.4 lists the major ozone-depleting CFCs.

Additional evidence of the effect of CFCs on the ozone layer emerged in the mid-1980s. In 1985, British scientists, using satellite data of ozone concentrations over Antarctica, observed that during each spring (September and October), ozone concentrations dropped to nearly half of what they had been 20 years before! This discovery of reduced ozone concentrations, popularly called the "ozone hole," spurred major international research efforts, which revealed that ozone losses were also occurring over middle latitudes and the Arctic. High levels of chlorine monoxide, the chemical link between ozone destruction and CFCs, were also observed.

Although some features are still vague, what seems to be occurring in the stratosphere over the Antarctic are reactions that need special conditions of darkness and cold. Ice particles in the atmosphere provide a surface for chemical reactions that release the chlorine from the CFCs, which then attacks the ozone in the spring when triggered by sunlight. In the fall, the hole is drastically reduced in size. The CFC gases take from seven to ten years to rise into the stratosphere, making effects difficult to observe for years.

The EPA estimates that for every 1% decrease in ultraviolet radiation absorbed in the atmosphere, there will be a 2% increase in skin cancer. Ultraviolet light not only causes skin cancers, but also damages our immune system, mak-

Table 9.4 USE AND EMISSIONS PROFILES OF COMMONLY USED CHEMICALS CONTRIBUTING TO OZONE DEPLETION

Chemical	1985 Emissions (thousand tons)	Atmospheric Lifetime* (y)	Applications	Annual Growth Rate (%)	Percent Share of Contribution to Ozone Depletion
CFC-11	238	76	Foams, aerosols, refrigeration	5	26
CFC-12	412	139	Air conditioning, refrigeration, aerosols, foams	5	45
HCFC-22	72	22	Refrigeration, foams	11	0
CFC-113	138	92	Solvents	10	12
Halon 1211	3	12	Fire extinguishers	23	1
Halon 1301	3	101	Fire extinguishers	Not applicable	4
Methyl Chloroform	474	8	Solvents	75	5
Carbon Tetrachloride	66	67	Solvents	1	8

*Time it takes for 63% of the chemical to be washed out by the atmosphere. (World Watch Institute [www.worldwatch.org], State of the World, 1989)

ing us more vulnerable to some infections and diseases. Ultraviolet light also seems to affect marine life. One of the basic foods for marine life is the single-celled phytoplankton, which floats near the surface. Ultraviolet light makes such plants less efficient in harvesting sunlight (through photosynthesis). A 1992 study found that phytoplankton in the waters around Antarctica had declined by 6 to 12%. Consequently, less food will be produced and less CO_2 absorbed. This latter situation points out a coupling between ozone depletion and global warming. The decrease of a terrestrial mechanism for CO_2 absorption can lead to increased CO_2 concentrations in the atmosphere. CFCs also contribute directly to the absorption of infrared radiation.

Two other major ozone-depleting compounds are carbon tetrachloride and methyl chloroform (known also as trichloroethane). By mass, carbon tetrachloride is more ozone-depleting than any of the CFCs listed in Table 9.4. Even if all CFCs were completely phased out, stratospheric chlorine levels could still increase because of the increased use of these chemicals. Both of these chemicals are used primarily as solvents, so substitution is relatively easy for some uses (with such things as soap and water). Since carbon tetrachloride is a cheap solvent, and less expensive than other replacements for CFCs, these chemicals are quite attractive to developing countries.

To replace CFCs, we must find chemicals that either contain no chlorine or are unstable enough that they will break down in the lower atmosphere before they reach the ozone layer. The most promising substitutes are hydrofluorocarbons (HFCs) and hydrochlorofluorocarbons (HCFCs). The HFCs do not contain chlorine and thus will not deplete ozone. HCFCs contain chlorine, but will break down in the lower atmosphere. However, both HFCs and HCFCs can contribute to global warming. Thus we must be concerned about producing chemicals that might bring about wholly unexpected consequences decades from now.

It's only when CFCs escape into the atmosphere that they cause a problem to the ozone layer. Even though car air conditioners contain less than half of all CFCs, they account for three quarters of CFC emissions. Older car air conditioning systems mainly use CFC-12. Eventually all of it is expected to leak out. The main substitute being pursued is HFC-134a. However, major changes in an older car's air conditioning system would have to be made to use HFC-134a.

In 1987, many of the nations that produce and use CFCs met in Montreal. An international treaty was signed that called for a freeze in CFC production at 1986 levels, a 50% reduction by the year 1999, and full phase-out of CFCs by the year 2000. This "Montreal Protocol" is important not only for the magnitude and the speed of the changes agreed on but also for being the "first truly global treaty that offers protection to every single human being" (U.N. Environment Director, Dr. M. Tolba). The United States ended the production of ozone depleting CFCs in 1996.

Like many other environmental problems, there exist wide differences among nations on the appropriate balance between economic development and protection of the global environment. The Montreal Protocol allows participating Third World countries to increase CFC use for ten years before they must reduce consumption by 50%. Many such countries make extensive use of CFC-based refrigeration and cannot afford the replacement chemicals. Some substitutes also may prove to be less durable and less energy efficient. Many developing countries feel that they need the help of the rich countries to use appropriate substitutes, not just an international protocol with lofty objectives. Therefore the protocol set up a fund to help developing nations pay for new technologies and equipment conversions. Large demand for these chemicals still exists in the developing world. For example, China, in the midst of strong economic growth, has seen the sales of refrigerators jump from 2 per 1000 households in 1981 to 423 per 1000 in 1990 to 750 per 1000 in 1998. Obviously, broader Third World participation is needed if worldwide consumption of CFCs is to be reduced.

D. Thermal Pollution

Thermal pollution is defined as the addition of unwanted heat to the environment, in particular, to natural waters. "Pollution" in this case is not the visible "dirtying" of water but the impairment or modification of a lake or river's environment. It might take some time before these effects become visible. The greatest source of heated water is from steam-electric generating stations. As discussed in Chapter 4, a condensing unit is needed after the turbine to complete the steam cycle and to improve the efficiency of the power plant (Fig. 3.3). In the condenser, thermal energy is removed from the hot steam by passing cooling water through the condenser coils. This cooling water is usually taken from, and discharged to, a body of water such as a lake or river.

The quantity of water passing through the condenser is very large. Recall from Chapter 4 that the addition of heat Q to a substance of mass m leads to an increase in temperature given by the equation

$$Q = mc\Delta T$$

where c is the specific heat of the substance. The amount ΔT by which the temperature increases is related to the mass, m, of water flowing through the condenser and the amount of heat that is added. A higher water flow rate will reduce the temperature rise ΔT. For a standard 1000-MWe power plant, a flow of about 10,000 gallons of water per second (1200 ft^3/s) must be maintained through the cooling coils to limit the temperature rise of the return water to 8°C (15°F). This volume is equivalent to about one quarter of the daily water needs of New York City.

Figure 9.9 shows the past and projected uses of water in the United States. The water demands for electrical generating plants account for about 50% of current water usage. Most of this water can be used again, since it passes through the condenser only once, but heat energy has been added with a subsequent 6°C to 17°C (10°F to 30°F) rise in temperature. The total freshwater runoff in the United States, which includes times of floods, is about 1200 billion gallons per day. The current usage of electrical plants is about 200 billion gallons per day! Increased use could put a severe thermal strain on some rivers in the near future, especially during normal or low flows. To meet this problem, all new power plants completed after July 1977 were required to use closed cooling systems (see the section on "Cooling Towers and Ponds" later in this chapter), for example, their water would not be drawn directly from a lake or river.

Fossil- and nuclear-fueled power plants have different impacts on the environment, but one common feature is their emission of waste heat. A nuclear plant emits about 40 to 50% more waste heat into the water than a fossil-fueled plant of the same electrical output. This is because fossil-fueled plants have a higher efficiency, since they can use higher steam temperatures, and because some of the waste heat leaves the fossil-fueled plant via stack gas emissions. Some of the waste heat characteristics of typical steam electric power plants are given in Table 9.5.

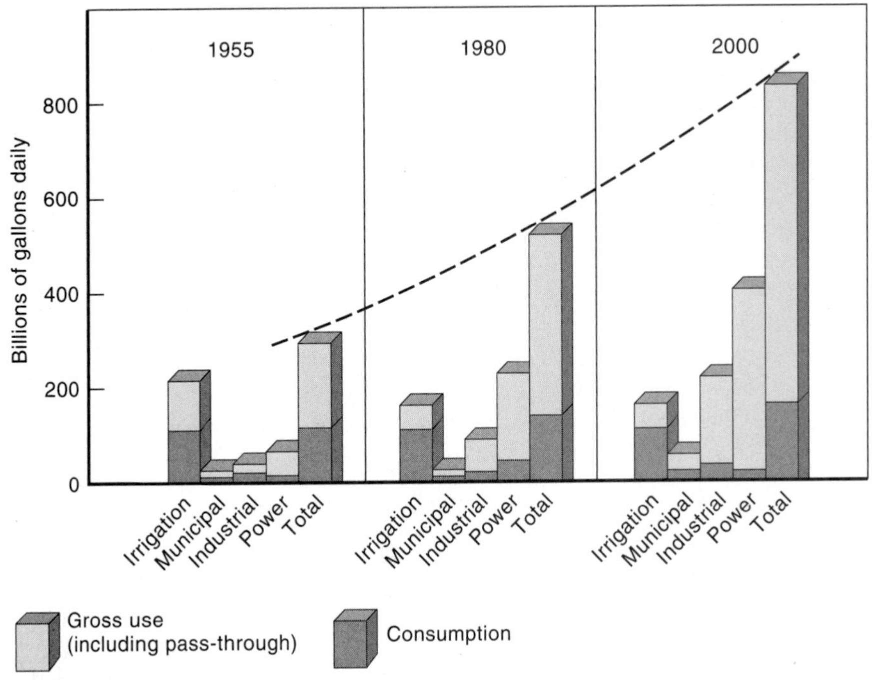

FIGURE 9.9
Present and projected water uses in the United States. (UNITED STATES DEPARTMENT OF THE INTERIOR)

Table 9.5 HEAT CHARACTERISTICS OF TYPICAL STEAM ELECTRIC PLANTS (VALUES IN Btu PER kWh)

Plant Type	Thermal Efficiency (%)	Required Heat Input	Total Waste Heat	Lost to Boiler Stack*	Heat Discharged to the Condenser	Cooling Water Requirement (ft^3/s/MW of capacity†)
Fossil fuel	33	10,500	7100	1600	5500	1.6
Fossil fuel (recent)	40	8600	5200	1300	3900	1.15
Light water reactor	33	10,500	7100	500	6600	1.9

*Approximately 10 to 15% × required input for fossil fuel; approximately 3 to 5% × required input for nuclear.

†Based on an inlet temperature in the range of 70°F to 80°F and a temperature rise across the condenser of 15°F (8°C).

Source: R. Rimberg, "Utilization of Waste Heat from Power Plants," William Andrews Pub. LLC, 1974.

E. Ecological Effects of Thermal Pollution

Aquatic Life

The effects on the aquatic environment as a result of dumping waste heat into a river or lake are numerous. Increased water temperatures lead to

- decreased ability of water to hold oxygen
- increased rate of chemical reactions
- changes in reproduction, behavior, and growth patterns throughout the food chain
- long-term damage (including possible "death") to natural bodies of water (including eutrophication, discussed in the next section)

Temperature is one of the most important factors governing the occurrence and behavior of life. In cold-blooded animals such as fish, body temperatures are linked quite closely to the environmental temperature. Warm-blooded animals such as humans maintain a uniform body temperature, usually higher than the environment, and are less dependent on the surrounding temperature. The insulation of the body by fat, hair, or feathers is much greater for warm-blooded animals. The metabolism (chemical reactions within the body) of cold-blooded animals can be affected strongly by temperature changes. As temperatures rise, animals become more active. There seems to be an exponential dependence between temperature and metabolic rate, with a doubling every 10°C. An increase in metabolism brings about a need for more oxygen. However, the dissolved oxygen concentration in water is inversely proportional to temperature. As the water temperature increases from 16°C to 35°C (60°F to 95°F), the saturation concentration of oxygen in water drops from 10 mg/L to 7 mg/L.

Some animals adapt more easily to increased temperatures, although gradual changes are more tolerated than sudden temperature changes. Figure 9.10 shows the preferred temperature ranges for several species of fish and shows upper lethal temperatures. Drastic declines in fish population occur when temperatures rise even a few degrees above the limit. A temperature of 34°C (93°F) is usually taken as an upper limit for aquatic life.

Growth and reproduction of fish as a function of water temperature vary among species. Usually young fish grow faster with increased temperatures because of their increased metabolism. Figure 9.11 shows the effects of temperature on growth rates for several animals and fish. Shrimp growth is increased by 80% when water is maintained at 27°C instead of 21°C, and catfish growth is nearly three times faster at 28°C than at 24°C. Clearly one of the uses for the controlled use of waste heat is in the production of fish for food.

For fish, reproduction rates are not necessarily changed in mildly warmer waters. However, there are critical temperatures above which fish will not reproduce, and the temperature interval suitable for this is narrow. Higher

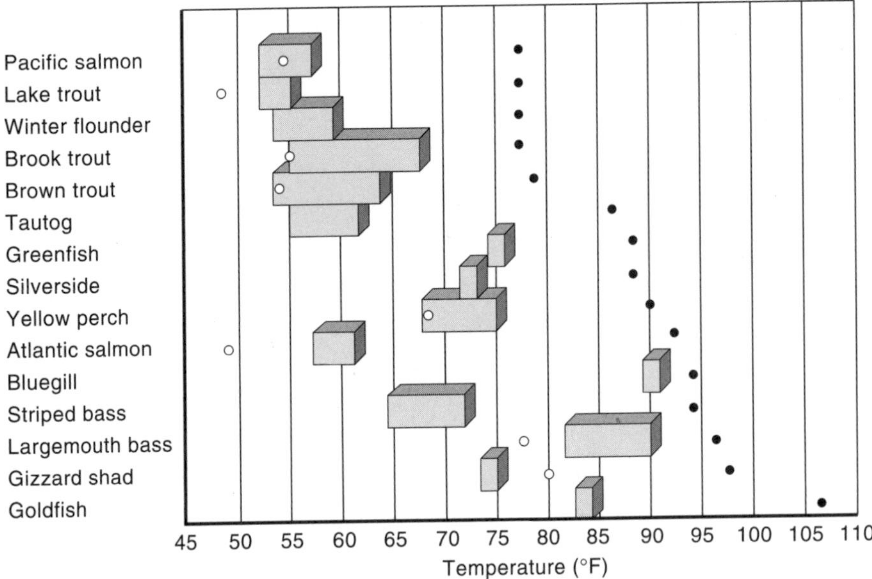

FIGURE 9.10

Sensitivity of fish to temperature. Preferred temperature ranges for some species, as determined in the field and laboratory, are shown as blocks ▪. The solid dot • indicates the upper lethal limit. The open dot ◦ is the temperature found to be best for spawning. (J. STARWOOD. *SCIENTIFIC AMERICAN*, VOL 220 [MARCH], p 24).

temperatures may cause some fish to move prematurely into a stream to spawn before conditions suitable for spawning prevail. Bluegills exhibit this symptom, and are often attracted to waters in which they cannot survive. Yellow perch are attracted to warm water but then swim at a reduced speed. This could be a significant factor in the spawning behavior of those fish that must swim against strong currents to reach their spawning grounds. Slower moving fish might not be able to escape their predators in time. Bacterial diseases also increase as the temperature increases. One famous case is that of the sockeye salmon of the Columbia River (Washington). A series of hydroelectric dams built on the river changed it from a cool, fast-flowing river to a series of warmer, slower-moving lakes. Consequently, bacterial diseases drastically reduced the salmon population.

The changes in an entire aquatic community because of temperature increases are difficult to observe or decipher because there are so many variables. Most of the work done so far on thermal effects on aquatic life has been done in the laboratory and not in the field, and not within a total community. Recent studies of the effects of thermal effluents on large lakes and rivers have not shown the dramatic changes that were observed initially on

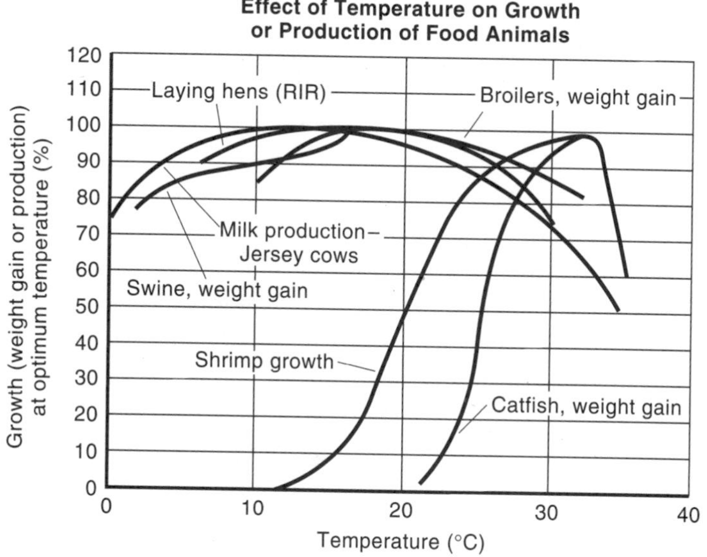

FIGURE 9.11

Effect of temperature on growth and production of food animals. (OAK RIDGE NATIONAL LABORATORY, JULY 1972, ORNL-4797)

smaller bodies of water, where the temperature remained high after heated releases. In general, long-lasting temperature changes will have effects throughout the food chain, from the plankton through algae to fish. Any change in one of these links can affect drastically the abundance of game fish. Hot water usually will lead to the elimination of thermally sensitive species, without replacement. The diversity of species within the community will decrease, although there may be a population increase in one or two dominant species.

Another problem arises from the mechanics of the cooling system itself. Fish are killed by passing through the cooling system or by hitting the intake screens. Also, chlorine is released into the cooling waters, because it is added to reduce the build-up of slime on the condenser coils. Many of the ecological damages once thought to be a result of waste heat discharges are now being traced to chlorination.

Lake Processes: Eutrophication

There are other permanent changes in aquatic communities on a large scale that can be brought about by waste heat discharges. Indeed, the very life of a lake can be destroyed. Under natural conditions, the water temperatures of a lake undergo two distinct stages. In the summer, the lake is thermally stratified

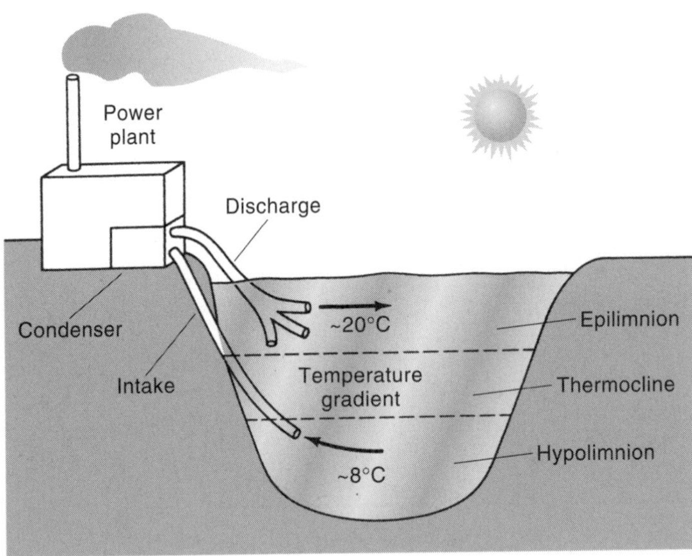

FIGURE 9.12
Stratification, or layering, of a lake during the summer. The average temperature of each layer is shown.

(Fig. 9.12); on top, water warmed by the sun forms a warm layer called the epilimnion. Because cold water is more dense than warm water, it remains at the bottom in a cold layer called the hypolimnion. The layer in between is called the thermocline, where the water temperature changes between the top and bottom layers. During the winter, mixing takes place between the warmer and colder water layers, as the water on top becomes colder and then descends because of its increased density. Heat transfer occurs through the process of convection. Mixing brings up nutrients from the bottom to enhance biological growth in the upper layers and supplies oxygen to the bottom layers.

The presence of a power plant can disturb this natural process. Colder water is taken from the hypolimnion for the condenser and hot water is discharged to the epilimnion (see Fig. 9.12). This results in an increase in the temperature of the upper water layer and consequently a lengthening of the period of stratification, which means a shorter mixing period and a reduction in the oxygen supplied to the animals in the bottom layer. Water taken from the bottom layer also brings up nutrients such as nitrogen and phosphorus that are discharged to the surface, increasing the growth of plant life, especially for those species with greater temperature adaptations. Several species of algae grow quite rapidly and can produce a green scum or a mat of algae covering the water's surface. Instead of providing additional food supplies for animal life, these aquatic plants appeal to only a small number of species and are toxic to many. As the algae die and sink to the bottom, they become food for the microscopic organisms that decompose the material. These "decomposers" demand oxygen, diminishing its availability for the fish. Further decay can lead to the creation of gases with unpleasant odors and products that are toxic to some species.

Eutrophication is the name given to the process in which a body of water is enriched by the addition of extra nutrients, stimulating the growth of algae. This process occurs in the natural aging of lakes, but is accelerated by the addition of pollutants such as phosphorus from municipal sewage (including household detergents), nitrogen from the runoff of fertilizers used in agriculture, and waste heat released from power plants.

F. Cooling Towers and Ponds

Because of the ecological impacts of thermal pollution, recent laws dictate that methods other than direct dumping into the aquatic environment must be used to dispose of the waste heat. **Cooling towers** are one of the most common means used to dispose of waste heat without putting it directly into a water system. Currently, facilities generating more than a third of our electricity use cooling towers or cooling ponds. Their use will undoubtedly increase in the years ahead.

In a wet-cooling tower, hot water from the condenser enters the tower near the top and is sprayed downward. The small droplets are cooled by evaporation as a stream of air is drawn in from the outside and circulates up through the tower. The cooled water is collected at the bottom of the tower and pumped back to the condenser. In the most common type of tower, using an open or wet

Natural-draft wet-cooling tower of the (now closed) Trojan Nuclear Power Plant, Oregon. (United States Environmental Protection Agency)

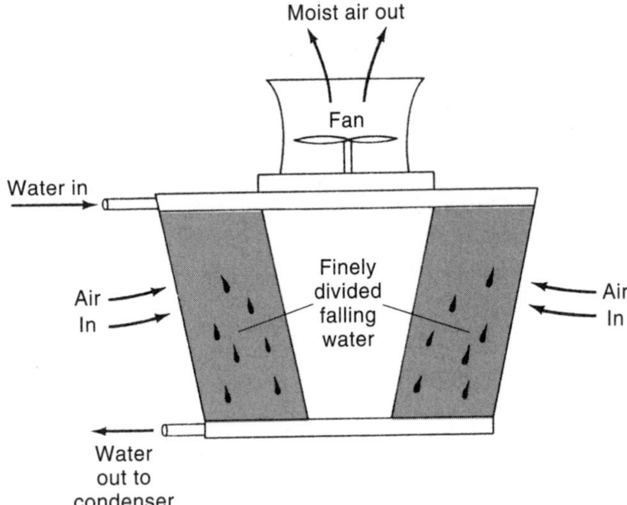

FIGURE 9.13
Mechanical-draft wet-cooling tower.

cycle, outside air comes into direct contact with the water (Fig. 9.13). The air is moved upward either by a mechanical fan or by natural draft, as in a chimney. In the mechanical-draft type of tower, the air enters from the sides of the tower and is drawn upward by the fans located on top. In a natural-draft type of tower, the warmer, moister air rises out the top and draws in cooler, drier air from the bottom. Natural-draft towers are generally larger and more expensive than the mechanical-draft types. The first natural-draft wet tower, in Kentucky, is 98 m (320 ft) high and 75 m (245 ft) wide at its base. It is capable of cooling 120,000 gal/min. Some water (about 3% of the circulating water) is lost through evaporation out the chimney.

One of the disadvantages of a wet tower is the possible alteration of local weather. In cooler, moister areas, fog may be formed. The evaporative water loss from a wet tower is comparable to a 1-in. rainfall over a square mile area per day. Another disadvantage of the wet tower is the continual emission, with the evaporated water, of chemicals that had been used to prevent slime buildup and corrosion. These emissions can cause damage to vegetation.

An alternative to the wet tower is a dry or closed type cooling tower (Fig. 9.14). In this system, as with the radiator in a car, there is no loss of water from the system. Air is moved past the cooling coils by natural or mechanical means. However, the capital cost of a dry tower is about four times as expensive per kilowatt as a natural-draft wet tower; in the wet-cooling towers, evaporation of the water itself is a cooling process. Energy costs are also important: With mechanical-draft cooling towers, 1 to 2% of the energy output of the power plant must be used to run the fans.

Another cooling device is a closed body of water like a reservoir, called a **cooling pond.** These artificial lakes are shallow to allow a maximum surface

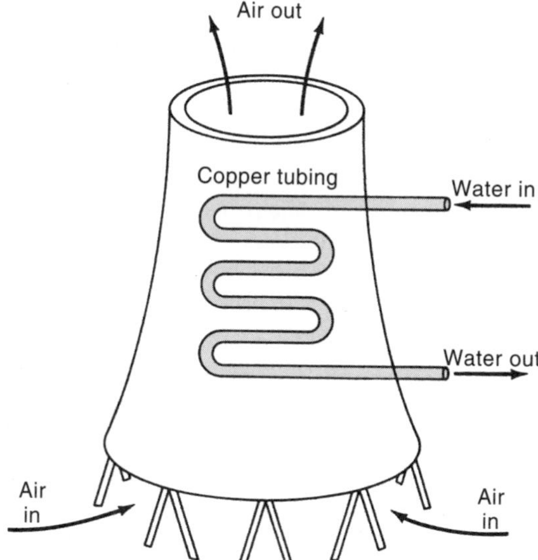

FIGURE 9.14
Natural-draft dry-cooling tower, which is much like a car radiator.

area to volume ratio for heat loss via evaporation. It is estimated that a 1000-MWe plant would need about 1000 to 2000 acres (about 5 km^2, or 2 mi^2) for a cooling pond for an allowed temperature increase of 8°C. Cooling ponds can be expensive if the power plant is located close to a city, where the cost of land is relatively expensive.

G. Using Waste Heat

In an energy conversion process, the efficiency of conversion is always less than 100% (the second law of thermodynamics). The energy that is not converted into a useful form (as into mechanical energy in the case of an electrical power plant) is "waste heat." Are there ways in which at least part of this waste heat can be used for other benefits? The answer is yes. Present developments in using waste heat include

- hot water for industrial use—cogeneration (see Section 11E)
- aquaculture, with increased fish growth through warm water cultivation
- greenhouse heating
- desalination of seawater
- increased crop growth and frost protection
- air preheating

Higher yields of finfish and shellfish through increased water temperatures appears to be a very desirable use of waste heat. The Long Island Lighting Co. (New York) ran a commercial operation for oyster cultivation for several years, and commercial shrimp farms are found in several cities in Florida. Catfish have been successfully grown in the warm effluents of the TVA's Gallatin steam plant in Tennessee. Catfish account for more than half of all aquaculture-raised fish in the United States. The Japanese have done much with aquaculture. Their results with shrimp show a 20% increase in growth rate in the summer and a 700% increase in winter if heat is added.

The use of waste heat for the heating of buildings or hot water systems is difficult because of the low temperatures of heat available. Cooling water at most power stations has an exit temperature of only 27°C to 38°C, so it is not economical to transport the water any distance at those temperatures. Consequently, in places where waste heat is used for the heating of buildings, the power plant must be very near. This is more likely to be the case if the building is a factory or a greenhouse.

On a smaller and non–power-plant scale, a good deal of heat is exhausted from a building through vented air, steam, or hot water (including sewer water). Energy can be recovered from these waste products through the use of heat exchangers, as illustrated in Figure 9.15, in which heat is transferred to a

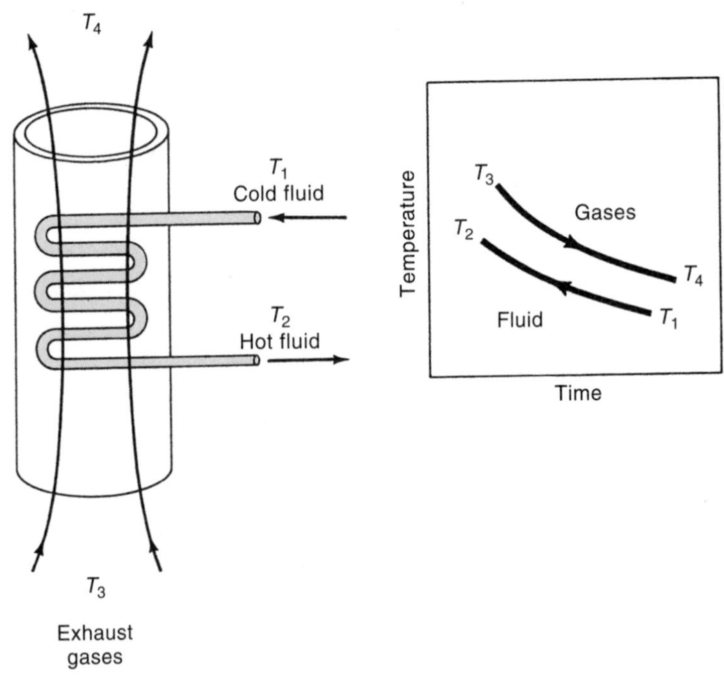

FIGURE 9.15
Air-to-liquid heat exchanger.

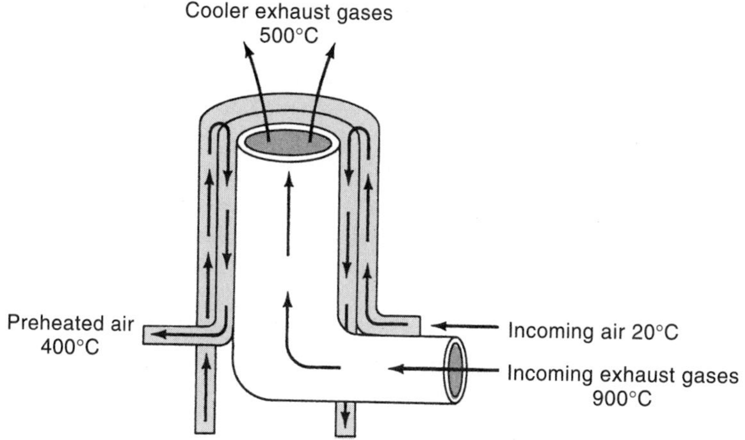

FIGURE 9.16
Industrial furnace recuperator to extract waste heat from exhaust gases.

liquid. These exhaust gases also can be used to preheat combustion air (that might enter as cold air from outside the building) for boilers and furnaces through a "recuperator," recovering maybe half of the waste energy that usually would go up the stack (Fig. 9.16). The economic payback period for these units can be less than one year.

In a residential situation, a lot of heat is wasted as hot water literally goes down the drain. Some of this energy can be recovered if the water from sink, shower, and laundry drains (termed "gray water") can be used to preheat cooler city or well water entering the water heater. A water tank with a heat exchanger is needed to hold up this gray water before it leaves the house.

H. Summary

Evidence for human impact on our climate is becoming clearer. The chemical composition of our atmosphere is changing through the emission of greenhouse gases. These gases, primarily carbon dioxide, methane, and nitrous oxide, trap heat radiated from the earth. Carbon dioxide is released to the atmosphere primarily through the combustion of fossil fuels, accounting for about 80% by weight of greenhouse gas emissions. Methane is emitted during decomposition of organic waste in landfills. Nitrogen oxide is emitted during agricultural and industrial processes. While changes in technology and industrial demands in the last 50 years have brought about a shift in the use of energy sources from a high carbon fuel like coal to lower carbon fuels like natural gas and oil, the total amount of carbon emitted into the atmosphere continues to rise.

In the last 200 years, the atmospheric concentration of CO_2 has increased nearly 30%. Exactly what impact the rising atmospheric concentrations of greenhouse gases will have upon our climate is uncertain due to the complexity of the atmospheric system. A predicted warming trend of 2°F to 6°F in the 21st century might cause increased sea levels as a result of melting of the polar ice caps. This warming will also produce changes in precipitation patterns that will affect prime agricultural regions and modify productivity. National plans to reduce emissions of greenhouse effect gases include achieving gains from increased energy efficiencies, technological fixes, fuel shifting, carbon sequestration, and economic disincentives (such as a carbon tax). Implementing such plans will not be free. What will reductions in emissions bring in terms of reduced environmental damage? And what will be the price tag? The contributions of developing countries to global warming will be even more important in the 21st century. By 2010, they might contribute 45% of all CO_2 emissions.

In our upper atmosphere, ozone (O_3) absorbs much of the sun's ultraviolet radiation. A 5 to 10% depletion of ozone concentrations in the stratosphere has been observed over the past 20 years. The consequences of this are increased skin cancer and damage to our food chain. In the Antarctica, an ozone hole, or region of depleted ozone, has been observed every spring for the last 15 years. Chlorinated compounds called CFCs (chlorofluorocarbons) are largely responsible for this depletion. They are emitted primarily from refrigerants, aerosol propellants, and cleaning solvents. The long active lifetimes of these chemicals in the atmosphere implies their destructive effects on ozone will be felt for years to come. Developed nations have agreed to cease production of CFCs by 2000 and are introducing clean substitutes.

In the operation of any steam electric generating plant, waste heat is discharged to the environment. Increased water temperatures can lead to detrimental effects on fish reproduction and growth, as well as long-term damage to the lake. Waste heat can accelerate the process of eutrophication, in which algae growth is stimulated by the addition of extra nutrients to the lake. Cooling towers can be used to dispose of waste heat without directly dumping it into a lake. Waste heat can be used for aquaculture, greenhouse heating, and heating of buildings in the vicinity of the power plant.

Internet Sites

For an up-to-date list of Internet resources related to the material in this chapter, go to the Harcourt College Publishers website at **http://www.harcourtcollege.com**. The links are in the *Energy: Its Use and the Environment* site on the Physics page. General energy related sites and some guidelines for using the World Wide Web in your class are on the inside front cover of this book.

References

Chapter 9

Ausubel, J. and H. Sladovich. 1989. *Technology and Environment.* Washington, D.C., National Academy Press.

Berger, J. 2000. *Beating the Heat: Why and How We Must Combat Global Warming.* Berkeley, CA, Berkeley Hills Books.

Brower, M. 1990. *Cool Energy: The Renewable Solution to Global Warming.* Cambridge, MA, Union of Concerned Scientists.

Cavanaugh, R. C. 1989. Global Warming and Least-Cost Energy Planning. *Annual Review of Energy,* 14.

Graedel, T. E., and P. J. Crutzen. 1989. The Changing Atmosphere. *Scientific American,* 261 (September).

Gribbin, J. 1990. *Hothouse Earth: The Greenhouse Effect and GAIA.* New York, Grove Weidenfeld.

Gyftopoulous, E. P., and R. F. Widmer. 1982. Cost-Effective Waste Energy Utilization. *Annual Review of Energy,* 7.

Khalil, M. 1999. Non-Carbon Dioxide Greenhouse Gases in the Atmosphere. *Annual Review of Energy,* 24.

Lyman, F. 1990. *The Greenhouse Trap.* Boston, Beacon Press.

Miller, T. 1999. *Living in the Environment.* 11th ed. Belmont, CA, Brooks/Cole.

Raven, P., L. Berg, and G. Johnson. 1998. *Environment.* 2nd ed. Philadelphia, Saunders College Publishing.

Rifkin, J. 1989. *Entropy: Into the Greenhouse World.* New York, Bantam.

Schneider, S. H. 1989. The Changing Climate. *Scientific American,* 261 (September).

Scientific American. 1990. *Managing Planet Earth.* New York, W. H. Freeman.

World Resources Institute. 2000. *World Resources.* New York, Oxford University Press.

QUESTIONS

1. Why does deforestation contribute to global warming?
2. What correlation has been observed between changes in the earth's temperature and increasing carbon dioxide emissions?
3. If the temperature of the earth rises enough so that the polar ice caps begin to melt, would this melting cause the global temperature to remain the same, drop, or rise higher? Why? (Consider feedback mechanisms.)
4. What strategies are available to reduce global warming? What are the detrimental consequences of such actions?
5. It has been suggested that the waste heat from a utility be used to provide heat for buildings in a city some distance away. What difficulties are involved? Why not raise the temperature of the outgoing cooling water?
6. What features of your lifestyle contribute to ozone depletion? What things would you be willing to give up to reduce this problem?
7. What reasons would developed or developing countries have for not agreeing with the conditions of the Montreal Protocol? The Kyoto Protocol?

8. Why do ozone concentrations pose both a problem and a benefit in the atmosphere?
9. Define thermal pollution.
10. What is the approximate temperature of the outgoing water from the condenser of an electrical generating plant?
11. Why do temperature changes adversely affect the life of fish?
12. What processes are responsible for lowering the temperature of river water into which waste heat had been dumped?
13. What processes in a lake are upset by the addition of hot water from a condenser?
14. What suggestions would you make to reduce CO_2 emissions in the United States? Worldwide? What disadvantages would there be to such plans?
15. Design a device for the recovery of waste heat from the combustion gases of a wood-burning stove or from the vent of a gas-fired clothes dryer. How could this heat be used?

PROBLEMS

1. How much CO_2 is produced by driving 100 mi, drying a load of clothes, and watching 5 hours of television? (See Tables 9.3 and 10.2.)
2. An automobile emits 20 lb of CO_2 for each gallon of gasoline it burns. Calculate the amount of CO_2 your vehicle emits for a round-trip to the grocery store.
3. Using the numbers of Table 9.2, at what time will the yearly emissions of CO_2 from China pass those from the United States? (Assume a rate of growth of fossil-fuel use in China to be 5% per year.)
4. If the efficiency of a geothermal plant is half that of a fossil-fueled plant, then how much more waste heat will be discharged to the environment from a geothermal facility than from a fossil-fueled plant with the same electrical output?
5. Using estimates of your residential consumption of energy (space heating, domestic hot water, lighting, appliances) for a month, estimate the size of your contribution of direct heat to the environment. Express your answer in watts/m^2 or Btu/hr/ft^2, where the area involved is the size of your building lot or apartment. Compare with local insolation.
6. If the temperature of the water leaving the condenser in a steam turbine cycle were raised from 20°C to 30°C, what would be the decrease in the Carnot efficiency? Assume a steam temperature of 500°C. (See Chapter 4.)

10

Electricity: Circuits and Superconductors

A. Introduction to "Electrification"

Electricity is accepted today so matter-of-factly and is so intertwined with our way of living that we seldom think about its fuel source or worry about its conservation. (A 1999 Roper poll concluded that less than one third of consumers know where their electricity comes from or that electricity generation is responsible for one third of all greenhouse gas emissions.) Electricity's convenience and availability make it very popular. Electricity consumption has the largest rate of growth of any major energy-use sector. Thirty-six percent of our energy resources today are used to produce electricity. In the 1950s, 1960s, and early 1970s, increased demand for electrical appliances and electric space heating, new enclosed shopping centers and sports arenas, and conversions to electrical processes in industry (in place of coal or natural gas) caused electricity

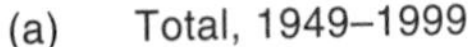

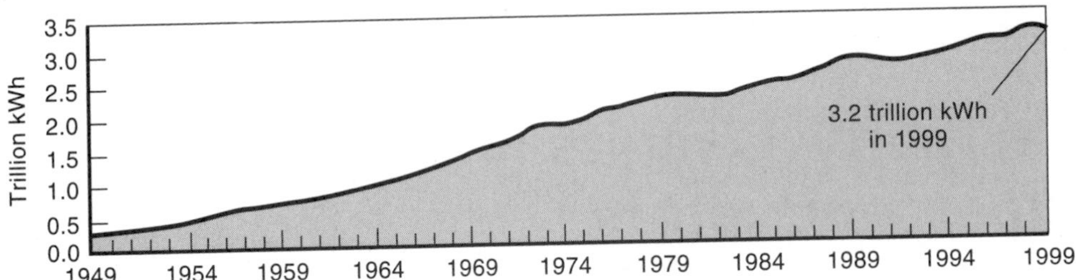

(b) By source, 1949–1999

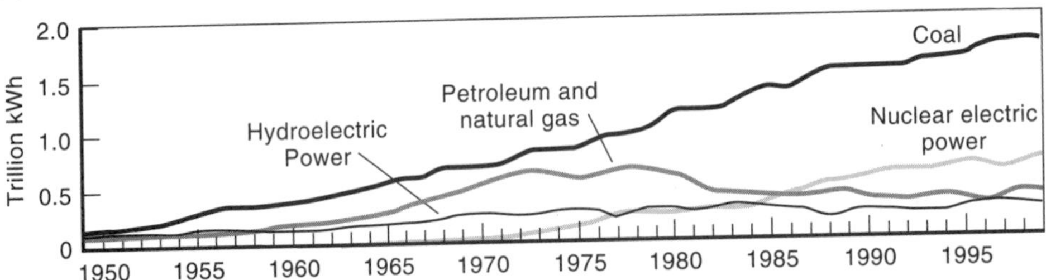

FIGURE 10.1

(*a*) U.S. electric utility net generation of electricity. (*b*) Production of electricity by the electric utility industry by type of generation, 1999. (UNITED STATES ENERGY INFORMATION ADMINISTRATION)

consumption to increase at a rate of almost 7% per year (Fig. 10.1). Such increased electrification required a doubling in electrical generation capacity every 10 years! Per capita consumption of electricity was six times higher in 1998 than in 1948. However, this growth rate has slowed considerably as most households already have acquired the basic electrical appliances. In 1997, 99% of U.S. households had a color TV and 47% had central air conditioning. In

Dual-fired (oil and natural gas) 850-MWe electrical power plant across the Hudson River from Manhattan. (NEW YORK POWER AUTHORITY)

1978, 8% of U.S. households had a microwave oven, but in 1997, 83% had microwaves. A slowing of the growth rate was also due to rising electricity costs as a result of increasing fuel prices and higher costs of new power plants (especially nuclear ones). The Clean Air Act also forced utilities to invest in expensive air pollution control devices. In the 1990s, electricity consumption grew by about 1.7% per year, not much more than the growth rate of the GDP.

About one half of our electricity comes from coal combustion. Other sources used are shown in Figure 10.2. If coal is used as the fuel, then one should be prepared to use large quantities. A 1000 MWe plant uses 9000 tons of coal per day, which is the equivalent of one trainload per day (90 cars of 100 tons each).

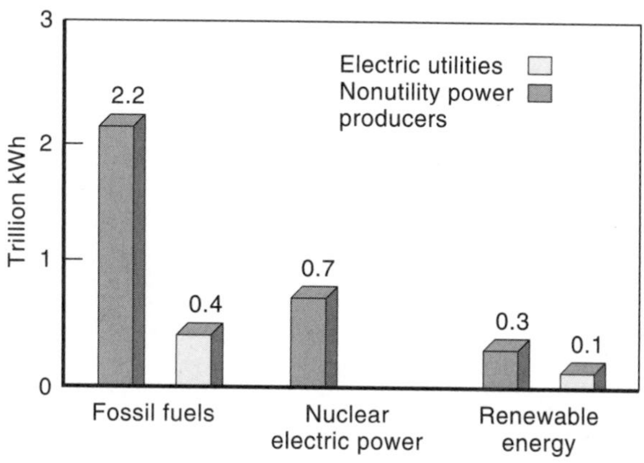

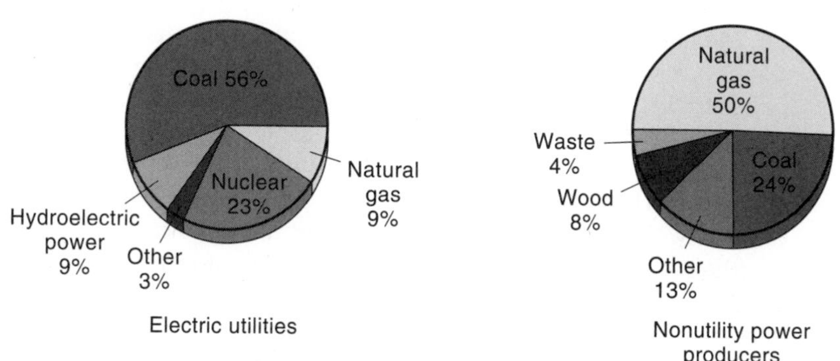

FIGURE 10.2

Electric power industry generation, 1999, for utilities and independent power producers. (UNITED STATES ENERGY INFORMATION ADMINISTRATION)

In this chapter and the next, we will examine several aspects of the electrification of America. However, we need to establish the principles behind the generation of electricity, as well as develop an understanding of electrical circuits. We begin with a discussion of the rapid changes that are now going on in the electric power industry.

B. Restructuring of the Electric Utility Industry

As we enter the 21st century, we are finding the electric utility industry undergoing massive changes in its operation, termed **restructuring.** Beginning in 1997 and spreading from state to state, consumers now have the ability to buy their electricity (and in some cases their natural gas) from suppliers other than their local utility in what is becoming a very competitive market. Suddenly, the electricity industry is in the news much more often and customers can exercise their right to choose their energy supply company, much like they can choose their long distance telephone company.

Historically, electricity has been generated primarily in central-station power plants that use chemical, nuclear, or gravitational potential energy from the sources of coal, natural gas, oil, uranium, or water and convert it into electrical energy. The first such power plants for supplying electric energy went into operation in 1882 under the direction of Thomas Edison. The first two plants were run by steam turbines and the third by hydropower. One of these systems, in New York City, initially serviced 59 customers, with about 1300 lamps. Larger and more efficient turbine generators made it cheaper to produce power at large plants rather than at small ones. Giant utilities serving entire regions, even states, dominated the power industry. Electric utilities were "vertically integrated" in that they owned and operated everything from the turbine to the customer meters. Investor-owned utilities were granted franchise areas in which they were given exclusive rights to provide all aspects of electric service. In exchange for this monopoly control, the electric utilities were regulated by the states.

The highly regulated structure of the 20th century is now changing to one in the 21st century that relies on competition and little regulation to set the price of electricity generation. Some of these reforms came about from public unhappiness with high electricity prices and a growing concern for the environment. In 1978, the U.S. Congress passed the **Public Utility Regulatory Policy Act (PURPA),** which opened the way to limited competition in the generation of electricity. This law required utilities to compare the cost of adding new capacity (the so-called "avoided-cost") with the cost of purchasing it from independent power producers using renewable energy or cogeneration (producing heat as well as electricity—see the next chapter) and choose the least expensive option. In some states, such purchases were required at an "avoided cost" as determined by the state's public service commission, at around $.06/kWh. Such legislation allowed the independent power producers to grow rapidly. In the early 1990s, they accounted for about half of all new additions to generating capacity. Most of this addition came through natural gas-fired turbines, wind, geothermal, biomass, and solar energy. Today, independent power producers generate about 11% of the electricity in the United States. Their net generation of electricity has doubled in the last ten years, compared to a 14% increase by electric utilities.

In this new emerging era, the electricity industry has been divided into three functions: generation, transmission, and distribution. Presently, only the first function is open to competition. This part represents about 25 to 40% of your electric bill. Your utility will continue to use (and maintain) existing wires for delivering electricity to homes and industry. The federal Energy Policy Act of 1996 made transmission lines more available for anyone to use, like a toll road. This law allowed an independent power producer in one area to sell electricity to an industry or utility in another area by using the transmission lines owned by a utility in between. Previously, the utility in between could prevent this sale by refusing to let their transmission lines be used.

Restructuring and retail competition are leading to widespread financial changes in the electric industry itself. Early retirement of some nuclear power plants, closure of less-competitive coal mines, and an increased use of natural gas are some probable changes. A major concern of the electric utilities are **stranded costs**—costs that have been incurred by utilities to serve their customers but cannot be recovered if the consumers choose other suppliers. It is argued that these stranded costs should be allowed to be recovered by the utility because the plant (probably nuclear) was built in a regulatory climate in which the state guaranteed a return on their investment. Some ways to meet these stranded costs during the transition to a competitive market might be to delay the start of retail competition or to charge the departing wholesale customers.

Competition is expected to lower the price per kilowatt-hour of electricity. It might also open the door wider for companies selling "green power," or electricity produced from renewable sources. Many states require that your electric bill disclose what sources were used. In this new environment of inter- and intrastate competition, we are seeing numerous mergers of utility companies, the selling ("divestiture") of noneconomical power plants, and many applications for new power plants, usually using gas-turbine technology. Increased efficiency from gas turbines and lower natural gas prices (due in part to the ability to get more gas out of the ground) have allowed some electric companies to generate electricity at about one-half the price of what people thought just a few years ago. This certainly is not the universal case across the country at the start of the 21st century.

With a competitive electricity market opened up through deregulation, a robust economy, tight supplies, and increased use, electricity prices have dramatically increased. In the hot summer of the year 2000, demand was close to supply, and prices shot up more than ten times for some periods. The power crisis in California that winter brought about rolling blackouts and the near

Focus On 10.1

MANAGING ENERGY DEMAND

On a hot, early summer day in the year 2000 in Milwaukee, Wisconsin Electric Power was about to experience a power shortage. They could have purchased electricity from another company on the open market, but increased demand that day placed the price at 10 to 20 times normal. So, the utility asked one of its largest industrial customers to shut down for the day, and the utility would pay the workers for their vacation. It worked. It ended up that Wisconsin Electric paid the industry about 30¢/kWh, *not* to use electricity (about six times the going rate), but it was still cheaper than buying power on the open market.

Today, due to deregulation, electricity is often bought and sold wholesale, and the cost can temporarily soar to $10/kWh. No utility can buy much at that price, resell it at 5 to 10¢/kWh, and make a profit. Everyone's costs would be much lower if customers had a choice (or one was made for them) to cut consumption when demand (and prices) are high.

Other utilities have managed energy demand by installing internet-controlled thermostats in homes to raise air conditioning settings by several degrees when it needs the power. Radio-controlled switches have also been installed on air conditioners that can be activated remotely when the need arises.

bankruptcy of the two largest utilities in the state. Utilities have a reluctance to build more power plants due to the uncertainty of the market and environmental legislation. President Clinton, speaking for people on fixed incomes, voiced concern for those who had to decide between medicine and air conditioning. Focus On 10.1, Managing Energy Demand, provides an example of one utility's method of dealing with large price fluctuations in their supply.

C. Electrical Charges and Currents

Electrical charges are of only two types. They are equal in size and opposite in character and are denoted as positive and negative. If a plastic rod is rubbed with a piece of fur, the rod acquires a negative charge. You can give a glass rod a net positive charge by rubbing it with nylon. Scientists in the 18th century observed that there was a force between charged objects, and they concluded that an "electric force" exists between any objects that have a *net* electric charge. This force is repulsive if the net charge on both objects is of the same type, and attractive if the objects have different charges (Fig. 10.3). The basic law of electrical charge is

Like charges repel; unlike charges attract

The unit of charge is the **coulomb** (C). An electron has a negative charge, while a proton has a positive charge, equal in size to the charge on the electron. The size of charge of these particles is quite small. It takes 6.25×10^{18} electrons to have a total charge of 1 coulomb. All neutral atoms have the same number of electrons as protons, so they have no net charge. In the process of electrically charging something, electrons are transferred from one material to another. No charge is created or destroyed; electrical charge is conserved. An object has a negative charge if there is an excess of electrons on the object. It has a positive charge if electrons are removed from the object, leaving behind an excess number of positively charged protons.

Further illustrations of electrical charge are discussed in the Special Topic on Electrostatics at the end of this chapter. It would be good to read this if you are unfamiliar with static electricity or wish to explore further some of the phenomena illustrated in the Further Activities section also at the end of the chapter. The following example shows some of the characteristics of the basic laws of electrostatics.

FIGURE 10.3

An example of electric forces showing that like charges repel. Just before this picture was taken, the plastic rod transferred negative charge to the pith ball, so now both objects have a net negative charge. (G. BURGESS)

ACTIVITY 10.1

You can construct a small device to observe electrical charging. Using a straw, a thread, a Styrofoam cup, and two pieces of aluminum foil rolled into small balls, construct the object shown. Rub a balloon (or plastic spoon) on your hair. Approach the aluminum foils slowly and observe what happens. Try putting some salt and pepper on the table and place the charged balloon about 1 cm above. What do you observe? Explain.

Replace the aluminum spheres with small pieces of Styrofoam or packing material and repeat the experiment with the balloon and the hair or with a plastic spoon and plastic food wrap. Were there any changes? Explain.

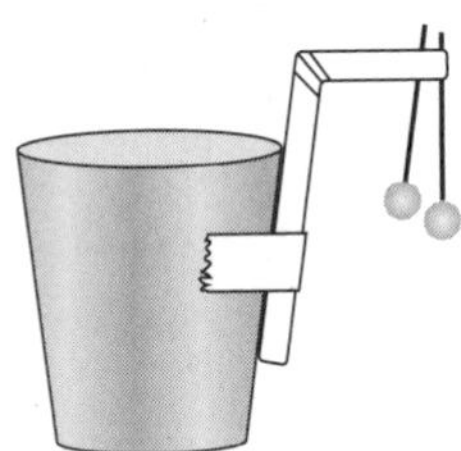

If you bring a copper wire near an object that has a net charge and touch it, the net charge on that object will decrease. If a piece of glass makes contact with the charged object, the net charge will remain about the same. This phenomenon has to do with the large variation in the ability of materials to conduct, or transfer, electrical charge. Metals are generally excellent electrical *conductors* while materials such as glass are very poor conductors and are called *insulators.* The atomic electrons in insulators are strongly bound to their particular atoms, while the electrons in metals are much freer to move and can easily migrate through the metal.

This flow of electrons in a conductor is called an **electrical current.** Electrical current is expressed in terms of the amount of charge flowing past a point in a given time and has units of coulombs per second, called *amperes,* A, or "amps." If two plates, one charged positively and the other negatively, were connected by a conductor such as a copper wire, electrons would flow from the negative material to the positive one. As long as net positive and negative charges remain on the plates, we say there is a **potential difference** between the plates and so the flow of charge will be maintained. The potential difference between two points A and B is defined as the work per charge that must be done to move that charge from point A to point B. The unit of potential difference is the **volt,** where $1 \text{ V} = 1 \text{ J/C}$. You might think of this concept of potential difference as analogous to the difference in gravitational potential energy between two points as a result of a difference in height. It takes work to move an object from the lower point to the upper point. Batteries and electrical generators are devices used for producing electrical potential differences.

For there to be a current between two points, there must be a potential difference between those points (e.g., that supplied by a battery) *and* a path between those points through which the charges can flow. This is called a circuit. If you examine the two slots of a house electrical outlet, the potential difference or voltage between them is 120 V. One of the slots (the larger one) is at a "ground" or zero potential while the other is 120 V higher. If a light was plugged into the outlet, one would have a path through which electrical charges could flow and so you would have a complete circuit. Figure 10.4 shows a cutaway view of a flashlight containing two 1.5-V D-cell batteries. Follow the diagram and observe that when the switch is on, there is a complete circuit from the positive terminal of the battery to the negative terminal through the filament; the lamp is lighted. The total potential difference across the lamp is 3 V, since the two batteries are in series, with the + terminal of one in contact with the − terminal of the other.

There are two types of electrical currents: direct current (DC) and alternating current (AC). In DC, the current is always in the same direction as, for example, in a circuit with a battery as the source of potential difference. In AC, the current continuously changes from one direction to another and back again. The frequency of commercial AC in the United States is 60 cycles per second (60 hertz [Hz]), while in most of Europe the standard frequency is 50 Hz. Alternating current is the type of current produced by our large electrical generating plants and thus is the type we find in our homes. The advantages of AC will be studied in Chapter 11.

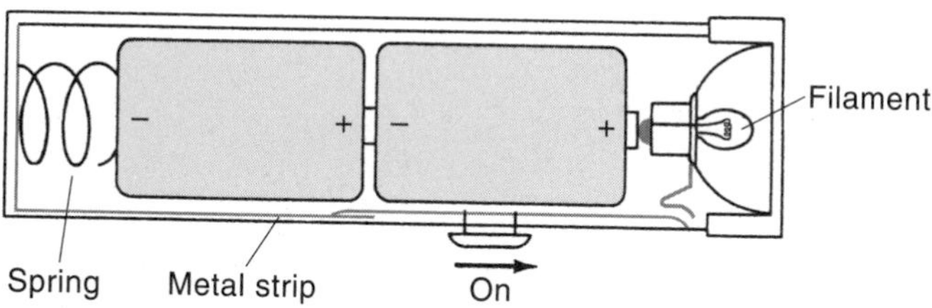

FIGURE 10.4

Diagram of the inside of a flashlight with a plastic case. When the switch is on, the metal slide makes contact with the metal ring around the bulb. This makes a continuous circuit through the metal strip on the inside of the casing to the spring and then to the negative terminal. The positive end of the battery is in contact with the filament of the bulb.

D. Batteries and Electric Vehicles

A battery is like a "pump" exerting force on the electrons in a wire. It is an energy converter as it transforms chemical energy into electrical energy by providing the potential difference between the two terminals of the battery. In general, a battery has two electrodes (or terminals) of different materials submerged in a chemical solution called the **electrolyte.** This is called a cell; a battery is a combination of cells. Small amounts of the compounds making up the electrodes go into solution in the electrolyte as free ions (electrically charged atoms). These create negative and positive terminals. When the two terminals are joined together with an external circuit, electrons can flow from one electrode to the other. The electrolyte can be either a liquid (such as the dilute sulfuric acid in a car battery) or a moist paste (as in a dry cell). The potential difference of the battery is maintained by the continued chemical action at each electrode.

The first battery was invented by the Italian scientist Alessandro Volta (1745–1827). His battery consisted of a small plate of zinc and one of copper, separated by a piece of cardboard that had been moistened in a salt solution. As a zinc atom entered into the electrolyte as an ion, $Zn \rightarrow Zn^{2+} + 2e^-$, two electrons were left behind. These electrons flowed as current through an external wire to the other electrode, joining with a copper ion that was in solution and so plating out the copper on the negative electrode: $Cu^{2+} + 2e^- \rightarrow Cu$.

A common battery today is the lead-acid battery used in automobiles. Focus On Common Batteries describes its operation, as well as that of a common dry-cell battery. An advantage of storage batteries is that they can be recharged, unlike dry cells. This is done by running an external current through the battery in the direction opposite to the current flow during regular operation. Lead-acid batteries are popular because they can be recharged thousands of times. They are also relatively low in price and can supply large currents for short periods.

Focus On 10.2

COMMON BATTERIES

The lead-acid battery usually consists of six cells connected in series. Each cell has a positive electrode made of lead dioxide and a negative electrode made of pure sponge lead, both immersed in an electrolyte of sulfuric acid, H_2SO_4, and water. When the two terminals are connected through an external circuit, two electrons at the negative electrode leave a lead atom as it goes into solution as Pb^{2+}. These positively charged lead ions combine with the sulfate ions in the electrolyte to form lead sulfate ($PbSO_4$). At the positive terminal the lead dioxide, PbO_2, is converted to lead sulfate and water, using the two electrons that "flowed" through the external circuit. As lead sulfate is deposited on *both* terminals during cell discharge, the sulfuric acid concentration of the battery decreases. Since the density of the acid is greater than that of water, a marked decrease in electrolyte density indicates that the cell is discharged. When the $PbSO_4$ buildup on each electrode is so great that Pb and PbO_2 are not available to the electrolyte, you have a "dead" battery. The potential difference generated across the two terminals in each cell is about 2 V, so a car battery with six cells connected together in a series yields $6 \times 2\,V = 12\,V$.

A dry-cell battery uses a carbon electrode for the positive terminal and a zinc container for the negative terminal. A wet paste serves as the electrolyte; as it dries out, the cell's voltage decreases. The dry cell also wears out as the zinc is used up.

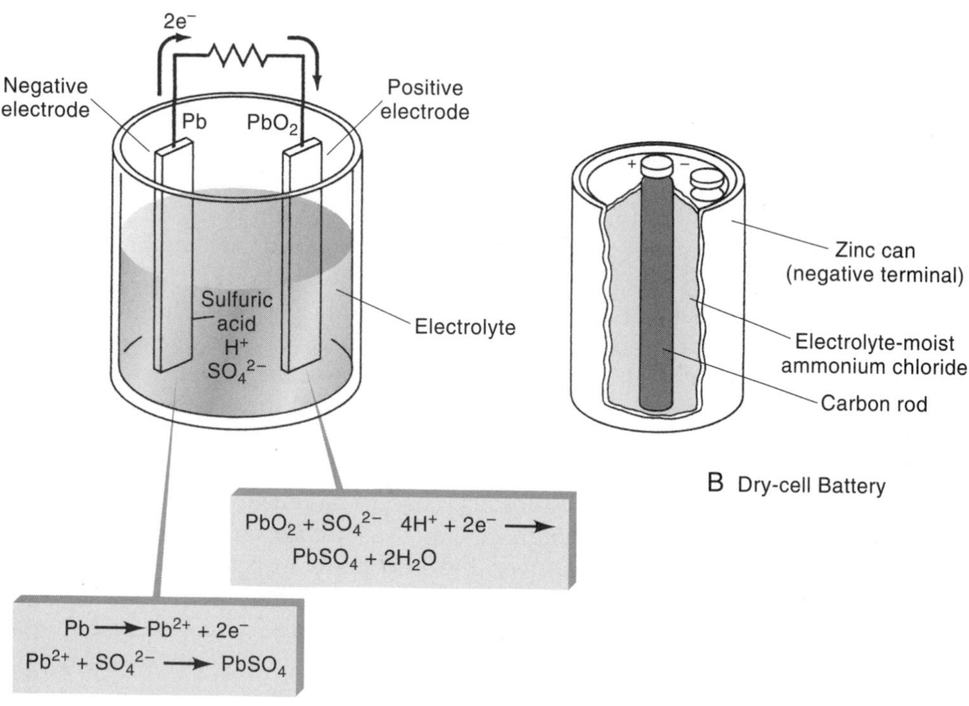

Chemical-to-electrical energy converters.

Two characteristics of a battery are its voltage and discharge capacity. The discharge capacity is equal to the current supplied by the battery (in amps) times the number of hours that the battery can supply that current. For example, a good 12-V car battery will have a discharge capacity of 60 amp-hours, meaning it can supply 3 amps of current for 20 hours, or 10 amps for 6 hours. The energy stored in a battery is very small compared to that stored in more conventional fuels. The lead-acid battery stores about 2% of the energy available in 1 gallon of gasoline! It is also very heavy.

The battery has been called the most reliable source of power known. It is convenient, portable, and reliable (no moving parts). Some important applications are electric vehicle propulsion and storage of solar electric energy (generated by solar cells). **Electric-powered vehicles (EVs)** have been built for many years; they were even as popular as gasoline powered cars at the turn of the century. The first speeding ticket for a car was given to an EV—for going 15 mph! In 1914, there were 20,000 EVs on the road.

Today's conventional EVs have about eight 12-V standard lead-acid batteries—weighing more than 230 kg (500 lb)—and taking six to eight hours to recharge. Generally they have a range of only 60 to 160 miles between recharges and a top speed of 60 to 70 mph. All of Detroit's Big Three auto manufacturers have been involved with EV development at one time or another, as well as Toyota, Honda, and Nissan. In December 1996, GM was the first manufacturer to make EVs commercially available; in two years its EV1 (with 26 lead-acid batteries) had only sold 550 models. It later used nickel metal hydride batteries, which gives it a longer range. The EV1 was built from the ground up, while Ford is producing an electric version of its Ranger pickup. U.S. Electricar removed the gas engines from Geo Prizm sedans, and installed electric motors and lead-acid battery packs. The electric Prizm cost $30,000, twice that of the standard model. The total number of EVs sold by all manufacturers in 1999 was about 1200. At a sticker price of $30,000 to $40,000, it's easy to understand why. It's like asking the customer to buy a car with a $15,000 gas tank that holds the range equivalent of three gallons of gas and takes six to eight hours to refuel.

Research and development to provide batteries that are lightweight, inexpensive, and capable of thousands of recharges has been very active, but slow. The automobile industry spent more than $150 million on EV research and development in the 1990s. However, breakthroughs in battery technology are still being awaited. Nickel metal hydrides, NiCad, zinc, sodium-sulfur, and lithium batteries are possible replacements for the lead-acid battery. Potential short lifetimes and economic considerations have hurt these alternatives, even though their energy densities (energy stored per mass of battery) are high (Table 10.1.).

There are over 2000 EVs in use today in the United States, mainly by government agencies or corporations. They have niche uses (as city delivery vans, postal vans, etc.) and a potential for most routine driving where distances are small and speeds not excessive. About 75% of private cars are driven less than

Table 10.1 BATTERY CHARACTERISTICS

Battery Type	Energy Density (W-h/kg)	Range, City (km)	Notes
Lead-acid	30–50	110–150	Reliable, low cost, heavy
Nickel-cadmium	55	180–200	Established technology, expensive
Sodium-sulfur	80–140	300	Good storage, high-temperature operation (350°C)
Lithium	150	450	Inexpensive, R & D needed
Zinc-air	180–200	400	Expensive, low life cycle
Nickel metal hydride	60	180–200	Popular, lightweight

(*Scientific American*, July 1999, p 92)

80 km per day! While capital costs are high, operating expense is small—about \$0.012/km (\$.02/mi). Gasoline costs alone for an automobile are from \$0.02 to \$0.06 per km (\$.03 to \$.10 per mile).

The EV can help reduce urban air pollution. California recently put into effect auto-emission laws that will require 10% of the new cars sold to be ZEV (zero-emission vehicles) by 2003. There is only one kind of ZEV (besides the bicycle). It is the electric vehicle, powered by batteries or fuel cells. The power plants that generate the electricity for these cars do add pollutants (including CO_2) to the atmosphere at their source point. However, it is easier to control the emissions at the power plant than at the tailpipe of a fossil-fueled car.

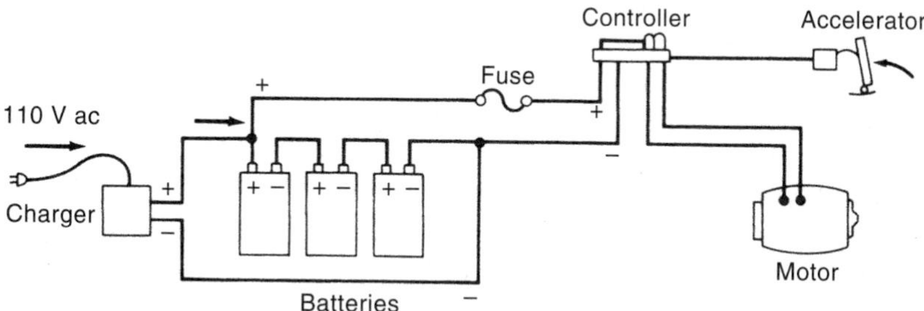

Basic circuit for an electric vehicle. The charger feeds the batteries, which can then supply power to the motor to move the car. The speed and power of the motor are regulated by the controller, which is in turn controlled by the accelerator.

FIGURE 10.5
The Honda *Insight* is a new hybrid car that gets 70 mpg in highway driving. (HONDA)

Interest in EVs has been displaced recently by **hybrids** and fuel-cell powered cars (see the section entitled Fuel Cells in this chapter), although both have a strong electric component. Honda's *Insight* is a recent entry into the EV field (Fig. 10.5) and is a hybrid. The vehicle uses a lightweight 1-L, 3-cylinder gasoline engine with a 10-kW DC motor mounted directly to the engine's crankshaft. Power to this motor comes from a 144-V nickel metal hydride battery pack. The battery pack never needs recharging by an outside power source since the motor also acts as a generator during deceleration and braking. The gasoline engine can be small and more fuel-efficient because the electric motor supplies low-end torque and power assist as needed. The engine shifts from electric to gas mode quietly and will shut itself off when you stop at a stoplight! Aerodynamically unique, the car has a rounded nose, low height, teardrop shape, and a plastic undercar cover to reduce drag. It has achieved the best EPA mileage rating in history, reaching 61 mpg in city driving and 70 mpg highway. The *Insight* is an ultra-low emission vehicle, producing 84% fewer hydrocarbons and 50% less NO_x than a typical car. Another hybrid, Toyota's *Prius,* was introduced into the American market in the year 2000. This car has a 274-V battery, a 33-kW electric motor, and a 1.5-L gasoline engine. It gets better mileage in the city than on the highway (52 mpg versus 45 mpg).

E. Ohm's Law

With a voltage source and a continuous path through which charge can flow, we have an electrical circuit. First, let's consider the size of the electrical current between two points across which there is an electrical potential difference. To formulate an expression for this current, consider the analogy of water flowing down a hill through a pipe. The amount of water flowing (current) in a certain time is proportional to the gravitational potential difference between the two ends of the pipe (the difference in elevations) and to the size of the pipe. A very small diameter pipe presents a large *resistance* to water flow, while a larger diameter pipe will allow water to flow at a greater rate. The water's flow rate is inversely proportional to the pipe's resistance. So it is with electric current. The **electrical resistance,** R (not the heat transfer coefficient) of a circuit element, expressed in units of ohms, is a property of the type of material and its size—both length and diameter. As with water flowing through a pipe, the larger the cross-sectional area of a wire, the smaller will be its resistance. The longer the length of a wire, the greater its resistance. The resistance also will vary with temperature. In general, the resistance of metals increases with an increase in temperature. Some materials have zero resistance at very low temperatures, and will be discussed in a later section on superconductors.

The quantities of resistance, potential difference, and current are related in most circuits by an important relationship called **Ohm's Law**:

$$I = \frac{V}{R} \text{ or } \mathrm{V} = I \times R$$

This formula states that the current I through a device is proportional to the potential difference V, or voltage across that device, and inversely proportional to the resistance R of the device. (Ohm's law is only an experimental relationship that is true for most metals; it does not hold for solid-state devices such as transistors.) Note that for there to be a current I in a circuit, we need both a potential difference and a complete path for the electrons.

EXAMPLE

Figure 10.6 shows a typical household circuit with a toaster plugged in. If the voltage at the plug (the potential difference between the terminals) is 120 V and the toaster coils have a resistance of 15 ohms, the current in the toaster will be

$$I = \frac{V}{R} = \frac{120\text{ V}}{15\text{ ohms}} = 8\text{ amps}$$

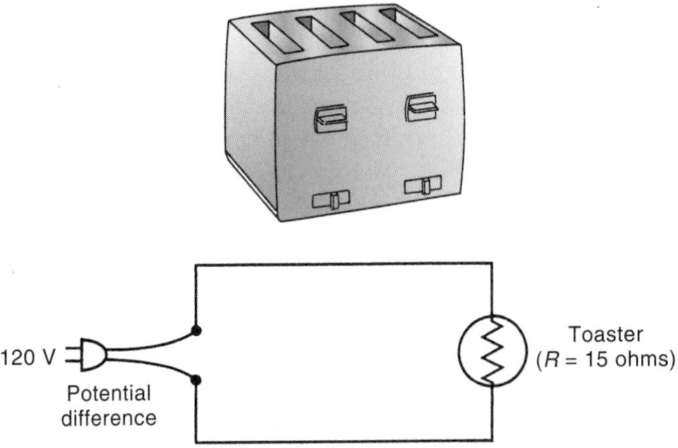

FIGURE 10.6
Household circuit with toaster.

An example illustrating Ohm's law deals with *electrical shock.* When you are working around electrical circuits, you must use extreme caution because a current of only about 0.1 amp through the heart can cause death. Two things should be noted about electric shocks. First, for there to be a current, there must be potential difference. A bird can land on a high-voltage wire and not get shocked, as long as it is only touching *one* wire: There is no difference in potential across the bird's body. Second, the amount of current is inversely proportional to the resistance. Since injuries occur because of excessive electric *current,* damage from a *voltage* source can be reduced if the *resistance* in the "circuit" is large. Standing on dry ground in tennis shoes and touching a voltage source with your hand would

ACTIVITY 10.2

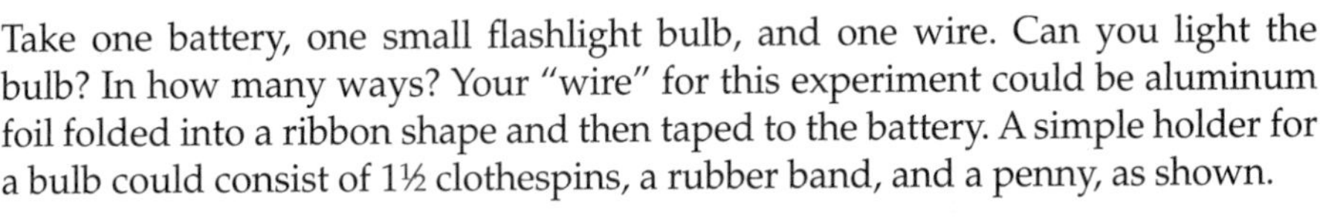

Take one battery, one small flashlight bulb, and one wire. Can you light the bulb? In how many ways? Your "wire" for this experiment could be aluminum foil folded into a ribbon shape and then taped to the battery. A simple holder for a bulb could consist of 1½ clothespins, a rubber band, and a penny, as shown.

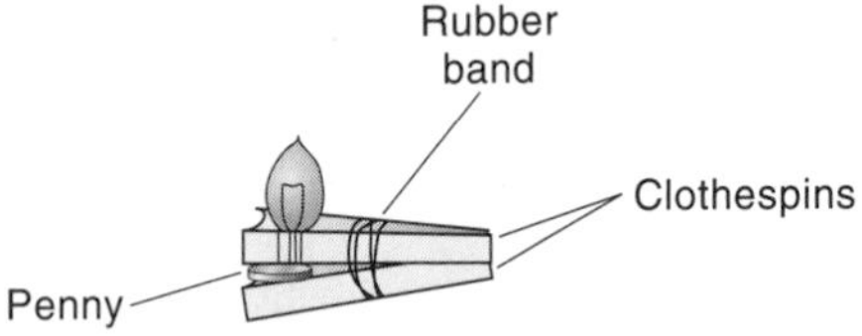

produce a much smaller current because of the larger resistance than if you were standing in a pool of water. (When working around high voltages, you might remember to keep one hand in your pocket: if that free hand was holding a metal pipe when the other hand touched a voltage source, an unhealthy amount of current would pass through your chest because this would be the path of much smaller resistance than one through your legs and shoes to ground.)

F. Superconductivity

One of the most exciting developments in recent times has been the discovery of "high-temperature" superconductors. Few scientific events have captured public attention as quickly and as thoroughly. Much of the excitement is a result of the many applications that can come from high-temperature superconductivity. These include magnetically levitated trains, high-voltage transmission lines with no electrical resistance, and high-speed, miniaturized electronic computer chips.

Superconductivity has been known since 1911, when the Dutch physicist Heike Kamerlingh Onnes discovered that the electrical resistance of mercury and other metals abruptly dropped to zero at temperatures below about 4 K, the temperature of liquid helium. (Such temperatures are called critical temperatures.) Advances over the next 50 years led to the discovery of superconducting states in other metals at higher temperatures (up to 20 K). Alloys of niobium were found to be superconducting at 23 K and are able to support large electrical currents, so they served as the basis for one of the most important applications of superconductivity (until recently)—superconducting magnets. However, progress in this area was slow and little advancement was made in attaining superconductivity at higher temperatures. Then in 1986, two physicists in Switzerland (K.A. Muller and J. G. Bednorz) announced the creation of a new class of superconducting materials—ceramics—able to superconduct at significantly higher temperatures (35 K). What made this so interesting was that ceramics are normally insulators. These results stimulated intense research leading to the discovery in 1987 of superconductivity at even higher temperatures (100 K) with a different ceramic material—an yttrium-barium-

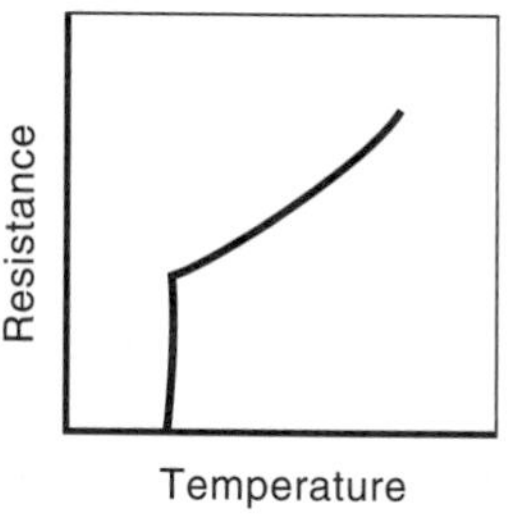

Resistance versus temperature for a superconductor. The resistance is like that of a normal metal for higher temperatures but drops to zero at the critical temperature.

copper oxide ($YBa_2Cu_3O_7$). The existence of a superconducting state at these high temperatures allowed the material to be cooled with relatively cheap liquid nitrogen (at 77 K), at ½5th the cost of liquid helium. Today, ceramic superconductors at temperatures up to 135 K have been achieved.

The loss of resistance at temperatures below the critical temperature leads to numerous applications. Perhaps the first use of superconductors in electrical power systems will be for underground high-voltage transmission lines. At this time, about 10% of the electricity carried by transmission lines is lost as heat because of the wire's resistance. These heating losses would be eliminated if superconducting cables were used. Because of the refrigeration required for superconductivity, the cables would have to be underground. However, underground transmission lines are presently 10 to 20 times more expensive than overhead lines. Another possible utility application is the use of superconducting coils to store electricity. Such coils would be charged during off-peak hours by using power from baseload generators, then discharged during times in the day when demand was largest (peak demand times). The conductor coils are envisioned as large circles one-half mile in diameter and buried underground. Other applications of superconductors include use in computers for higher speeds and denser arrays of electronic components. Superconductors also can be used in medical imaging devices, such as nuclear magnetic resonance (NMR) units, that are used to image the body's soft tissue.

Another property of superconducting material is that it excludes magnetic fields from its interior; this is called the "Meissner effect." This can lead to levitation. When a magnet is placed near a superconductor, the magnet will hover in the air (Fig. 10.7). This levitation occurs because the magnetic field induced

FIGURE 10.7

Demonstration of magnetic levitation. A magnet "floats" above a superconductor, which is in a bath of liquid nitrogen at 77 K (−196°C). (United States Department of Energy)

FIGURE 10.8
Magnetically levitated (MAGLEV) trains, such as this one in Japan, might eventually make use of superconductors. Speeds in excess of 340 mph have been achieved on a test track. Germany is constructing a commercial MAGLEV line between Hamburg and Berlin. (CORBIS/BETTMAN)

by the current in the superconductor and the field of the regular magnet repel one another, as do the north poles of two conventional magnets. Magnetically levitated trains can make use of superconducting materials. Prototype trains in Japan and Germany equip the train with conventional or superconducting magnets and place electromagnets on the track. The train moves as the magnetic field travels along the track, energized by current from the substation, pushing the train forward or reversing the direction to provide braking. Once the vehicle reaches cruising speed, little energy is needed to keep it going because there is no friction between the train and the track (Fig. 10.8).

As with any new technology, it is often a long way from the lab to large-scale applications such as those we have mentioned. Superconductivity is a fragile process and depends on more than just the temperature of the material. High currents and strong magnetic fields can cause the material to return to its non-superconducting state. Also, fabrication of these new ceramic superconductors might be difficult because they are brittle. As with any development, economic considerations play a central role. Whether we build magnetically levitated trains depends not so much on the energy savings that high-temperature superconductors might make possible, but on whether such high-cost transportation options are in the public interest. Is the energy savings enough to justify increased capital costs? What about the value of less air pollution because fewer people are driving to work?

G. Elementary Circuits

A simple electrical circuit uses a source of potential difference connected via wires to different devices (termed the load) to convert the electrical energy to other forms of energy, such as heat and light and work (as in a motor). Each electrical device has its individual resistance *R*. (We use the symbol -ww- for a resistor to represent the electrical resistance of the load.) Each device can be combined in a circuit in one of two ways: in series or in parallel with other elements.

Series Connections

The devices can be put one after the other in **series,** as shown in Figure 10.9a; for this situation, the *same current* flows through each one. No electrons are lost. As more devices are added to this circuit, the total resistance of the circuit *increases.* The total resistance is the sum of the resistances of the individual devices. As the resistance in the circuit increases, there will be a *decrease* in the amount of current, according to Ohm's law. Light bulbs in a series circuit will grow dimmer as more are added on. Unfortunately, if one device burns out, you have an "open circuit" (like a switch being opened) and none of the remaining devices will receive electricity since there is no closed path through which charge can flow.

Parallel Connections

Another way for the resistors to be arranged is in **parallel,** as shown in Figure 10.9b. This is how circuits in your house are wired. In this arrangement, the incoming current divides between the devices. The *potential difference* across each resistor *is the same;* each device in a parallel circuit receives the same voltage supplied by the source and so is independent of the others. The amount of current through each "leg" of the circuit depends on the value of the resistance of that leg. The total current leaving the source is the sum of the currents through all the legs of the parallel circuit. In this arrangement, a burned-out device (such as a light) will create a break in that part of the circuit, but the other devices will continue to receive the same current and function as usual. The total

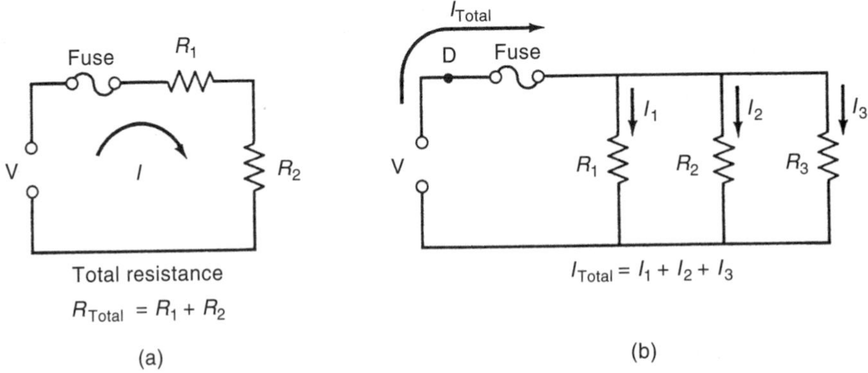

FIGURE 10.9
(*a*) A circuit containing resistors in series. As more devices are added, the total resistance increases, and so the current decreases. (*b*) A circuit containing resistors in parallel. As more devices are added, the total resistance decreases. However, the current through each device (I_1, I_2, etc.) remains the same.

resistance of the parallel circuit *decreases* as more devices are added. This is analogous to the case of more water being able to flow through three pipes in parallel than through the same three pipes in series. As the total or equivalent resistance of a parallel circuit decreases, the total current *increases.*

If a circuit has too many devices connected in parallel, and all are operating at the same time, there is a fire hazard because of excessive heating of the wire from the large I_{total} current. (This current I_{total} exists at point D in Figure 10.9b, a location between the voltage source and the point at which the current divides into the other devices.) As protection from overheating that could cause an electrical fire, a *fuse* or *circuit breaker* is used in series with the source. This device will open at a selected value for I_{total}, limiting the current to a safe value. (A fuse can be thought of as an intentionally placed "weak link" in a circuit.) It is better that overheating occur in the fuse than in the circuit's wiring or components.

EXAMPLE

A 12-ohm resistor and a 24-ohm resistor are connected in series across a 12-V battery. What is the current in the circuit?

Solution

Since they are in series, the total resistance will be 36 ohms. From Ohm's law,

$$I = \frac{V}{R} = \frac{12\ V}{36\text{ ohms}} = 0.33\text{ A}$$

A perennially observed advantage of parallel circuits over series circuits is found in the different types of Christmas tree lights. In the older, less expensive sets, the lights are all placed in series, so one bulb burning out will spell disaster for the rest of the string. The more expensive light sets have the bulbs arranged in parallel so that if one bulb goes, the rest remain lit with the same intensity.

H. Electrical Power

The electrical energy in a circuit, supplied by a generator or battery, is either converted into work (as in a motor) or dissipated as heat energy (as in a resistor) (Fig. 10.10). The *rate* at which electrical energy is delivered to a device is given by the equation.

Power delivered = volts × current or $P = VI$

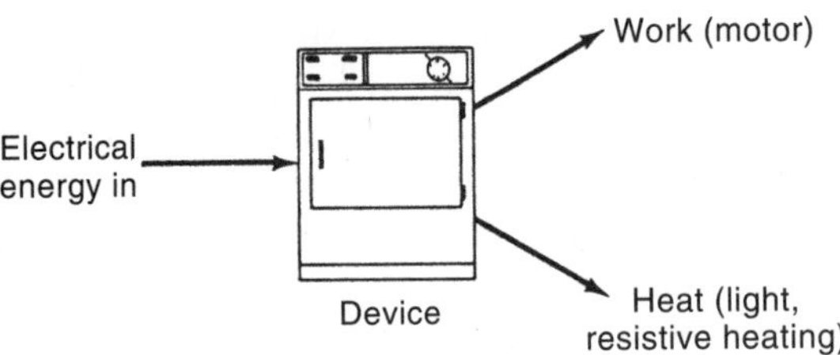

FIGURE 10.10
Electrical energy goes into work or heat.

The unit of power is the watt, where 1 watt = 1 V × 1 amp. For the previous example of a toaster, the electrical power delivered is $P = 120\text{ V} \times 8\text{ A} = 960\text{ W}$. When you buy a 100-W light bulb, you might not be aware of another number printed on the bulb (besides the price)—120 V. The power dissipated by the bulb will be 100 W *if* the source is 120 V. If the source voltage is less than 120 V, the bulb will draw less current, since $I = V/R$, and so the power output $P = VI$ will be less. During a "brown-out," when there is excessive demand, the voltage generated by the utility is reduced, resulting in dimming of the lights.

EXAMPLE

An electric iron has a resistance of 18 ohm and is connected to a 120-V wall socket. What is the power consumed by the iron?

Solution

Since $P = VI$, we need to find the current I.

$$I = \frac{V}{R} = \frac{120}{18} = 6.7\text{ A}$$

Now,

$$P = 120\text{ V} \times 6.7\text{ A} = 804\text{ W}$$

The rate at which electrical energy is converted into heat is related to the resistance of the device. Since the potential difference across a resistor is $V = IR$, the rate at which energy is dissipated as heat is given by

$$P = VI = (IR)I = I^2R$$

To prevent the overheating of wires, household circuits use heavy copper wires (about 1⁄16 in. in diameter); the large cross-sectional area of this wire gives it a low resistance. As the resistance of this wire is small (about 0.003 ohms per linear foot), the current I that is flowing is primarily determined by

Focus On 10.3

EDISON MAZDA LAMPS, CIRCA 1925

From an ad in the November 1925 Ladies Home Journal:

"In the days of Gov. Bradford, light was so expensive that the frugal Puritan family extinguished its single candle during prayer. The early settlers had to learn to make candles themselves.

Your light comes at a finger's touch—and it's more than 100 times cheaper than candle light.

So use light freely. A 75-W Edison Mazda Lamp will give more than twice as much light as a 40-W Edison Mazda Lamp—but will average only a third of a cent more per hour for current."

the resistance of the appliance or load, and not by the resistance of the wire. Therefore, heat dissipation I^2R in the wire itself is kept small by having a small R.

In your home's electrical circuits, the appliances are put in parallel (Fig. 10.11). The maximum power that can be delivered in one circuit with a 20-amp

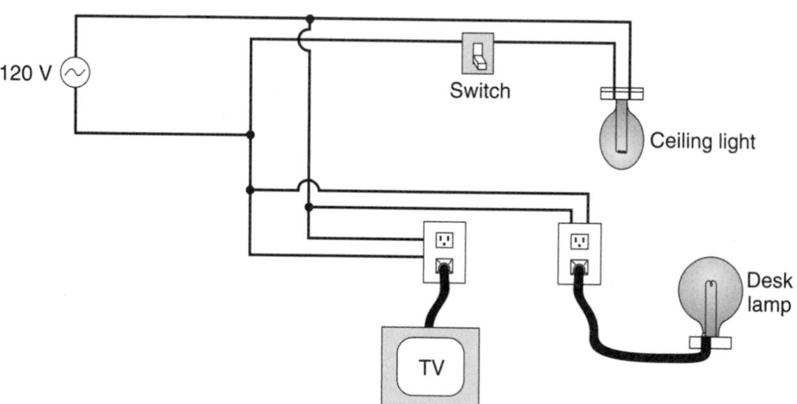

FIGURE 10.11

Parallel connections in a household circuit. (Wiring connections are shown by a •.)

fuse is 120 V × 20 amps = 2400 W. Considering that a hair dryer might use 1000 W and a refrigerator 400 W, you can see that you need more than one circuit for a house. Consequently, a home might have ten individual circuits, each allowing a maximum current of 20 amps.

EXAMPLE

A household circuit at 120 V is fused for 20 amps. How many 100-W light bulbs can be used simultaneously in this circuit?

Solution

Each bulb will draw a current of $I = P/V = 100\ \text{W}/120\ \text{V} = 0.83$ A. The maximum number of bulbs in this circuit will be 20 A/(0.83 A/bulb) = 24 bulbs.

I. Pricing Electrical Energy Use

Let us review the computation of electrical energy costs. We pay for the *energy used*, not the power delivered. The energy used is the power (in watts) expended times the period of use (in hours), expressed in units of kilowatt-hours (kWh). For example, if ten 100-watt light bulbs were left burning for 24 hours, the total electrical energy expended would be 10 bulbs × 100 W/bulb × 24 h = 24,000 W-h = 24 kWh. The cost for this "oversight," at $.08/kWh, is $1.92. (Unfortunately, this small cost of energy has been one of the reasons we have largely ignored electrical energy conservation and have continued to increase our per capita electricity consumption.) The average home uses about 500 kWh of electricity per month (excluding space heating). A more detailed breakdown of annual use for different appliances is presented in Table 10.2. The important item is not the wattage rating itself but the power consumption *times* the average number of hours of operation per year, that is, the kilowatt-hours used annually. Notice the advantage of microwave ovens—1450 W versus 12,000 W for a conventional oven. The microwave oven heats only the food, not the dish and surrounding air, and does it in considerably less time.

A new system of setting electricity rates as well as increasing system efficiency is called "time-of-use" pricing. This method charges customers according to what time of day they use electricity; the cheapest rates are for the use of electricity during off-peak hours, usually 9 PM to 7 AM. Higher-than-standard rates would apply during times of peak load. Electricity demand for the day shows peak-load periods in the morning and early evening (see Fig. 11.20). Leveling the demand curve would reduce the need for less efficient and more costly plants that are normally needed just to meet peak-load demand. In a deregulated electricity market, tight supplies during high demand mean very high prices per kilowatt-hour.

Time-of-use metering is done with special meters. You can lower your personal electricity cost with this system if you use major electric appliances mostly from 9 PM to 7 AM weekdays (or all day Saturday and Sunday) rather than at other times. Computers and electronic controls can be used to turn on appliances when the demand is low and electricity is less expensive. Some utilities are experimenting with two-way communication between home and utility that will permit optimum energy use by allowing the utility to turn off your air conditioning, for example, for a short period of time.

Energy conservation in the residential sector can be practiced through judicious appliance purchases. Life-cycle costs (initial cost + maintenance + energy costs) should be taken into account. The initial cost of such an appliance might be larger, but the total lifetime cost can be less since appliance manufacturers must display energy cost information on most major appliances. This helps the consumer conserve energy and reduce household costs (Fig. 10.12).

Substantial savings in electrical energy use, especially in the commercial and industrial sectors, can be accomplished by using new energy-efficient light bulbs. We have already noted that fluorescent lights are four times more efficient than conventional incandescents, although their use in homes has not been large because many people don't like their flicker, shape, coolness/color, or high initial investment. However, significant improvements in both fluorescent and incandescent bulbs have been made in recent years.

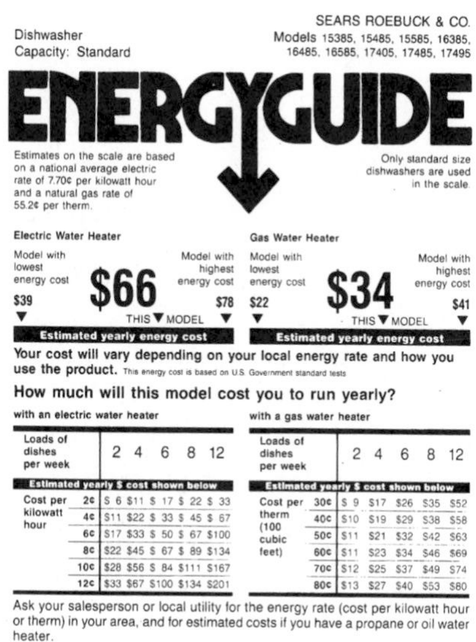

Dishwasher
Capacity: Standard

SEARS ROEBUCK & CO.
Models 15385, 15485, 15585, 16385, 16485, 16585, 17405, 17485, 17495

ENERGYGUIDE

Estimates on the scale are based on a national average electric rate of 7.70¢ per kilowatt hour and a natural gas rate of 55.2¢ per therm.

Only standard size dishwashers are used in the scale.

Electric Water Heater

Model with lowest energy cost $39 — **$66** THIS ▼ MODEL — Model with highest energy cost $78

Estimated yearly energy cost

Gas Water Heater

Model with lowest energy cost $22 — **$34** THIS ▼ MODEL — Model with highest energy cost $41

Estimated yearly energy cost

Your cost will vary depending on your local energy rate and how you use the product. This energy cost is based on U.S. Government standard tests

How much will this model cost you to run yearly?

with an electric water heater

Loads of dishes per week		2	4	6	8	12
Estimated yearly $ cost shown below						
Cost per kilowatt hour	2¢	$ 6	$11	$ 17	$ 22	$ 33
	4¢	$11	$22	$ 33	$ 45	$ 67
	6¢	$17	$33	$ 50	$ 67	$100
	8¢	$22	$45	$ 67	$ 89	$134
	10¢	$28	$56	$ 84	$111	$167
	12¢	$33	$67	$100	$134	$201

with a gas water heater

Loads of dishes per week		2	4	6	8	12
Estimated yearly $ cost shown below						
Cost per therm (100 cubic feet)	30¢	$ 9	$17	$26	$35	$52
	40¢	$10	$19	$29	$38	$58
	50¢	$11	$21	$32	$42	$63
	60¢	$11	$23	$34	$46	$69
	70¢	$12	$25	$37	$49	$74
	80¢	$13	$27	$40	$53	$80

Ask your salesperson or local utility for the energy rate (cost per kilowatt hour or therm) in your area, and for estimated costs if you have a propane or oil water heater.

Important Removal of this label before consumer purchase is a violation of federal law (42 U.S.C. 6302).

Part No 154031912A

FIGURE 10.12

Sticker displaying energy costs for an appliance, a dishwasher in this case.

Table 10.2 RESIDENTIAL APPLIANCE USAGE

Appliance	Average Wattage	Estimated kWh Used Annually
Food Preparation		
Carving knife	92	0.8
Coffee maker	894	106
Deep fryer	1448	83
Dishwasher	1201	363
Frying pan	1196	186
Hot plate	1257	90
Mixer	127	13
Microwave oven	1100	462
Range with oven	12,200	1175
Roaster	1333	205
Sandwich grill	1161	33
Toaster	1146	39
Trash compacter	400	50
Waffle iron	1116	22
Waste disposer	445	30
Food Preservation		
Freezer (15 ft^3)	250	660
Freezer (frostless 15 ft^3)	440	1761
Refrigerator (12 ft^3)	110	460
Refrigerator (frostless 12 ft^3)	160	600
Refrigerator/freezer (14 ft^3)	280	1275
Refrigerator/freezer (frostless 14 ft^3)	400	1829
Home Entertainment		
Computer	500	390
Radio	20	25
Radio/turntable	80	100
Color television	115	165

Comfort Conditioning		
Air cleaner	50	216
Air conditioner (room)	320	320*
Bed covering	177	147
Dehumidifier	257	377
Fan (attic)	100	144
Fan (circulating)	88	43
Heater (portable)	1322	176
Heating pad	65	10
Humidifier	177	163
Housewares		
Clock	2	17
Floor polisher	305	15
Sewing machine	75	11
Vacuum cleaner	630	46
Laundry		
Clothes dryer	4856	993
Iron (hand)	1008	144
Washing machine (automatic)	512	103
Washing machine (non-automatic)	286	76
Water heater	2475	4219
(quick recovery)	4474	4811
Health & Beauty		
Hair dryer	1000	52
Heat lamp (infrared)	250	13
Shaver	14	1.8
Sun lamp	279	16
Toothbrush	7	.05
Vibrator	40	2

*Based on 1000 hours of operation per year. This figure will vary widely depending on area, home insulation, and size and efficiency of unit.

Source: Snohomish Public Utility District, Everett, WA, and local appliance retailers.

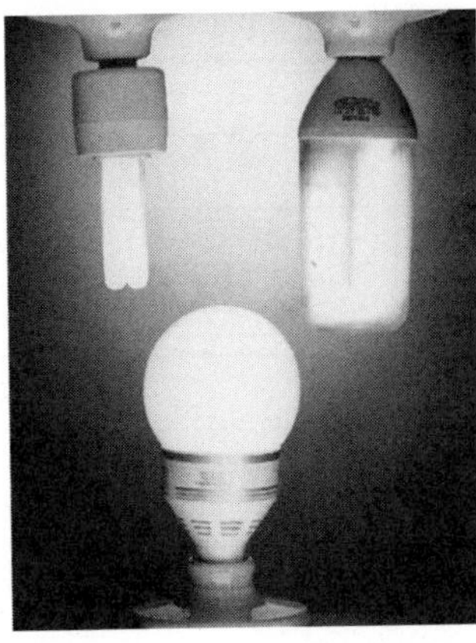

FIGURE 10.13
Energy-efficient fluorescent light bulbs with standard bases. (NEW YORK POWER AUTHORITY)

For fluorescent tubes, flicker can be eliminated by the use of solid-state ballasts, which are also more efficient than the older wire-wound devices. (Ballasts are necessary for all fluorescents to run the tubes at the correct voltages and to heat the filament so it will start.) New compact fluorescent bulbs have a bent tube inside a globe (Fig. 10.13), with the ballast in the base. The screw-in base fits standard incandescent light sockets. Phosphors that coat the insides of the tubes correct the color and produce warm, soft light that is very close to that of incandescents. These compact fluorescents will deliver the same amount of light as a 75-W bulb but use 75% less energy and last up to nine times longer! Payback will be less than a year for average use (see the following example).

EXAMPLE

Payback

A fluorescent bulb that can replace an incandescent bulb in a table lamp may be expensive to buy. Yet it can save energy by its higher efficiency. If a 22-W fluorescent bulb costs \$8 (on sale) and a regular 70-W bulb costs \$1, and if electricity costs \$.08/kWh, what will be the simple payback?

Solution

The 22-W fluorescent bulb puts out as much light as a 70-W incandescent bulb. If the bulb is assumed to be on 4 hours per day, the cost of operating the incandescent bulb for 1 month will be

$$4 \text{ hr/day} \times 30 \text{ days/month} \times 70 \text{ W} \times \frac{1 \text{ kW}}{1000 \text{ W}} \times \$.08/\text{kWh}$$

$$= \$0.67/\text{month}$$

For the fluorescent bulb, the cost will be

$$4\ \text{hr/day} \times 30\ \text{days/month} \times 22\ \text{W} \times \frac{1\ \text{kW}}{1000\ \text{W}} \times \$.08/\text{kWh}$$

$$= \$0.21/\text{month, a } \textit{savings} \text{ of } 46¢ \text{ a month.}$$

For the first year, the total cost (operating + investment) will be \$9.04 for the incandescent and \$10.52 for the fluorescent. This difference in total cost (\$1.48) will be made up in 148¢/46¢ per month = 3 months of the second year, so the payback time is 1.25 years. (Because most incandescent bulbs last only about 1000 hours—less than one year at the assumed usage—the fluorescent bulb will last many times longer; thus the payback time will actually be less than this.)

New incandescent light bulbs are now available that use 30 to 50% less energy and have three times the lifetime of older types. These bulbs are called "tungsten-halogen lamps," and first came to prominence in their use as car headlights. The tungsten filament is encased in an inner capsule that contains halogen gas—produced by iodine vapor—which slows filament wear and increases efficiency. (Standard incandescents contain a tungsten filament surrounded by argon gas.) Additional bulb cost is easily recovered by decreased energy cost over the lifetime of the bulb (3000 hours).

J. Fuel Cells

The fuel cell is a unique power converter that is efficient, nonpolluting, and flexible. It combines a fuel (usually natural gas or hydrogen) with oxygen by an electrochemical process to produce electricity. The fuel cell was invented more than a hundred years ago, but came of age in the 1970s when it was first used on space missions. Today, there is a strong renewed interest in the fuel cell. Its large power-to-weight ratio, compactness, and reliability (there are no moving parts) make it a popular power source for such uses as vehicle propulsion and commercial and residential small-scale electric facilities. However, the high cost of fuel cells and unknowns about their lifetimes have hindered their commercialization. The present price of \$3000 to \$4000 per kilowatt might be reduced by one third by using mass-production techniques.

The fuel cell is similar to a battery, providing direct current through an electrochemical process. However, a battery uses up the materials that are stored at the electrodes (Pb and PbO_2 for a storage battery), while in a fuel cell the chemical reactants are fed to the electrodes on demand. The two chemical reactants in a fuel cell are generally hydrogen and oxygen, which are fed into the cell through porous electrodes (Fig. 10.14).

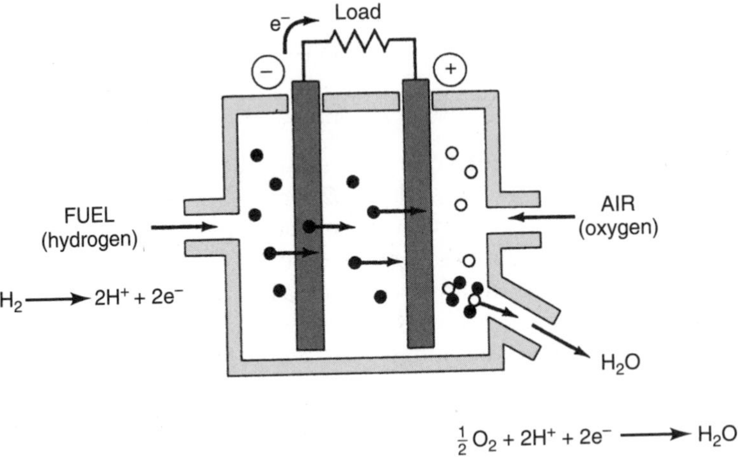

FIGURE 10.14
Cross section of a fuel cell. The two carbon electrodes are immersed in the electrolyte. Other fuels can be used.

The fuel cell reaction can be thought of as a very slow hydrogen gas combustion process. Normally in combustion, the H_2 fuel molecules are oxidized as $H_2 + ½O_2 \rightarrow H_2^{2+} O^{2-}$ + energy. In this very rapid process, electrons pass directly from the fuel to the oxidizer. In a fuel cell, however, the electrons are transferred slowly to the oxygen via an external circuit. At the positive electrode, the hydrogen atoms are stripped of their electrons and enter into the electrolyte (usually potassium hydroxide [KOH] or phosphoric acid). At the negative electrode, the hydrogen ions combine with the oxygen atoms and electrons to form water and heat. The electrons flow through an external circuit connecting the electrodes. The net result, as shown in Figure 10.14, is the reaction of oxygen and hydrogen to form water.

Five different types of fuel cells are in research, testing, or development stages (Table 10.3). The proton exchange membrane (PEM), the molten carbonate cell ($NaCO_3$), and solid-oxide ceramic cells are all in the demonstration stage. The phosphoric acid cell has been operated in a commercial size of 200 kW and tested with an 11 MW unit. The alkaline (KOH) process has been used by NASA in its space program, in the Gemini space capsules in the 1960s.

The phosphoric acid cell is the most commercially developed type of fuel cell. The proton exchange membrane cells have high power density and may be applicable for light-duty vehicles, and for smaller applications such as replacements for rechargeable batteries in video cameras and cell phones. Carbonate cycle fuel cells use methane as fuel, and molten sodium carbonate as the electrolyte. They have the ability to use coal-based fuels making them suitable for large-scale electricity generating facilities. Solid oxide fuel cells use hard ceramic material instead of a liquid electrolyte, allowing high operating temperatures.

Table 10.3 TYPES OF FUEL CELLS

Type of Cell	Efficiency	Operating Temp	Unit Size
PEM	40–50%	80°C	50 kW
Phosphoric acid	40–50%	200°C	200 kW
Molten carbonate	60+%	650°C	2000 kW
Solid-oxide ceramic	60+%	800°C	100 kW
Alkaline	70%	60°C	2–5 kW

Several experiments with thin films are being explored that could result in small, lightweight miniature fuel cells that would replace conventional batteries. A solid metal-hydride ($LaNi_5H_6$) source performs slightly better than a lithium-ion battery. Another approach uses a 25-micron-thick film of plastic to which an electrolyte is added to react directly with methanol diluted to 2% in water. This system generates a power-density level at 30 W/L. Another thin-film cell generates power from a gas mixture that contains both hydrogen and oxygen rather than having two separate gas supplies. This approach uses a gas-permeable electrolyte less than one-micron thick, placed between two layers of platinum. Hydrogen ions diffuse through the membrane and combine with oxygen in an electrochemical reaction called reduction, producing water and a potential of 1 V. In theory, this design, with a surface area of 16 cm^2, could generate up to 0.85 W of power. A stack of six cells would develop 5 W and a current of 1.7 amps. The result would be to deliver up to 100 W-h/L from a solid hydride and up to 1400 W-h/kg from methanol. By comparison, lead-acid batteries deliver 30 to 50 W-h/kg or 80 W-h/L. Nickel-cadmium batteries provide about 50 W-h/kg, or 130 W-h/L. Lithium-ion batteries are rated at 130 W-h/kg or 300 W-h/L.

The Mail Processing Center in Anchorage, Alaska uses one of the nation's largest fuel cell systems, 1 MW. (Courtesy of International Fuel Cells)

The fuel cell has many benefits. First, its efficiency can be as high as 50 to 70%. Second, its nonpolluting nature means it can be located in cities where the power is needed, rather than far away. Such power plants are relatively simple and can be constructed quickly from factory-made modules. Third, a variety of fuels can be used, including natural gas, hydrogen, methanol, and biogas. Fuel cells are finding many uses in stationary facilities, mainly businesses. The units are about 200 kW. Fuel cells for homes are relatively new. A 5- to 10-kW unit might meet all the needs of a home except during peak demand.

The fuel-cell can be used to power an electric drive motor. There are a number of buses now running on fuel cells. (The Canadian company Ballard Power Systems delivered 25 fuel cell buses to Los Angeles in 2000 for testing.) General Motors and Daimler/Chrysler are planning to market such passenger cars by 2004. These would satisfy the ZEV mandate in California. They will utilize a 50-kW system. Honda has developed hydrogen and methanol fuel cell prototype models and is committed to introducing a fuel cell-powered vehicle by 2003. A fuel cell-powered car would use a stack of cells connected in series that would occupy about the same volume as a conventional gas tank. Operated on hydrogen gas, they would be three times as efficient as gasoline powered internal combustion engines. Costs might be comparable to those of conventional cars, and the net release of greenhouse gases (from resource recovery to end uses) would be much less for fuel-cell cars than for gasoline cars. Like conventional cars, they could be fueled in minutes. If the fuel is hydrogen gas stored at high pressures, a range of 400 km would be possible using a 35-gal tank. Metal hydrides can also be used to store the H_2 gas, reducing the volume by a factor of three for that needed to hold enough gas to provide a range of 400 km; however, they are expensive. Batteries might be used in some systems to provide help during times of peak demand. They could then apply regenerative braking—generating power when slowing down—to recharge the batteries. Some automobile manufacturers are considering the use of a processing system to make H_2 directly on board from gasoline or methanol. However, overall efficiencies will drop.

K. Summary

There are two types of electrical charges found in nature: negative and positive. The flow of charge constitutes an electrical current. A potential difference is needed for there to be a current between two points. The current I (measured in amps) is equal to the potential difference across a device (in volts) divided by the resistance (in ohms) of that device. This is Ohm's law: $I = V/R$. The devices in a circuit can be arranged either in series or in parallel. In a series circuit the current through each device is the same, while in a parallel circuit the voltage across each device is the same. The rate at which electrical energy is converted into work or dissipated as heat is the power, which is given by $P = VI$. The cost of running a device is equal to the power delivered times the time of use times the cost per unit of energy (usually in dollars per kilowatt-hour).

Internet Sites

For an up-to-date list of Internet resources related to the material in this chapter, go to the Harcourt College Publishers website at **http://www.harcourtcollege.com**. The links are in the *Energy: Its Use and the Environment* site on the Physics page. General energy related sites and some guidelines for using the World Wide Web in your class are on the inside front cover of this book.

References

Chapter 10

Appleby, J. 1999. The Electrochemical Engine for Vehicles. *Scientific American*, 281, (July).

Hewitt, P. 1998. *Conceptual Physics.* 8th ed. New York, Harper Collins.

Hobson, A. 1999. *Physics: Concepts and Connections.* 2nd ed. Englewood Cliffs, NJ, Prentice-Hall.

Joskow, P. L. 1988. The Evolution of Competition in the Electric Power Industry. *Annual Review of Energy*, 13.

MacKenzie, J. 1994. *The Keys to the Car: Electric and Hydrogen Vehicles for the 21st Century.* Washington, D.C., World Resources Institute.

Ostdiek, V., and D. Bond. 1995. *Inquiry into Physics.* 3rd ed. Minneapolis, West.

Patterson, W. 1994. *Transforming Electricity: The Coming Generation of Change.* Cambridge, MA, Earthscan Publishing.

Schafer, L. 1992. *Taking Charge: An Introduction to Electricity.* Washington, D.C., National Science Teachers Association.

Srinivasen, S. 1999. Fuel Cells: Reaching the Era of Clean and Efficient Power Generation in the 21st century. *Annual Review of Energy*, 24.

Zubrowski, B. 1991. *Blinkers and Buzzers: Building and Experimenting with Electricity and Magnetism.* New York, Beech Tree.

QUESTIONS

1. Tollbooth stations on roadways and bridges usually have a piece of wire stuck in the ground before them that will touch a car as it approaches. Why?
2. In regions of low humidity, such as the southwestern United States, one develops a special "grip" when opening car doors, or touching metal door knobs, which involves placing as much of the hand on the device as possible, not just the ends of one's fingers. Discuss the induced charge and explain why this is done.
3. Why would you expect to get a more potent shock if you touched a high-voltage source with your shoes off rather than with them on?
4. Why are some people killed by 120 V while others are only shocked by the same voltage?
5. If you want a particular circuit to carry a large current, is it better to use a large-diameter or a small-diameter wire?

6. Why are 220-V circuits used for such devices as electric clothes dryers and stoves? What differences do you expect to find in the wire used for these circuits compared to 120-V lines?
7. What is the purpose of a fuse or a circuit breaker?
8. If you plug a stereo system into one wall socket and a lamp into another socket in that room, are you connecting these loads in series or parallel? What limits the number of devices that can be plugged into the outlets of the same house circuit?
9. What is the disadvantage of using an inexpensive extension cord (with a small-diameter wire) to run an air conditioner or a refrigerator?
10. What conditions are necessary for there to be a current between two points in a wire?
11. In the following sketch of a three-way switch (Fig. 10.15) used to control a light from two different places, trace the circuit and conclude whether the light is on or off. (The blackened parts in the switch are metal conductors.)

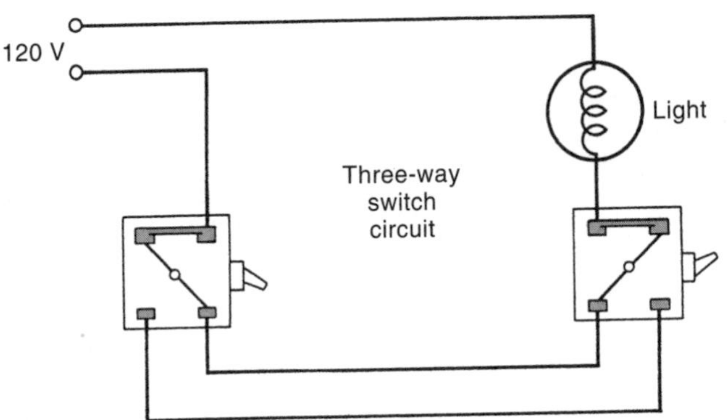

FIGURE 10.15
Three-way switch.

12. What are some of the advantages to the development of electric vehicles (EV)? In what situations would they be at their best?
13. Suggest methods by which EVs could be charged so that the owner would have minimum inconvenience.
14. Large-scale development of EVs takes massive investments on the part of the auto industry. How would you begin to convince a manufacturer that the market will justify such a large investment?
15. Suggest a public policy for offering incentives to further transportation technologies that will have low emissions of greenhouse gases.
16. What do we gain from high-speed rapid-transit systems?
17. If your house were subject to time-of-use pricing for electricity, what price structure would you think reasonable to make this work? Describe the concept and explain your rationale.

18. What are some of the advantages of using a fuel cell? What fuels can be used?
19. Investigate recent installations of fuel cells in your state.

PROBLEMS

1. A light bulb draws a current of 2 amps when connected to a wall outlet (120 V). What is the resistance and wattage of the bulb?
2. What is the cost of running a 1500-W hair dryer for 30 minutes, when the cost of electricity is $.12/kWh?
3. A 12-V battery is connected across a device whose resistance is 24 ohms. Calculate the current in the wire. What is the power dissipated by this load?
4. How much current is drawn by a 1000-W electric skillet operating on 120 V?
5. With a 1200-W toaster, how much electrical energy is needed to make a slice of toast (cooking time = 1 minute)? At $.08/kWh, how much does this cost?
6. Using the appliance ratings on Table 10.2, estimate the electric energy (in kWh) used by your household in a day. Make a table showing appliance, time of use on an average day, and energy used.
7. Estimate the cost of the electricity used (at $.08/kWh) for one month for each of the following (use Table 10.2).
   (a) ten 100-W bulbs for 6 hours per day
   (b) color television set for 5 hours per day
   (c) night light of 1 W left on continuously
8. A refrigerator is rated at 6 amps at 120 V. How much power is used by the refrigerator when operating at its rated voltage?
9. A 50-ohm resistor is connected across 120 V. How much heat is dissipated in the resistor?
10. Resistors of 10 and 30 ohms are connected in series to a 120-V source. What is the current flowing in the 30-ohm resistor? What is the current in this resistor when the same resistors are connected in parallel?
11. You might challenge the assertion that batteries are inexpensive energy converters. Calculate the price per kilowatt-hour for a 12-V automobile battery with a 50 amp-hour capacity that sells for $40. If its weight is 45 lb, what is the energy density in watt-hours per pound?
12. Show that a 60 amp-hour battery stores the equivalent of 2% of the energy available in 1 gal of gasoline. Use Table 3.4 for conversion factors (1 gal gasoline = 125,000 Btu).
13. The common exit sign in a public building uses two 20-W incandescent bulbs. These could be replaced with light-emitting diodes (LEDs) at 2 W per sign. If there are 40 million signs in use in the United States, what savings in electricity use for an entire year could be secured with this changeover?

14. Calculate the price of electricity (per kWh) in 1925 using the information found in the "Edison Mazda" ad.

FURTHER ACTIVITIES

1. Inflate a balloon, tie the end, and rub it on your hair. Try to stick the balloon to a wall, placing the part of the balloon that touched your hair against the wall first. Rotate the balloon by 180° and again try placing it against the wall. Was there a difference? Why?
2. Experiment with static charge by taking a large plastic comb and rubbing it through your hair or on a piece of fur or wool cloth. Turn on a water faucet to obtain a slow stream of water. Move the comb close to the water. Why does the water react as it does?
3. Examine a flashlight and observe its electric circuit. Draw a diagram representing this circuit. Include the switch.
4. Examine the differences between parallel and series circuits. Using two batteries (wired in series to give 3 V), put two small bulbs in series and then in parallel. Note the bulb brightness in each connection. (A battery holder might make successful completion of each circuit a little easier.)
5. Find out what is the best battery to use in an electronic device or a toy with the following experiment. Place a set of batteries in the electronic device and run it until it stops, recording the time. Do the same for other brands of batteries. Figure the cost per hour of use by dividing the cost of the battery by the number of hours the battery worked. Compare results.
6. Investigate your house wiring by following the circuit connected to one particular circuit breaker or fuse. Throw the breaker or take out the fuse and see what outlets are affected. Draw this circuit and label the devices. What is the maximum power allowed for this circuit?
7. Ask your local electric utility about the variation of electrical consumption with time of day. Graph this information. Find out whether the utility's electric rate is a function of the time of use. Discuss how the pricing of electricity relates to energy conservation. Give your recommendations for using pricing more effectively to conserve energy.
8. When two different metals (copper, galvanized steel, or aluminum) are inserted into a piece of fruit or a vegetable, a potential difference exists between them. A sensitive DC ammeter connected between the electrodes will measure a current, although not large enough to light a bulb. The fruit or vegetable acts as the electrolyte in this battery. Try different fruits and vegetables with combinations of electrodes. This is the principle behind the commercial "two-potato clock."

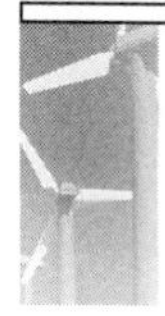

SPECIAL TOPIC

Electrostatics

It was known to the Greeks, about 600 BC, that a piece of amber (a fossil resin) rubbed with fur would attract small pieces of straw and wood. We say that the amber was "electrified," or electrically charged. (The word "electricity" comes from the Greek name of this material, "elecktron.") Although originally thought unique to amber, the process of electrification can be done to many other substances by rubbing them with other materials. If a plastic rod is rubbed with a piece of fur and brought into contact with a piece of cork (called a pith ball) suspended from a thread, the pith ball flies away from the rod (Fig. 10.16). If a glass rod is rubbed with a silk cloth and brought into contact with another pith ball, again the ball flies away from the rod. After such contact, if the two pith balls are placed near each other, they will swing toward each other.

These phenomena can be understood in terms of a property of matter called **electrical charge.** There are two kinds of charges, one positive and the other negative. When the plastic rod was rubbed it acquired a charge that arbitrarily was named a negative charge. Some of this charge was transferred to the pith

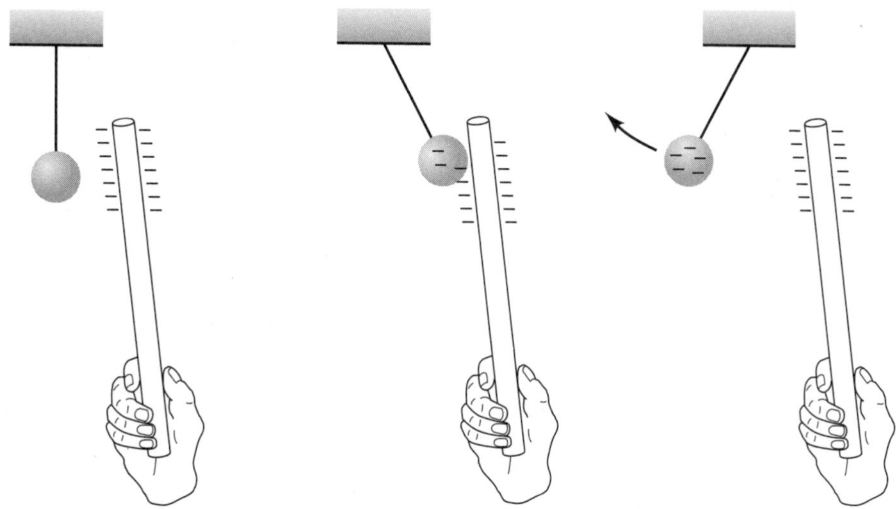

FIGURE 10.16

The pith ball is at first attracted to the plastic rod; after making contact, it flies away in the opposite direction.

ball during contact. The flying away of the pith ball suggests that negative charges repel each other. For the glass rod, rubbing produced a positive charge on it. The pith ball became positively charged because negative charges were transferred from it to the positive glass rod on contact. Now both pith ball and glass rod have a net positive charge. This experiment suggests that positive charges repel each other. Finally, the motion of the two pith balls toward each other suggest that unlike charges attract each other. We can summarize the results of these mini-experiments by stating that an "electrical force" exists between any objects that have a *net* electric charge; this force is repulsive if the net charges on both bodies have the same sign and attractive if the bodies have opposite signed charges. *Like charges repel; unlike charges attract.*

Further experiments would show that the magnitude of the electrical force varies inversely as the square of the distance between the charged objects and directly as the product of the magnitudes of the two charges. This result is expressed quantitatively by Coulomb's law:

$$\text{Electric force} = \text{K}\,\frac{Q_1 \times Q_2}{R^2}$$

where K is a constant, and Q_1 and Q_2 are the charges (in units of coulombs) separated by the distance R.

The origin of electric charges is found in the atom (discussed in detail in Chapter 13). The building blocks of the atom are negatively charged electrons, positively charged protons, and uncharged (neutral) particles called neutrons. The atom can be visualized as a miniature solar system, with a small positively charged nucleus consisting of protons and neutrons, surrounded by a cloud of electrons. The electrical attraction between the electrons and the nucleus provides the force needed to keep the electrons in orbit, analogous to the gravitational force that holds the planets in orbit as they move around the sun. The negative charge of the electron has exactly the same magnitude as the positive charge of the proton. Since the atom normally has the same number of electrons as protons, it is electrically neutral (i.e., its net charge is zero). If an electron is removed, the remaining fragment has a net positive charge and is called a positive **ion.** If, instead, an electron is transferred to the atom, it becomes a negative ion. When two materials are rubbed together, as in the case of amber and fur, electrons are transferred from one material to the other, giving one object an excess number of electrons and the other a deficiency of electrons. The first object has a net negative charge and the other object a net positive charge. No electrons were created or destroyed; they were just transferred from one thing to another. This is an important point and is a fundamental principle of physics: the conservation of charge.

You can make an object charged by passing a comb through your dry hair. The comb can then be used to pick up small pieces of paper or lint (Fig. 10.17). You might wonder how the charged comb could attract the small pieces of paper, since the paper was originally uncharged. The explanation depends on the variation of the electrical force with distance. When the negatively charged

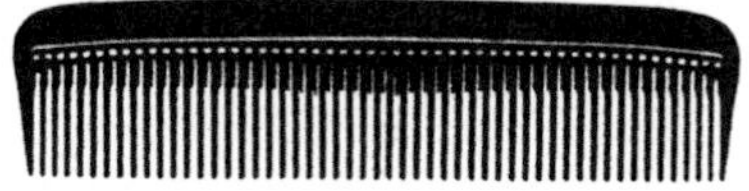

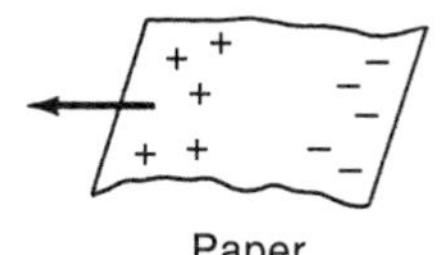

FIGURE 10.17
The comb has acquired a negative charge by being run through hair. It will attract a small piece of paper.

comb is brought near the paper, some of the negative charges in the paper move away from the edge of the comb, leaving an excess of positive charges closer to the comb. We have a separation of charges in the paper; we say that the individual molecules are electrically **polarized**—the comb has induced a net positive charge in the side of the paper closest to the comb, leaving the other side negative. Because the electrical force varies inversely with separation, the positive charges will feel more of an attractive force from the comb than the negative charges will feel the force of repulsion; consequently, the pieces of paper will be attracted to the comb. (Observe what happens to some of the pieces of paper after they have been on the comb for several seconds.) This electrical "polarization" is why the pith ball is first attracted to the charged rod (Fig. 10.16).

Water is a fair conductor because of the presence of ionic impurities that have a net charge. Where the air is unusually dry (a low humidity or low water content), as in the southwestern United States or indoors on a winter day, you must take care to avoid shocks from "static electricity." Dry air acts as a good insulator to prevent any acquired net charge from leaving your body. Sliding your feet across a furry rug or moving across the seat of a car under such conditions can provide you with irritating shocks on touching a grounded object (a doorknob or a friend) and discharging (Fig. 10.18). To avoid sparks in an operating room where oxygen is used, personnel wear slippers with metallized bottoms to ground themselves. A car can pick up a net charge as it moves; consequently, grounding rods sticking up from the pavement are sometimes used to discharge the car as it approaches a tollbooth.

An example of using electrostatic concepts is the electrostatic precipitators that are found at some fossil-fueled electrical power plants. The electrostatic

FIGURE 10.18
Personal electrification. Sliding on a car seat to get out can give you a net charge, which is discharged to the car door after you step out onto the ground. The shock results as the charge imbalance between you and the car is equalized.

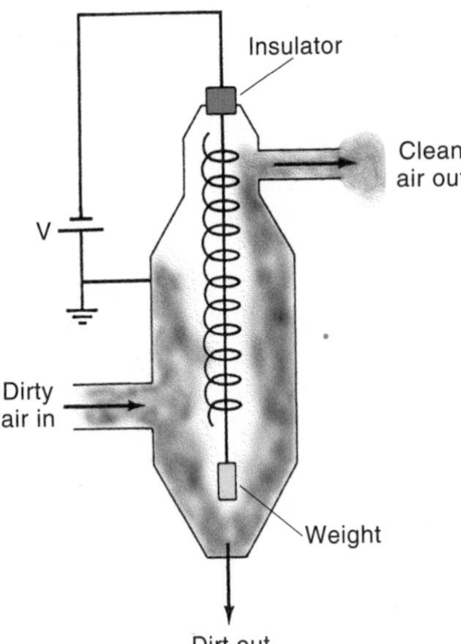

FIGURE 10.19
Schematic diagram of an electrostatic precipitator for the removal of particulate matter from combustion gases. Note the large negative voltage V on the center wire.

precipitator has been very successful in reducing particulate emissions from power plants' stack exhaust. The gases of combustion exiting the plant pass through a set of metal wires and plates situated as shown in Figure 10.19. There is a large negative voltage on the wires, and the plates are at ground potential. Electrons emitted by the wire can be captured by an ash particle, forming a negatively charged piece of ash that will move toward the plate, where it sticks and can be removed mechanically. The electrostatic precipitator can remove as much as 99% of the particulates present in the combustion gases, eliminating unsightly black smoke from the stack. (See Fig. 8.17.)

11

Electromagnetism and the Generation of Electricity

To make use of electricity in our homes and factories, we must have a way of providing that energy. It is not convenient to have batteries or fuel cells in your home or (currently) economically viable to use a large array of solar cells. It is presently less expensive and more convenient to generate electricity at centrally located power plants and then to transmit this energy to homes and factories. However, as discussed in the previous chapter, things are changing rapidly in the electric power industry as a result of deregulation and increased competition between independent power producers and conventional, centralized utilities. Many independent power producers are specializing in building small-scale generators that can provide both electricity and heat for a locality or industry. Recent legislation in the use of transmission lines also has contributed to such changes. In this chapter we will examine the relationship between electricity and magnetism and the principles behind the generation and transmission of electrical energy as well as changes in the electric power industry.

A. Magnetism

Simple Magnets

The first observation of a magnetic phenomenon was that of a particular rock that had the property of attracting (and repelling) similar rocks. The rock, an iron ore known as lodestone or magnetite, was known by the Greeks as early as 500 BC. It could pick up and cluster together tiny pieces of iron around its ends. A magnet acts as if there are two centers of force, which are called **poles,** labeled north (N) and south (S). The force between two magnets varies inversely as the square of the distance between them, and can be attractive or repulsive, depending on which poles face each other: In general, like poles repel each other and opposite poles attract (Fig. 11.1).

For forces that act over a distance, such as magnetic, electrical, and gravitational forces, it is convenient to speak of their respective "fields,"—for example, the gravitational field. A gravitational force will be exerted on a mass when it is in a gravitational field. The lines in Figure 11.1 from the N to S poles of a bar magnet denote "magnetic field lines," which are imaginary lines indicating the direction a magnet or small compass needle will point when placed in the vicinity of the bar magnet. Iron filings sprinkled on a sheet of paper (which lies on a bar magnet) similarly will align themselves to map out the magnetic field lines seen in this figure. The earth has its own magnetic field, similar to that produced by a bar magnet.* (See Fig. 11.5.) A compass needle will align itself with the earth's magnetic field, and so can help us keep track of our direction when navigating through the woods or on the sea. It is thought that the Chinese, about 100 BC, used permanent magnets to navigate. Recent archaeological finds suggest that the Olmec Indians of Central America also used magnets for navigation, but years earlier than the Chinese.

The Magnetic Field of a Current

As electricity and magnetism were studied, common features appeared, but few connections were made between the two subjects until the 19th century. Similarities observed between two charged bodies and two magnets were that (1) both created forces that could act in a vacuum, (2) both forces were inversely proportional to the square of the distance between the two charges or poles, and (3) the forces could be both attractive and repulsive. However, almost any substance could be charged or electrified, while magnetism was a property of iron and only a few other substances. The link discovered between electricity and magnetism established the field of "electromagnetism," which has played a crucial role in shaping the technological world in which we live.

---

*One set of units for magnetic fields are "gauss." The earth's magnetic field is about 0.5 gauss. In SI, the unit is the tesla; 1T = 10,000 gauss.

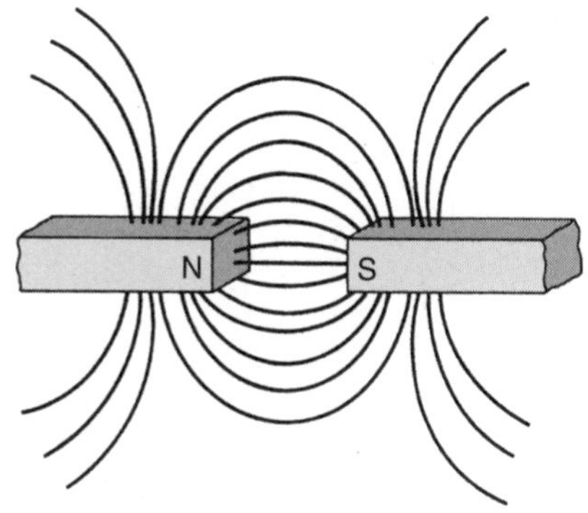

FIGURE 11.1
Two bar magnets: opposite poles attract. The lines are magnetic field lines, which indicate the direction a compass needle will point when placed there.

ACTIVITY 11.1

MAGNETS

Small magnets are available from many sources, such as electronic stores or hardware stores (where they are sold for use as kitchen-cabinet door holders).

1. Obtain two small magnets and explore how they interact with each other.
2. From a collection of objects in your room, find and list which ones are attracted by a magnet.
3. Investigate through which materials a magnet will attract a paper clip.
4. Examine the strength of a magnet by seeing how many paper clips, or small weights, can be suspended by a large paper clip. (See figure.)
5. Put two magnets together and investigate the question "Which will hold more paper clips or weights: two separate magnets or one two-unit magnet?"

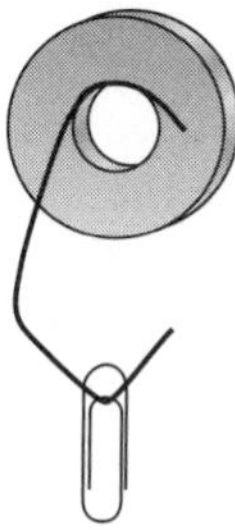

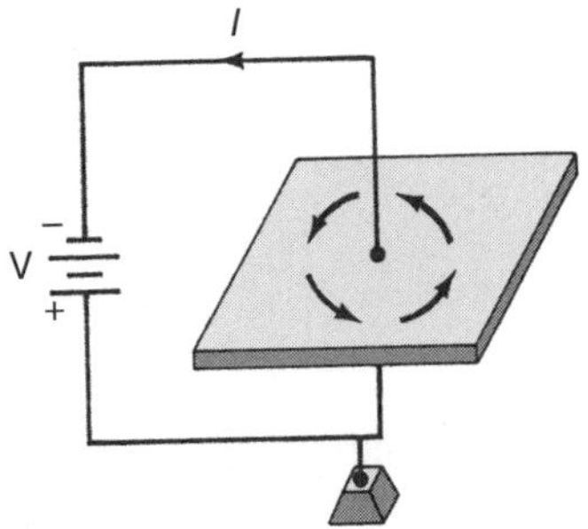

FIGURE 11.2
Oersted's experiment showed that a current-carrying wire gave rise to its own magnetic field. The arrows indicate the magnetic field lines as a result of the current-carrying wire.

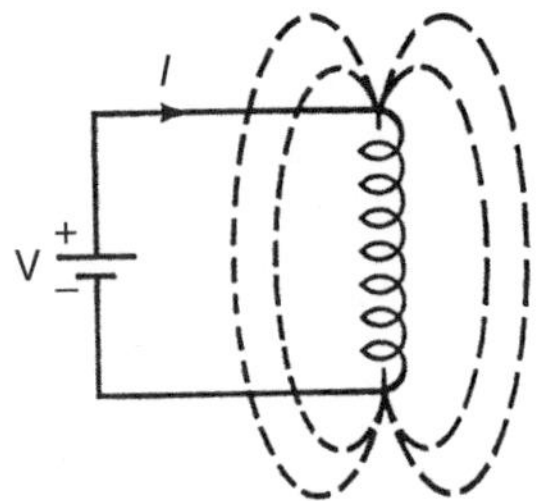

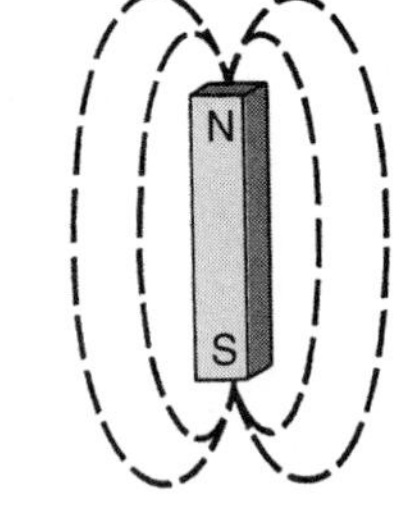

FIGURE 11.3
A solenoid, or coil of current-carrying wire, gives rise to a magnetic field (indicated by the dashed lines) similar to that of a simple bar magnet, as shown.

The connection between electricity and magnetism was discovered by a Danish physicist, H. C. Oersted, in 1820. He observed that a compass needle (which is a small magnet) experienced a force when placed near a wire that was carrying a current (Fig. 11.2). From this Oersted concluded that *an electric current creates a magnetic field.*

A coil of wire with a current passing through it also acts as a magnet, similar to a bar magnet (Fig. 11.3). This coil, given the name "solenoid," is an example of an **electromagnet.** The strength of the electromagnet can be increased by wrapping the wire around a core of soft iron (e.g., an iron nail), since the magnetic field of the solenoid makes the iron core magnetic also. Electromagnets are used in many places today, such as motors and relays. A magnetic field can be created or removed by closing or opening a simple switch that connects the electromagnet to a voltage source.

Motion of Charged Particles in a Magnetic Field: Cosmic Rays, Magnetic Bottles

Since a magnet will experience a force when placed near a current-carrying wire, Newton's third law states that there should be a reaction force felt by the current as a result of the magnet. Because a current consists of moving charges, another way of stating this is that *a charged particle will experience a force when moving in a magnetic field.* The charged particle will experience this force when moving at an angle to the magnetic field, but will feel no force if its velocity is parallel to the field. The force will deflect the particle in a direction perpendicular to its original motion (Fig. 11.4).

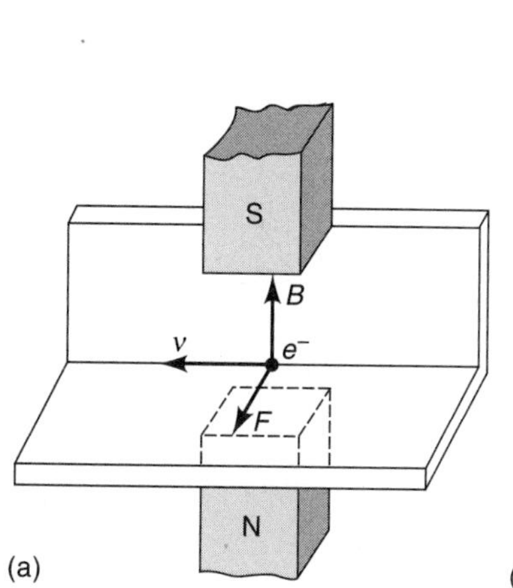

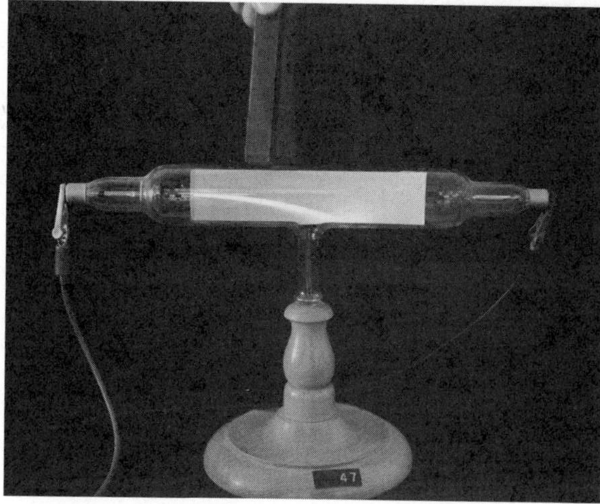

FIGURE 11.4

(*a*) A charged particle (e^-) experiences a force F when moving in a magnetic field B. The direction of the force is perpendicular to the velocity v and the magnetic field B. (*b*) A beam of electrons in the evacuated tube is deflected by a magnetic field. The beam is visible as it hits a fluorescent screen.
(G. BURGESS)

An example of this phenomenon is as old as the earth. The earth is continually subject to bombardment from cosmic rays (fast-moving charged particles) from outer space. This bombardment provides the majority of the natural ionizing radiation we receive (see Chapter 15). This incoming radiation on the earth is reduced below what it would be otherwise, because the incoming charged particles are deflected by the earth's magnetic field. Many of these charged particles that are deflected end up in radiation bands, called Van Allen belts, that surround the earth, extending from a few hundred kilometers to about 40,000 km above the earth. The particles remain "trapped" in these doughnut-shaped shells by the earth's magnetic field (Fig. 11.5). The pretty phenomenon of northern lights (*aurora borealis*) originates from the occasional interaction of the particles in these belts with air molecules in the atmosphere over the North Pole.

Another application of this phenomenon occurs in "magnetic bottles" (boxes without physical walls) used for the confinement of charged particles in thermonuclear fusion reactors (Fig. 11.6). (This subject will be discussed in detail in Chapter 16.) The object of confinement is to contain a hot, highly ionized gas (called a plasma). If a solid container were used, the gas would cool when making contact with the sides and lose some of its energy. However, suitable magnets can be placed so that their magnetic fields can trap the moving charged particles in a ring or doughnut-shaped geometry (as in Fig. 11.6).

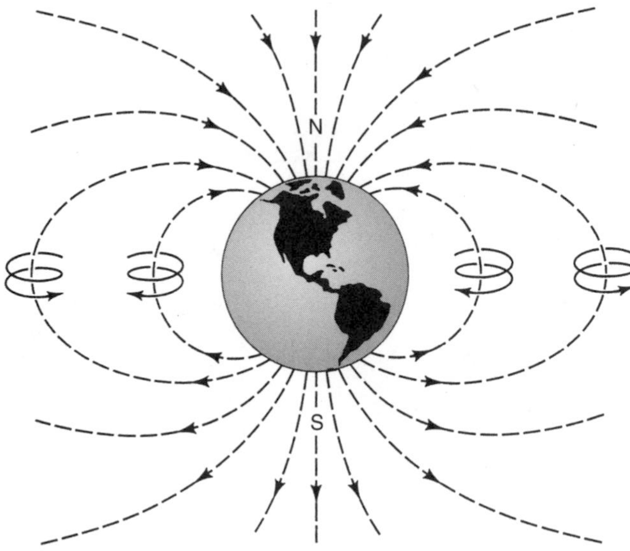

FIGURE 11.5
The earth's magnetic field and the Van Allen belts. The Van Allen belts are collections of charged particles that are trapped in doughnut-shaped shells by the earth's magnetic field. The particles spiral in these shells from N to S to N, being reflected at each end as if by "mirrors" as a result of the shape of the earth's field.

FIGURE 11.6
Fusion Test Reactor at Princeton University: the "Tokamak." In the center is an evacuated ring (torus) containing an ionized gas of deuterium and tritium at a very high temperature. The gas is held in place by the electromagnets surrounding the torus. This test facility was closed in 1998. It achieved 10 million watts of fusion power. (PRINCETON UNIVERSITY, PLASMA PHYSICS LABORATORY)

Electric Motors

A current-carrying wire placed in a magnetic field will experience a force, causing it to move. If this wire is connected to a shaft, useful work can be obtained from this motion. We call this device an electric motor.

Figure 11.7 shows the essential components of an electric motor. Coils of wire (called an armature) are placed in a magnetic field. (The magnetic field is usually produced by an electromagnet, although some motors use permanent magnets.)

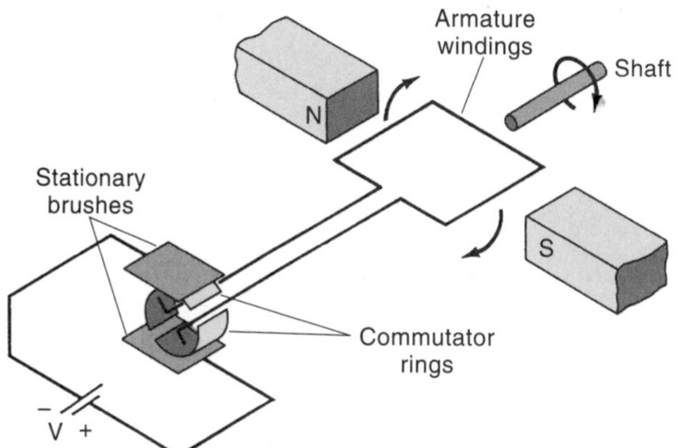

FIGURE 11.7
Schematic of a DC electric motor. For the continuous rotation of the shaft, the current in the armature windings must be alternating. With a DC source such as the battery, current reversal is possible with a "commutator"—two connecting rings on the armature shaft that make contact alternately with the positive and negative terminals of the battery.

The armature turns a shaft; the wires from the armature are brought out to sliding contacts called commutator rings. To understand how a motor operates, consider one coil of the armature, shown in Figure 11.7 in the shape of a square, with a current flowing in the wire. There will be a force on one side of the wire (near one pole) directed upward and a force on the other wire (near the other pole) directed downward, as a result of the different current directions. (There is no force on the wires that lie parallel to the magnetic field lines.) As the wire loop goes through half a turn, the directions of these two forces change, which would cause the loop to rotate back. For the shaft to continue to rotate in the *same* direction, the current in the wire has to change direction. If the wire carries alternating current, then the reversal is automatic. For wires carrying direct current, a split commutator ring must be used to send current through the coil first one way, then the other, as the shaft rotates. Figure 11.8 shows a very simple DC motor,

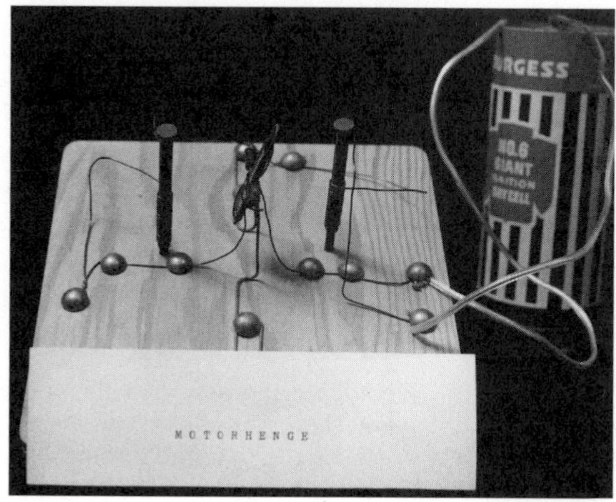

FIGURE 11.8
"Motorhenge"—the makings of a simple DC motor. (G. BURGESS)

using nothing more than a battery, wire, and iron nails. The wire-wrapped nails are connected to a battery and provide the magnetic field. Another very simple DC motor is shown in the "Further Activities" section at the end of this chapter.

B. The Generation of Electricity

Consider the following question: If an electric current creates a magnetic field, then could a magnetic field create a current? Suppose a loop of wire is placed around a magnet, as in Figure 11.9. Would a meter that measures current (an ammeter) give a reading for this circuit? The experiment was tried, but the answer was no. However, a variation of this experiment was tried in 1831 by Michael Faraday in England and Joseph Henry in the United States with surprising results. A magnet *at rest* next to or inside the wire loop will *not* cause any deflection of the ammeter's needle. However, if the magnet (or the wire) is *moved*, a potential difference is produced between the ends of the wire and thus a current is produced in the wire. The faster the magnet moves, the greater the induced voltage. This phenomenon, known as **Faraday's law of induction,** is the principle behind electrical generators today.

> A potential difference will be induced across the ends of a coil of wire if the magnetic field through the loop is changing; the amount of voltage induced is directly proportional to the rate at which the field through the coil changes.

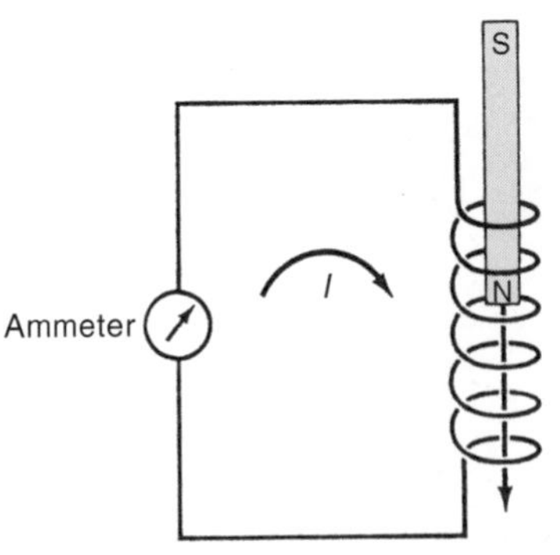

FIGURE 11.9
Faraday's law of induction: A moving magnet will induce a current in the surrounding wire.

Electrical generator at nuclear power plant. The pipes shown carry high-pressure steam to the turbine. (NEW YORK POWER AUTHORITY)

A model of a simple electrical generator is shown in Figure 11.10. (Its construction will be one of the activities at the end of this chapter.) In this model, a coil of wire is positioned between two stationary magnets and rotated by turning a shaft. The magnetic field through the rotating coil is constantly changing because of the motion of the coil, and so a varying voltage is induced across the ends of the coil (Faraday's law) (Fig. 11.10 insert). Alternating current is generated.

Conversion to direct current (DC) is desirable for many applications, such as charging a battery. In a generator, DC output can be obtained with the help of a split-ring commutator, as in Figure 11.7. Contact is made to the commutator with brushes, and the current is removed moving in one direction. A generator also can provide DC output by the use of a solid state device called a **rectifier.** Automobiles have an alternator (like a generator) that charges the battery, providing electricity for the ignition. The shaft is rotated by a fan belt. An alternator works on the same principle as a generator except that the armature coils are fixed and the magnetic field rotates.

There is a great similarity between the motor of Figure 11.7 and the generator of Figure 11.10. In the former, a current is supplied to the wires of the coil or armature, which experience a force in the magnetic field, causing them to rotate and turn a shaft. In a generator, the same shaft is turned by an external force—moving steam or water—causing the wires to rotate in the magnetic field, inducing a current in the loop. The homemade device of Figure 11.10 can be used in either way to become a motor or a generator. A large electrical generator is shown in Figure 11.11.

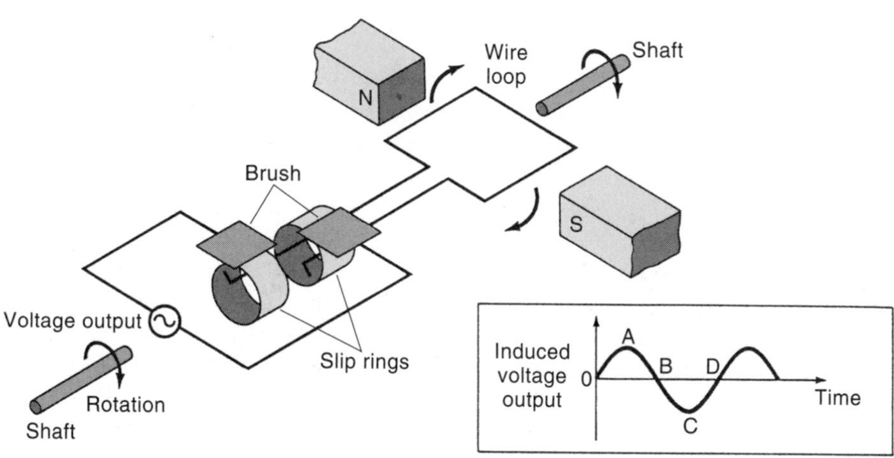

FIGURE 11.10
Simple electrical generator.

FIGURE 11.11
Cutaway view of a modern nuclear steam-turbine generator. On the left (with housing removed) are the turbine blades; the turbine shaft is connected to the generator—the cylinder on the right. (GENERAL ELECTRIC POWER SYSTEMS)

To review, the main concepts of electromagnetism are

1. A current-carrying wire produces a magnetic field.
2. A current-carrying wire experiences a force when in a magnetic field (as with an electric motor). Similarly, a charged particle experiences a force when traveling in the proper direction in a magnetic field.
3. A conductor moving in a magnetic field will have a voltage induced across its ends (as in an electric generator).

C. Transmission of Electrical Energy

Transformers: Voltage Changers

So far we have discussed the use of electricity in the home, simple circuits, and the generation of electricity. The transmission of electricity from plant to home could in principle be accomplished by the use of simple conducting wires of copper or aluminum, much like those we use in the home. Electricity could be generated at 120 V and then transmitted directly to the community. However, the transmission efficiency would be very poor. To understand this, suppose that ten households constituted our community and each had a need for 1000 W. Then the total power that must be supplied by the electric company would be $10 \times 1000 = 10{,}000$ W. If the power was transmitted at 120 V, and since $P = IV$, then the current carried in the line would be $I = P/V = 10{,}000/120 = 83.3$ amps. However, some of the electrical power is dissipated as heat according to the expression $P = I^2R$. If the transmission line from the power company to the community and back again had a total resistance of 1 ohm, then the energy lost per unit time would be $P = I^2R = (83.3)^2(1) = 6944$ W. To deliver the necessary power to ten homes, the plant would have to generate $10{,}000 + 6944 = 16{,}944$ W.

FIGURE 11.12
High-voltage (345-kV) transmission lines. Each system transmits power on three wires, and each set of towers carries two systems. (NIAGARA MOHAWK)

This means we have a transmission efficiency of only 59%. In practice, most transmission lines (Fig. 11.12) lose only 10% of the generated power as heat.

One can avoid such extreme losses by increasing the voltage at which the electricity is transmitted. (Another way would be to eliminate the resistance of the line with the use of superconducting wires.) If the transmission voltage is increased to 1200 V, the current would be reduced by a factor of 10 and the power dissipated as heat reduced by a factor of $(10)^2$, or 100. The power lost to heating, I^2R, is 69.4 W now, and so the (ideal) efficiency of this line rises to better than 99%.

To change or transform voltage to a higher or lower value, a **transformer** is used. The principle of its operation is a corollary to Faraday's law of induction. Figure 11.13 shows two loops of wire in close proximity, with only the left loop attached to a voltage source. The current in this coil gives rise to a magnetic field. If this magnetic field varies with time, as would happen if the voltage source provided alternating current (AC), the coil on the right will experience a changing magnetic field and so have a current induced within it.

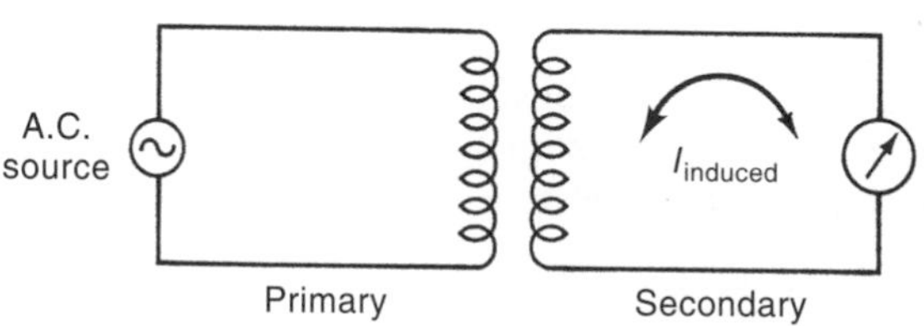

FIGURE 11.13
Corollary to Faraday's law: The current in the primary loop produces a magnetic field that varies with time, since we have an AC source. This varying field induces an alternating current in the secondary loop.

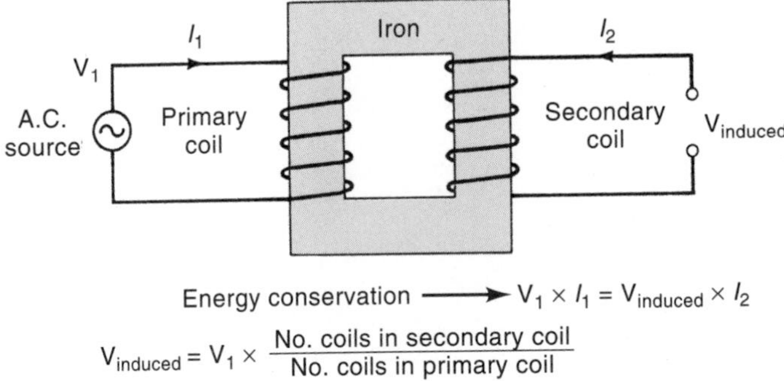

FIGURE 11.14
Transformer model: a device for stepping voltage up or down.

A better example of a transformer is shown in Figure 11.14. Two coils of wire are connected by an iron yoke. A change in the current in the first coil (or primary) causes a voltage to be induced in the second coil (or secondary). The voltage generated across the ends of the secondary coil depends on the ratio of the numbers of turns in the coils. If there were two turns of wire in the primary coil and two turns in the secondary, then the voltage induced in the secondary coil would be the same as the voltage across the primary leads. However, if there were twice as many turns in the secondary as the primary, then the voltage induced in the secondary would be twice the primary voltage. We have "stepped-up" the incoming voltage with this transformer. To "step-down" the voltage the opposite arrangement is made, with the number of the coils in the secondary being less than the number in the primary loop.

At first glance you might think that we are getting something for nothing with a step-up transformer. However, energy conservation still prevails; even though we can increase the voltage output with a transformer, the power transmitted will remain the same (ideally, in the absence of heating effects), that is,

$$\text{Power in} = \text{Power out}$$

Since $P = VI$, this expression can be written as

$$V_{\text{in}} \times I_{\text{in}} = V_{\text{out}} \times I_{\text{out}}$$

An increase in the secondary voltage is balanced by a decrease in the secondary current. The factor by which the output voltage is increased is equal to the ratio of turns between the secondary coil N_s and the primary coil N_p:

$$V_{\text{out}} = V_{\text{in}} \times \left(\frac{\text{turns in secondary coil}}{\text{turns in primary coil}}\right) = V_{\text{in}} \times \left(\frac{N_s}{N_p}\right)$$

EXAMPLE

An electrical generating plant produces electricity at the rate of 1000 MW at a voltage of 24,000 V. To transmit this power at 360,000 V, what must be the ratio of turns in the transformer? What current is transmitted at this voltage?

Solution

$$\frac{N_s}{N_p} = \frac{V_s}{V_p} = \frac{360{,}000\ \text{V}}{24{,}000\ \text{V}} = 15$$

The power transmitted is 1000 MW. Since $P = V \times I$, the current is

$$I_s = \frac{1{,}000{,}000{,}000\ \text{W}}{360{,}000\ \text{V}} = 2780\ \text{A}$$

The transformer works only if the current in the primary is changing; thus AC must be available. The secondary output is also AC. Some generators use rectifiers to change the outgoing AC to DC for use in charging batteries (e.g., to store energy generated by wind generators) and other applications. Examples of a few transformers are pictured in Figure 11.15, from a small one used to

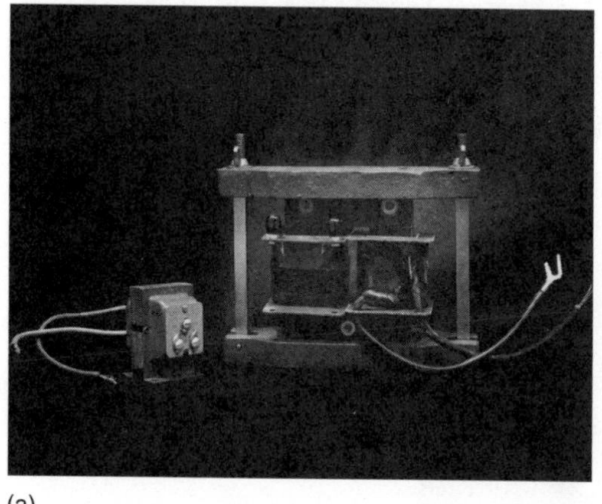

(a)

(b)

FIGURE 11.15

Examples of transformers: The smaller one (*a*) is for a doorbell system (120 V stepped down to 12 V), and the larger one (*b*) is a step-up transformer at a power plant. (NIAGARA MOHAWK)

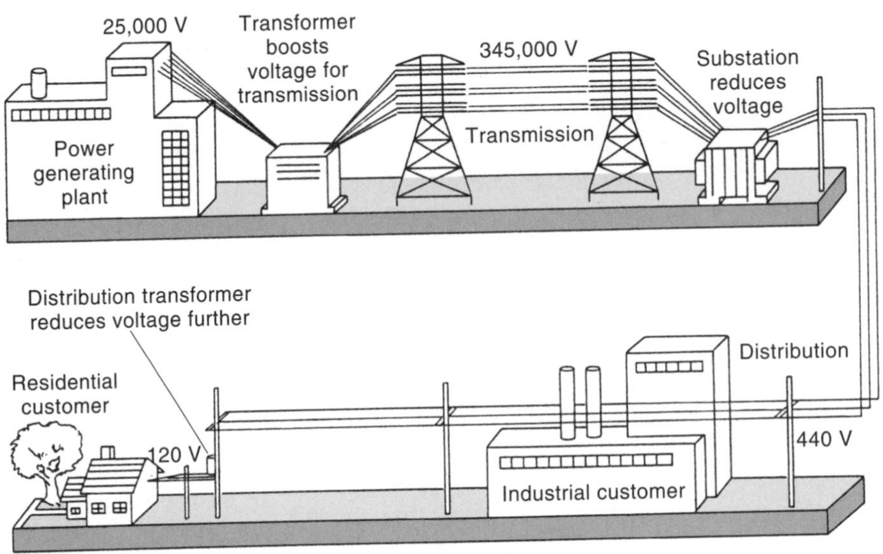

FIGURE 11.16

Electrical transmissions system. Electricity passes through a number of steps on its way from the power plant to the customer.

change 110 V to 6 V for use with a doorbell to large transformer stations located at a commercial electrical plant.

To minimize the I^2R power loss in transmission, electricity is transmitted at very high voltages from the generating plant to the receiving community. A city has several transformer substations to reduce the voltage for use in industry and the home (Fig. 11.16). The final transformer might be small enough to be placed on a telephone pole in that part of the community to reduce the volt-

FIGURE 11.17

First line crew on a high-voltage transmission line, Niagara Falls to Buffalo, New York, 1896. (NEW YORK POWER AUTHORITY)

age to 120 V. Electricity is usually generated at about 25,000 V (25 kV) and raised to 115 kV, 345 kV, or 765 kV. Most newly installed transmission lines today are at a voltage of 345 or 765 kV; work is progressing on lines carrying electricity at more than 1000 kV, with such lines already in use in Russia.

The first transmission of electricity by Thomas Edison used direct current and was restricted to a service area of only a few square miles. The first use of alternating current and higher transmission voltages (with transformers) was in 1896 for the transmission of electricity from Niagara Falls to Buffalo, a distance of 20 miles. A picture of one of the first line crews is shown in Figure 11.17. Today there are some two million miles of power lines across the United States.

Environmental and Health Impacts of High-Voltage Lines

Electricity at high voltages is usually carried by steel-cored, bare-aluminum conductors that are mounted on steel towers using large ceramic insulators (Fig. 11.18). The diameter of the conductor is proportional to the voltage carried by that line. High voltages can cause ionization of the surrounding air, resulting in a glow visible at night (a corona) and an ever-present high frequency buzzing; radio reception is also affected. Farmers living near such power lines have registered complaints about electric shocks from houses (e.g., while installing rain gutters or aluminum siding), loud electric noises, and TV and radio interference.

The maximum magnetic field directly beneath a 765-kV transmission line is less than 1 gauss. The earth's magnetic field is about 0.5 gauss. However, the earth's field is static, while the electromagnetic fields (EMFs) from power lines

FIGURE 11.18

Close-up view of a high-voltage transmission line crossing the Hudson River, New York. Shown are the large ceramic insulators. (NEW YORK POWER AUTHORITY)

are 60 Hz. Typical magnetic field levels in a home are between 0.1 to 50 milligauss, but the values several centimeters away from TVs and hair dryers can be 10 to 20 times higher than these (Table 11.1). Fortunately, the fields drop off rapidly with distance from the source.

The electric fields in homes are about 1 to 10 V per meter (V/m); within 20 cm of small appliances the electric fields reach 20 to 300 V/m, and right next to an electric blanket they approach 10,000 V/m, the maximum level one would experience while standing on the ground directly beneath a 765-kV line. Fortunately, electric fields are blocked easily by vegetation and buildings and are attenuated strongly in the human body. However, magnetic fields can pass through most objects easily. (See Activity 11.1 earlier in this chapter.)

The biological effect of EMFs from power lines has become a controversial issue in recent years, with large differences of opinions among scientists. In 1979, N. Wertheimer et al. studied the correlation between childhood mortality records and the proximity to high-voltage power lines in the Denver area. They proposed a correlation between long-term exposure to weak EMFs and increases in cancer. Critics pointed out that actual field measurements were not taken in the homes, and the study was not "blind"—the researchers knew which homes were those of cancer victims. A study by David Savitz several years later eliminated both of these problems and found a modest statistical correlation between children with cancer and the proximity of their homes to high-voltage power lines. However, there seemed to be no correlation between the magnitude of the magnetic field and cancer.

The weak EMFs near power lines do not have the energy themselves to initiate cancer, because radiation-induced cancers are usually caused by more energetic radiation breaking or rearranging DNA bonds. One hypothesis on the biological effects of EMFs suggests that when fields interact with a cell, they initiate a chemical change on its surface, possibly disrupting the flow of ions through the cell membrane. Inside the cell, a sequence of chemical reactions is triggered that distort the normal flow of biological information and may result in a cell whose growth is out of control.

Table 11.1 TYPICAL ELECTRIC AND MAGNETIC FIELDS ENCOUNTERED IN DAILY LIFE*

Situation	Electric Field (V/m)	Magnetic Field (mG)
Home wiring	1–10	1–5
Electrical appliances	30–300	5–3000
Under distribution lines	10–60	1–10
Under high-voltage transmission lines	1000–7000	25–1000

Notes: Background from earth is 120 V/m and 500 mG (milligauss); values for power line frequencies of 60 Hz.

*Committee on Interagency Radiation Research and Policy, 1992.

A 1995 University of North Carolina study (David Savitz and Dan Loomis) looked at case histories of 139,000 utility workers over a 36-year period. Estimates were made of each worker's cumulative magnetic field exposure, and actual measurements of exposure to magnetic fields were taken for some workers. The UNC researchers concluded that the data did *not* support an association between occupational magnetic field exposures and leukemia, but did suggest a weak link to brain cancer. A 1994 Canada–France study of utility workers also estimated magnetic field exposure. This study *did* find a statistically significant association between magnetic field exposure and at least one type of leukemia. Such inconsistencies in results are frustrating. Some recent Scandinavian studies on the effect of EMFs on childhood cancers offered inconclusive results. Laboratory studies using rodents at the Illinois Institute of Technology found no evidence that EMFs adversely affected the animals' reproductive or immune systems.

Much more must be learned about the biological effects of EMFs to interpret the inconclusive findings that have been gathered to date. Many studies have been conducted over the last two decades in this area. While there is uncertainty in some of the findings, there has been a failure to substantiate or reproduce those studies that have reported adverse health effects from exposure to EMFs. Unfortunately, policy options in a climate of scientific uncertainty pervade this issue. The American Physical Society issued a statement in 1995 on power-line fields. They said that "the scientific literature and the reports of reviews by other panels show no consistent, significant link between cancer and power-line fields." Furthermore, relative to the costs of mitigation relating to the power lines–cancer connection they said that "the diversion of resources to eliminate a threat which has no persuasive scientific basis is disturbing. More serious environmental problems are neglected for lack of funding and public attention, and the burden of cost placed on the American public is incommensurate with the risk, if any."

Alternatives to Transmission Lines

Electrical transmission lines are the most expensive method for transporting energy. Coal by barge or rail and natural gas by pipeline are much cheaper. The costs of electrical service are broken down in Table 11.2. More than *half the cost* goes to transmission and distribution. However, the cost of transmission per kWh decreases as the voltage increases because of reduced rights-of-way needed for transmission of the same amount of power and reduced heating losses.

An alternative to conventional transmission lines is underground cables, which are used primarily in congested urban areas. About 1% of the total transmission mileage in the United States is underground. Underground lines use negligible land, but can cost 6 to 20 times as much as overhead lines of the same length and power. A problem with lines under the ground is the dissipation of the heat produced. Wires above the ground are usually bare (uninsulated) and

Table 11.2 COSTS OF ELECTRICAL SERVICE*

Capital Costs (Original Costs, No Depreciation)	
Production	44%
Transmission	22%
Local distribution	34%
Operating Costs	
Production (including fuel)	89%
Transmission	3%
Local distribution	8%

*Typical utility costs.

Source: Niagara-Mohawk Power Corporation.

the circulating air easily removes the heat. Underground, insulation is needed for the wires; this is usually an oil-coated paper. The earth cannot transfer the heat as well as air, so there is a limit on the power that can be transported with such wires.

One potential solution to energy losses from heating is the use of superconducting cables. Superconductors, at low enough temperatures, lose their resistance ($R = 0$ ohms) and so no energy is lost as heat. (See Chapter 10.) Some of the common superconducting materials are alloys of the rare metal niobium, which becomes superconducting at temperatures less than −264°C (−445°F). To reach such temperatures, large refrigerating systems must be used. Liquid nitrogen can take temperatures down to −196°C, but liquid helium is necessary to reduce temperatures down to almost absolute zero, −273°C. Even though the requirement for refrigeration equipment is large, superconducting cables can carry much more electricity than conventional lines and so their cost may be competitive with other types of underground lines. Much research is being done to find commercially usable materials able to carry large currents that are superconducting at higher temperatures, perhaps eventually even at room temperature.

D. Standard Steam–Electric Generating Plant Cycle

The majority of electricity used in the world today is produced by generators that transform mechanical energy into electrical energy. As studied in the last section, electrical generators operate on the principle of electromagnetic induction; that is, a conductor, such as a wire, moving in a magnetic field has a voltage difference induced across its two ends. The source of mechanical energy that turns this wire may be falling water (hydropower), steam produced by the burning of fuels, or the kinetic energy of wind.

Electrical energy also can be produced in direct conversion processes, rather than the conventional heat to mechanical to electrical conversion processes. Examples of direct energy conversion devices are solar cells and fuel cells.

Fifty-six percent of the electricity generated in the United States is produced from coal. The combustion of coal for the production of steam is a multifaceted process. First, the coal is unloaded from the train cars by picking them up with large cranes, turning them upside down, and shaking out the contents. The coal is then ground to a fine powder and blown into the furnace connected to a steam generator. Here it mixes with preheated air and burns to release heat for steam production.

Oil is shipped to a power plant via train, truck, pipeline, or tanker, and then stored in large tanks. Most plants burn residual fuel oil, that part left over at the refinery after lighter fractions have been removed. The oil is quite viscous, and so is usually heated for easy handling prior to combustion. The oil is sprayed into the furnace, where it mixes with the entering air for burning. Natural gas is received via pipeline and is used immediately, at a rate determined by electrical demand. No storage is necessary.

A block diagram for a standard fossil-fueled electrical generating plant was given in Figure 3.3, from fuel in to electricity out. Let us examine some of these components in more detail. The steam generator or boiler consists of miles of tubing carrying the water (and/or steam). These tubes receive heat by radiation from the fire or by convection from the gases of combustion as they flow through the boiler. Steam is produced and reheated between turbine stages to increased temperatures for higher power-plant efficiencies. The steam leaves the boiler at more than 1000°F (538°C) and enters the turbine unit, passing through nozzles to increase its velocity to about 1000 mph. The high-speed steam hits the blades of the turbine, turning a shaft or rotor on which wire coils are mounted (Fig. 11.19). As the steam proceeds through the turbine, its pressure and density decrease, so progressively larger blade surfaces must be used to capture the steam's waning energy. As the figure indicates, the blade surfaces, or buckets, become larger. The steam rushes through the system of buckets in about one thirtieth of a second, spinning the rotor at 3600 rev/min. The temperature drops from 1000°F (for a fossil-fueled plant) to about 100°F during this time. The steam enters at a pressure of 2000 lb/in^2 and expands to a final pressure less than atmospheric pressure. The steam is "really exhausted" by the time it leaves the turbine.

Leaving the turbine, the steam enters the condenser chamber, which consists of some 400 miles (667 km) of tubes that carry cold water from an external source such as a lake or cooling tower. The steam transfers heat to the cold water and so is cooled and condensed to liquid water. The cycle of a steam electric plant is completed with the return of this water to the feedwater pump, where it is pumped to a high pressure and fed back to the steam generator. The

FIGURE 11.19
Cutaway view of a steam turbine, showing 16 blades and the shaft, which connects to the generator at the top of the picture. Steam enters by the small blades and exits at the larger blades.
(NIAGARA MOHAWK)

warmed condenser water is returned to the lake or river or goes to a cooling tower, with an increase in temperature (ΔT) of about 20°F (11°C) over ambient temperatures. The purpose of the condenser is to increase the efficiency of the power plant and to complete the steam cycle by returning water to the boiler. The effects of thermal pollution and alternative uses of waste heat were discussed in Chapter 9.

After the cooled gases of combustion have left the furnace (carrying with them part of the ash remaining after the coal or oil has been burned), they pass through a series of air-pollution control devices before being released through a tall stack into the atmosphere. The effects and control of air pollution from the burning of fossil fuels were discussed in Chapter 8.

The electricity generated at a central power station is in the form of AC. The frequency of this current, 60 Hertz in the United States, must be maintained within close limits. Since the frequency depends on the shaft speed of the turbine, steam flow must match the electrical energy needs at that time. If the electric load increases, the shaft slows down so the steam flow must be raised, and vice versa.

One of the problems in generating electricity is fluctuating demand. During the day, with industries churning and households active, the demand is much higher than at night. High demand in the day also leads to high prices. A power plant must be built with the capacity to supply the maximum needs of its customers, yet during the evening hours part of the plant capacity is idle, not pro-

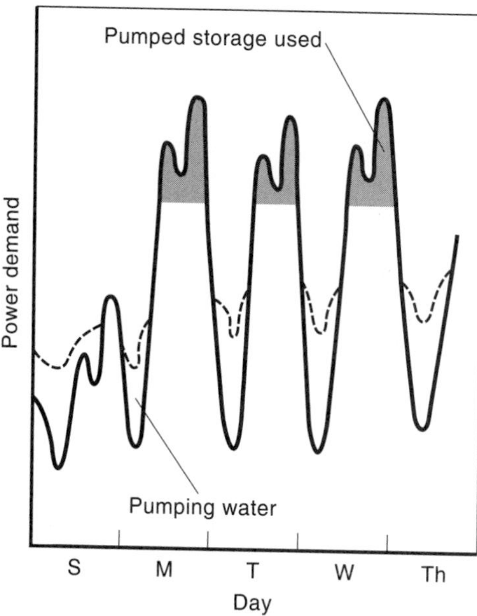

FIGURE 11.20
Electrical power demand during the week. During periods of reduced demand, electricity is used to pump water into a reservoir. The dashed line indicates energy use when this is done. During the day, pumped storage meets the demand shown by the shaded part.

ducing revenue for the utility. Some utilities build smaller power plants to reduce their capital costs for a plant, and then use less expensive gas turbines to supply the extra daytime demand. However, more fuel must be used to provide the same output, as gas turbines are less efficient than conventional steam turbines (25% efficient compared to 35%). One partial solution to the problem of fluctuating demand is rate restructuring, with cost incentives to use electricity at night. (See Chapter 10.) Another solution is energy storage systems, such as pumped water. In **"pumped storage,"** excess power generating capabilities at night are used to pump water up into large storage reservoirs. During the day this reservoir water can be used to provide hydroelectric power to meet the increased demands. Figure 11.20 shows electrical demand over a work week and the use of pumped storage to reduce peak demands on the steam plants. Pumped storage systems are expensive to build, however, as large areas of land are needed for the reservoirs.

E. Cogeneration

One area in which there are large opportunities for saving fuel, especially in the industrial sector, is cogeneration. **Cogeneration** is the production of both electricity and useful heat from the same fuel source (Fig. 11.21). Two types of cogeneration projects have been pursued. The first generates electricity from a turbine generator and uses the exhaust gases or high-pressure steam for

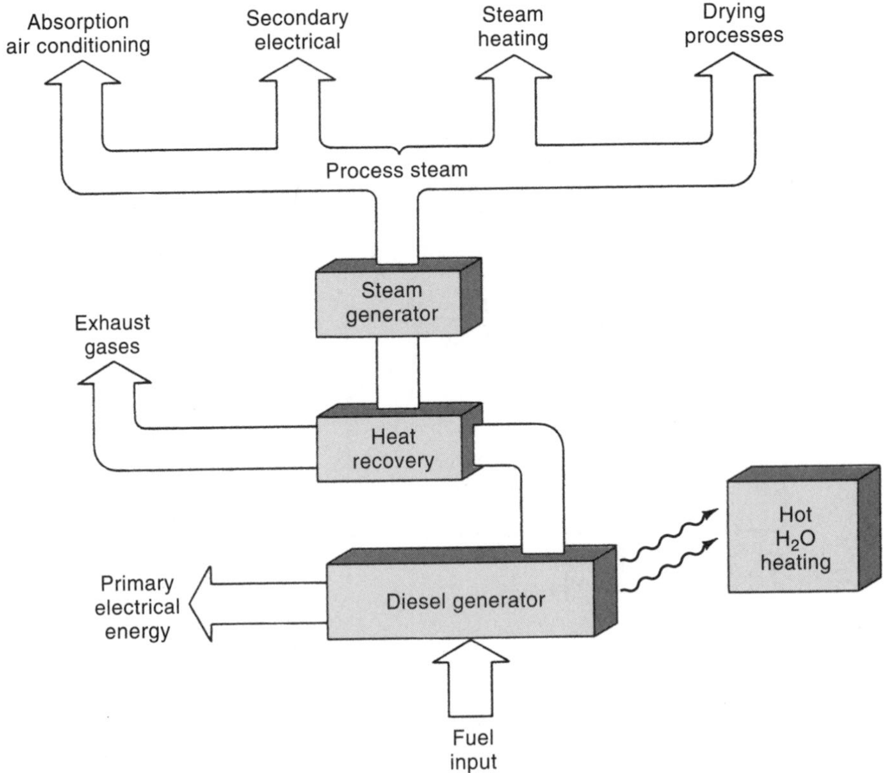

FIGURE 11.21
Cogeneration is the simultaneous production of various forms of energy from one power source.

process steam or district heating and/or additional electricity. This type is called a "topping cycle." The second scheme takes steam that already has been used in an industrial process and runs it through a low-pressure steam turbine for the generation of electricity. This type is called a "bottoming cycle." In both cases the overall conversion efficiency (the ratio of useful energy output to fuel energy input) is improved substantially, since some of the waste heat has become useful.

Cogeneration has made dramatic steps in recent years. More than 5000 projects were undertaken in the 1980s and 1990s, with nearly 100,000 MWe of electrical capacity put on line. The most commonly used fuel in these cogeneration plants is natural gas, in part because of its low price and relatively small impact on the environment. Facilities that burn solid fuels, primarily municipal waste, are increasing because of the high cost of refuse disposal and lack of adequate landfill space.

Historically, self-generation of industrial power was the rule rather than the exception in the United States for the first half of the 20th century. Electric utilities made technological advances in power generation and transmission after World War II, and in the 1960s and 1970s utilities were installing 40 new large central-station units per year. Pacing this growth was an average 7% annual increase in electric-power demand. Things changed in the 1980s because of a slower growth rate in power demand, the demise of the nuclear power industry because of burdensome costs and the Three Mile Island accident, and increased concern for the environment. The change was accelerated also by the passage of the Public Utility Regulatory Policies Act (PURPA), in which utilities were required to purchase power from qualified sources at the utilities' avoided cost. Consequently, on-site electric power generation by small units and cogeneration became economically favorable. One approach that has become popular is the development of a cogeneration project by an outside firm, which sells the steam and/or electric power to an industrial customer and the balance of the power to the local utility. Natural gas is the favorite fuel for most of these projects. The biggest user of cogeneration units is the chemical industry.

Some cogeneration units use "combined-cycle" systems (Fig. 11.22). They use gas turbines to generate electricity and the high-temperature exhaust gases (1000°F to 1200°F) to generate additional electricity with a conventional steam turbine or use the high-pressure steam for industrial processes. Such systems are enormously popular due to the present abundance and cheapness of natural

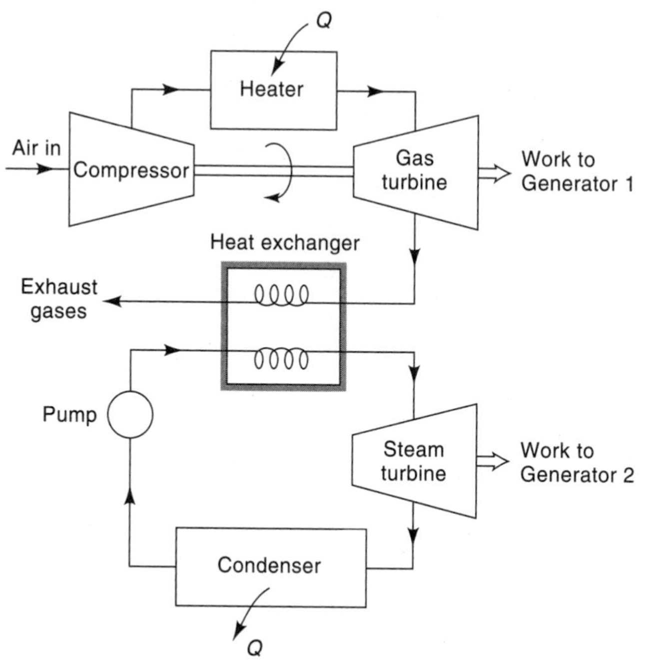

FIGURE 11.22

Combined-cycle system for the production of electricity by a gas turbine and a steam generator.

FIGURE 11.23
This 1040-MWe cogeneration power station in upstate New York uses natural gas and provides thermal energy to a nearby industry. Water vapor emerges from the cooling towers on the left, behind the plant. The 4 stacks are associated with the gas turbines. In the background is Lake Ontario. (SITHE ENERGIES, INC.)

gas and their high efficiencies. One of the largest such cogeneration facilities in the United States is a 1040-MWe plant in Oswego, New York, completed in 1995 (Figure 11.23). Its overall efficiency is 54%! The world's largest combined-cycle facility, in Korea, is pictured at the beginning of Chapter 2. A cutaway view of a high-performance gas turbine is shown in Figure 11.24.

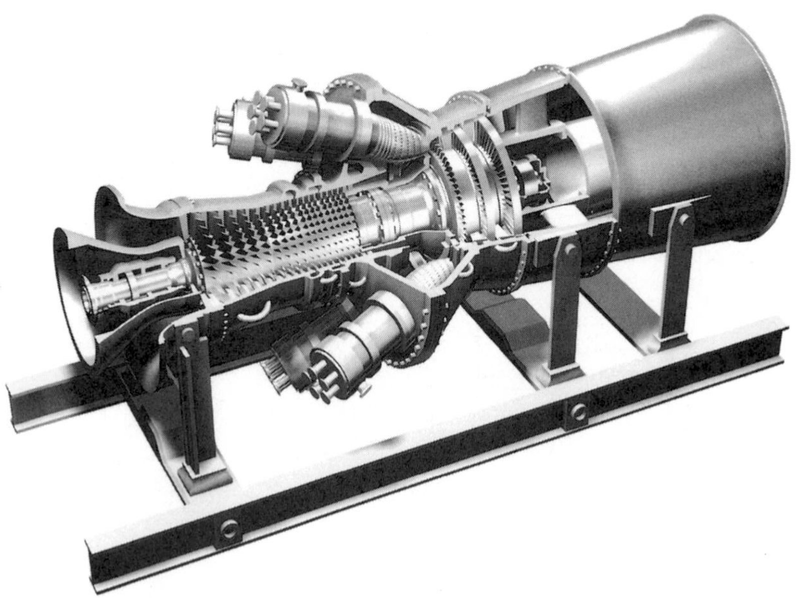

FIGURE 11.24
Cutaway view of a high-efficiency gas turbine. Shown from left to right are the compressor, the combustion chambers (four are shown), the three-stage turbine, and the exhaust system. The electrical generator would be connected on the left. (GENERAL ELECTRIC POWER SYSTEMS)

F. Summary

The production of electricity accounts for about 25% of the total energy consumed in the United States. This share probably will continue to grow. Most of the alternative energy sources discussed in this book are directed to the production of electricity.

The interaction between electricity and magnetism can be summarized as:

1. An electrical current produces a magnetic field.
2. A moving charged particle can experience a force in a magnetic field.
3. A wire moved in a magnetic field has a potential difference induced across its ends. (This concept is the principle behind electric generators.)

After generation, electrical energy is transmitted at very high voltages to reduce losses because of heating. The rate of heat loss because of heating is given by $P = I^2R$. Transformers, requiring alternating current, are needed to step up the voltage from the generator to the transmission line, and again to reduce it for household use. The electrical power transmitted is given as $P = VI$.

Cogeneration is the production of both electricity and useful heat from the same fuel source. This process either can generate electricity first and use some of the exiting steam from the turbine for industrial purposes, or make use of steam that was used for an industrial process to generate electricity in a turbine generator. In both cases, the overall fuel conversion efficiency has been increased. Typically a standard steam-cycle power plant has an efficiency of 35 to 40%, while a combined cycle plant can attain efficiencies of 55 to 60%.

Internet Sites

For an up-to-date list of Internet resources related to the material in this chapter, go to the Harcourt College Publishers website at **http://www.harcourtcollege.com**. The links are in the *Energy: Its Use and the Environment* site on the Physics page. General energy related sites and some guidelines for using the World Wide Web in your class are on the inside front cover of this book.

References

Chapter 11

Boebinger, G., A. Passno, and J. Bevk. 1995. Building World-Record Magnets. *Scientific American*, 272 (June).

Hughes, T. 1993. *Networks of Power: Electrification in Western Society, 1880–1930.* Baltimore, Johns Hopkins University Press.

Nadel, S. 1992. Utility Demand-Side Management Experience and Potential. *Annual Review of Energy,* 17.
Sathaye, J., A. Ghirardi, and L. Schipper. 1987. Energy Demand in Developing Countries: A Sectoral Analysis of Recent Trends. *Annual Review of Energy,* 12.
Schramm, G. 1990. Electric Power in Developing Countries. *Annual Review of Energy,* 15.
Serway, R., and J. Faughn. 1999. *College Physics.* 5th ed. Philadelphia, Saunders College Publishing.
Stix, G. 1992. Air Trains. *Scientific American,* 267 (August).
Zubrowski, B. 1991. *Blinkers and Buzzers: Building and Experimenting with Electricity and Magnetism.* New York, Beech Tree.

QUESTIONS

1. If you hold a charged particle in a magnetic field, what force is there on the particle? Suppose you drop the particle so that its path is perpendicular to the magnetic field. What do you think will happen?
2. How can you use a small electric motor as a generator in a homemade wind generator? Why can't this substitution always be made? (Think of an electromagnet.)
3. Suppose you are hiking in the woods and using a compass to get your bearings. Suddenly a bolt of lightning strikes nearby. Would you expect this to have an effect on your compass? Explain.
4. Briefly describe how an electric motor works.
5. Why do utilities transmit electricity at high voltages?
6. A transformer is used to step down the voltage to 240 V for residential use. Does this mean that the electrical energy is also reduced? Explain.
7. What issues do you think need to be further studied or investigated in the controversy over health effects of high-voltage transmission lines?
8. Electric resistance heating of new homes is very popular. How do you explain this trend, considering that the cost of electricity is so high?
9. With the increased production of coal and its use in electric generating plants, proposals have been made for "mine-mouth" electricity generation: power plants located adjacent to the source of the coal. Considering the location of coal mining areas in the United States, what are the advantages and disadvantages of this plan?
10. Cogeneration is a popular means to provide electricity and heat. How would your school use this technology to save money? Be specific.
11. In cogeneration, would the overall efficiency of a steam-turbine power plant generating electricity and using the process steam be greater if the steam was expelled from the turbine at a higher temperature?

PROBLEMS

1. A power plant is capable of producing electricity at a rate of 1000 MWe. If the electricity is sold for $.08/kWh, what revenue is lost per day if the plant is not operating?
2. A small transformer used for a doorbell steps down the voltage from 120 V and 0.5 A to 12 V. What is the current flow to the doorbell?
3. Ten megawatts of electricity are to be transmitted over a power line of resistance 4 ohms.
   (a) If the electricity is generated at 10,000 V and is to be transmitted at 130,000 V, what should be the ratio of turns of the transformer?
   (b) At what current will the power be transmitted?
   (c) What percentage of the original power will be lost in the line because of resistive heating?
4. If a pumped storage system for a fossil-fuel electric generating plant uses a pump of overall efficiency 70% and a turbine generator of efficiency 80%, what is the overall efficiency? What are the advantages of pumped storage?

FURTHER ACTIVITIES

1. Construct an electromagnet by wrapping about 1 meter of small-diameter (24-gauge) insulated wire around a bolt or a nail. Connect the ends to a D-cell battery as shown. See how many staples or paper clips you can pick up with the electromagnet. How would you increase the strength of the magnetic field?

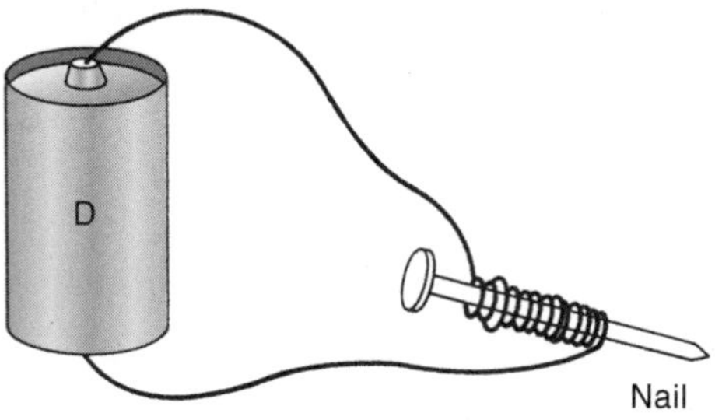

2. Construct a simple meter for measuring electrical current as follows: Place a compass in the middle of a cardboard cylinder and wrap about 2 m of small-diameter insulated wire around the cardboard, leaving both ends free with 1 cm of insulation stripped from each end. Use the meter to observe the differences in current for one and two flashlight bulbs in series with a D-cell battery. (The amount of rotation of the compass needle will not be proportional to the current.)

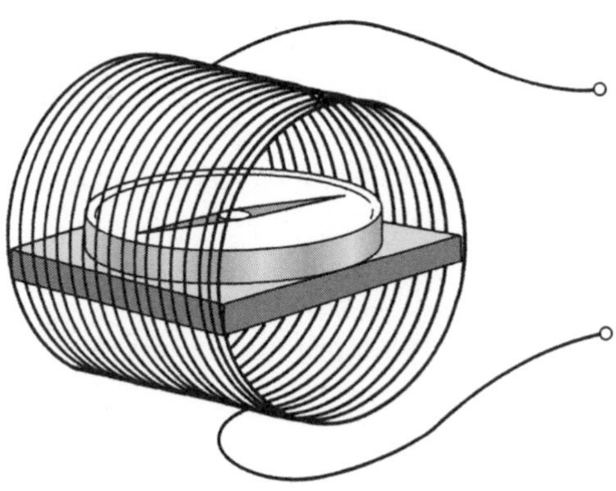

3. Illustrate Faraday's law of induction with the following activity. Make a coil of wire large enough for a bar magnet to pass through. Attach the ends of the coil to the current meter constructed in Activity 2. Move the bar magnet back and forth through the coil and observe the direction in which the needle moves. Does the needle's deflection depend on how rapidly the magnet moves through the coil?
4. You can use some simple motors as electric generators. Find a small, inexpensive motor that uses permanent magnets (not electromagnets). Hook the leads up to a flashlight bulb. Turn the shaft of the motor with your fingers or by pulling a string wound on the shaft. Does the brightness of the bulb depend on the speed of the shaft?
5. Construct a direct current electric motor similar to the one shown in Figure 11.8.

6. Another DC motor that is very portable is shown below. You can construct this using wire, two paper clips, a D-cell battery, and a small magnet. Use the battery as a form around which to wrap 10 to 15 turns of magnet wire. Both ends of the wire should have the insulation stripped off of one side. Put the motor together as shown. Start the coil moving by hand and it should continue on its own.

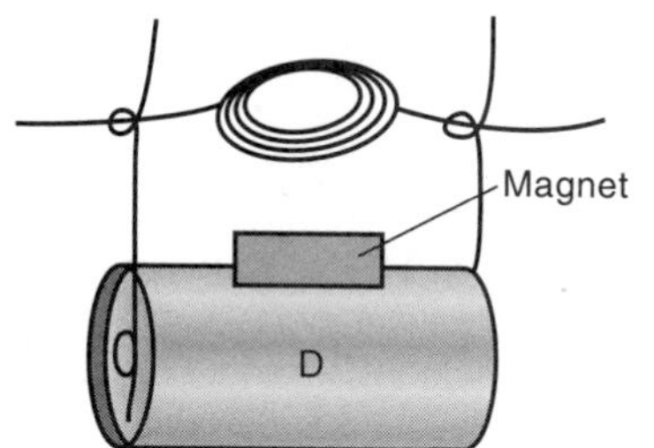

Portable motor.

7. We can make a magnet from material with iron in it by stroking it (in one direction) with a permanent magnet. See how many staples a pair of metal scissors can pick up after they have been stroked different numbers of times. Hammer the scissors several times and see if they will still pick up the same number of staples.

12

Electricity from Solar, Wind, and Hydro

A. Introduction

Chapter 6 introduced the many dimensions of renewable energy: from space heating to photovoltaics to wind energy to hydroelectric power. However, it covered in detail only the uses of solar energy for domestic hot water and space heating. We now direct our attention to the generation of electricity through photovoltaics, wind energy, hydropower, and thermal electric systems. Each of these areas can play a substantial role in helping us meet our energy needs and in fact are some of the fastest growing energy technologies today, although their total role is still small. For example, wind energy capacity has grown in the United States by 150% over the last 15 years, but still provides less than 1% of U.S. electricity. Table 12.1 looks at contributions of solar technologies to our nation's energy supply over the last 20 years. Research and development, as well as the world price of oil and public policy, will have a strong bearing on the future contributions of solar energy.

Table 12.1 SOLAR CONTRIBUTIONS TO U.S. ENERGY SUPPLY*

Source	1980	1990	1999
Solar heating, photovoltaic, thermal electric	Small	0.063	0.076
Wind	Small	0.032	0.038
Biomass	2.4	2.6	3.5
Hydroelectric	3.0	3.1	3.4
Total Solar	5.4	5.8	7.0
Total Consumption	78	84	97

*Units: quads (10^{15} Btu) per year (United States Department of Energy)

Photovoltaic (PV) generation, the conversion of sunlight directly into electricity, has been and will continue to be one of the more glamorous technologies in the energy field. This technology began many years ago, and received a boost in the 1950s through its use in the U.S. space program. The prices for solar cells have dropped by a factor of almost 1000 since those days. Yet solar cells remain relatively expensive, and the degree of future market penetration is highly dependent on lower production costs and increased cell efficiency. There appear to be no remaining technical obstacles to widespread use of solar cells. Over the past several years, in fact, there have been significant advances in the development of low-cost PV materials, and efficiencies of almost 30% have been obtained. Figure 12.1 shows the decrease in PV cost per watt over time, as well as the strong increase in world PV sales. Costs in the late-1990s were about 25 to 30¢/kWh.

In spite of high relative costs, the market for PV continues to grow. Tens of thousands of PV systems are already providing power for a variety of applications, including utility scale power plants, lighting, communications, water pumping, battery charging, vaccine refrigeration, and so on (Fig. 12.2). In many remote areas, stand-alone PV systems are the only viable sources of power. Solar cells have the advantage that no pollution (or very little) is associated with their use. Because they directly convert light into electricity, they are not constrained by the fundamental limitations of the second law of thermodynamics as are heat engines. They can be assembled very rapidly; construction time for a PV power plant is 1 to 2 years, compared to 5 to 8 years for a fossil-fuel plant. Their principal material is silicon, which is abundant on the earth, so no resource limitations appear likely.

The market for PV in remote applications, utility power, and consumer products (calculators and watches, for example) is growing at about 15% per year in the United States. From 1986 to 1998, PV shipments in the United States of solar cells and modules for all uses grew by 780%. In 1998, the United States produced PV cells and modules with a peak output of 51 MW, about one third of the world's total. In 1999, PV manufacturing was up again, by 52%. Since 1982,

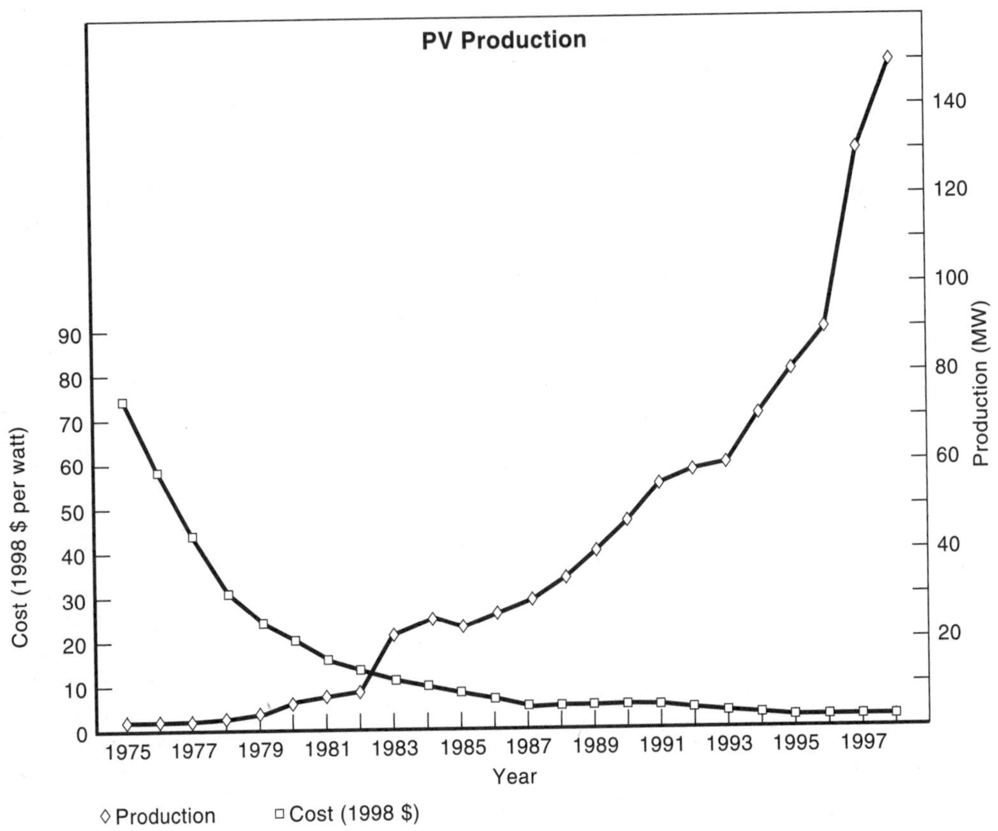

FIGURE 12.1
Cost of PV cells has decreased significantly over the years and world annual production has risen above 150 MW. (NATIONAL RENEWABLE ENERGY LAB)

about 140 peak MW of PV has been installed in the United States for all uses. As module costs decline from present prices of about $3.50/peak watt to $2.50/peak watt,* PV will become very competitive with diesel-powered generators (estimated to have a market of thousands of megawatts). When PV modules drop to about $1.50/peak watt, or a total system cost of $2.50 to $3.00/peak watt, PV electricity can be produced at a cost of $0.12/kWh, which will allow PV to make a more substantial entry into the U.S. utility market. Utility deregulation linked with consumer choice for "green power" will also be a factor in increasing PV contributions. The goal for the PV industry is to produce energy for $0.06 to $0.09/kWh, which is expected to occur early in the 21st century.

*PV arrays are rated in peak watts (W_p), which refers to its maximum power output when operating at 25°C under insolation of 1000 W/m^2.

FIGURE 12.2
Photovoltaic modules can power vaccine refrigeration units in remote locations.
(SIEMENS SOLAR)

B. Solar Cell Principles

The principle behind the direct use of the sun's energy for the production of electricity was discovered in 1887 by Heinrich Hertz and explained in 1905 by Albert Einstein. It was observed that when light strikes certain metals, electrons are emitted. This phenomenon, known as the **photoelectric effect,** can be studied with an apparatus such as the one shown in Figure 12.3. When light shines on the negative plate, electrons are emitted with an amount of kinetic energy inversely proportional to the wavelength of the incident light. Originally this effect was not considered surprising, and was thought to be consistent with a classical understanding of nature. However, for certain colors of light, no electrons were emitted. In classical physics, the only thing that determined whether electron emission would occur or not was the light intensity reaching the surface, not its color or frequency.

Einstein explained this effect by assuming that light behaves like a particle in this situation, rather than as a wave. The energy of each light particle, called a

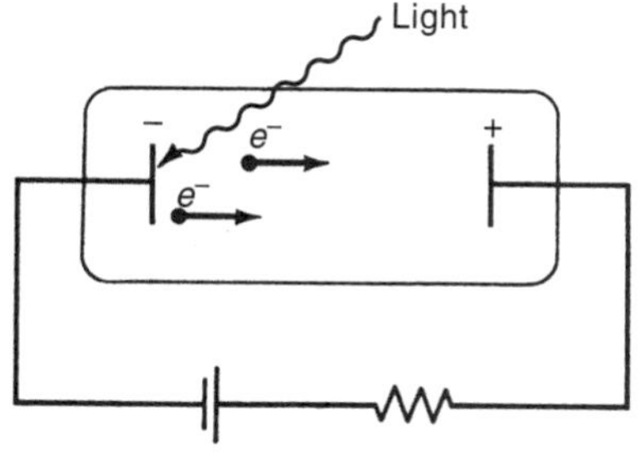

FIGURE 12.3
Apparatus for observation of the photoelectric effect. Light hits the metal plate (in the evacuated tube) and electrons are emitted.

photon, depends on its frequency only and is equal to $h \times f$, where h is a constant known as Planck's constant and f is the frequency of the light. An electron in a metal atom is able to "capture" a photon and obtain the energy necessary to escape *if* the energy of the photon exceeds the binding energy of the electron in the metal. This will happen if the frequency of the light is large enough or if the wavelength λ is short enough, since the wavelength is inversely proportional to the frequency. Recall that $\lambda \times f = c$, and the speed of light c equals 3×10^8 m/s.

Most solar cells are made by joining two very thin layers of crystalline silicon that have been treated in a special way. Normally there are no free electrons in silicon, and therefore silicon is a good insulator. By a process called "doping," impurities are added to the silicon to change its properties and make it a better conductor. If a small amount of phosphorus is added, there will be extra electrons in the crystal, producing an "n-type" (negative) semiconductor in which the current carriers are negative electrons. If boron is added, there are fewer electrons than in the silicon, so there are empty "holes" in the crystal—places in which electrons should be but are not—producing a "p-type" (positive) current-carrier semiconductor. These holes act just like positive charges. When these two types of semiconductors are joined together, they form a "p–n junction." The rearrangement of electrons and holes at this junction forms a barrier to the flow of electrical charge.

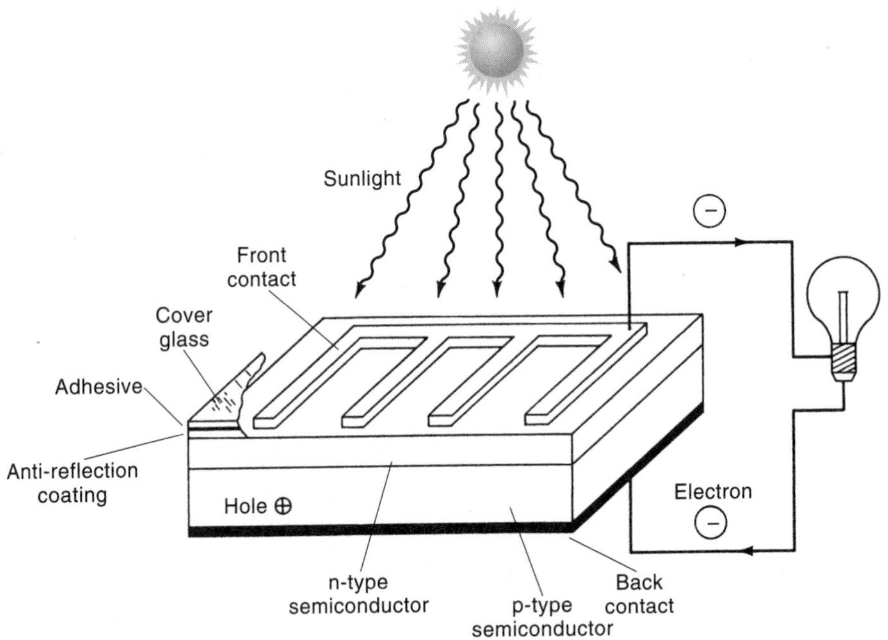

FIGURE 12.4
Solar cell construction. (SOLAR ENERGY RESEARCH INSTITUTE)

When light strikes a solar cell, electrons and holes are created by the photoelectric effect. These charges are separated by the potential barrier at the p–n junction. If the n-type and p-type sides of the solar cell are connected by an external circuit, the electrons will flow out of the electrode on the n-type side, through a load to do useful work, and into the p-type side, where they recombine with holes.

Figure 12.4 shows the construction of a solar cell. The thin top layer is usually made of n-type silicon about 1 μm (10^{-6} m) thick. On this layer a conducting grid has been attached, arranged as fingers to avoid blocking out too much of the light. The bottom p-type layer of silicon is about 400 μm thick; a metal electrode is attached on its back.

The potential barrier at the p–n junction produces a voltage of about ½ V in single-crystal silicon. Like a battery, the output is direct current. The output current from a solar cell is directly proportional to the amount of incident light and the cell area. Under bright sunlight of 1000 W/m^2, a current of about 100 milliamps per cm^2 of cell surface area is produced by typical single-crystal cells. A cell 10 cm in diameter will produce about 1 W under insolation of 1000 W/m^2. A cell 5 cm in diameter will produce about ¼ W under the same insolation.

Much of the solar energy incident on a cell is lost before it can be converted into electrical energy. While conversion efficiencies can range up to 30%, typical efficiencies are 10 to 15% (and even less with thin-film cells). The energy losses occur because some of the light is not energetic enough to separate electrons from the atomic bonds in the crystal; about 55% of the solar spectrum has light with wavelengths too long to excite electrons in silicon. Some light is too energetic, and the extra energy of the electron-hole pair becomes heat. Reflection from the cell's surface and electron-hole recombination also contribute to decreased efficiency. The use of several different thin films placed back to back (Fig. 12.5) allows the absorption of solar energy at

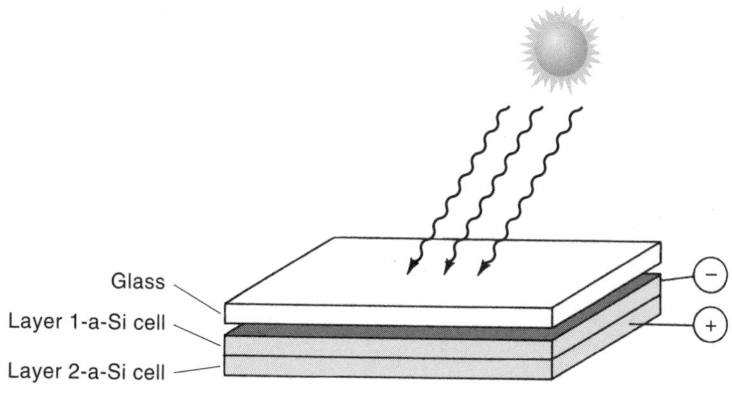

FIGURE 12.5

Multilayered solar cell. A second thin film is used so the stack can respond to a broader spectrum of light, thus increasing its efficiency.

Focus On 12.1

PV CAR: "THE SUNRAYCER"

In 1987, the General Motors "Sunraycer" entered and clearly won the World Solar Challenge Race across Australia—more than 3100 km. The aerodynamic (and expensive) car was covered with 8600 cells, with an area of about 8 m^2. The average electric output supplied to the motor during the race was about 1000 W. Silver-zinc rechargeable batteries were used early and late in the racing day, and for acceleration when needed. The cruising speed was 67 km/h, or 41.6 mph. It weighed 390 lb without driver. It is now on display at the Smithsonian Museum in Washington, D.C.

different wavelengths; this multilayered technique has produced demonstration cells with efficiencies as high as 28%. Some of these materials have efficiencies that do not decrease with increasing temperatures, such as silicon cells. Mirrors or lenses can then be used to concentrate the sun's rays on the cell and increase the power output.

C. Cell Manufacture

Most PV cells in use today are made from **single-crystal silicon,** although other manufacturing processes are fast becoming economically and technically competitive. To make a solar cell, silica (SiO_2) is first refined and purified. It is then melted and solidified in such a way that the silicon atoms are arranged into a perfect lattice. One way this is done is to introduce a seed of crystalline silicon into a molten mass of pure silicon and slowly draw it out (called the Czochralski process). The cylindrical ingot so formed is then sliced into wafers about 0.5-mm thick and doped with impurities of phosphorus (to create the n-type layer) and boron (for the p-type layer) to form the p–n junction.

More advanced crystal growing techniques can reduce the cost of cell manufacture. One method grows cells automatically in continuous ribbons. The thin ribbon is pulled from a furnace and cut to size. Two other types of PV cells that are cheaper to produce are polycrystalline and amorphous silicon. Polycrystalline cells are made of many grains of single-crystal silicon that are randomly packed. These cells are easier and cheaper to make and have been fabricated with efficiencies of more than 10%.

Amorphous silicon (Fig. 12.6) has quite different properties from crystalline silicon because its atomic structure is disordered. Amorphous cells are used in calculators, watches, and similar applications. Under fluorescent light they are more efficient than single-crystal cells. Their efficiency is currently about 5 to 10%. One of the major problems with amorphous silicon cells concerns "dangling" bonds, which can trap free electrons before they can get into

(a)

(b)

(c)

FIGURE 12.6

Amorphous silicon accounts for about 40% of worldwide PV sales, with products as diverse as calculators (*a*), go-carts (*b*), and PV cells (*c*). [(*A*) ENERGY SCIENCES, INC.; (*B*) SANYO CORPORATION]

an external circuit. The addition of hydrogen to the silicon removes some of the dangling bonds. This alloy can be doped with impurities to form a p–n junction (albeit different from crystalline junctions). Another problem is that the electrons do not move through amorphous silicon as rapidly as through crystals. What makes up for these limitations in amorphous silicon is that it absorbs light 40 times better than crystalline silicon. A very thin cell can be manufactured (about 1 μm thick) with good performance. Both material and money are saved. One problem with these cells is that their efficiency drops over time when exposed to light.

Materials other than silicon also are being used, such as gallium arsenide, cadmium telluride, cadmium sulfide, and (very recently) copper indium gallium diselenide. Most of these compounds are manufactured as thin films. They have promise for increasing cell efficiency because they provide energy gaps appropriate for solar energy conversion and they have high optical absorption coefficients.

D. Photovoltaic Systems and Economics

Individual solar cells are connected electrically in flat plate arrays to meet electrical energy needs. One such array, shown in Figure 12.7, provides 47 W at 12 V under full sunlight. These arrays are wired together to form a PV system. In the system shown in the illustration, the total output will be $6 \times 47 = 282$ W_p. These arrays cost about \$350 each, or about \$7.50/peak watt. Prices continue to decline in this market.

On a smaller scale, individual solar cells can be wired together in series or in parallel (Fig. 12.8). Each cell is capable of delivering only a specific

FIGURE 12.7
This 282-watt, 12-V DC PV unit replaced a noisy, high-maintenance diesel generator. It is used for communication in remote areas by the Wycliff Bible Translators. (SIEMENS SOLAR)

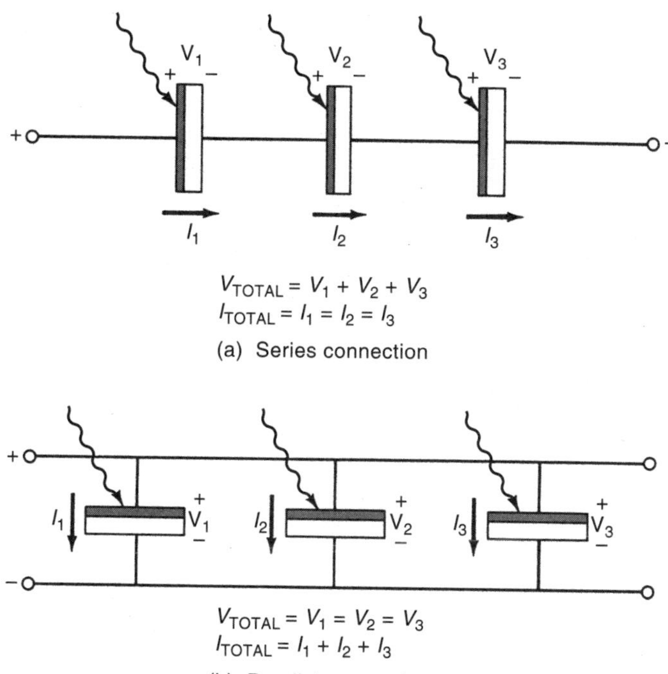

FIGURE 12.8
Solar cells can be wired in series (*a*) or parallel (*b*) to provide increased voltage or current, respectively.

amount of power. Additional cells must be added to increase the power output. When three silicon solar cells are wired together in series, the voltages add to yield an output of 3×0.5 V = 1.5 V. The current remains the same as from one cell. If the three cells are put together in parallel, the voltage of the combination remains at 0.5 V while the current is the sum of the currents from the three cells. An array usually consists of combinations of series and parallel groups of cells.

A typical arrangement of cells for a residential or commercial dwelling is shown in Figure 12.9. The output of the solar cells is direct current. There are many loads in the house that can use DC, such as incandescent lights. To store energy for backup, batteries can be used and a regulator is needed to prevent overcharging. For loads that require AC, such as TVs or motors, an inverter must be used to convert DC to AC. Any extra output can also then be passed to the outside electrical grid for sale to the utility (at a price determined by the utility).

Much larger PV systems are functioning throughout the world with power outputs on the order of megawatts. These either provide electricity directly for

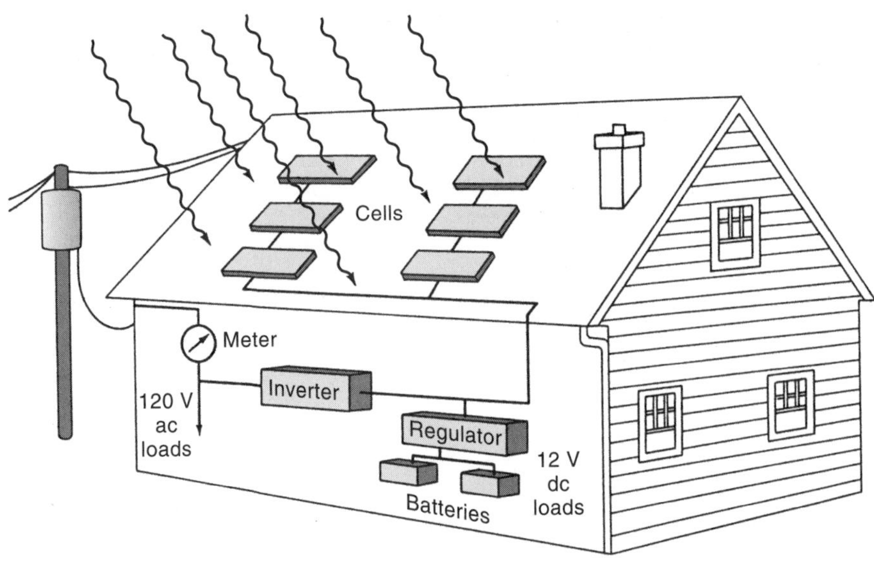

FIGURE 12.9
Photovoltaic system for a residential dwelling.

a commercial establishment or feed the output directly into the local electrical grid. Figure 12.10 shows what was the largest U.S. utility-scale PV power system, rated at 6.5 MW, located in Carrissa Plains, California; it was completed in 1985 and dismantled in the mid-1990s. Costs of large PV systems have been reduced by a factor of 40 over the past decade to about $0.30/kWh, making them comparable to peak load prices on a conventional system. (Operation and maintenance are usually very low, at about $0.005/kWh.)

FIGURE 12.10
The 6.5-MW Siemens Solar PV central station on the Carrissa Plains in California provided electricity to meet the annual needs of more than 2300 average homes. (It was dismantled and sold to private parties in the mid-1990s.) (SIEMENS SOLAR)

Photovoltaic-powered EV recharging station at the University of South Florida.
(UNITED STATES DEPARTMENT OF ENERGY)

Water pump powered by a PV unit.
(SIEMENS SOLAR)

PV is making important inroads into developing countries, with markets for PV growing fastest in the Third World. With two billion people without access to an electric utility grid, Shell Renewables, Inc. has begun to market solar home systems to meet electricity needs of rural homes in the developing world. Albeit on a small scale, many NGOs (nongovernmental organizations) are assisting in the installation of PV systems. For example, Sri Lanka, with 8000 solar home systems, is tackling first-cost financial disincentives (i.e., high initial costs), with help from NGOs. Communications is the largest and fastest growing area for remote applications. It is estimated that there are 250,000 homes worldwide that have rooftop PV systems, usually in remote areas. More than 10,000 systems are added annually worldwide. Systems range in size from kilowatts down to 160-W single modules for health care applications. In remote locations, PV at $0.25 to $0.35/kWh competes favorably with diesel generators at $0.30 to $0.50/kWh.

One of the important uses of PV is for powering water pumps. Because a majority of the world's rural populations live in the sunny tropics or subtropics, the use of the sun's energy is a good alternative to hand- or diesel-powered methods of pumping water. For example, in the 1990s Kenya installed more than 100,000 water-pumping stations powered by PV. Focus On Pumping Water explores the sizing of solar arrays that could be used for water pumping.

Besides remote locations, another use of PV is for buildings in which the solar cells are integrated into the structure, as you might have with a roof module. Several countries are investigating the integration of PV into roof tiles to replace traditional roofing materials. Japan provides subsidies to 10,000 homeowners per year to install rooftop grid-connected solar electrical systems. In Europe, several countries are developing building-integrated PV systems for roofing and facades. Utilities are assessing the use of PV in several areas of grid support. One application is to support the utility's transmission and generation systems by improving the quality of service and reducing peak loads enough to put off upgrading its transformers for several years. PV also can be used for battery-charging stations for electric vehicles.

Focus On 12.2

PUMPING WATER

Suppose we wish to lift 60 m^3 (16,000 gal) of water through a height (H) of 5 m in a period of 8 hours. What are the electrical energy requirements of the pump?

The potential energy we must give to a mass m of water is given by (work) $W = PE = mgH$. The mass of water is equal to its density times its volume, or $m = pV$, where p is the density of water, 1000 kg/m^3. We are interested in a pumping rate, which is the work done divided by the time taken. This is the power,

$$P = \frac{W}{t} = \frac{pVgH}{t} = \frac{(1000\ \text{kg/m}^3)(60\ \text{m}^3)(9.8\ \text{m/s}^2)(5\ \text{m})}{8\ \text{h} \times 3600\ \text{s/h}} = 102\ \text{W}$$

If the pump is 60% efficient, the required power will be 102/0.6 = 170 W, which could be met by about four 40-W arrays.

The worldwide market penetration of PV will be determined primarily by economic and policy decisions. Significant markets (remote as well as power generation) have developed, and no technical obstacles to widespread utilization remain. Global warming concerns may also accelerate the rate of PV used for meeting both peak- and base-load utility requirements.

While the price of PV has been declining steadily over the past decades, government–industry partnerships are at work to identify and solve manufacturing problems that affect module cost and production capacity. PV will be economic for conventional systems when system prices fall to \$2 to \$3/peak watt (they are presently \$5 to \$7). This will result in electricity costs of \$0.10 to \$0.15/kWh. Recent partnerships between the government and the PV industry have seen direct manufacturing costs for modules reduced from an average of \$4.50/W to about \$2.70/W. Increased competition and reforms taking place in the electric industry (see Chapter 10) are causing utilities to consider a broader array of generation options, such as PV and wind.

E. Wind Energy

The extraction of energy from the wind, especially in the form of electricity, has enjoyed renewed interest among both utilities and governments. Wind energy is the fastest growing form of energy today, up 75% in installed capacity in the

United States since 1990. Today, there are more than 30,000 wind turbines worldwide, with a capacity of 13,000 MW. It has been estimated that wind energy could supply anywhere from 5 to 15% of the total electrical needs of the United States by the year 2020. Currently this number is about 0.1%. Wind power's environmental impact is almost insignificant, its main problem being visual pollution, although concerns about noise, television interference, and raptors have been expressed. Other positive features of wind turbines are that they have short construction times, they are small units relative to other types of electrical generators and therefore provide greater adaptability in responding to electrical demand, and they can be tailored to specific uses and locations. Another advantage of wind power, especially for residential use, is that it is a fine complement to radiant solar: Days with little sun are usually those days of better-than-average winds.

Interest in harnessing the winds is certainly not new; it was one of the first natural energy sources to be used. There are indications that windmills were used in Babylon and China around 2000 to 1700 BC to pump water and grind grain. Windmills were introduced into Europe about the 12th century, and by 1750 Holland had 8000 windmills and England had 10,000. Their use declined after the introduction of Watt's steam engine in the late 18th century and further declined in the early 20th century as a result of the availability of cheap, reliable fossil fuels and hydropower. The windmill was (and is still) very important in the economic development of the United States since it offered a means to pump and supply water on remote farms for beef and agricultural production. Today there are about 150,000 windmills in operation in the United States, the large majority of which are used for pumping water. Thousands of 2- to 3-kW units were installed in the 1930s and 1940s to generate electricity in rural areas, but the Rural Electrification Administration and the Tennessee Valley Authority phased out such uses when they encouraged electrification through loans and construction.

The development of the current wind energy industry in the United States began after the energy crisis of 1973 with the construction of multi-kilowatt demonstration machines by NASA and the Department of Energy. One of the first large-scale demonstration machines is shown in Figure 12.11—the 100-kW NASA wind turbine near Sandusky, Ohio. This horizontal-axis machine had two blades with a diameter of 125 ft and was designed for a rated wind speed of 18 mph. The next demonstration-sized machines were 200-kW ones, with similar designs, built in Puerto Rico, New Mexico, and Rhode Island. Several megawatt-sized machines were built in Boone, North Carolina (200-ft diameter blades) and Washington State (2.5 MW, 300-ft blade diameter). These machines proved to be uneconomical and displayed component failures caused by metal fatigue. The largest windmill ever run prior to 1980 was located on Grandpa's Knob in Vermont in the 1940s (Fig. 12.12). This Putman-Smith windmill had a capacity of 1.2 MW, and a cost of $1000/kW. Large-scale production might have lowered this cost. The two steel blades had a diameter of 175 ft and weighed 8 tons each. In 1945, one

FIGURE 12.11
NASA experimental wind-turbine generator in Ohio. This first large-scale unit had an output of 100 kW in 18-mph winds. It was put into operation in 1975. Blades were 125 ft in diameter. (UNITED STATES DEPARTMENT OF ENERGY)

FIGURE 12.12
Grandpa's Knob wind generating station. This unit operated near Bennington, Vermont, in the 1930s and 1940s, producing up to 1.2 MW. (UNITED STATES DEPARTMENT OF ENERGY)

blade broke off (flying several hundred meters), shutting the windmill down; it was never replaced. Today's blades are lighter and more durable, with carbon fiber and fiberglass replacing the metal.

Commercial efforts in the area of wind energy increased quite dramatically in the 1980s, brought about largely by energy tax credits, new technologies in wind energy systems, and the passage of the Public Utilities Regulatory Policies Act (PURPA), in which utilities were required to purchase energy from qualified sources at the utilities' avoided cost (Chapter 10). Wind development in the United States fell off in the late 1980s and early 1990s as oil prices dropped (precipitously in 1986), natural gas-fired generating plants increased, and tax credits expired. Presently, the United States is seeing a significant growth in wind energy development—a 15% growth in 1998 and 38% in 1999. Federal Production Tax credits, paying 1.5¢/kWh for the electricity generated, have been extended to 2002. Other countries are also seeing such rapid growth. From 1995 to 1999, worldwide wind capacity had an average growth rate of 29% per year—a sevenfold increase in the 1990s! While the United States possessed 95% of the world's installed wind capacity in the early 1980s, that share has dropped to about 20%.

Table 12.2 lists installed wind energy capacity for the top worldwide wind energy markets at the end of 1999. Germany overtook the United States in 1997 as the country with the largest installed wind capacity. Denmark supplies 10%

Table 12.2 TOP WIND ENERGY MARKETS

	Installed Capacity (MW)	
Country	**1998**	**1999**
Germany	2872	4072
United States	1770	2502
Denmark	1433	1733
Spain	822	1722
India	1015	1077
United Kingdom	334	534
Netherlands	375	428
China	224	300
Italy	199	249
Sweden	176	216

(U.S. Energy Information Administration)

of its electricity with the wind. Spain almost doubled its wind energy capacity in both 1998 and 1999. High energy prices, fuel security issues, renewable energy subsidies, and concern for the environment has fueled this growth in many European countries. Growth in developing countries has slowed from the mid-1990s due to underperformance of wind units, economic instability, and changes in tax credits.

Wind turbines increased in size in the 1990s from 100 to 1000 kW units. Increasing turbine size has also dramatically lowered cost. A 1981, 25-kW model cost $2600/kW, while today's 1-MW model costs about $800/kW. Today there are more than 17,000 intermediate-sized wind turbines in the United States, averaging about 100 to 200 kW. Ninety percent of these are on "wind farms," mainly in California, with an installed capacity of more than 1600 MW (Fig. 12.13). California still has about 73% of the U.S. installed wind capacity. Here, thousands of wind turbines (of many different types, both horizontal and vertical axes) are running, feeding their output directly into the power lines of the state's electrical grid. Currently, 5% of the electrical energy of California's largest utility, PG&E, is supplied by wind energy.

The basic parts of a wind energy system are shown in Figure 12.14. Wind pressure turns a rotor made of blades or vanes. This rotor is attached to a shaft, which is connected through various gears to an electrical generator. For smaller, residential systems, the DC output of the generator can be stored in batteries or used to run devices that use resistive heating (e.g., lights, toasters, and heaters). For large generators, such as those found on wind farms, a system that has revolutionized the wind energy industry is the "synchronous inverter." The synchronous inverter converts DC power from the wind generator into AC and feeds it into the utility grid or to a local load at the correct frequency (60 Hz in the United States). The electricity is sold to the utility at a rate determined by each state or by the market.

FIGURE 12.13 Horizontal-axis wind machines (100–200 kW) generating power in the Altamont Pass area, northern California. (U.S. WINDPOWER; PHOTOGRAPHER ED LINTON)

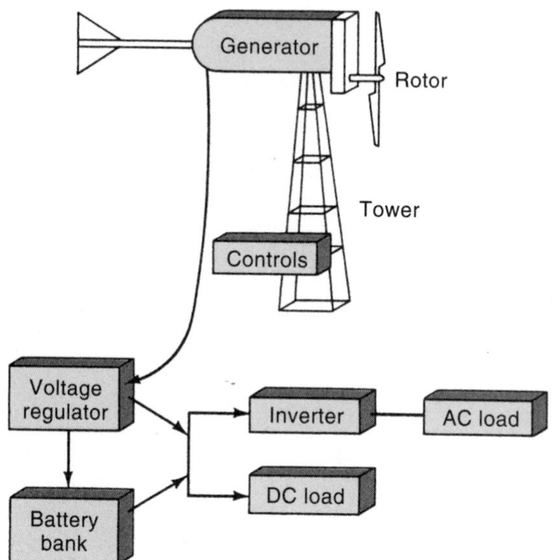

FIGURE 12.14
Residential wind energy system.

The power that can be extracted from the wind is proportional to the *cube* of the wind's velocity v and the area swept out by the blades. To understand the cubed relationship, remember that kinetic energy is expressed as $\frac{1}{2}mv^2$. The mass m in this expression is the mass of air that hits the wind-generator blades in a unit of time, and so depends on the wind velocity v. The greater the velocity, the more wind will impact on the rotor blades every second. Therefore, the wind energy transferred in a certain time (the power output) is proportional to $v \times v^2 = v^3$. The power output is also related to the area swept out by the blades; the area is proportional to the blade diameter D squared. Since the air has a small mass-to-volume ratio (i.e., density), the turbine blades must sweep out a large area to produce significant output. Putting all these factors together, with the use of appropriate conversion factors, and a factor of 0.59 for the maximum energy conversion efficiency achievable by wind turbines, the maximum power output can be written as

$$P = 2.83 \times 10^{-4}\, D^2 v^3 \text{ kW}$$

with D in meters and v in meters per second, or

$$P = 2.36 \times 10^{-6}\, D^2 v^3 \text{ kW}$$

with D in feet and v in miles per hour.

EXAMPLE

What is the maximum achievable power output of a wind turbine of blade radius 2 m in a 25-mph wind?

Solution

We will use the second of the power equations given and convert the diameter to feet. Also note that a *radius* of 2 m means a *diameter* of 4 m.

$$D = 4\text{ m} \times \frac{3.28\text{ ft}}{1\text{m}} = 13.08\text{ ft}$$

$$P = 2.36 \times 10^{-6}\,(13.08)^2\,(25)^3 = 6.3\text{ kW}$$

Using these equations, a tabulation of power output for several wind velocities and blade diameters is given in Table 12.3. An increase in wind velocity from 5 mph to 15 mph will yield $(3)^3$ or 27 times more power from the wind turbine. Consequently, wind turbine siting is very important. While some locations are clearly better than others for wind availability, there is a considerable variation in wind velocity from day to day and from year to year. Local topographical features influence the winds dramatically, sometimes with large variations in wind velocities over a small area. (For example, 3 hours of 20-mph winds and 3 hours of 10-mph winds in one location will generate twice as much electricity as a site at which the wind velocity is 15 mph for the entire 6 hours.) The support tower for the turbine must be as high as is possible, because the wind velocity in-

Table 12.3 WINDMILL OUTPUT AS FUNCTION OF WIND SPEED AND BLADE DIAMETER

	Power Extracted (kW)*			
Wind Speed (mph)	*D* = 12.5 ft	*D* = 25 ft	*D* = 50 ft	*D* = 100 ft
10	0.37	1.48	5.9	23.7
20	2.95	11.8	47	189
30	9.96	39.8	159	637
40	23.6	94.4	378	1510
50	46.1	184	738	2950

*Maximum theoretical output, assuming the windmill converts 59% of the wind's energy into usable output. Because of aerodynamic imperfections and mechanical and electrical losses, these numbers would have to be multiplied by about 0.5 to 0.7.

creases with distance from the ground. Good wind turbines are those that can use high-speed winds efficiently because the power output is related to the velocity cubed. Generators with fewer blades (two or three) are much more efficient than multivane rotors, which are good at low wind speeds. In strong winds, the rotor must spill or dump the excess power the generator cannot handle without damage. This is usually done by "feathering" or tilting the blades so less of their area faces the wind and therefore less power is extracted.

Wind turbines are classified by the orientation of the rotor shaft. There are *horizontal-axis* wind turbines and *vertical-axis* machines. The most common types are those with horizontal axes and vertical blades. Figure 12.15 shows three types of horizontal-axis windmills. The Dutch four-arm rotor was used for pumping water

FIGURE 12.15

Three types of windmills: (*a*) "Dutch" type windmill. Thousands were used for centuries in Holland, but few are in use today. They had small efficiencies (7%) and output (10 hp). (*b*) "American multivane" windmill. Dependable and able to operate in winds with small velocities. Extremely important during the past century for lifting water. (*c*) Two-bladed wind turbine: prototype of many in use today.

and grinding flour, but has a small (7%) efficiency for electrical energy conversion. The American multivane type is still in use for pumping water, but will generate only about 4 hp (3 kW) in a 15-mph wind. The two- (or three-) bladed propeller is the most efficient of these types for generating electricity; it is the least massive and most efficient for its size. Even this model will not extract all the energy of the wind. This is because if the rotor *did* extract all the wind's energy, the wind velocity would be zero after hitting the blades and thus would be stationary behind the blades; then air would pile up! It can be shown that the maximum efficiency of wind-to-electrical energy conversion for an ideal rotor is about 59%.

An example of a vertical-axis machine is the eggbeater-shaped Darrieus rotor (Fig. 12.16). Vertical-axis rotors have the advantage that they don't have to shift with changes in wind direction. They also can have the gearbox and generator mounted on the ground and not at the top of the tower, which reduces tower costs. However, they are difficult to put high up on a tower to take advantage of the higher wind speeds, so their popularity is small. Figure 12.17 compares horizontal- and vertical-axis wind turbine configurations.

A wind-driven generator has a rated output of so many watts, which occurs for a given *rated wind speed*. If the rated power output for a particular system is achieved with a 20-mph wind, then for higher wind velocities the pitch of the blades has to be changed, or feathered, to prevent the wind turbine from delivering more mechanical energy than the generator can handle. If the rated out-

FIGURE 12.16

Darrieus rotor (250 kW) in Altamont Pass, California, wind farm. (FLO-WIND CORP)

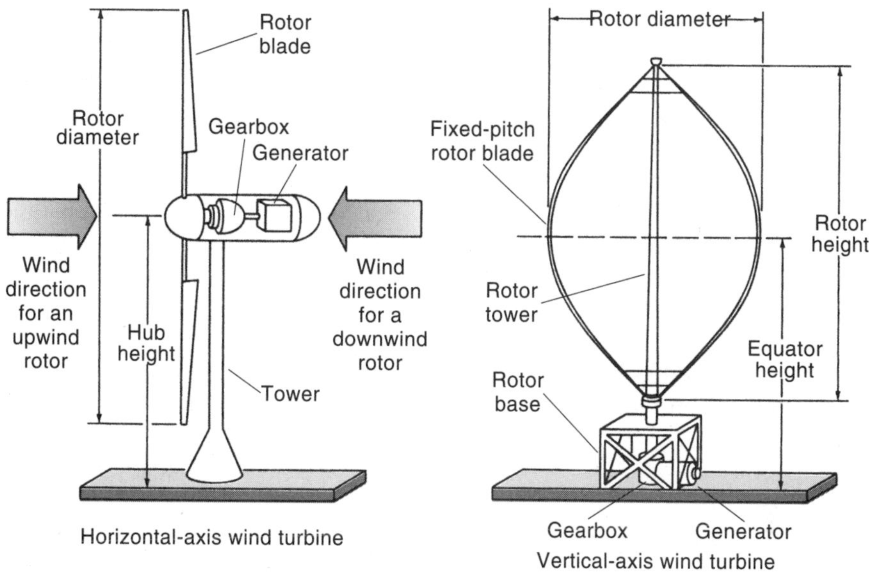

FIGURE 12.17
Horizontal- and vertical-axis wind turbine configurations.

put, say 10 kW, is for 20-mph winds, then a good part of the time a smaller output will be delivered. If the wind velocity drops to 15 mph, then the power output will be reduced to $(15/20)^3 \times 10$ kW = 4.2 kW.

To establish the *size* of wind generator needed for a single residence, you must determine your electrical energy needs. This is about 600 kWh per month or 20 kWh per day for a typical household. The electrical energy generated in a month with a wind turbine depends on its rated wind speed (i.e., the wind speed at which the generator achieves its rated output) and the wind velocity profile (i.e., the number of hours of wind at a particular velocity). Because of the velocity-cubed dependence of the power output, we cannot use the average wind velocity. To calculate the monthly energy output, we could calculate the power at each wind speed, multiply by the number of hours at that speed per month, and sum to get the number of kilowatt-hours per month. This would be very tedious, so calculations for localities with better-than-average winds use an empirical relationship that one can expect about 70 kWh per month per rated kilowatt for a unit with a rated wind speed of 25 mph. Thus a 6-kW wind generator will provide $6 \times 70 = 420$ kWh per month, adequate for a residence. However, the cost might be prohibitive. Present prices for a complete wind-generator system are about $3000 for each 100 kWh of monthly output. Unless power lines have to be brought a long distance from the road to your house at your expense, it is now less expensive to buy your electricity from the local utility. Consequently, most construction today is for large wind-farm units where economy of scale reduces the overall cost. These units feed the output directly to the grid.

Since winds are intermittent and cannot be held back, *storage* of energy for future use is very important, if one cannot tie into the existing electrical grid and sell the excess generated electricity to the local utility. For residences, electrical energy is usually stored in 12-V, lead-acid batteries. Ten of these can be put together in series to provide 120 V DC output. The storage capacity of a battery is measured in amp-hours (see Chapter 10). Since energy can be expressed in terms of watt-hours, and watts = volts × amps, the energy stored can be written as the product of the storage capacity of the battery and the rated voltage. For example, a set of ten 200-amp-hour 12-V batteries has a storage capacity of 200 amp-hours × 120 V = 24,000 watt-hours = 24 kWh, about one to two days' supply for an average home.

Installed costs of \$1000/kW are now obtainable, competitive with large-scale steam power plants. Total costs of \$0.05 to \$0.06/kWh make this one of

Table 12.4 U.S. WIND ENERGY POTENTIAL AND INSTALLED CAPACITY

State	1999 Installed MW	Rank	Potential MW	Rank
North Dakota	0.4		138,000	1
Texas	188	4	136,000	2
Kansas	0		122,000	3
South Dakota	0		117,000	4
Montana	0.1		116,000	5
Nebraska	2.8	10	99,000	6
Wyoming	73	5	85,000	7
Minnesota	272	2	75,000	8
Iowa	242	3	63,000	9
Colorado	21	8	55,000	10
New Mexico	0.7		50,000	11
Michigan	0.6		7500	12
New York	0		7100	13
Illinois	0		7000	14
California	1840	1	6800	15
Wisconsin	23	7	6400	16

(American Wind Energy Association, 2000)

the lowest cost options for new generating capacity. Although the use of wind energy was confined almost entirely to California initially, wind energy development has emerged in the Pacific Northwest, the Midwest, and the South. Table 12.4 lists installed and projected wind energy projects for many states, ranked according to their wind energy potential. Note that California is number 15 on this list for potential. Presently under construction in Texas is a 208-MW wind facility that is designed to be the largest single wind-power installation in the world. There will be 160 wind turbines, each with a capacity of 1.3 MW. What has been helpful in such growth is that some states require their utilities to purchase so many MW of renewable energy per year. Restructuring of the electric utility industry has also given a boost to wind energy. More than 20 electric utilities in the United States are offering wind energy as the primary or only resource in "green power" programs. Customers can choose such electricity over electricity from conventional sources. Continued strong growth is expected in this century as improved wind turbine designs, higher fuel prices, increased capacity needs, and environmental considerations make wind energy very attractive.

F. Hydropower

Historically, hydropower has been harnessed to control water to do useful work—to grind grain, saw lumber, and provide power to do other tasks. Power was transferred to a variety of rotary motion machines via shafts, pulleys, cables, and gears. The Greeks used vertical-axis waterwheels as early as 85 BC, and horizontal-axis wheels from about 15 BC. Water power was the only source of mechanical energy (other than wind) until the development of the steam engine in the 19th century.

Another invention of the 19th century was the development of electric generators. Hydropower was a natural source of power for such generators. Hydropower converts potential energy into kinetic energy by virtue of elevation changes. While rivers with widely varying rates of flow were found to be unsuitable for installation of generators, dams in rivers provided an easy means of adjusting the flow of water to meet varying demands for electricity. The first hydroelectric power facility to be built in the United States was in Appleton, Wisconsin, in 1882. The generator produced direct current primarily for local industries.

Figure 12.18 gives a very simple model of a hydroelectric plant. Water flow to the power plant from the dam is through a large pipe called the penstock, where it drives a reaction or impulse turbine. The output is a function of both the "head" and the rate of water flow. The head is the distance from the highest level of dammed water to the power-producing turbine. In low-head dams this distance is smaller than 30 m (100 ft), while in high-head dams this distance can be 300 m (1000 ft) or more. A large output can be achieved using either a high

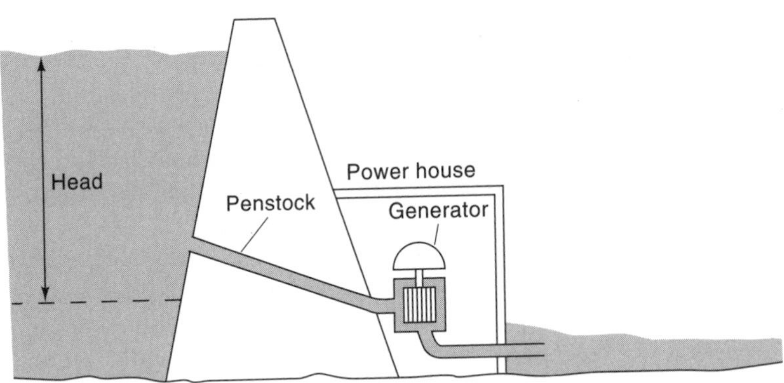

FIGURE 12.18
Model of a medium- to high-head hydroelectric power plant.

head or a low head with a large volume of water flow. Hoover Dam on the Colorado River is the highest dam in the United States at 221 m (725 ft) with an output of 1300 MW. The Robert Moses/Robert Saunders power dam, an international facility that stretches across the St. Lawrence River, has a generating capacity of 1800 MW but a head of only 9 m (30 ft).

Turbines have been developed for different rates of flow and pressures created by the head. A basic type of waterwheel is the undershot wheel (Fig. 2.2); this wheel can have a diameter of 5 m or more with flat blades around the perimeter of the wheel. It operates at a low head with an efficiency of 20 to 40%. The breast wheel (Fig. 12.19a) can be used with a higher water level on the wheel and uses both the flow of the water and the weight of the water to increase the efficiency to about 65%. The overshot wheel (Fig. 12.19b) uses a flume to create a force to the top of the wheel as well as using the weight of the water, yielding efficiencies up to 85%.

Modern turbines are of two major types—the impulse and the reaction. The Pelton turbine (see figure in Focus On 12.3) uses nozzles aimed at cupped blades to develop rotational speeds up to 1300 rpm. This is like a garden hose hitting your hand—the impulse pushes your hand back. The Pelton is similar to the old fashioned waterwheel on a grist mill. The reaction turbine is used on many of the larger hydro facilities today. One type is the Francis, which is used in large facilities with high heads (Fig. 12.19c). Like a hose on the rim of a wheel or a rotating lawn sprinkler, the wheel turns due to the third law. A more modern reaction turbine is the Kaplan (Fig. 12.19d), which uses a variable-pitch system and is like a ship's propeller.

The United States produces about 9% of its electricity today from hydropower. While the total amount of electricity produced from hydropower has increased over the years, the percentage of electricity has decreased from

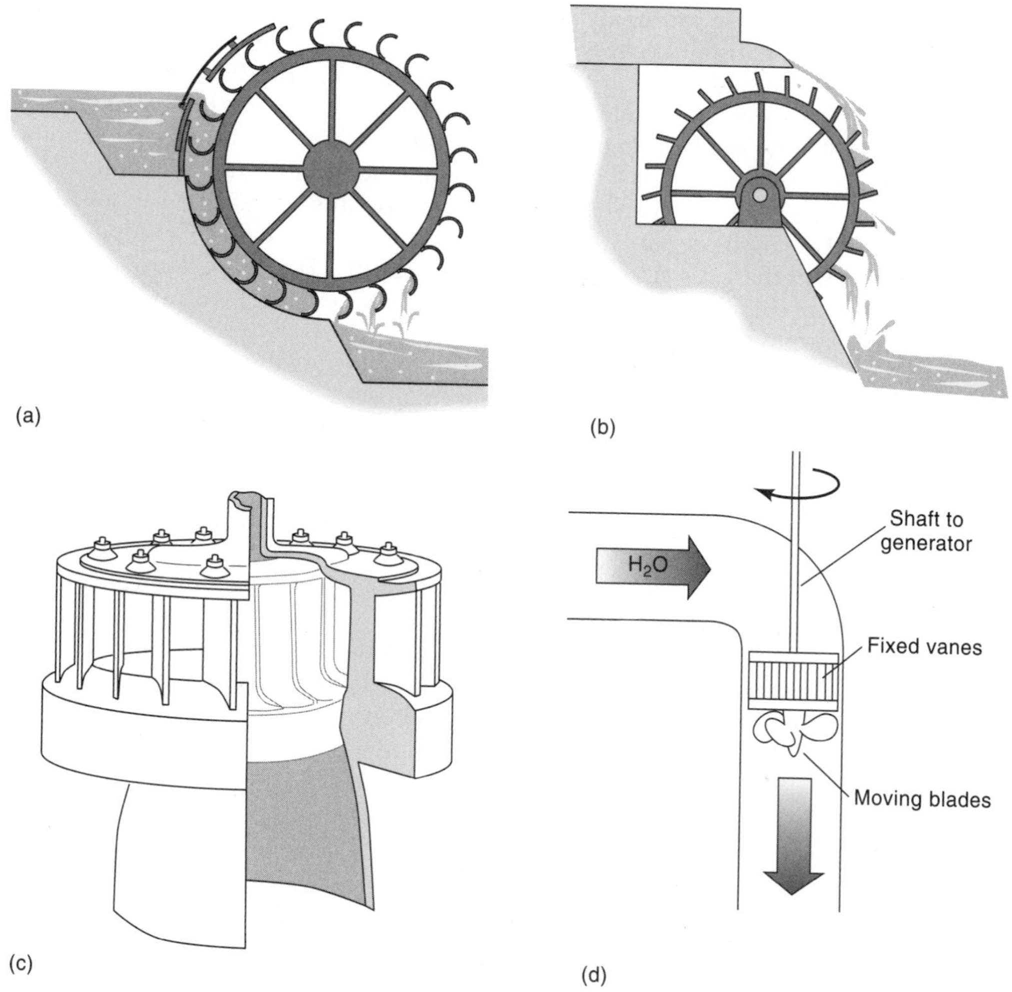

FIGURE 12.19

Models of waterwheels or turbines: (*a*) Breast wheel, (*b*) Overshot wheel, (*c*) Francis turbine, (*d*) Kaplan or Propeller turbine. (USED WITH PERMISSION FROM MICRO-HYDRO POWER: *REVIEWING AN OLD CONCEPT.* ©1979. BUTTE, MT, NATIONAL CENTER FOR APPROPRIATE TECHNOLOGY)

about 35% to 9% over the last 50 years. About 45% of U.S. conventional hydroelectric sites have been developed. The largest facility in the United States is Grand Coulee Dam, with a capacity of 6495 MW; plans are underway to increase its output to 10,800 MW.

About 19% of the world's electricity is produced by hydropower. This percent varies from 75% in Latin American (in 1998) to 16% in Africa, where there is immense potential. Table 12.5 gives both the electrical energy generated and the installed capacity for the top countries for 1998. In the last ten years, world hydroelectric output has increased by only 15%. In recent years, there has been a substantial increase in the amount of electricity that can be generated by small-scale or low-head units. This is sometimes referred to as **micro-hydro**. If there is a sufficient volume of falling water available, heads of as little as 2 to 3 m can be used. Some micro units are as small as 200 to 500 W, and are used primarily for battery charging. These units are cheaper than a PV unit of the same size, with greater availability. They are finding use in developing countries to provide needed electricity for medical clinics and telecommunications. The power output is a product of the head and the flow. A smaller flow will require a larger head to generate the same power. An example of the power obtainable from a small-scale hydroelectric unit is given in Focus On 12.3.

The largest hydroelectric plant in the world currently operating is in Venezuela, with a capacity of 10,000 MW. Russia has a facility planned for 20,000 MW, and the Three Gorges Dam in China, on the Yangtze River, will have a capacity of 18,600 MW when completed in 2009 (Fig. 12.20). This dam will be 2.3 km (1.4 miles) wide and 185 m (607 ft) high and will create a reservoir 625 km (375 miles) long.

While hydroelectric plants do not pollute, they do affect the environment. A dam will result in the flooding of large areas of land. The Three Gorges project will displace about 1.2 million people, and inundate national

Table 12.5 HYDROPOWER OUTPUT (1998)

	Electricity Generated (Billion kWh)	Installed Capacity (Thousand MW)
United States	350	99
Canada	330	67
Brazil	289	54
China	203	60
Russia	150	44
Norway	115	27
Japan	90	21
India	76	22
Sweden	73	16

(U.S. Energy Information Administration)

Focus On 12.3

SMALL-SCALE HYDROELECTRIC SYSTEMS

Small-scale hydroelectric installations, sometimes called micro-hydro, have capacities on the order of 1 to 100 kW and can provide enough electrical energy to meet the needs of 1 to 100 homes, excluding electric space heating and electric water heating. In such installations, some water is diverted from a source upstream and transported through a conduit or pipe to a turbine, which turns a generator to produce electricity. This electricity might be stored in batteries, especially for smaller units.

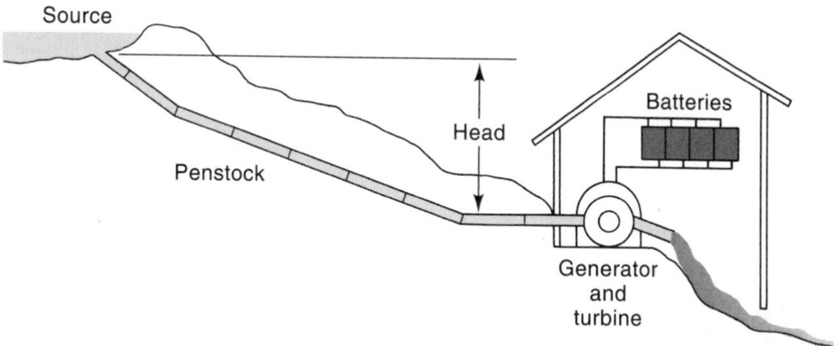

Low-head hydroelectric installation.

The power produced from this water is equal to the rate of loss of gravitational potential energy as the water drops from the source. The change in gravitational potential energy is equal to the weight of water × the vertical height through which the water drops (called the head).

$$\text{Power} = \frac{\Delta(\text{PE})}{\text{time}} = \text{weight} \times \frac{\text{vertical drop}}{\text{time}}$$

If 20 kg of water per second (32 gal/min) undergoes a vertical drop of 5 m, then the energy change each second is 20 kg $\times$ 9.8 m/s^2 $\times$ 5 m = 980 J. The electrical power available is

$$\text{Power} = 980\ \text{J} \times \frac{\text{eff.}}{1\ \text{s}} = 980\ \text{W} \times 0.8 = 780\ \text{W},$$

(continues on p. 414.)

where eff. is the conversion efficiency of the turbine (0.8 here). If this power can be obtained for 24 hours per day and 30 days per month, the electrical energy obtainable for a month is

$$\text{Energy} = \text{power} \times \text{time} = 780\text{ W} \times 24\text{ h/d} \times 30\text{ d/month} = 560\text{ kWh/month.}$$

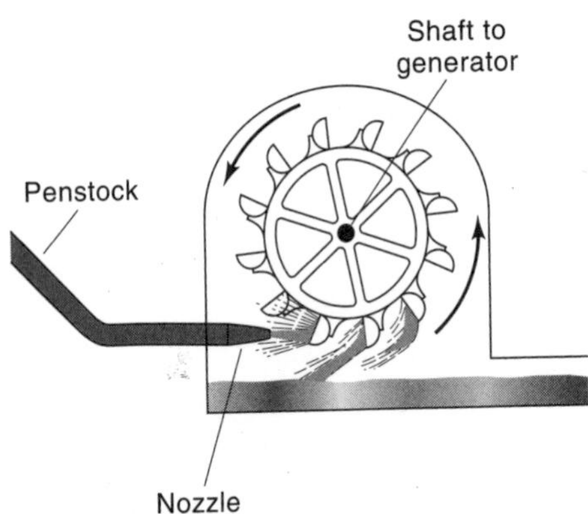

Pelton turbine.

One common type of turbine is the Pelton turbine, designed for high-head use (vertical drop greater than 10 m). Low-head systems (2 to 6 m deep) use propeller-type turbines, which are similar to boat propellers in a tube (see Fig. 12.19d).

treasures that are centuries old. While the water behind a dam will provide a large recreational area, it eliminates the habitat for some endangered plants and animals. Another concern is the potential for health problems arising from the trapping of raw pollutants produced by the large cities upstream. There also will be a reduction in the sediment and nutrients that usually flow downstream. After the construction of the Aswan Dam in Egypt in 1964, fishing in the eastern Mediterranean was harmed for this reason. Stagnant or slow-moving warm waters can also lead to other health problems. In Egypt, snails that breed in the mud of the Aswan Dam and carry a parasite have infected millions of people with schistosomiasis, a very debilitating disease.

FIGURE 12.20
Site of Three Gorges Dam on the Yangtze River. This is China's largest construction project since the 19th century. When completed in 2009, it will provide 10% of China's electricity. (ADRIAN BRADSHAW/ LIASON AGENCY)

G. Solar Thermal Electric Facilities

Solar thermal systems use concentrating collectors to focus direct sunlight for the production of high-temperature fluids. Solar thermal technology has a broad range of applications that includes electrical power generation, industrial process heat, and metallurgical and chemical production. There are three types of concentrating collector systems (Fig. 12.21): parabolic troughs, which are used for mid-temperature applications, parabolic dishes, and central receivers, which are capable of achieving high temperatures.

Trough systems work by using parabolic reflectors in a trough configuration to concentrate sunlight up to 100 times onto a fluid-filled tube positioned along the line of focus. The main benefit of concentration is that one can achieve high fluid temperatures, up to about 400°C. Through a heat exchanger, the fluid produces steam, which is then used to drive a turbine to generate electrical power. Figure 12.22 shows the Solar Electric Generating System (SEGS) located in southern California. Built in 1984, the initial output was about 13 MW of electrical power. SEGS now supplies 350 MW and generates about one half of the world's electricity that is generated directly from the sun. The overall efficiency of converting sunlight to electricity with this technology is about 25%. The cost for the first unit was about $4000/kW. Today, prices for electricity

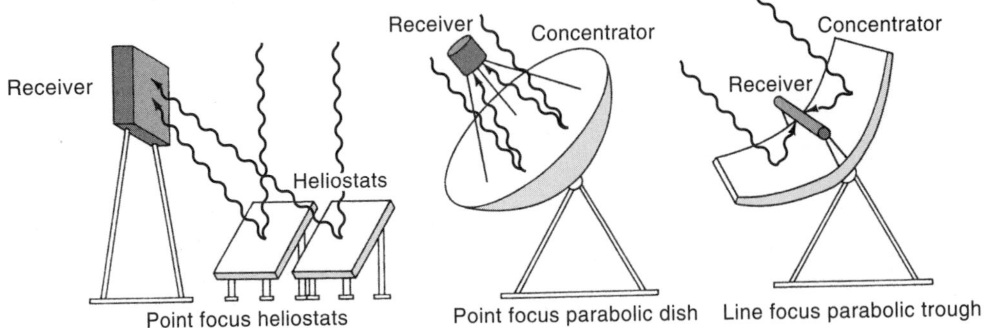

FIGURE 12.21
Three types of concentrating collector systems.

run between \$0.08 and \$0.12/kWh from this facility. Today there are more than 15 million ft^2 of concentrating collectors, mainly in California, representing 1200 MW of peak-power potential.

Central receivers (sometimes called power towers) use heliostats (highly reflective mirrors) which track the sun and reflect its energy to a central receiver atop a (100 m high) tower. The sun heats a fluid in the receiver to temperatures up to 650°C (1200°F). Figure 12.23 shows the demonstration 10-MW Solar One system near Barstow, California. The cost of this system was about \$12,000/kW. It was modified (called Solar Two) to heat molten salt rather than direct steam,

FIGURE 12.22
Solar Electric Generating System, Kramer Junction, California, provides 350 MW from concentrating collectors. (LUZ INTERNATIONAL)

FIGURE 12.23
The Solar One (and later Solar Two) 10-MW plant in Barstow, California, served as a test facility until 1999. Its 1926 moveable mirrors were used to heat a fluid at the top of the tower, which was used to generate electrical power. (SOUTHERN CALIFORNIA EDISON)

and was shut down in 1999. Estimated costs of $0.08/kWh now appear possible, especially with facilities producing 100 to 200 MW. Central receivers seem suitable for electricity peaking and industrial process heat applications, although low natural gas prices can be a critical factor.

H. Summary

There are a number of popular technologies for the direct production of electricity from solar energy. Photovoltaics uses solar cells that are usually made of pure silicon. Using the photoelectric effect, incident light frees electrons in the crystal, which are forced through a load before recombining with positive charges in the crystal. A single solar cell acts like a battery with a potential difference across it of about 0.5 V. The amount of current produced depends on the insolation as well as the area of the cell. Efficiencies of 10 to 28% are obtainable.

Wind turbines used for the generation of electricity have found large-scale use in California. Worldwide, wind is the fastest growing energy technology. Their power output increases as the cube of the wind velocity and the square of the diameter of the blades.

Hydropower provides about half of the electricity generated through renewable resources. The other half is provided by biomass, which will be discussed in Chapter 17. While development of large-scale hydroelectric facilities has reached a plateau in the United States, there is much potential for further growth of hydropower in developing countries.

Thermal electric plants use the sun's energy to produce a very hot fluid by concentrating the sunlight on a central receiving area. High-temperature fluids then can be used to run a conventional steam-turbine electrical generator.

Internet Sites

For an up-to-date list of Internet resources related to the material in this chapter, go to the Harcourt College Publishers website at **http://www.harcourtcollege.com**. The links are in the *Energy: Its Use and the Environment* site on the Physics page. General energy related sites and some guidelines for using the World Wide Web in your class are on the inside front cover of this book.

References

Chapter 12

Barlow, R., B. McNelis, and A. Derrick. 1993. *Solar Pumping*. Washington, D.C., World Bank.

Braun, G. W., and D. R. Smith. 1992. Commercial Wind Power: Recent Experience in the U.S. *Annual Review of Energy*, 17.

Brown, N. L. 1980. Renewable Energy Resources for Developing Countries. *Annual Review of Energy*, 5.

Chalmers, B. 1976. Photovoltaic Generation of Electricity. *Scientific American*, 236 (October).

Cuff, D., and W. Young. 1986. *The United States Energy Atlas*. 2nd ed. New York, Macmillan.

Davidson, J., and R. Komp. 1983. *The Solar Electric Home: A PV How-To Handbook*. Ann Arbor, aatec Publishing Co.

Dracker, R., and P. de Laquil. 1996. Progress Commercializing Solar-Electric Power Systems, *Annual Review of Energy*, 21.

Eldridge, F. 1980. *Wind Machines*. New York, Van Nostrand Reinhold.

Gipe, P. 1999. *Wind Energy Basics* (Preface). White River Junction, VT, Chelsea Green Publishing Co.

Gipe, P. 1995. *Wind Energy Comes of Age*. Wiley Series in Sustainable Design. New York, John Wiley & Sons.

Holdren, J. P., G. Morris, and I. Mintzer. 1980. Environmental Aspects of Renewable Energy Sources. *Annual Review of Energy*, 5.

Kenna, J., and B. Gillet. 1985. *Solar Water Pumping*. London, Intermediate Technology.

Kozloff, K., and R. Dower. 1993. *A New Power Base: Renewable Energy Policies for the Nineties and Beyond*. Washington, D.C., World Resources Institute.

Moretti, P. M., and L. V. Divone. 1986. Modern Windmills. *Scientific American*, 254 (June).

Ogden, J., and R. Williams. 1989. *Solar Hydrogen*. Washington, D.C., World Resources Institute.

Robinson, S. 1978. *The Energy Efficient Home*. New York, New American Library.

Smith, D. R. 1987. The Wind Farms of the Altamont Pass Area. *Annual Review of Energy*, 12.

Strong, S., and W. Scheller. 1993. *The Solar Electric House*. Emmaus, PA, Rodale Press.

QUESTIONS

1. What is meant by the statement that the efficiency of a solar cell is 15%?
2. Would you expect all wavelengths of light incident on a solar cell to be equally good in emitting electrons? Would a light filter over the cell help?

3. Discuss some of the limitations to the use of PV to produce power for a home. What reasons would someone have for installing PV devices for a residence?
4. How would you wire together individual solar cells to produce enough voltage to run a 1.5-V motor?
5. Using solar cells, design a simple circuit that allows a small 1-V light to turn on when the sun goes down.
6. Cost is a problem to the widespread use of PV. To lower manufacturing costs, larger markets are needed, which require lower prices. Why is this a "catch-22" problem, and how might it be addressed?
7. List as many applications of PV as you can. What are the benefits to the use of PV?
8. What is a micro-hydroelectric system? What factors determine the amount of power one can extract from a stream? Sketch how you might put together such a facility.
9. Why do we need concentrating collectors to produce high-temperature fluids for the generation of electricity? How is electricity produced by this fluid? (Provide a sketch.)
10. Why are solar thermal facilities located primarily in the southwestern United States?
11. What are some recent advances in wind energy technology? Examine Figure 12.13 of the Altamont Pass (California) area. Estimate the power output per square km of the windfarm shown.
12. What sites in your geographical area might be suitable for wind turbines? How might you evaluate the potential for this? What wind development is taking place in your state?
13. What are some of the reasons that wind energy capacity is growing by 30% per year in Europe?
14. Investigate present prices for PV modules and single cells. How much is this per kilowatt?
15. What is the largest capacity hydroelectric plant now operating in the world?

PROBLEMS

1. If a commercial solar cell with a maximum output of 2 amps at 0.5 V costs $30, what will be the cost of a PV plant per peak kilowatt?
2. We have discussed a solar cell "farm" for the conversion of sunlight into electrical energy. If the solar cells used have an efficiency of 10%, what area of land is needed to produce an output of 1000 MW? Assume that the mean insolation is 500 W/m^2.

3. A commercially available solar cell module puts out 3 V and 0.1 A, and measures 5 cm by 8 cm. How large an array is necessary to provide an output of 40 W?
4. What maximum output would you expect from a wind turbine with a blade of diameter 20 ft in a 15-mph wind?
5. A stream flowing at a rate of 12 kg/s has a vertical drop of 4 m. What is the maximum power that one can obtain from this stream?
6. How many square meters of PV would it take to provide 1 kW (about ⅓ of the power needed per person for health and comfort) for a world population of 6 billion? If these cells were crystalline silicon of 200 microns thick (200×10^{-6} m), what is the minimum mass of silicon material needed for this venture? (Density of silicon = 2330 kg/m^3.)
7. If the U.S. growth rate for wind energy continued at 22% per year, when would it provide 5% of our present demand for electricity?

FURTHER ACTIVITIES

1. A simple example of a windmill was considered in a Further Activity in Chapter 2. A pattern was given for its construction. You might wish to experiment with this model now.
2. The following device can be used to measure wind speed. You'll need a protractor, some fishing line, and a ping-pong ball. Check out the calibration curve by having a friend hold the device outside a car window and noting the angle the fishing line makes with the protractor as you drive at different speeds.

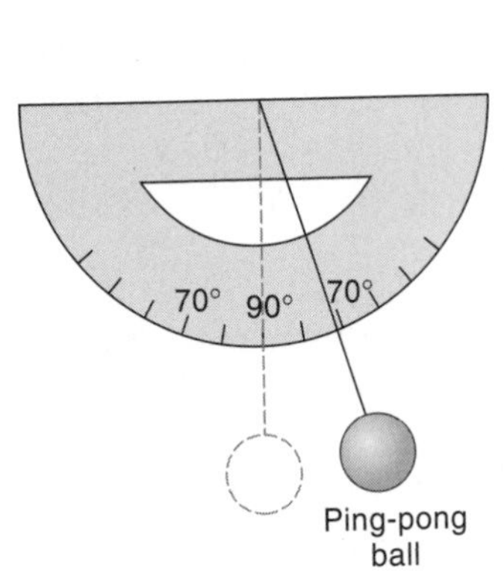

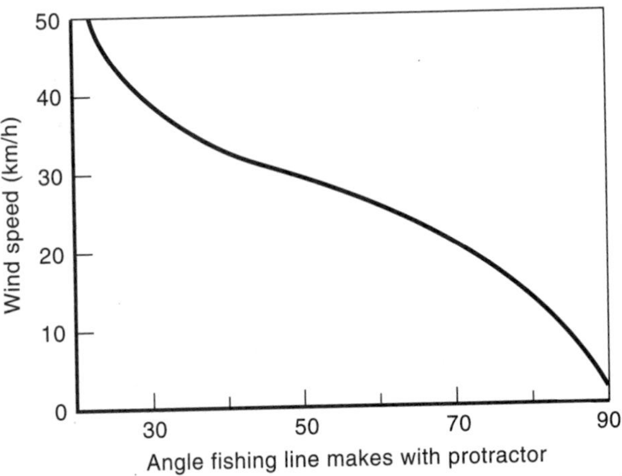

3. Solar cells can be wired together to produce an output capable of lighting a small bulb. PV cells can be purchased from such places as Radio Shack or Edmond Scientific Co. (New Jersey). Individual cells can be wired together in series to produce the desired output. Each cell produces about 0.45 V. The current output (in intense sunlight) will be about 0.1 A/cm^2.

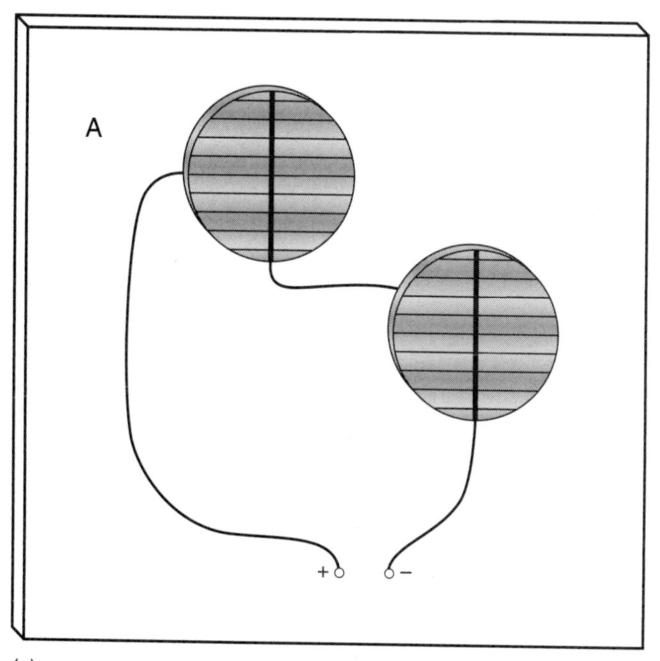

(a)

(b)

(*a*) Wires can be attached to two solar cells in series to provide an output of almost 1 volt. (*b*) Care must be taken when soldering the fragile cells.

If the cells are not in plastic with wires already attached, one will need to do some soldering. To wire the cells in series, one needs to solder a wire from the (+) back of one cell to the (−) grid on the front of the next cell. The end of the wires can be attached to a screw and nut that are secured to a piece of board. For protection, a piece of Lucite to cover the cells can be hot-glued on top of the board. (Don't press too hard on the Lucite, as the cells are fragile.)

13

The Building Blocks of Matter: The Atom and Its Nucleus

For better or for worse, we find ourselves today well into the "nuclear age." The symbol of the atom, with its nucleus surrounded by orbiting electrons, has become the trademark for "progress" in the eyes of many. Humankind has discovered how to tap the vast stores of energy contained within the nucleus and has put this energy to use in many ways, both destructively and constructively. For many people, a study of nuclear physics implies only a study of "atomic" bombs and nuclear reactors. However, the field is much broader than this; an understanding of radioisotopes and their many uses and of fusion reactors are only a few of the topics that require a knowledge of the nucleus and its properties.

We might be at the dawn of a new era of alternative energy sources, such as solar energy, but the importance of conventional nuclear energy for the intermediate term also must be considered. Arguments for and against nuclear power are made sometimes solely from emotional grounds. It is the purpose of this and the following chapters to examine the nucleus, radioactivity, nuclear reactions, and nuclear reactors so as to provide the background necessary to evaluate these arguments. Clearly, more than technical knowledge is needed to develop an energy policy for the first part of this century, but these nontechnical factors will be discussed later.

A. The Atomic Hypothesis

For thousands of years, people have been interested in the structure of matter and its ultimate composition. With the many different kinds of substances known—salt, water, stone, air, hair, and so forth—people wondered whether there were an infinite number of different substances in nature or only a small number of fundamental ones that made up the many different things around us. If we subdivide a piece of coal into many small pieces, how far can we take this process until the identifying properties of the substance are lost? What are the basic building blocks of nature? What common properties of all objects can be found?

Today, almost regardless of our backgrounds, we know about atoms and readily accept their existence, even though we cannot easily see, feel, or even smell them. We may know there are more than 100 different kinds of atoms or elements, and that combinations of these elements—such as letters in the alphabet—form the substances that we see around us. The idea that all matter is constructed out of tiny building blocks called **atoms** (from the Greek *atomos*, meaning indivisible or uncuttable) goes back to the Greek philosopher Democritus (ca. 420 BC). He believed that there was some common feature underlying all matter, some basic structure from which all matter is made. However, like many theories, this description of nature was apparently too early, and lacked evidence, to be universally accepted. This theory was superseded by the view of Aristotle (ca. 340 BC) that matter was composed of four elements: air, fire, water, and earth. For 2000 years, this view was the dominant "model" of the physical universe.

The atomic theory of matter was not revived in earnest until about the beginning of the 19th century, mainly as a result of the work of the English chemist John Dalton. From quantitative analyses of the way in which various elements combine to form chemical compounds, Dalton arrived at the conclusion that every element is made up of atoms, the indestructible and indivisible basic units of matter. He proposed that each element consists of only one kind of atom, which is different from the atoms of any other element. Each element also has a different mass and its own particular set of properties. With the aid of an accurate balance, Dalton showed that whenever water is formed from hydrogen and oxygen, no matter how the elements are mixed or in what quantities, there is a definite proportion of one mass of hydrogen to eight masses of oxygen. The best explanation of this "law of definite proportions" was the existence of elementary particles, or atoms.

Dalton identified about 20 different kinds of atoms in his work. By studying chemical reactions between these elements, he was able to determine their relative masses. His list went from hydrogen with 1 mass unit to gold with 190 mass units. Later work determined that 1 atomic mass unit (amu) was equal to 1.66×10^{-27} kg. The classification of all known elements appears in the periodic table, a brief overview of which is given in the Special Topic at the end of this chapter.

B. Building Blocks of the Atom

It was philosophically difficult for some scientists to believe that the universe had at its most fundamental level as many elementary "particles" as there were different elements. Firm evidence that the atom was not a hard, indivisible sphere nor an elementary particle finally became available at the end of the 19th century with the discoveries of the electron and of radioactivity. In the case of the electron, scientists were investigating the passage of electricity through gases at low pressures. An evacuated tube (at a pressure of about one thousandth of an atmosphere) contained a negative cathode and a positive anode and a collecting cup (Fig. 13.1a). A voltage difference between these

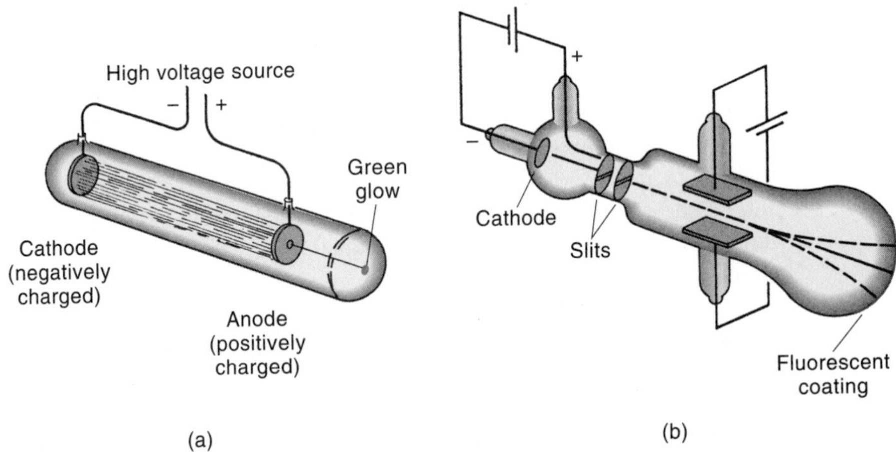

FIGURE 13.1
(*a*) Evacuated tube used in observation of cathode rays. (*b*) Apparatus used by J. J. Thomson (1897) to measure the charge-to-mass ratio of the electron. The evacuated tube is similar to a TV picture tube. The negatively charged particles emitted from the cathode are deflected by either an electric field or a magnetic field. The parallel plates connected to a battery provide the electric field. Two current-carrying coils (not shown) produce a magnetic field perpendicular to the electric field. The sizes of the deflections, as noted on the fluorescent screen, can be used to determine the charge-to-mass ratio of the electron.

two plates gave rise to a visible ray inside the tube from the cathode to the anode, which also caused the collecting cup to become negatively charged. These rays streamed from the cathode irrespective of the material used, leading J. J. Thomson (1897) to reason that all materials contain **electrons,** the name given to these "cathode rays."

More elaborate apparatus was used to study the deflections of the cathode rays in electric and magnetic fields. The apparatus used (Fig. 13.1b) was similar to our present TV tubes, with a high voltage applied to one end. Electrons were emitted from the negatively charged cathode, accelerated to the positive anode, and then continued on to a fluorescent screen, where light was emitted. By deflecting this beam with electric and magnetic fields, Thomson determined the charge-to-mass ratio of the electron. He concluded that the electron has a negative charge and a mass 1837 times smaller than that of the hydrogen atom.

These experiments indicated that, because the electron came from an atom, the atom was neither indivisible nor fundamental. Because the atom is electrically neutral, the negative electrons have to be balanced electrically by particles of positive charge. Thus the atom itself has structure. The form of this internal structure was the subject of much speculation in the beginning of the 20th century. Thomson envisioned the atom to be a sphere in which equal amounts of positive and negative charge were distributed evenly. His model could be likened to a bowl of oatmeal (a sphere of positive charge) in which were embedded raisins, the lumps of negative charge.

The discovery of radioactivity also helped make it clear that the atom was not an indivisible particle. Although radioactivity will be discussed in more detail later in the chapter, a brief outline will be given here. During the last part of the 19th century, because of the work of Henri Becquerel and of Marie and Pierre Curie, it was observed that several of the most massive elements found in nature, such as uranium and radium, emit radiation spontaneously. These atoms were "radiation-active" or **radioactive.** It was observed that three types of radiation are emitted from these radioactive materials. The names given for these three distinct types of radiation were alpha (α), beta (β), and gamma (γ) rays. These three radiations are not emitted simultaneously from radioactive substances. Some elements emit alpha rays and some emit beta rays, while gamma rays sometimes accompany α and sometimes β radiation. If these radiations are directed between two charged plates, they behave in three different ways (Fig. 13.2). Scientists learned that the alpha rays were positively charged particles, beta rays were negatively charged particles of very small mass, and gamma rays were similar to what we know as X-rays.

It was also discovered—to the surprise of many—that after the emission of radiation, the original radioactive atom had been transformed into an entirely different chemical atom. Furthermore, the radiation of a given sample was not affected in any way by physical or chemical processes, such as changing the temperature or chemical composition of the substance. As a consequence, it became clear that radioactivity is a *nuclear process,* resulting from nuclear decay.

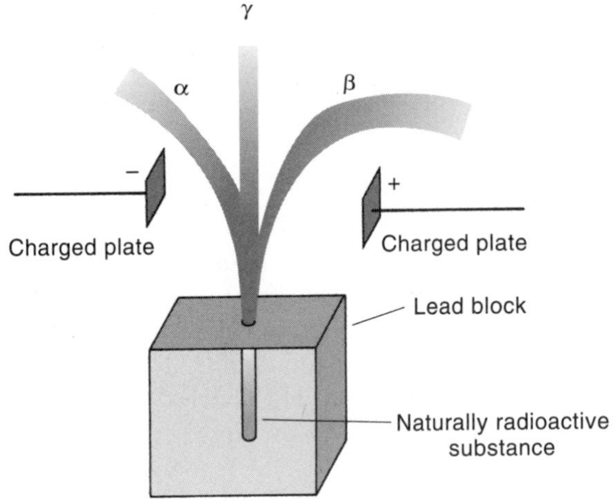

FIGURE 13.2
Radioactive elements may emit three types of radiation: electromagnetic radiation called gamma rays; fast-moving electrons called beta particles; and alpha particles, which are the nuclei of helium atoms. If radioactive material is placed at the bottom of a hole in a lead block, radiation will be emitted through the top. If the beam passes through an electric field, it will separate into the three types of radiation.

The atom's internal structure was probed by the scattering experiments of Hans Geiger and Ernst Marsden in England at the beginning of this century. In these experiments, alpha particles (positively charged particles, about 7400 times the mass of an electron) from a naturally radioactive substance bombarded a very thin foil of gold. The alpha particles scattered through different angles and were observed as they hit a fluorescent zinc sulfide screen (Fig. 13.3). If the positive charge was distributed uniformly throughout the atom as suggested by Thomson's model, then the alpha particles would only be scattered through small angles (i.e., suffer small deflections). There was not enough of a concentrated positive charge present in this model of the atom to allow its interaction with the massive alpha particle to cause more than a small deflection. However, Geiger and Marsden observed that some alpha particles were deflected through very large angles.

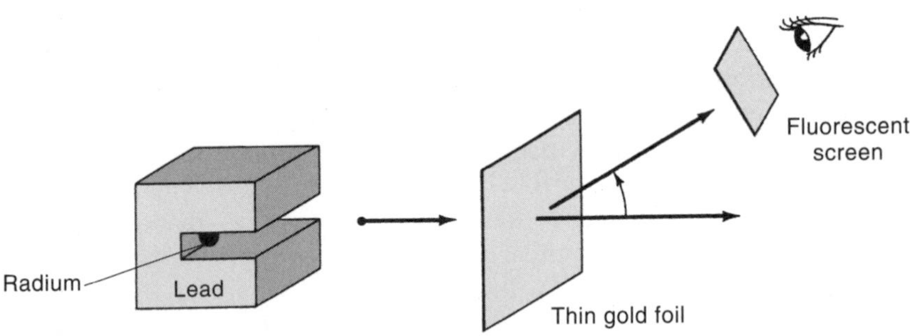

FIGURE 13.3
Scattering of alpha particles from a thin gold foil.

(It was likened to a cannonball being shot at a target of tissue paper and bouncing backward.)

The British physicist Ernest Rutherford analyzed the data and concluded that the positively charged particles must be concentrated in a volume much smaller than the entire atom, forming a central core or **nucleus** around which the electrons moved. These experiments indicated that the nucleus is extremely small, about 20,000 times smaller than the atom. (To picture this large difference in size, imagine that the atom is the size of a football stadium; then the nucleus would be about the size of a ball bearing.) Despite its small size, the nucleus contains more than 99.9% of the mass of the atom!

The nucleus is also composite, consisting of positively charged particles called **protons** (from the Greek for "first") and neutral particles of about the same mass, called **neutrons.** It has even been established that these three particles (electrons, protons, and neutrons) are themselves not the fundamental building blocks of all matter. Experiments in recent years with large particle accelerators (Fig. 13.4) have shown that the proton and the neutron seem to be composed of smaller particles called **quarks.** These quarks have a fractional charge (either plus or minus ⅔ or ⅓ the electron charge). Three quarks combine to form a proton of net charge plus one ($\frac{2}{3} + \frac{2}{3} + (-\frac{1}{3}) = +1$). (The standard model of matter holds that there are six different kinds of quarks, named up and down, charm and strange, top and bottom. Particle discoveries in the past several decades confirmed the existence of five of these quarks, but only recently (1995) has evidence been attained for the existence of the (heaviest) "top" quark. Large accelerators that slam together particles were necessary to create this top quark. This finding was extremely important because it appears to validate a theoretical model central to understanding the nature of matter, time, and the universe.)

For an isolated, electrically neutral atom, the number of electrons surrounding the nucleus is equal to the number of positive charges in the nucleus

FIGURE 13.4
Aerial view of the Fermi National Accelerator Laboratory in Batavia, Illinois, the world's highest energy particle accelerator. The accelerator ring is 6.3 km (3.8 miles) in circumference. Protons can be accelerated up to 99.99% the speed of light. (FERMILAB)

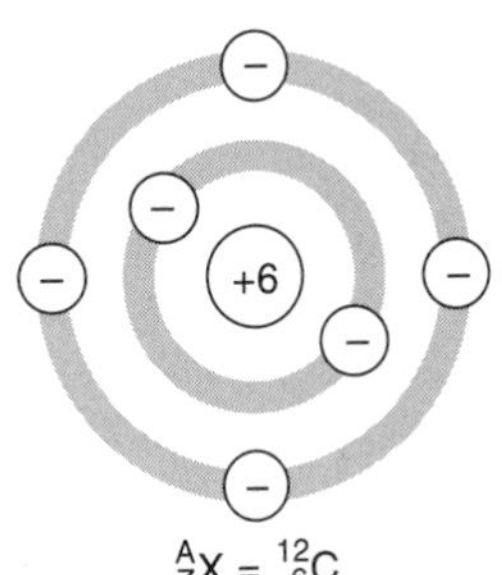

FIGURE 13.5

The nucleus of the carbon atom has a positive charge of 6. It is surrounded by six electrons, arranged in two major shells. The number of protons gives the element its atomic number.

(Fig. 13.5). The number of protons is a very important property of the atom, called the **atomic number.** It defines the position occupied by the element in the periodic table (see Fig. 13.15). The atomic number also determines the chemical properties of the element.

The building blocks of most of the matter around us are a combination of two or more atoms. These atoms, which can be different elements, form a **molecule** by sharing their electrons with each other or exchanging them in such a way that the different elements can be held together by electrostatic attraction. The force holding the elements together in a molecule is electrical, caused by the force of attraction between positive and negative charges. Such molecules range in size from the simplest diatomic ones (such as carbon monoxide or oxygen) to the extremely large molecules found in biological systems.

In the combustion of coal, paper, natural gas, and other fuels, energy is released. These are chemical reactions in which the bonds between the atoms of the molecule are either formed or broken, releasing energy that was stored in the bonds. We shall see in Chapter 14 that when the "bonds" between the particles in the nucleus are rearranged, a much larger amount of energy is released than occurs in purely chemical reactions. This energy is called **nuclear energy.**

C. Energy Levels

One factor in the radical break between modern physics and classical physics was the discovery that the properties of matter in the microscopic world have distinct or **"quantized"** values. It is not possible for these properties to have arbitrary values between the "allowed" ones. Until the end of the 19th century, it had been assumed that the energy possessed by any object was "continuous," that is, could assume any value. A ball rolling down a ramp has both kinetic and potential energies and can assume any value for the sum of its energies. This is not true in the atomic world. The electron "orbiting" the nucleus possesses both kinetic energy from its motion around the nucleus and potential energy from the electrical force between it and the nucleus. Yet this electron's energy cannot be just any amount, but must have particular discrete values. We say that the electron can exist only in certain definite "energy levels" or quantized states within that atom. In everyday experience one deals with quantized quantities: The number of atoms in a container is a whole number, as is the number of coins in my pocket. I don't have 14⅓¢ with me.

An early and still helpful model of the atom (Niels Bohr, 1913) postulated that electrons are permitted to move around the nucleus only in orbits of fixed radii. The electrons in each orbit possess a quantized amount of energy, depending in part on the atomic number of the atom. The number of electrons allowed in each orbit is also limited. The electrons can gain or lose energy only in amounts corresponding to the differences in energy between these levels. An analogy of this situation is a multilevel parking garage. The floors in the garage correspond to the different energy levels the electron can possess. Electrons, like cars, can be placed only on one floor, not in between (Fig. 13.6). Each floor also has a given capacity. An electron on one floor can move up to a higher (if unfilled) floor if energy is given to it—for example, by the heat of a fire or by an electric discharge. However, in the atom the electron will not remain in its new orbit more than a fraction of a second, and will drop back almost immediately to a lower orbit. In the transition or "fall" from a higher floor to a lower one, energy is released in the form of electromagnetic radiation, such as visible or ultraviolet light or X-rays. Even though the Bohr model of the atom has been superseded by the more complex quantum mechanical picture of the atom, the concepts of discrete energy levels and transitions between them remain the same in both models.

This simple model of the atom permits one to understand **atomic spectra**—the colors of light emitted from various elements that have been excited by either heat or electric discharges. Notice the red light coming from a neon tube, the blue-purple light from mercury, the yellow light from a sodium-vapor lamp. Each chemical element has its own unique set of energy levels, so the transitions of electrons from an excited state back to another energy level results in the emission of unique energies with characteristic colors. Although

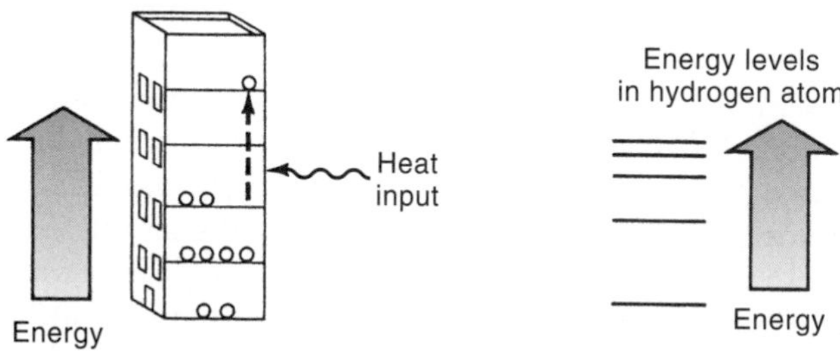

FIGURE 13.6

Energy levels of electrons within atoms are analogous to floors in a building. Here, one electron has been excited to a higher state by the addition of heat to the atom.

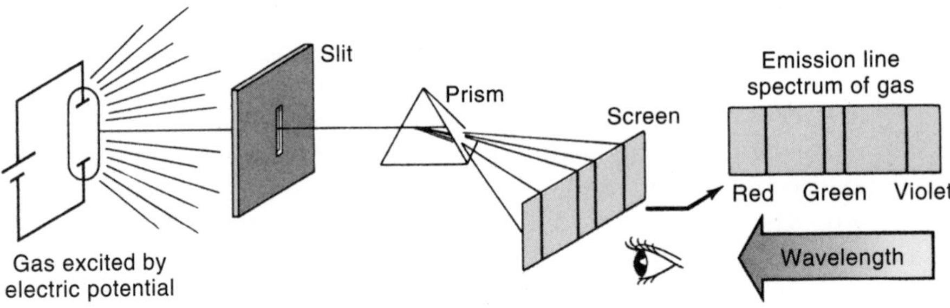

FIGURE 13.7
Spectrum of light emitted by a gas that has been excited by electrical discharge or heat.

there appears to be only one color, the visible light emitted from each element is a mixture of many emission lines. If the light from an excited gas is passed through a narrow slit and separated into its different colors with a prism or grating, we do not see a continuous band of colors but a set of discrete, sharp lines (Fig. 13.7). (Have you ever worn a pair of diffraction glasses?) By measuring the wavelengths of these lines, we can determine the composition of the elements in the source, since each element has its own set of characteristic colors. Using this spectral "fingerprint," the chemical compositions of the stars can be determined. In fact, it is such line spectra of stars that convince us that the universe is composed only of those elements in the periodic table.

D. Nuclear Structure

In the early part of this century it was observed that atoms of the same element were not all identical but had as many as ten different masses. We call these similar atoms **isotopes,** which is Greek for "the same place" (that place being the same position in the periodic table). Isotopes of a given element have the same number of protons but different numbers of neutrons, and thus different masses. All isotopes of a substance behave nearly the same chemically, but each have different nuclear properties. Most of the elements that occur in nature have at least two isotopes. An example of a series of isotopes for one element is that for hydrogen (Fig. 13.8). The nucleus of the simplest and most common isotope (99.985% of all hydrogen atoms) has one proton and is called hydrogen. If one neutron is added to the nucleus, we have an atom with a mass of 2 atomic mass units; we call this *deuterium,* given the symbol D. This isotope exists in nature (0.015%) and combines with oxygen to form what is called heavy water, or D_2O. A very rare isotope of hydrogen is *tritium,* which has one proton and two neutrons, and a mass of 3 amu. It is important as a fuel for fusion reactions.

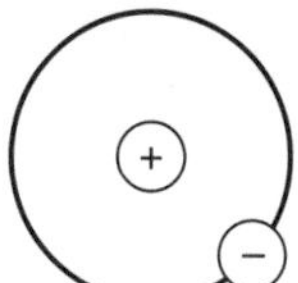

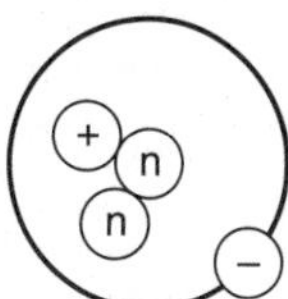

Name:	Hydrogen	Deuterium	Tritium
Symbol $^{A}_{Z}X$:	$^{1}_{1}H$	$^{2}_{1}H$	$^{3}_{1}H$
Abundance:	99.985%	0.015%	Unstable

FIGURE 13.8
Isotopes of hydrogen.

When speaking of different isotopes, we will use the notation of Fig. 13.8, $^{A}_{Z}X$. The subscript Z denotes the number of protons in the nucleus, and is called the atomic number. (Most of the time the subscript Z is omitted, since the chemical symbol X specifies the number of protons. However, for problems where this information is necessary, we will include Z.) The superscript A is the mass number of the isotope and is equal to the number of protons plus the number of neutrons, $A = Z + N$. Protons and neutrons behave much the same if the proton charge is not considered. They are collectively referred to as "nucleons." When referring to a particular nuclear species (characterized by the name of the element and its mass number, such as cobalt-60 or uranium-238), the name "nuclide" is used.

The existence of isotopes explains why the atomic masses of some elements are not whole numbers. Because isotopes behave nearly the same chemically, elements are mixtures of isotopes in differing amounts. An example is chlorine with an atomic mass of 35.5; this is a mixture of two isotopes with atomic masses of 35 and 37, in relative amounts of 76% and 24%. Thus, the weighted average atomic mass is

$$0.76 \times 35 \text{ amu} + 0.24 \times 37 \text{ amu} = 35.5 \text{ amu}$$

which is what appears for the atomic mass of chlorine in the periodic table.

E. Radioactivity

A radioactive nuclide is a nuclide that spontaneously undergoes nuclear decay, resulting in the emission of nuclear radiation in the form of particles or rays. Important in the description of a radioisotope is the type of nuclear radiation emitted and its "half-life." The Greek letters α, β, and γ are given to the three types of radiation, which are characterized by their mass, their charge, and their abilities to penetrate matter (see Fig. 13.2). The ranges of these rays—that is, the distance the radiation can travel before it loses all its energy—

Table 13.1 PROPERTIES OF NUCLEAR RADIATIONS

Type of Radiation	Range
α-particles	A sheet of paper, a few centimeters of air, or thousandths of a centimeter of biological tissue
β-particles	A thin aluminum plate, or tenths of a centimeter of biological tissue
γ-rays	Several centimeters of lead, or meters of concrete

in different types of matter (for energies typically found in radioactive decay) are given in Table 13.1.

We have since learned that alpha particles are the nuclei of helium atoms, with two protons and two neutrons. Many radioactive nuclei that are more massive than lead will emit alpha particles during their decay. Beta decay is the most common type of radioactive decay. A beta particle can be either an electron (β^-) or its antiparticle, called a positron (β^+); the latter has the same mass as an electron but a positive charge. A β^- particle is emitted from the nucleus during the decay of a neutron into a proton, and a β^+ in the decay of a proton into a neutron. Gamma rays are electromagnetic waves that accompany alpha or beta decay. They arise from the fall of a nucleus from an excited state to a more stable state, similar to the emission lines of chemical elements caused by electrons moving from one quantized energy level to another.

In the α and β decay of a radioactive nucleus (Fig. 13.9), "transmutation" of elements occurs—an atom of one element becomes an atom of another element, as the number of protons in its nucleus changes. The resulting atom may also be radioactive, and decay again into yet another atom. By knowing the types of radiation emitted and the rules for nuclear decay, you can follow the decay scheme through a sequence of radioisotopes to the final stable nucleus. This will be discussed later.

Another very important property of a radioactive nuclide is its half-life. The **half-life** of an isotope is defined as the time it takes for half of the original

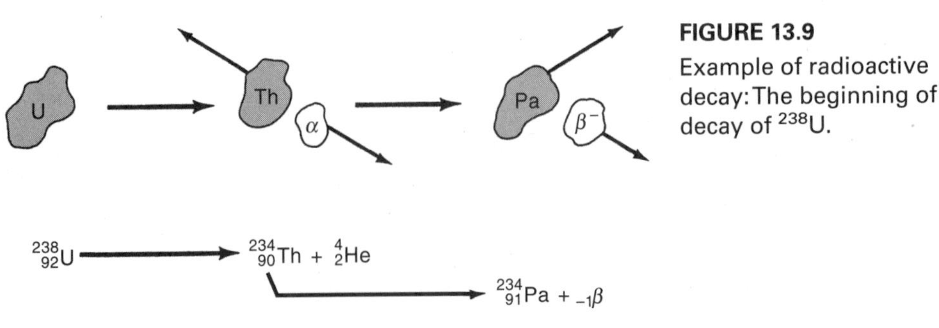

FIGURE 13.9
Example of radioactive decay: The beginning of decay of ^{238}U.

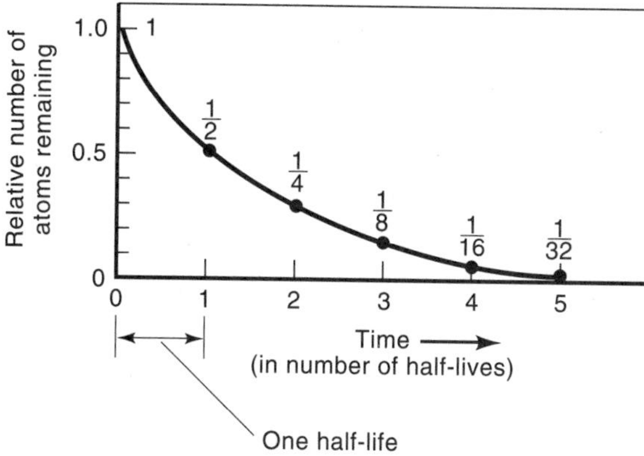

FIGURE 13.10
The half-life of a nucleus is the time it takes for one half of the original amount of that substance to decay. Radioactive decay is an exponential process.

amount of that isotope to decay into another element. For example, if we start with 100 g of cobalt-60, which has a half-life of 5.3 years, then 5.3 years later we will have 50 g of cobalt-60 left. The other 50 g has decayed into nickel-60, which is stable. After another 5.3 years we will have 25 g of cobalt-60 left; after 5.3 more years we'll have 12.5 g, and so on. We say that this substance is decaying exponentially with time, as shown in Figure 13.10. (Anything that has a fixed doubling time or halving time changes exponentially.) The half-life is fixed and does not depend on the environmental temperature, the chemical composition of the source (i.e., whether it is a pure element or part of a chemical compound), or its previous decay history. The symbol $\tau_{1/2}$ is used for half-life.

The half-lives of radioactive nuclei cover an enormous range of values. Uranium-238, found naturally, is radioactive with a half-life of 4.5 billion years, decaying by the emission of alpha particles. Bismuth-208 can be made in a nuclear reactor and has a half-life of 3 milliseconds (0.003 s). A rule of thumb is that after ten half-lives have elapsed, the amount of the initial element left will be quite small compared to the original amount (about 0.1%).

EXAMPLE

If 100 g of a radioactive element were present at 9 AM on Monday and 25 g of that element were present at 9 AM the following Friday, what is the half-life of the source?

Solution

Because nuclear decay is exponential, 50 g would remain after one half-life, and 25 g would remain after two half-lives. Since it took four days for the amount to decay to 25 g, the half-life is

$$\tau_{1/2} = \frac{4 \text{ days}}{2} = 2 \text{ days}$$

The intensity of radiation from a radioactive source depends on the number of nuclei present and the half-life. A small amount of a substance with a short half-life may have a greater intensity than a large quantity of a long-lived isotope. The number of atoms that disintegrate per second is called the **activity,** and it is measured in units of Becquerels (Bq) or Curies (Ci). A Becquerel is equal to one disintegration per second. A quantity of radioactive substance in which 37 billion atoms decay per second is said to have an activity of 1 Ci.* Because this is much larger than the decay rates of many radionuclides, the activities of substance are often expressed in microcuries (μCi), or millionths of a Curie. An expression for the activity A is

$$A = \lambda N$$

where $\lambda = 0.693/\tau_{1/2}$ and N is the number of atoms of the radioactive source present at any time. From this expression you can see that a short half-life can yield a large activity. Also note that the activity decreases exponentially with time, since A depends on N. The damage that radiation does to a person's cells depends not only on the activity of the radioactive source and the person's distance from it, but also on the type of radiation emitted, the exposure time, and the part of the body affected. We shall study these factors in Chapter 15.

Half-lives of radioisotopes can be used in the **radiometric dating** of organic and inorganic materials. Radiocarbon dating has been important for many years in estimating how long ago the source of a sample of organic material died. Radioactive ^{14}C is produced in the earth's upper atmosphere by nuclear reactions induced by high-energy gamma rays. Since this isotope behaves chemically identically to the abundant stable isotope ^{12}C, it can form such common compounds as carbon dioxide. In the atmosphere the ratio of ^{14}C to ^{12}C as found in CO_2 is about 1.3×10^{-12} to 1. The same ratio occurs in living organisms because plants use CO_2 during photosynthesis and animals then eat the plants. When an organism dies, this ratio will decrease because ^{14}C decays by beta emission and is not replenished, since the biological processes have stopped. The half-life of ^{14}C is 5730 years. Assuming the $^{14}C/^{12}C$ ratio in the atmosphere has remained constant for thousands of years, it is possible to estimate the age of organic material by measuring its activity. Today, you would find an activity of about 15 disintegrations per minute per gram of living material. A 1-g sample that is 5700 years old would have an activity of about 7.5 disintegrations/minute. Radiocarbon dating is considered reliable back to about 5 half-lives, or 25,000 years.

---

*This somewhat odd-looking number was selected because it is (approximately) the number of disintegrations per second in 1 g of radium.

Geologic dating of rocks must use radioisotopes with very long half-lives. Since the activity of such a source is very small (remember that activity is inversely related to the half-life), this type of radiometric dating is done by measuring the amount of decay product (called the "daughter") relative to the original radioisotope (called the "parent") present in the sample. Since the ratio of parent and daughter elements in the rock is a function of time, the age of the sample can be determined. Such measurements must assume that a "closed system" exists for the rock—in other words, none of the parent or daughter atoms have been added to or escaped from the rock during its history. As a consequence, samples must be free from weathering influences and other contamination.

One such method of geologic dating uses potassium and argon. The isotope ^{40}K, found in many minerals, decays to ^{40}Ar by the capture of an electron from the innermost shell around the nucleus, and has a half-life of 1.2×10^{10} years. Argon is not found in rock initially, because it is an inert gas and not bound chemically with other atoms. Measurements of the ratio of ^{40}Ar to ^{40}K have yielded rock ages of up to 4.7 billion years. Another dating method makes use of the ratio between the parent nucleus rubidium-87 (^{87}Rb, $\tau_{1/2} = 5 \times 10^{10}$ years) and its daughter nucleus strontium-87 (^{87}Sr).

F. Nuclear Glue, or Binding Energy

Today we know of 115 different elements, 81 of which have at least 1 stable isotope. Among the other 34 elements, some have radioactive isotopes with long half-lives, which are found in nature, while others have been produced artificially and have very short half-lives (e.g., several microseconds). Naturally occurring radioactivity is limited primarily to heavy nuclei, but artificial transmutations from stable to radioactive nuclei are brought about today in many laboratories with particle accelerators and nuclear reactors (see the next section). If we consider isotopes instead of elements, more than 1500 nuclides are known, 279 of which are stable. Many elements have only 1 stable isotope, but some have as many as 10 (e.g., tin). Clearly, not all combinations of neutrons and protons within a nucleus are stable. A general rule for most stable nuclei with mass numbers less than 40 is that the numbers of neutrons and protons will be about equal (e.g., $^{16}_{8}$O, $^{24}_{12}$Mg, $^{40}_{20}$Ca). For massive stable nuclei, there are more neutrons than protons (e.g., $^{90}_{40}$Zr, $^{208}_{82}$Pb).

Whereas the electrons in an atom are bound by the electrical force of attraction between them and the positively charged nucleus, the protons in the nucleus exert an electrical repulsive force on each other. What holds nuclei together in spite of this is a much stronger nuclear force, acting as a "nuclear glue." The character of the nuclear force is responsible for the general rules for relative numbers of neutrons and protons within light nuclei described previously.

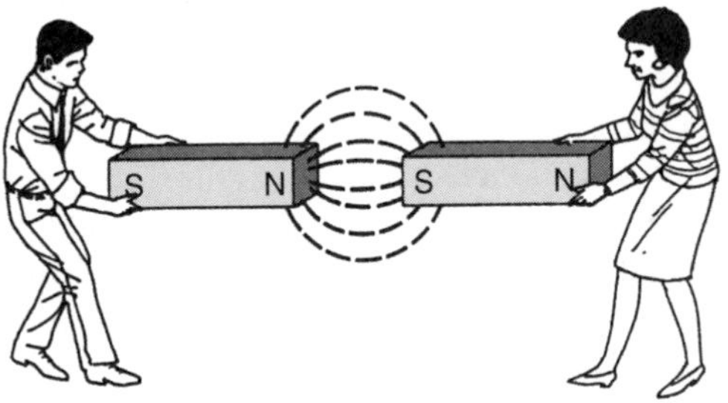

FIGURE 13.11
Just as it takes energy to pull two magnets apart, energy is also necessary to pull apart the nucleons that are bound together in the nucleus. The total binding energy is the energy required to disassemble the entire nucleus.

If you have tried to pull two magnets apart, you know that it takes a force (Fig. 13.11). Similarly, it takes energy to pull apart the nucleus. The **binding energy** of a nucleus is the energy needed to disassemble the nucleus into its constituent neutrons and protons; conversely, it is the energy released when that nucleus is assembled from its component parts. The larger the binding energy, the more stable the nucleus.

EXAMPLE

Which of the following isotopes might you expect to be radioactive?

$$^{6}_{3}\mathrm{Li},\ ^{20}_{10}\mathrm{Ne},\ ^{29}_{13}\mathrm{Al},\ ^{35}_{17}\mathrm{Cl}$$

Solution

Recall that the superscript is A, the atomic mass, and the number of protons is Z, the atomic number; the number of neutrons is N, where $N = A - Z$. Referring to the periodic table at the end of the chapter for the atomic numbers of the elements, the listed isotopes have the following numbers of protons and neutrons:

	Z	N
^{6}Li	3	3
^{20}Ne	10	10
^{29}Al	13	16
^{35}Cl	17	18

Only ^{29}Al has a relatively large difference between Z and N. In fact, it is radioactive, with a half-life of 6.6 minutes. The other three isotopes are stable.

If we look closely at the atomic masses of different nuclei, we see that they are nearly whole numbers, but not exactly, and this small difference is very important. The mass of the helium nucleus is 4.0016 atomic mass units, while the mass of a proton is 1.0073 amu and that of a neutron is 1.0087 amu. Therefore, the mass of two protons and two neutrons is $2 \times 1.0073 + 2 \times 1.0087 = 4.0320$ amu, or 0.0304 amu *more* than the actual measured mass of the helium nucleus. If we were able to join together these four nucleons to make helium, what would happen to the extra mass? It would be released as energy. Albert Einstein proposed in 1905 that mass and energy are equivalent, and that mass–energy is conserved in an isolated system. Mass and energy are related by the expression

$$E = mc^2$$

where m is the mass lost (or gained) in a reaction, c is the velocity of light in vacuum (186,000 mi/s or 3×10^8 m/s), and E is the energy released. The mass difference between one helium atom and its four constituents is only 0.0304 amu or 5.05×10^{-29} kg, but when this is multiplied by the square of the velocity of light (a large number!), the amount of energy released is 4.5×10^{-12} J (or 28.4 MeV)*. One gram of helium formed as a product of the fusion of neutrons and protons would yield 6.8×10^{11} J, or the energy equivalent of burning 23,000 kg of coal.

G. The Joy of Atom-Smashing, or Nuclear Reactions

The study of nuclear reactions dates back to the work of Rutherford during the first part of the 20th century. Using alpha particles (emitted from a polonium-210 radioactive source) as projectiles, Ernest Rutherford was able to induce the first human-made transmutation by bombarding nitrogen gas. This reaction can be written as

$$^{4}_{2}\mathrm{He} + ^{14}_{7}\mathrm{N} \rightarrow ^{17}_{8}\mathrm{O} + ^{1}_{1}\mathrm{H}$$

By observing the transmission of ^{1}H particles through a thin metal foil (which was thick enough to stop the alpha particles), Rutherford concluded that high-energy particles were ejected from the nitrogen nuclei after bombardment by the alpha particles (Fig. 13.12). He identified this particle as the proton, giving further evidence that it is a fundamental particle. The result of this nuclear reaction was the conversion of one element into another. Alchemists had

*A more convenient unit for such small amounts of energy is an **electron volt,** abbreviated eV. This is the energy that one electron would acquire when moving through a potential difference of 1 volt: 1 eV = 1.60×10^{-19} J. One million eV (or 1 MeV) = 1.60×10^{-13} J.

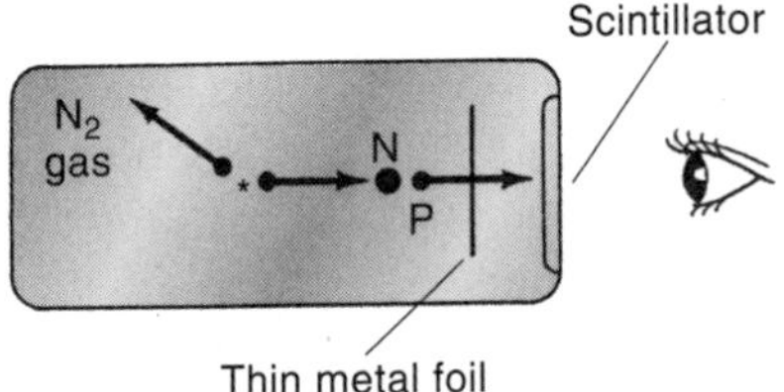

FIGURE 13.12

Rutherford's apparatus to study nuclear reactions. The protons p produced in the transmutation of ^{14}N are detected in the scintillator. The incident alpha particles are produced in the decay of the ^{210}Po.

searched for centuries for ways in which base metals could be converted or transmuted into gold. Now, with the help of particle accelerators and nuclear reactors, such transmutations can be carried out, although they are not economically practical.

As we discuss nuclear reactions, there are several rules that must be followed. First, since **electric charge is always conserved,** the total charge will remain the same between the reactants (on the left side of the equation) and the products (on the right side). For Rutherford's reaction, this equality is $2 + 7 = 8 + 1$. Second, the **mass number is conserved** (at least as far as the integers are concerned). Thus $4 + 14 = 17 + 1$. We have "balanced the equation." We can use these rules to follow the successive transformations of a radioisotope as it decays. If an alpha particle is emitted, the atomic number of the reactant will decrease by 2 and the mass will decrease by 4. For example, the isotope uranium-238 decays with the emission of an alpha particle:

$$^{238}_{92}\text{U} \rightarrow {}^{234}_{90}\text{Th} + {}^{4}_{2}\text{He}$$

Thorium will decay (with a 24-day half-life) by electron emission (β^-) to protactinium. Thus

$$^{234}_{90}\text{Th} \rightarrow {}^{234}_{91}\text{Pa} + {}_{-1}\beta^-$$

The emission of an electron from the nucleus leads to an increase in the atomic number by 1 (as $90 = 91 + (-1)$ for the charge balance). What has happened is the conversion of a neutron into a proton and an electron*: $n \rightarrow p + \beta^-$. Protactinium decays by electron emission to another isotope of uranium: $^{234}_{91}\text{Pa} \rightarrow {}^{234}_{92}\text{U} + \beta^-$. The product of this conversion (uranium-234 in this example) will continue to decay until a stable isotope is reached. This is called a "decay chain" or radioactive series. The most massive substance with a stable isotope is lead.

It is difficult for a projectile to get into the nucleus to cause a nuclear reaction. If the projectile is a charged particle, such as an alpha particle or a proton, it will experience an electrical repulsive force as it approaches the positively

*In electron emission (beta minus decay), an antineutrino is also emitted. This particle is massless, travels at the speed of light, and interacts very weakly with other particles. It is not necessary to discuss it any further at this juncture.

charged nucleus. As we learned in Chapter 10, the electrical force is proportional to the inverse of the square of the distance between charges, so the projectile needs a significant amount of kinetic energy to surmount the electrical repelling force and get into the nucleus, where the attractive nuclear force takes over. This electrical force is also proportional to the product of the atomic numbers of the target nucleus and the projectile. An alpha particle experiences twice as much electrical repulsion force as does a proton. The production of some radioisotopes requires large accelerators to impart the needed energy to the projectile.

The first production of an artificial radioisotope was accomplished by bombarding aluminum-27 with alpha particles:

$$^{27}_{13}\text{Al} + ^{4}_{2}\text{He} \rightarrow ^{30}_{15}\text{P} + \text{n}$$

The isotope phosphorus-30 has a half-life of about 3 minutes. Energy is needed to make this reaction occur, and it is obtained from the kinetic energy of the alpha particle. The kinetic energy of the incoming projectile is usually provided by a particle accelerator, such as the Van de Graaff accelerator shown in Figure 13.14.

The first reaction to produce a large release of nuclear energy (1932) used protons that were accelerated to bombard a lithium target:

$$^{1}_{1}\text{H} + ^{7}_{3}\text{Li} \rightarrow ^{4}_{2}\text{He} + ^{4}_{2}\text{He}$$

The energy released (30 times as much as was put into the reaction) appears as the kinetic energy of the alpha particles. Although a lot of energy is released in this reaction, it is of little practical use because of the low yields (small probability

FIGURE 13.14

Van de Graaff accelerator. Nuclei are accelerated by a high-voltage (9 million volts) terminal located within each of the cylindrical tanks. The accelerated particles travel within an evacuated beam tube (shown emerging from the tank). In the foreground is an electromagnet that deflects the beam of particles into a room to the right, where experiments are conducted.
(UNIVERSITY OF WASHINGTON)

Focus On 13.1

RADON

A very important example of a decay chain is the one that produces indoor air pollution by radon gas. Uranium-238 is found as a trace element in the earth's crust. A continuation of the radioactive series given earlier ($^{234}_{91}Pa \rightarrow ^{234}_{92}U + \beta^-$) is $^{234}_{92}U \rightarrow ^{230}_{90}Th \rightarrow ^{226}_{88}Ra \rightarrow ^{222}_{86}Rn \rightarrow ^{218}_{84}Po$. Chemically inert radon (^{222}Rn, $\tau_{1/2} = 3.8$ days) is a gas and can accumulate in tiny air pockets in the soil; from there it can be pushed into a house by pressure differences between outside and inside the house, created by the rising warm air inside. Once inside, radon decays into other radioactive isotopes that can lodge in the lungs, where the emitted alpha particles can cause cell damage and cancer. New, energy efficient, airtight homes can be particularly sensitive to this problem, because the low infiltration rates (less than 0.5 air changes per hour) can prolong the time of residence of the radon. Health effects of radon pollution have been calculated to be between 2000 and 20,000 cases per year of lung cancer, based on studies of lung cancer in uranium miners, who often have had high radon exposure. (See Chapter 15.) Detectors for measuring radon concentrations are shown in Figure 13.13 and can be purchased in most hardware stores.

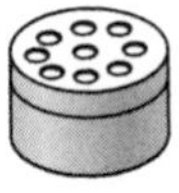

Charcoal canisters   Alpha track detectors

FIGURE 13.13
Two types of detectors for measuring radon concentrations. These devices are exposed to air in your home for a specified time, then sent to a laboratory for analysis.

of occurrence). To make such a reaction useful, we need to get a much larger percentage of the nuclei of a target involved in releasing energy, a situation achieved in fission.

H. Fission

As with the charged particles from radioactive sources or accelerators, neutrons are quite effective in forming new nuclei and in causing nuclear disintegrations. The use of neutrons has the advantage that the neutron does not carry an electric charge and so does not experience any electrical repulsive force when approaching the nucleus. However, there are no convenient sources of free neutrons in nature, so neutrons have to be produced by nuclear reactions with charged particles, such as deuterium plus tritium or helium plus beryllium:

$$^{2}_{1}H + {}^{3}_{1}H \rightarrow n + {}^{4}_{2}He \quad \text{or} \quad {}^{4}_{2}He + {}^{9}_{4}Be \rightarrow n + {}^{12}_{6}C$$

Studies in the 1930s used neutrons to create new radioactive products. A discovery was made in 1939 that changed the world. In the course of bombarding uranium with low energy (slow) neutrons (with the intention of producing a more massive nucleus), some barium was found in the product, an element much less massive than uranium. It soon was realized that some of the uranium nuclei had been split. The splitting of the uranium isotope $^{235}_{92}U$ left two lighter products, barium and krypton. This reaction is written as

$$n + {}^{235}_{92}U \rightarrow {}^{93}_{36}Kr + {}^{141}_{56}Ba + 2n$$

Energy is released in this process and is carried off by the products. The loss of mass between the product nuclei and the reactants has been converted into energy.

Another important feature of the fission reaction is that additional neutrons are emitted, which could be used to cause other neighboring uranium nuclei to fission, which in turn gives off neutrons to cause other fissions and release additional energy, and so on. This "chain reaction" overcomes the low-yield problem of other nuclear reactions. One can make almost any nucleus break apart by bombarding it with a projectile of sufficiently high energy, using an accelerator. However, the interesting feature of the uranium reaction is that ^{235}U fissions with the addition of very low energy neutrons. This does not happen for any other naturally occurring isotope.

In Chapter 14 we shall study the details of the fission process and how it can be controlled for practical uses in generating power. In Chapter 16 we shall study the other side of the binding energy curve and discuss the use of fusion for energy production. Since both of these processes have radiation associated with them, we shall study the biological effects of nuclear radiation, and important uses of radioisotopes, in Chapter 15.

I. Summary

In this chapter we have seen that the atom, instead of being indivisible and unchangeable, has a definite structure and is destructible. A simple model of the atom is one in which negative electrons orbit a positively charged nucleus. This picture should not be carried too far, because some of the tenets of modern physics are quite different from those of classical physics. In modern physics, the electron can occupy only discrete energy levels within the atom. We say that the electron's energy is "quantized." The addition of energy to the atom will cause the electron to jump up to a higher energy level; conversely, when an electron drops from one energy level to a lower one, energy is released in the form of visible light, infrared radiation, X-rays, or other radiation.

The nucleus of the atom accounts for more than 99.9% of the atom's mass. An element is characterized by the number of protons present in the nucleus (its atomic number). The combined mass of protons plus neutrons in a nucleus is approximately equal to the atomic mass. Nuclei that have the same number of protons, but differing numbers of neutrons, are called "isotopes" and behave the same chemically. Most elements have many isotopes. The majority of these isotopes are unstable, decaying into other elements with the emission of α, β and/or γ radiations. A radioactive isotope is characterized by the types of radiation emitted and by its half-life—the time it takes for half of the initial amount of the isotope to decay.

Nuclear forces hold the nucleus together. An element's binding energy is the energy required to separate the nucleus into its constituent protons and neutrons. If one considers the binding energy per nucleon, one finds differences between nuclei, with iron being the most tightly bound nucleus.

A radioactive isotope will undergo nuclear disintegration. A natural transmutation is the spontaneous change of one element into another element. Transmutations can also be brought about with the help of artificial nuclear reactions, such as ${}^{6}_{3}\mathrm{Li} + {}^{28}_{14}\mathrm{Si} \rightarrow {}^{2}_{1}\mathrm{H} + {}^{32}_{16}\mathrm{S}$. Fission is an example of a transmutation in which the addition of a neutron to a massive nucleus causes the nucleus to split into two less massive nuclei.

Energy is released in naturally occurring nuclear disintegrations and in many artificial nuclear reactions. (In other cases, energy may be required to initiate the reaction.) In nuclear decay and most reactions, there is a conversion of some mass m of reactants into energy E. This is expressed quantitatively by the equation $E = mc^2$, where c is the velocity of light. The use of this energy will be studied in Chapter 14.

Internet Sites

For an up-to-date list of Internet resources related to the material in this chapter, go to the Harcourt College Publishers website at **http://www.harcourtcollege.com.** The links are in the *Energy: Its Use and the Environment* site on the Physics page.

General energy related sites and some guidelines for using the World Wide Web in your class are on the inside front cover of this book.

References

Chapter 13

Bodansky, D. 1987. *Indoor Radon and Its Hazards.* Seattle, University of Washington Press.

Bodansky, D. 1996. *Nuclear Energy: Principles, Practices, Prospects.* New York, American Institute of Physics.

Cowan, G. 1976. A Natural Fission Reactor. *Scientific American,* 235 (July).

Hewitt, P. 1998. *Conceptual Physics.* 8th ed. New York, Harper Collins.

Serway, R., and J. Faughn. 1999. *College Physics.* 5th ed. Philadelphia, Saunders College Publishing.

QUESTIONS

1. What does the atomic number of an element represent?
2. Dalton showed that the masses of the elements found in a pure compound always occur in the same ratios. For the compound ZnS (used in making fuel for model rockets), what fraction of the mass is contributed by the zinc? How would you distinguish between this compound and mixtures of zinc and sulfur? See Figure 13.15.
3. List four isotopes of oxygen ($Z = 8$). Write down their masses. Which of these might you expect to be stable? (There are three stable ones.)
4. What mechanisms (other than heat from a Bunsen burner) can be used to excite the electrons in an atom to higher energy levels?
5. Describe the source of the red light emitted from a neon sign in terms of the excitation and relaxation of electrons in neon atoms.
6. In the text we used a multilevel parking garage as an analogy to quantization of energy levels. Think of another analogy from everyday life to describe quantization.
7. Why are the atomic masses of most elements not whole numbers?
8. Suppose you have 10 g of a radioactive isotope with a half-life of 5 days. How much of this isotope would have decayed after 20 days?
9. From the following nuclear reaction with heavy ions, find the mass and atomic number of the argon nucleus:

$$^{12}_{6}\text{C} + ^{30}_{14}\text{Si} \rightarrow ^{A}_{Z}\text{Ar} + ^{4}_{2}\text{He}$$

10. The nucleus ^{60}Co decays by the emission of an electron (β^-). (Energetic gamma rays also accompany this decay and are used for the treatment of cancer.) What is the resulting nucleus? (It is stable.)

11. In many cases one radioactive nucleus will decay into another radioactive nucleus, which will in turn decay, and so on. You have to know the radiations emitted not only from the parent but from the daughter nuclei in this decay chain. One typical series is that for uranium-235. The series is long and continues all the way down to a stable lead isotope. For the first three decays in this chain, write the atomic numbers and masses of the daughter nuclei that are formed:

$$^{235}_{92}\text{U} \xrightarrow{\alpha} \text{Th} \xrightarrow{\beta^-} \text{Pa} \xrightarrow{\alpha} \text{Ac}$$

12. Using Figure 13.15, calculate how many neutrons are present in the (stable) lead-206 nucleus.

13. Complete the following nuclear reaction:

$$^{235}_{92}\text{U} + \text{n} \rightarrow {}^{103}_{42}\text{Mo} + ? + 2\text{n}$$

SPECIAL TOPIC

The Periodic Table

After Dalton's hypothesis of the atomic nature of matter and his identification of 20 different elements, many other elements were discovered in the latter part of the 19th century. As the list of known elements grew, and their atomic masses relative to hydrogen were determined, there was a need to categorize this information. It was observed that there were similarities between some of the elements. There appeared to be families of elements that had very similar chemical properties, including the kinds of compounds they could form. One example is the halogens—fluorine (F), chlorine (Cl), bromine (Br), and iodine (I)—nonmetallic elements that are very active chemically. Another family is the alkaline earth metals—beryllium (Be), magnesium (Mg), calcium (Ca), strontium (Sr), and barium (Ba). These elements are shiny and fairly reactive with water, and they form compounds with very similar properties.

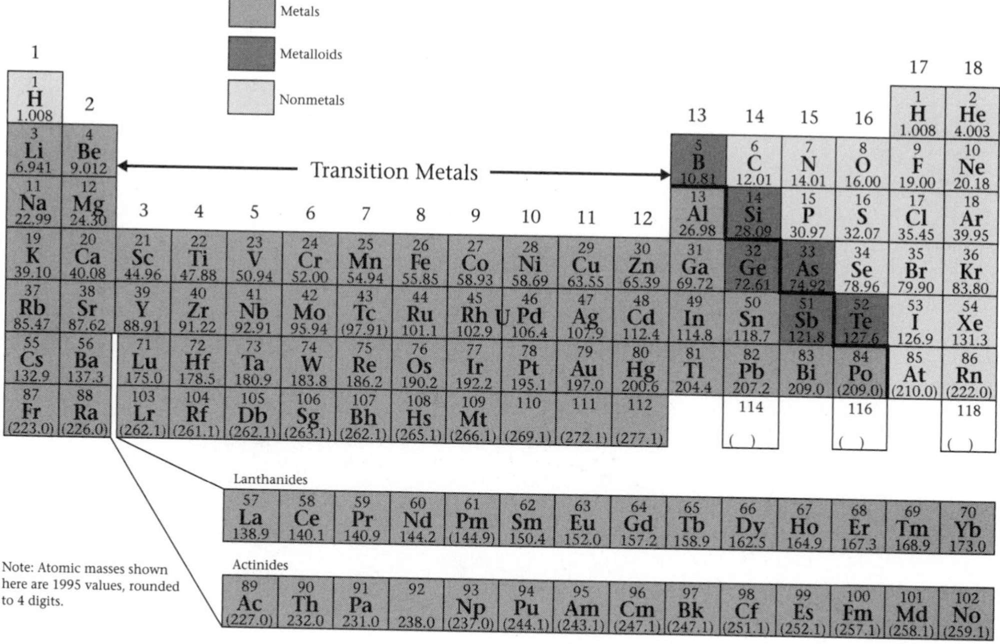

1	2	3	4	5	6	7	8	9	10	11	12	13	14	15	16	17	18
1 H 1.008																1 H 1.008	2 He 4.003
3 Li 6.941	4 Be 9.012											5 B 10.81	6 C 12.01	7 N 14.01	8 O 16.00	9 F 19.00	10 Ne 20.18
11 Na 22.99	12 Mg 24.30											13 Al 26.98	14 Si 28.09	15 P 30.97	16 S 32.07	17 Cl 35.45	18 Ar 39.95
19 K 39.10	20 Ca 40.08	21 Sc 44.96	22 Ti 47.88	23 V 50.94	24 Cr 52.00	25 Mn 54.94	26 Fe 55.85	27 Co 58.93	28 Ni 58.69	29 Cu 63.55	30 Zn 65.39	31 Ga 69.72	32 Ge 72.61	33 As 74.92	34 Se 78.96	35 Br 79.90	36 Kr 83.80
37 Rb 85.47	38 Sr 87.62	39 Y 88.91	40 Zr 91.22	41 Nb 92.91	42 Mo 95.94	43 Tc (97.91)	44 Ru 101.1	45 Rh 102.9	46 Pd 106.4	47 Ag 107.9	48 Cd 112.4	49 In 114.8	50 Sn 118.7	51 Sb 121.8	52 Te 127.6	53 I 126.9	54 Xe 131.3
55 Cs 132.9	56 Ba 137.3	71 Lu 175.0	72 Hf 178.5	73 Ta 180.9	74 W 183.8	75 Re 186.2	76 Os 190.2	77 Ir 192.2	78 Pt 195.1	79 Au 197.0	80 Hg 200.6	81 Tl 204.4	82 Pb 207.2	83 Bi 209.0	84 Po (209.0)	85 At (210.0)	86 Rn (222.0)
87 Fr (223.0)	88 Ra (226.0)	103 Lr (262.1)	104 Rf (261.1)	105 Db (262.1)	106 Sg (263.1)	107 Bh (262.1)	108 Hs (265.1)	109 Mt (266.1)	110 (269.1)	111 (272.1)	112 (277.1)		114 ( )		116 ( )		118 ( )

Lanthanides

57 La 138.9	58 Ce 140.1	59 Pr 140.9	60 Nd 144.2	61 Pm (144.9)	62 Sm 150.4	63 Eu 152.0	64 Gd 157.2	65 Tb 158.9	66 Dy 162.5	67 Ho 164.9	68 Er 167.3	69 Tm 168.9	70 Yb 173.0

Actinides

89 Ac (227.0)	90 Th 232.0	91 Pa 231.0	92 238.0	93 Np (237.0)	94 Pu (244.1)	95 Am (243.1)	96 Cm (247.1)	97 Bk (247.1)	98 Cf (251.1)	99 Es (252.1)	100 Fm (257.1)	101 Md (258.1)	102 No (259.1)

Note: Atomic masses shown here are 1995 values, rounded to 4 digits.

FIGURE 13.15

A modern form of the periodic table of the elements. Elements that behave the same chemically are in columns.

Credit for the classification of elements as we know it today goes to the Russian chemist Dmitri Mendeleev (1834–1907). Mendeleev listed the known elements horizontally in rows by increasing mass, but put elements of similar chemical properties in the same vertical columns. As he arranged the elements into these rows and columns, he left gaps or spaces where an unknown element might be but had not yet been identified. This classification of elements is known as the periodic table, a recent version of which is seen in Figure 13.15. The periodic table is important not only because of the groupings of elements that emerged but also because of its great predictive powers. Many of the gaps in the table were soon filled in, because the properties of the missing elements were predictable on the basis of known information about other members of the same chemical family (those in the same column). Today the mass of the most common isotope of carbon atom is assigned a value of 12.0000 amu. 1 amu is then exactly 1⁄12th the mass of a ^{12}C atom. The mass of hydrogen is 1.0078 amu. (Remember that 1 amu = 1.66×10^{-27} kg.) The periodic table today is arranged in order of increasing atomic number, not atomic mass, but this has resulted in only a handful of position changes. The chemical properties of an element are determined by its atomic number, as discussed in this chapter.

14

Nuclear Power: Fission

A. Introduction

The discovery and use of nuclear fission has been seen both as one of our greatest hopes for a growing, energy-dependent society and as the instrument of our destruction. Today we are well aware of the role nuclear weapons have played in shaping the history and attitudes of the world after World War II. The question that confronts us in this chapter is the role that controlled nuclear power can and should have in our society.

The discovery in 1939 of fission (see Chapter 13) with its release of large amounts of energy was a history-making event. Enormous, but as yet untapped, sources of energy seemed very close, if the technology could be developed. A war was raging in Europe, and so the development of an "atomic" bomb (more properly called a nuclear bomb) was the first goal of those familiar with fission. The splitting of a uranium nucleus yields two fission fragments as well as several neutrons. These neutrons make it possible for fission to continue in other uranium nuclei, rather than being just a single event. The first self-sustaining "chain reaction" was produced in 1942 in a small reactor constructed at the University of Chicago. To build a bomb, substantial amounts of either uranium-235 (^{235}U) or plutonium-239 (^{239}Pu) were necessary, and so methods were developed in the early 1940s for their production. A nuclear reactor was used for producing fissionable ^{239}Pu, while complex physical methods were used for separating the isotope ^{235}U from natural uranium. The "A-bomb" dropped on Hiroshima used almost pure ^{235}U, while plutonium was used in the bomb dropped on Nagasaki.

After the war, many people thought that the use of nuclear energy for peaceful purposes could provide a cornerstone for an energy-dependent economy. Indeed, the abundance of uranium fuel was thought to be large, there were methods for breeding additional fissionable fuel, and the technology was available. The University of Chicago reactor served as a prototype for the development of large reactors, and in 1951 the first electricity was generated from a nuclear reactor called the "Experimental Breeder Reactor," near Detroit. Further developments occurred on several fronts. In 1953 the nuclear-powered submarine *Nautilus* was built, and in 1957 the first commercial reactor to generate electricity was completed in Shippingport, Pennsylvania.

The 1960s were colored by strong optimism on the part of the nuclear industry, and nuclear power was envisioned as providing very cheap electricity compared to that obtainable from coal and oil. It was also thought to be the ideal substitute for oil and natural gas resources that were being depleted, and it was considered to have few environmental problems. However, in the 1970s concern grew over the safety of nuclear power. Although many of the fears were ungrounded, numerous protests surrounded the construction of nuclear plants. Then in March 1979, the first major accident at a commercial U.S. nuclear power plant occurred at the Three Mile Island (TMI) reactor near Harrisburg, Pennsylvania. Although no one was killed and the nuclear reactor safety systems worked, thousands of people were evacuated and even the experts were not sure for days after the initial event whether there would be a massive release of radioactive material.

Three operating nuclear power plants on the shore of Lake Ontario, New York: Nine Mile Point Units One and Two (with the cooling tower) and the James A. FitzPatrick reactor. (NIAGARA MOHAWK POWER CORPORATION)

In April 1986, a much more severe accident occurred at the 1000-MWe Chernobyl nuclear power plant near Kiev in what was then the Soviet Union. An ill-conceived experiment led to a large power surge, which resulted in a steam explosion and fire that virtually destroyed the plant, resulting in the release of massive amounts of radioactivity. More than 100,000 people were evacuated, and the spread of a radioactive cloud over part of northern Europe contaminated food supplies. Although the design of the Chernobyl reactor was significantly different from those of nuclear power plants in other countries, the accident continues to lead to reappraisals of nuclear power safety and emergency planning around the world. (The TMI and Chernobyl accidents are described later in the chapter.)

By the end of the 1990s, the United States had in operation 104 nuclear power plants, with a total output of about 97,000 MWe (Fig. 14.1.). This accounted for about 19% of our total electricity generated and 5% of our total energy output. Regionally, the New England states depend on nuclear power for one third of their electricity. Worldwide, the United States has more operating reactors than any other country. However, some countries currently supply more than 50% of their electricity from fission: France is one of the prime examples, with 77%. South Korea provides about 50% with nuclear, Germany about 30%. By 1999, there were 425 reactors in operation in the world. Twenty-nine reactors are under construction (Table 14.1.).

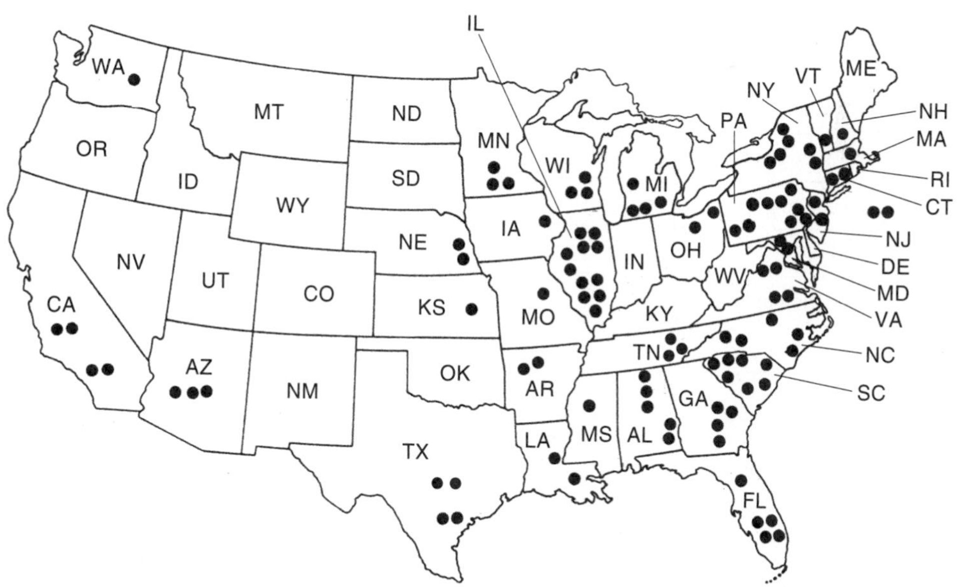

Due to space limitations, symbols do not represent actual locations.

FIGURE 14.1

Operable nuclear power plants in the United States, 1999. (UNITED STATES DEPARTMENT OF ENERGY)

Table 14.1 WORLD NUCLEAR POWER 1999

	In Operation		Under Construction
Country	**Number of Units**	**Total Net MWe**	**Number of Units**
Argentina	2	935	1
Armenia	3	380	
Belgium	7	5700	
Brazil	1	630	1
Bulgaria	6	3520	
Canada	14	10,300	
China	3	2100	4
Czechoslovakia	4	1630	2
Finland	4	2510	
France	58	61,700	1
Germany	19	21,000	
Hungary	4	1730	
India	10	1780	6
Japan	52	43,200	1
Korea, South	14	11,400	6
Lithuania	2	2500	
Mexico	2	1300	
Netherlands	1	450	
Pakistan	1	125	1
Romania	1	630	
Russia	29	19,800	3
Slovakia	5	2030	1
Slovenia	1	620	
South Africa	2	1842	
Spain	9	7400	
Sweden	12	9930	
Switzerland	5	3120	

Taiwan	6	4880	
Ukraine	14	12,150	2
United Kingdom	35	13,000	
United States	104	96,980	
World Total	425	342,390	29

(IAEA)

Today the nuclear power industry is beset with a variety of problems. No new reactors have been ordered in the United States since the TMI incident, and many plants that were on order (including some nearing completion) were canceled. Very large cost overruns (some by factors of 10), long delays, problems in quality control, and operating mishaps undermined public and investor confidence. Economics has been the chief reason for the decline in nuclear power. The last 20 reactors built in the United States cost $3 to $6 billion, or $3000 to $6000/kW. A gas-fired plant costs almost 10 times less, and can be built much faster. Wind turbines can be installed for about $1000/kW. However, in the present era of deregulation and competition in the electric utility industry, stranded cost* bailouts in some states have breathed new life into nuclear power by reducing the utility's capital debt and so strengthening their competitiveness.

The role of nuclear power in the United States is expected to decline precipitously as power plants are retired at the end of their lifetimes. (They have 40-year licenses.) More than half the U.S. nuclear capacity could close by 2020. Included in this forecast are some early retirements, due to the fact that possible major capital investments needed after 30 years of operation might be more costly than building new fossil-fueled capacity.

Worldwide, it is expected that nuclear electrical generating capacity will continue to rise for a few years, and then level off, as economics and public acceptance of nuclear power will affect new construction in the same way as it has in the United States. The Asian economic crisis that began in late 1997 may cause financing problems and delay or cancel orders of nuclear plants. An exception is China, which has ambitious plans to meet rapid growth in its electricity demand. By 2020, they are projected to have four times the current nuclear capacity!

Another country with ambitious plans for nuclear expansion is Japan, which wishes to achieve energy independence. However, the uncertainties in the financial market in Asia, and increased public opposition to nuclear power will affect

---

*Costs that a utility was obligated to pay for plant construction but may not be able to recover from the customer who no longer uses the utility's service.

these plans. A fire and explosion at a reprocessing plant in Tokai in October 1999, although resulting in negligible impacts, undermined public support.

Although western Europe relies heavily upon nuclear power, the trend is away from nuclear; most countries have frozen all new construction. Both Sweden and Germany have voted for the eventual phase-out of all nuclear power plants. The new German government (elected in October 1998) plans to shut down all 19 nuclear plants (which provide more than 30% of the country's electricity) without compensation.

With respect to the environmental problems raised in previous chapters, we need to understand the implications of the risks we are currently taking in all areas. Are we being overly skeptical and overlooking problems with other energy options? How concerned is the public with global warming? What would happen to our energy-dependent society without nuclear power? In this chapter the principles of nuclear physics introduced in the previous chapter will be used to study the construction and operation of nuclear reactors, the nuclear fuel cycle (including the disposal of radioactive wastes), and the potential environmental impact of nuclear power during both normal and abnormal situations.

B. Chain Reactions

As discussed in Chapter 13, any nucleus can be "smashed" or broken apart when bombarded by a nucleon of high enough energy. However, only a few naturally occurring isotopes will fission with the absorption of a neutron of low energy. The most common of these isotopes is ^{235}U, which constitutes only 0.7% of natural uranium. Its capture of a slow-moving neutron (kinetic energy 0.025 eV) to form ^{236}U provides enough extra internal energy to cause the nucleus to split apart into two isotopes of unequal mass (Fig. 14.2).

Uranium-238 will not fission unless it captures neutrons of energies greater than 1 MeV, and then the probability for fission (called the fission cross section) is 2000 times smaller than that for ^{235}U with low energy neutrons. Consequently, only the isotope ^{235}U can be realistically considered a fissionable fuel, and it is called "fissile" material.

One advantage of a nuclear power plant is that it uses only 35 tons of uranium dioxide fuel (which contains about 1 ton of fissile ^{235}U) to provide an output of 1000 MWe of electrical power for a year. One kilogram of ^{235}U, which

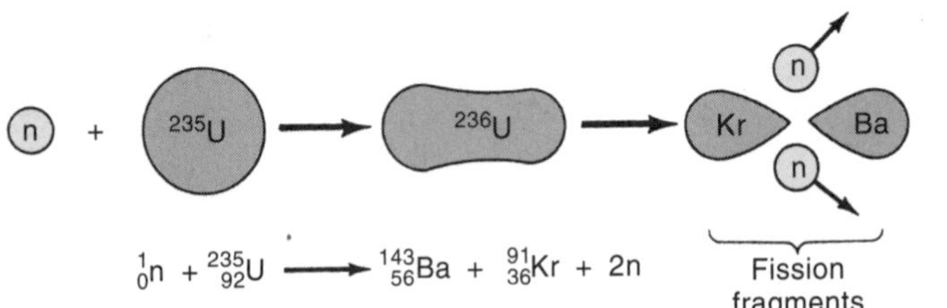

FIGURE 14.2
The fission process. From neutron capture to fission takes about 10^{-5} seconds.

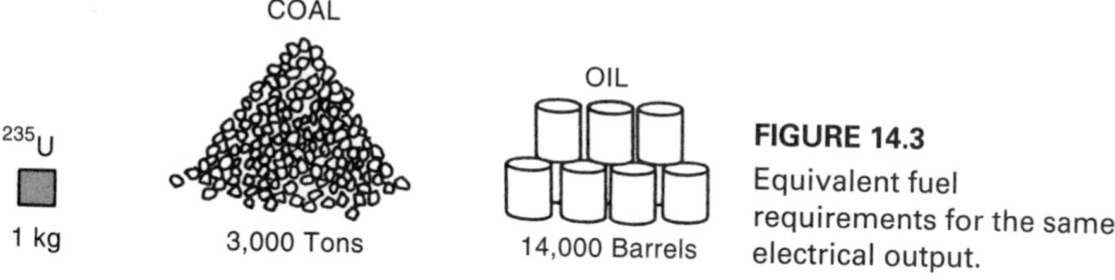

FIGURE 14.3
Equivalent fuel requirements for the same electrical output.

contains 2.6×10^{24} nuclei, can yield 8×10^{10} Btu of energy under complete fission, equivalent to the energy found in 3000 tons of coal or 14,000 barrels of oil. A nuclear plant requires only one shipment of fuel per year, compared to one trainload of coal per day for an equivalent coal-burning plant (Fig. 14.3).

A special feature of fission is the possibility that the process can be self-sustaining; it would be impractical to always have to add an external neutron to the ^{235}U fuel when fission was desired. A typical fission reaction (one of many possibilities) is

$$n + {}^{235}_{92}U \rightarrow {}^{142}_{56}Ba + {}^{91}_{36}Kr + 3n$$

If the reactor is properly designed, the three neutrons that are emitted in this example can cause other uranium nuclei to fission, which will emit other neutrons, and so on, producing a **chain reaction** (Fig. 14.4). A fission reaction will be

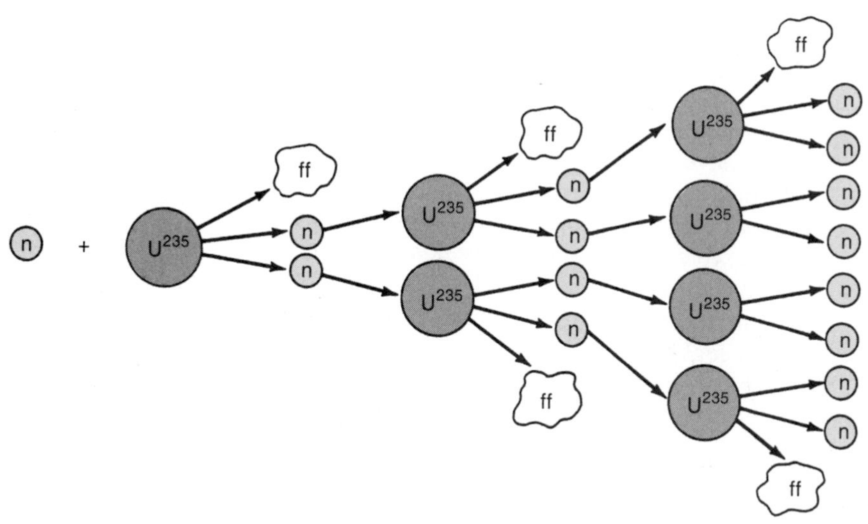

FIGURE 14.4
Schematic of a chain reaction such as would occur in a bomb. The number of fissions increases exponentially as more neutrons become available. Not all the fission fragments (ff) associated with each fission event are shown. In a nuclear reactor, the number of fissions per second is held constant.

ACTIVITY 14.1

A nuclear chain reaction can be simulated by using mousetraps and ping-pong balls. With the mousetraps set and ping-pong balls carefully placed on the latches, drop one ball onto a center trap and observe the action. This experiment works best in a cardboard box. Describe the analogy with a nuclear reactor, including the need for the box in this demonstration.

self-sustaining if at least one of the two to three neutrons given off per fission is captured by another ^{235}U nucleus and causes fission. An illustration of a chain reaction that is self-sustaining is the "dominoes" patterns that can go for hours as one domino after another falls down. Some of the fission neutrons also may be removed by capture in nonfissionable nuclei such as hydrogen and structural material in the reactor core. To ensure that the reaction will be self-sustaining, the fuel used in conventional reactors must be "enriched" to 2% to 3% ^{235}U.

The reaction can be more than self-sustaining. If more than one neutron per fission is captured by ^{235}U nuclei, the chain reaction will grow rapidly (multiply) with an ever increasing power output. This is what happens in a nuclear bomb, in which a significant portion of the bomb's material undergoes fission in a very short period of time. In assembling the material for a bomb, one needs a **critical mass** of fissile material, which allows the number of fission events to increase. The critical mass (about 10 kg for uranium) must be assembled into a sphere, a geometry that allows the smallest number of neutrons to escape. On initiation, the chain reaction occurs almost instantaneously, with a rapid release of energy. In a nuclear reactor, the density of fissionable material (3% ^{235}U) is much smaller than that used for weapons (95%) and so it is impossible for a reactor to explode, even in the event of a meltdown.

The level of power production in a reactor is controlled by **control rods.** Certain materials such as cadmium and boron have a very high cross section for the capture of neutrons. By inserting rods made of these materials at variable distances into the core, the number of free neutrons, and thus the number of fission events, can be controlled so that the desired amount of energy is released. These control rods are inserted from the top or bottom of the reactor vessel and are pulled out when a higher power output is desired. The reactor is shut down (or "scrammed") in an emergency situation by inserting the rods all the way into the reactor core.

C. Assembling a Nuclear Reactor

We now consider the design of a commercial nuclear power plant. Figure 14.5 shows the schematic of the general layout of one such reactor, called a **boiling water reactor** or BWR. The central part of the plant is the reactor "core," in which the uranium fuel is assembled for the generation of steam for the turbine

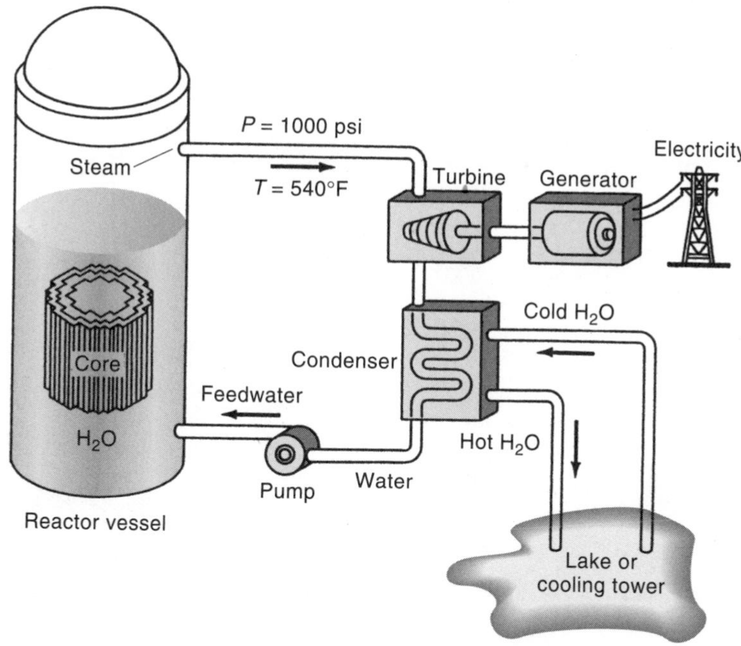

FIGURE 14.5
Block diagram of a boiling water reactor (BWR).

generator. The fuel is in the form of pellets of uranium dioxide, fabricated from uranium ore that has been enriched to about 3% ^{235}U. These pellets (about the size of a marble) are placed in long fuel rods (4-meters long and 1 cm in diameter) made of a zirconium alloy called Zircoloy. The fuel rods are assembled into more than 500 bundles, or assemblies, each bundle containing 50 to 70 rods.

In a typical BWR, there are about 35,000 fuel rods containing about 120 tons of uranium fuel. The bundles of rods are placed in a configuration within the core similar to that shown in Figure 14.6 and covered with water, like bottles or test tubes placed on a sterilizing rack. Between these bundles are from 130 to 180 control rods.

The core and water are contained in a reactor vessel (Fig. 14.7) made of special 6-inch-thick steel plate built to withstand an internal pressure of more than 1000 lb/in.^{2}. A primary containment structure, made of steel and surrounded by reinforced concrete 4- to 6-ft thick, encloses this vessel and much of the primary water cooling loop (Fig. 14.8). This primary containment is located within a large airtight building known as the secondary containment. This structure is composed of several feet of steel-reinforced concrete and is built to absorb an impact as great as that of a jetliner crash. In many cases the secondary containment is the visible exterior of the nuclear plant. A cutaway layout of a BWR and its containments is shown in Figure 14.9.

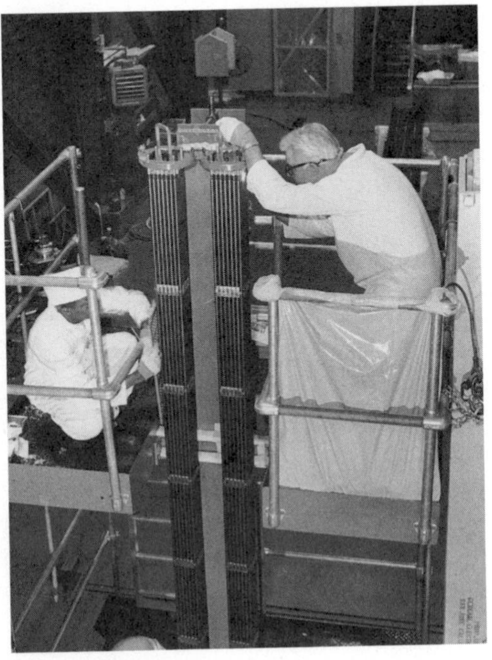

FIGURE 14.6
Inspecting fuel bundles in preparation for loading the reactor core. (NIAGARA MOHAWK POWER CORPORATION)

FIGURE 14.7
Reactor vessel for a BWR. (NEW YORK POWER AUTHORITY)

FIGURE 14.8
Construction of the James A. FitzPatrick Nuclear Power Plant, showing primary containment. The secondary containment is the circular structure. (NEW YORK POWER AUTHORITY)

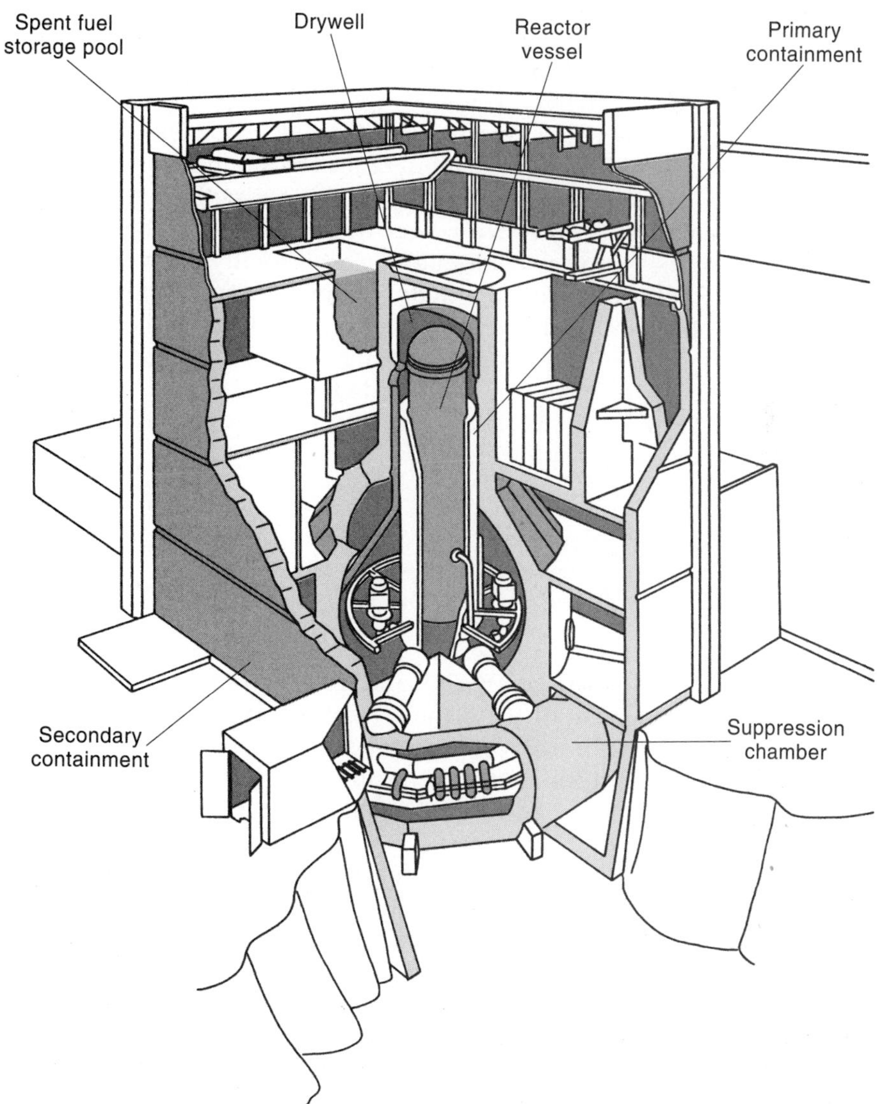

FIGURE 14.9

BWR reactor building, showing both primary and secondary containments. Excess steam would be condensed in the suppression chamber in the event of a loss of cooling water. Note the spent fuel storage pool. (WASH-1250, UNITED STATES DEPARTMENT OF ENERGY)

The water that surrounds the fuel rods in the reactor serves two purposes: (1) to carry away the heat energy produced in the fission process, and (2) to moderate (slow down) the neutrons produced in the fission reaction. The energy liberated in fission heats the fuel rods and the surrounding water, turning it to steam at about 280°C (540°F). The neutrons produced in fission are captured by ^{235}U nuclei in the same fuel rod or escape through the zirconium walls to interact in adjacent fuel rods. The second purpose of water in the core is to act as a **moderator.** The neutrons in the core have a better chance of being captured by ^{235}U and inducing fission if their kinetic energy is small. (The effective power output of the reactor is reduced if the energy of the neutrons is too high.) Consequently, the neutrons are slowed down by collisions with the hydrogen nuclei (protons) of the H_2O, losing much of their original energy. The relative masses of the two colliding objects are important: a tennis ball hitting another tennis ball will lose more of its kinetic energy than if it had collided with a bowling ball. A 2-MeV neutron will slow down to 0.025 eV in about 10^{-5} seconds after (on the average) about 18 collisions with water.

D. Types of Light Water Reactors

There are two common types of conventional light water reactors* or LWRs: boiling water reactors (BWRs) and pressurized water reactors (PWRs). Ninety-nine percent of the commercial nuclear power in the United States, and 80% of that in the world, is produced by these two types of reactors. Sixty percent of them are PWRs, including the one that failed at Three Mile Island. The schematic for a PWR is shown in Figure 14.10. The water in the core is heated to approximately 315°C (600°F), but it is not turned into steam because of the high pressure in the primary loop (2200 lb/in.^{2}, compared to 1000 psi for a BWR). A heat exchanger or "steam generator" is located between the primary loop and a secondary loop; it is used to transfer heat into water in the secondary loop, causing it to turn into steam. The secondary loop acts like the main loop in a BWR; the steam turns the turbine, passes through a condenser, and then is pumped back to the heat exchanger. The radioactivity that is present in the primary loop of the PWR is better contained than in a BWR, since the primary loop never comes into contact with the turbine or the condenser. However, because of the increased pressure, a sturdier reactor vessel is required for a PWR. Fuel data for both these types of reactors are given in Table 14.2.

*LWRs, as used in the United States, use regular water as the coolant/moderator. Canadian reactors use heavy water, or deuterium oxide.

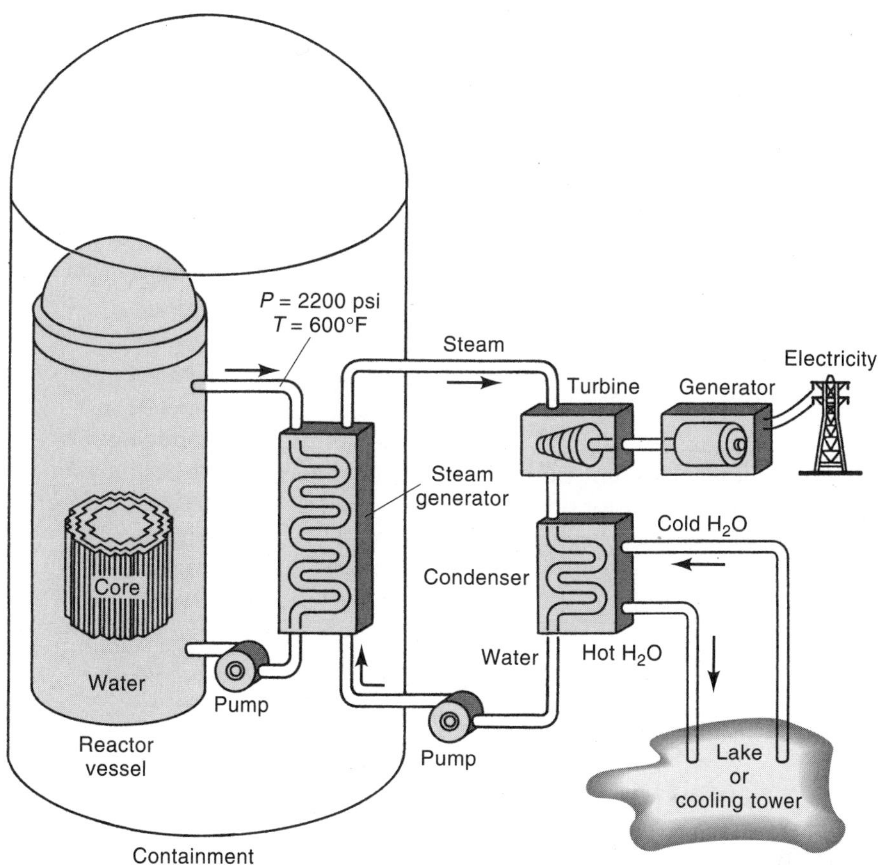

FIGURE 14.10
Block diagram of a pressurized water reactor (PWR).

Table 14.2 LIGHT WATER REACTOR FUEL DATA

	BWR	PWR
Electrical output (MWe)	1000	1000
Initial load (tons of uranium oxide)	135	80
Fuel rods per assembly	50	200
Fuel assemblies per core	750	180
Number of control rods	180	45

(WASH-1250, U.S. Department of Energy)

E. The Nuclear Fuel Cycle

The nuclear fuel cycle involves the physical and chemical processes necessary to produce the fuel used in nuclear reactors and to dispose of waste products and unused fuel (the spent fuel). The "back end" of the fuel cycle, or those processes that occur after the spent fuel has been extracted from the reactor, are particularly important today, and can include reprocessing of the spent fuel to extract the unused uranium and plutonium, and/or the storage of the high-level radioactive wastes. The steps in this cycle, although not all operating at present in the United States, are sketched in Figure 14.11.

Mining

In the "front end" of the fuel cycle, the first step in producing fuel rods for the reactor is the mining of uranium-bearing ores from the earth by methods similar to those used for other metal ores, such as open pit and underground mining. The major portion of U.S. uranium ore is found in Wyoming, Texas, Colorado, New Mexico, and Utah. Uranium mills extract the uranium from the ore by chemical methods and convert it to an oxide form called "yellowcake,"

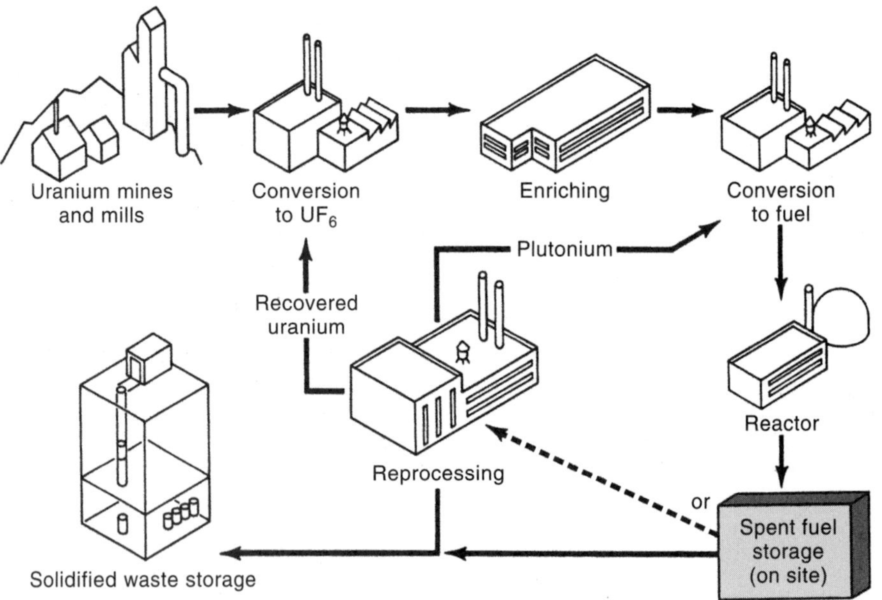

FIGURE 14.11
The nuclear fuel cycle from mining to waste disposal. There are no commercial reprocessing plants operating in the United States today. By law, plutonium cannot be recycled for use as fuel for a light water reactor.

which contains about 70 to 90% U_3O_8. These oxides are then shipped to a plant where enrichment of the ^{235}U takes place.

Fuel Enrichment

Enrichment is a major undertaking in the nuclear fuel cycle. The separation of different elements is not too difficult because of their different chemical properties. However, one cannot use chemical differences between isotopes of the same element since the chemical properties are determined by the electrons surrounding the nucleus, which are identical in number for isotopes. Therefore, physical methods of extraction must be used.

The main method of uranium **enrichment** to 3% ^{235}U is through the process of **gaseous diffusion.** In this method, the uranium oxide is treated with hydrogen fluoride to convert it to uranium hexafluoride (UF_6), which is a gas at high temperatures. The UF_6 gas is forced through a series of thin, porous barriers. The kinetic energy, $\frac{1}{2}mv^2$, of a gas depends only on its temperature. Since the ^{235}U-hexafluoride molecule has a smaller molecular mass than the ^{238}U-hexafluoride, it must have a larger velocity at the same temperature. It will therefore diffuse through the porous barriers at a slightly faster rate. By using many barriers (more than 1000 stages), a 2 to 3% enrichment of ^{235}U can be achieved. This enrichment process is quite energy intensive, requiring an electrical energy input equivalent to about 10% of the net output of the nuclear power plant.

Another enrichment method is with a **centrifuge**. As particles move in a circle, the more massive ones will move toward larger radii. Starting with uranium oxide, the ^{238}U compounds will move to larger radii than the ^{235}U compounds as they are made to spin at very high speeds, and then can be separated out. Centrifuge plants are being developed in the United States and Europe. The centrifuge method has a better uranium separation factor and is less energy-consuming to run than the diffusion method, but requires more precise engineering.

Recent developments in uranium enrichment have used lasers. As opposed to the previous physical methods of extraction, **laser enrichment** makes use of subtle differences in the electronic structures of atoms of different isotopes. As described in Chapter 13 an electron moves into a discrete energy state when it is excited from its ground state, as when an elevator rises to the third floor in a building. The energies of these levels are determined primarily by the number of protons in the nucleus, but there is a small effect as a result of the number of neutrons as well. Thus each isotope of an element will have a slightly different set of energy levels. An isotope can be excited to a particular energy level if it absorbs a photon of the correct energy (or correct wavelength). The project starts by producing an atomic vapor of uranium using an oven. By using a laser, which provides intense, monochromatic (of one wavelength) light, the ^{235}U atoms in the atomic beam can be moved to an excited state while the ^{238}U atoms are left unaffected. The laser light must be very monochromatic: a wavelength 0.1 angstrom

(10^{-10} m) shorter (one part in 60,000) can excite ^{238}U atoms instead. The excited ^{235}U atoms can be ionized by shining ultraviolet light on them, and then collected on an electrically charged device (Fig. 14.12). This enrichment method has an advantage over gaseous diffusion in that very high levels of enrichment (60%) can be achieved in a single step. With fewer stages, the overall cost of this process is expected to be much less. Gaseous diffusion plants are expensive to build and operate; a typical plant can cost \$2 to \$3 billion. Laser enrichment has been demonstrated only with microscopic amounts of ^{235}U, but intense research is in progress to evaluate the technical and economic feasibility of large-scale laser enrichment. (In 1999, technical challenges and economics brought about the closing of the largest demonstration facility in California.) Unfortunately, a disadvantage of this technology is that membership in the world's "nuclear club" might increase because of this simpler and cheaper method to produce fissionable fuel.

After enrichment in a gaseous diffusion plant, the UF_6 is converted to uranium dioxide (UO_2) and then sent to a fuel fabrication plant where ceramic pellets are made and sealed in tubes made of Zircoloy. The loaded tubes or fuel rods are then shipped to the nuclear reactor.

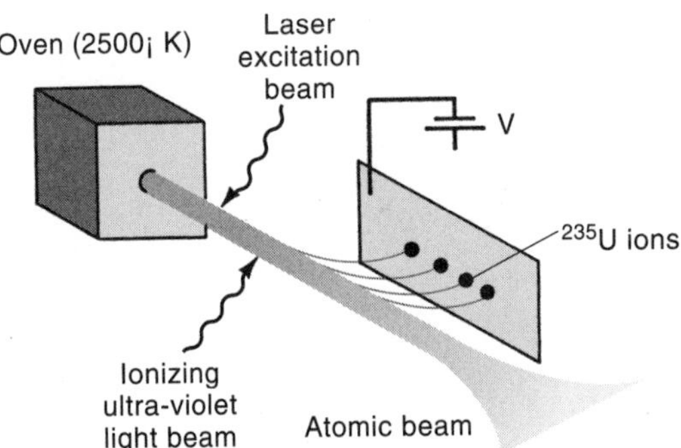

FIGURE 14.12
Schematic of laser enrichment apparatus, using an atomic beam of uranium.

Fuel Reprocessing

After about three years of use in a reactor, the fuel elements must be removed, even though they still contain some unused fissile material. Fuel changes in a reactor are made annually, with about one third (30 tons) of the core being removed each time. These **spent fuel** elements are stored under water in storage pools at the reactor site to let the short-lived radioactive isotopes decay. This storage should be only temporary, but is prolonged today because of U.S. government decisions against reprocessing and the lack of an acceptable plan for the permanent disposal of radioactive waste. Today all the spent fuel that has ever been generated by that reactor is stored on site. Storage pools at some reactors are running out of space, which might cause some reactors to shut down when all their space is filled. Storage on site in dry casks is being considered by several utilities so as to provide additional capacity for storing their spent fuel.

Focus On 14.1

JAPAN AND PLUTONIUM

Japan now provides as much electricity from nuclear reactors as it does from oil. Its goal is to have electricity self-sufficiency. One approach is to extract plutonium from the spent fuel rods and use it with enriched uranium. Plans are now underway to convert 12 of Japan's 52 reactors to run a U/Pu mixture. The plutonium is presently extracted at reprocessing plants in the United Kingdom and France and then shipped back to Japan. Much controversy greeted the first shipment (via ship) of plutonium from France to Japan. Self-sufficient fuel sources are very important in Japan, which has virtually no energy resources of its own, while electricity demand is growing at 4% per year.

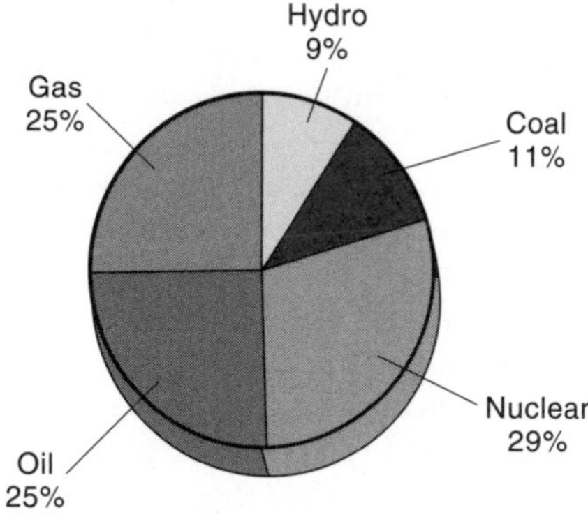

Japan's Electricity Sources: 1995.

The **reprocessing** of reactor fuel to extract unused uranium and plutonium has long been thought to be an integral part of the nuclear fuel cycle program. The plutonium and uranium extracted could be recycled back for use in the plant. However, fears of nuclear proliferation have posed questions about whether reprocessing is worth the risk. The United States now has a federal law prohibiting reprocessing, although a privately owned plant at West Valley, New York, did reprocess fuel between 1966 and 1971. (About 600,000 gallons of radioactive waste remained at this site for many years. The low-level radioactive wastes have been solidified into cement and placed into 20,000 steel drums, and are being shipped to a waste burial facility in Utah. The high-level waste is being solidified into glass logs to be eventually shipped to a permanent burial ground.) Government reprocessing facilities at Savannah River, South Carolina, and Hanford, Washington, have been operating for many years for the production of plutonium and other fuels for nuclear weapons. Many questions dealing with the safety of these plants have been raised in recent years. In Europe, France and England have reprocessing plants and have exported some of this technology to developing nations, a matter of great controversy.

In reprocessing, a series of chemical extraction processes is used to separate uranium and plutonium from the fission products produced in the fuel rods during irradiation, and then from each other. The fuel rods are chopped up into small pieces as they enter the reprocessing plant and then dissolved in nitric acid. The uranium and plutonium are separated from the fission products and other impurities by adding to the radioactive acid an organic solvent in which the plutonium and uranium are more soluble. The organic solvent is made to pass through the acid, dissolving the uranium and plutonium. To separate these two elements, another compound is added that precipitates out the plutonium. The remaining nitric acid solution is then neutralized with sodium hydroxide and stored (temporarily?) in large, double-lined steel underground tanks.

The benefits of reactor fuel reprocessing are (1) the increase (about 30%) in energy available from uranium by the recovery of unused uranium and plutonium; (2) a reduction in the cost of nuclear power; (3) the development of a technology important for the breeder reactor, in which reprocessing is necessary to recover the plutonium that is produced (see Section J, Alternate Reactor Designs); and (4) lessening the problem of radioactive waste disposal by a reduction in the amount of long-lived radioisotopes (as a result of U and Pu extraction).

Debate centers on whether these benefits are small compared to the long-term risks of introducing into the civilian sector a technology from which one product is weapons-grade plutonium. Would this step undercut efforts to limit nuclear proliferation, or can we safely guard the reprocessed fuel with the same security we have developed for other military operations?

F. Radioactive Wastes

One of the most crucial issues facing the nuclear power industry and the federal government today is whether we can develop an acceptable and safe method of isolating radioactive wastes from the environment for thousands of years. This question has been studied for more than 30 years, but to many people it is far from being solved satisfactorily.

The spent fuel elements taken from the reactor after refueling are called **high-level radioactive wastes.** The spent fuel removed from all U.S. nuclear power plants in any year totals about 2000 tons. These high-level wastes are very radioactive and contain many nuclides with half-lives of hundreds or thousands of years. Their health hazards come from both their radiation and their chemical toxicity. Almost all the high-level rad-wastes generated by civilian nuclear plants since their initial operation are now stored on site in large pools of water next to the reactor (see Fig. 14.9). More than 40,000 tons are now stored.

Because many of the radioisotopes have long half-lives, these concentrated wastes will remain thermally hot for a long time because of the radioactive decay heat. Since it is difficult to find conventional materials that will withstand elevated temperatures for a long time (1000 years or more), the long-term storage of high-level wastes presents a difficult problem. A principal hazard is that a leak will develop in the containment structure and that radioactive products will escape into the groundwater and eventually into food and drinking water. Consequently, a disposal site must provide multiple barriers to water or waste movement. The movement of any waste material away from the containment structure must be inhibited by the surrounding rock from reaching the surface for thousands of years.

Not all radioactive wastes are "high level." High level usually refers to those wastes generated in the nuclear reactor fuel cycle. **Low-level radioactive wastes** are generated at educational facilities, hospitals, industries, as well as at nuclear power plants; they include radiochemicals, gloves, papers, contaminated machine parts, and similar items (Table 14.3). These have primarily been buried in shallow trenches in the ground at a number of privately owned sites in Nevada, South Carolina, Utah, and Washington State.

Table 14.3 U.S. LOW-LEVEL RADIOACTIVE WASTES

Source	Waste Form	Percentage of Total Volume	Percentage of Total Radioactivity
Commercial nuclear power plants	Used filters, old equipment, paper, cloth	57	78
Academic and medical	Radioactive liquids used in testing, sealed sources, biological cultures	3	0.1
Industry	Radioactive equipment, machine parts, sources used in testing manufacture of radiopharmaceuticals	37	21
Government	As from industry	3	1

(United States Department of Energy)

There are two categories of high-level radioactive wastes. The first includes fission products radioactive atoms of medium atomic mass. The most important of these are ^{90}Sr and ^{137}Cs; these nuclei are gamma and/or beta emitters with half-lives of about 30 years. The second category of wastes includes those formed not by fission but by neutron absorption in the original uranium fuel. These elements are called actinides (with atomic numbers greater than 88) and are chemically very toxic. ^{239}Pu is the best example of an actinide, with a half-life of 24,000 years. It is formed when ^{238}U captures a neutron (yielding ^{239}U) followed by a series of beta decays to ^{239}Pu. If the spent fuel was reprocessed, 99% of the plutonium would be removed. Table 14.4 lists the most important radioisotopes found in radioactive wastes.

The level of radiation, or activity, from nuclear wastes decreases with time, as shown in Figure 14.13. After about 600 years, the activity has been reduced by more than 10,000 times. A reasonable time for the isolation of high-level waste would be at least 1000 years.

Radioactive wastes come from operation of both commercial and military reactors. The volume of the solid high-level waste generated by a single 1000-MWe plant during one year's operation is about 2 m^3, about the size of an office desk. If an individual's electricity needs were supplied solely by nuclear power plants, the maximum volume of solid waste accumulation per individual in 70 years would be about 300 cm^3, about the size of a soft drink can.

The largest amount of radioactive wastes in the United States today comes from the military defense program. These wastes (about 80 million gallons) are

Table 14.4 RADIOACTIVE WASTES FROM SPENT LWR FUEL

Fission Products		Actinides	
Nuclide	**Half-Life (y)**	**Nuclide**	**Half-Life (y)**
^{90}Sr	28.8	^{237}Np	2.1×10^6
^{99}Tc	2.1×10^5	^{238}Pu	89
^{106}Ru	1.0	^{239}Pu	2.4×10^4
^{125}Sb	2.7	^{240}Pu	6.8×10^3
^{134}Cs	2.1	^{241}Pu	13
^{137}Cs	30	^{242}Pu	3.8×10^5
^{147}Pm	2.6	^{241}Am	458
^{151}Sm	90	^{243}Am	7.6×10^3
^{155}Eu	1.8	^{244}Cm	18.1

(WASH-1250 U.S. Department of Energy)

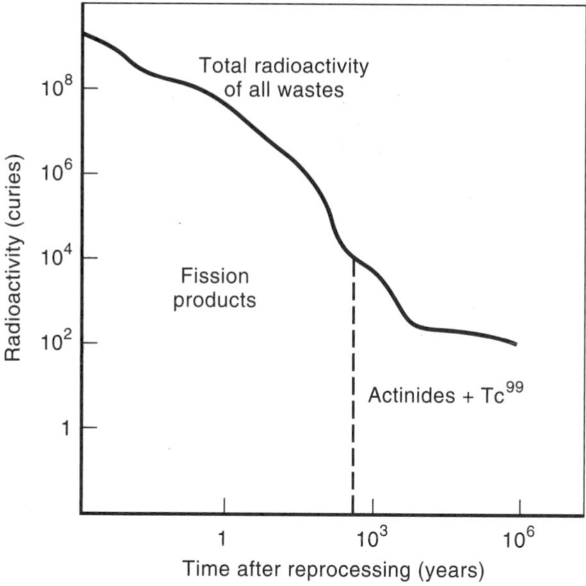

FIGURE 14.13

The total radioactivity of high-level nuclear wastes over time. (J. ZHU AND C. CHAN. RADIOACTIVE WASTE MANAGEMENT. *IAEA BULLETIN*, VOL. 31, 1989)

stored in large underground tanks at Hanford, Washington, and Savannah River, South Carolina. Some storage tanks at Hanford developed leaks in earlier years, with about 500,000 gallons discharged into the ground over a period of years, although with no known human casualties. Methods are being developed now to concentrate the liquid radioactive wastes from military reprocessing facilities into solid form, which will reduce their volume by a factor of 10 and make them easier and safer to store.

The selection of a nuclear waste repository is a very sensitive political issue. Although many experts agree that we now have the technical know-how to isolate such wastes for thousands of years, a "not-in-my-backyard" (NIMBY) sentiment is embraced by many people. The Nuclear Waste Policy Act of 1982 and its Amendment Act of 1987 established a national policy for the management of nuclear waste. A schedule was set for the siting, construction, and operation of a geologic repository. In December 1987, the Yucca Mountain, Nevada, location was selected by the U.S. Congress as the candidate repository site (Fig. 14.14). The site is about 70 miles northwest of Las Vegas.

The Department of Energy has begun a site characterization study of Yucca Mountain. This investigation will last several more years at a cost of about $6 billion, and is focusing on earthquakes, volcanoes, water movement through the rock, and performance over time of the canisters containing the waste. Present work is proceeding on a 14-mile-long tunnel in the side of Yucca

FIGURE 14.14
Yucca Mountain, Nevada: An artist's concept of the handling complex at the proposed site for the first geologic repository for high-level radioactive waste in the United States. (UNITED STATES DEPARTMENT OF ENERGY)

Mountain to investigate water movement in the rock and carry out seismological studies. If at any time during its studies the Department of Energy (DOE) finds Yucca Mountain is not suitable for a high-level waste repository, site characterization will stop and the site will be returned to its natural state.

If the site proves favorable, the DOE will recommend to the president that the repository be built. If the president approves, the DOE will apply to the Nuclear Regulatory Commission (NRC) for a license to build it. At this point the state of Nevada can submit a "notice of disapproval," which can be overruled only by a majority vote of Congress. The NRC review of the DOE's application is expected to take at least three years. If approved, construction could begin by 2010 and the facility could begin accepting spent fuel several years later.

The rock formations at the Yucca Mountain site consist of "tuff," which is a dense form of volcanic ash, produced more than 13 million years ago. The site

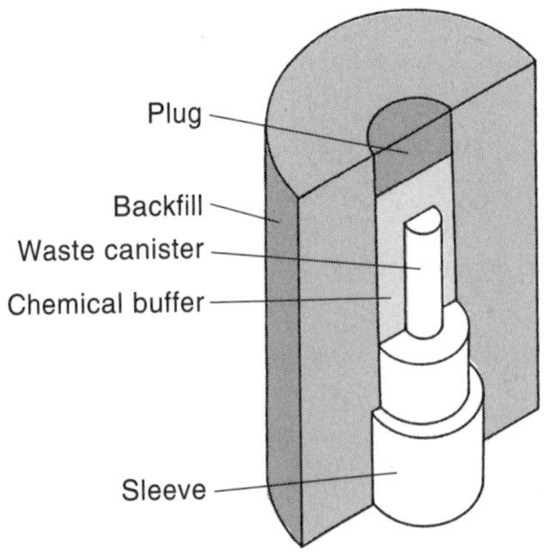

FIGURE 14.15
Possible container for storing radioactive wastes.

is in a very dry area (less than 6 in. per year of rainfall) and the water table is 1700 ft deep. The groundwater travels laterally only about a mile in 3400 to 8300 years; the nearest discharge point into surface waters is 30 miles away. Tuff also has the advantage that it can trap by adsorption (i.e., collecting on its surface) any radionuclides within the rock that might leak from the containers. In addition, the federal government already owns the site.

For most permanent isolation disposal schemes, a multiple-barrier approach for containment will be used (Fig. 14.15). The spent fuel from commercial power plants is in the form of metal tubes bundled together as fuel assemblies. The solidified high-level defense wastes will be first encapsulated in ceramic or glass containers (which can withstand the decay heat). These wastes will be placed inside stainless-steel canisters for placement underground in a large cave. The repository rooms will be surrounded by backfill materials to retard the penetration of water. These storage rooms are located in stable geologic formations, the third component of the multiple-barrier system.

Additional options proposed for the final management of high-level radioactive wastes are (1) isolation in other geologically stable regions, and (2) total elimination. Figure 14.16 summarizes these options. Representative of the

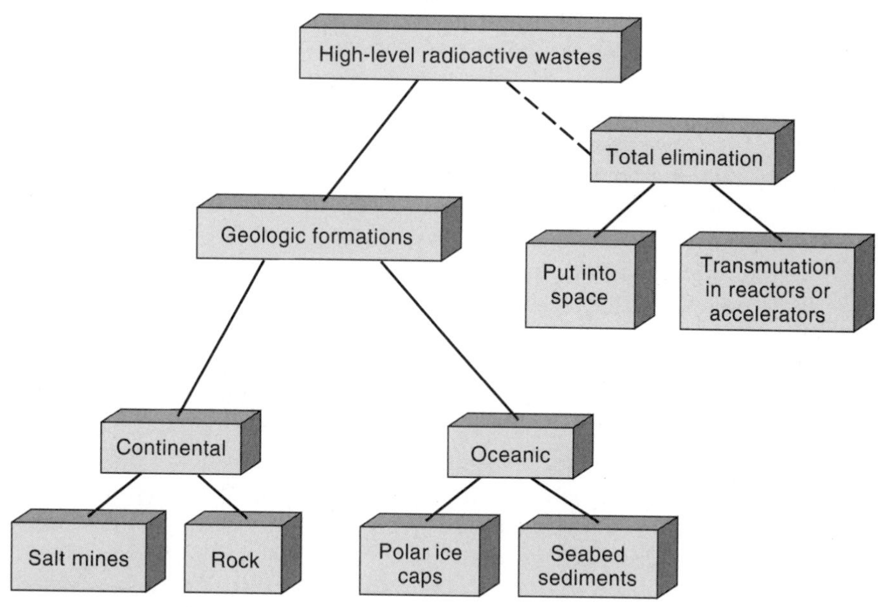

FIGURE 14.16

Summary of radioactive waste disposal options. Possible management and disposal technologies for high-level wastes are diagrammed. The total elimination schemes are not likely to be pursued. Granite or salt formations can be developed for either retrievable or nonretrievable disposal.

first alternative is disposal in other geologic formations such as salt mines, seabeds, or the polar ice caps. The second alternative includes transmutation or disposal into outer space. Although the United States is primarily pursuing disposal in rock, the other methods bear consideration since they are options for other countries.

One method that received much attention in the United States for many years, and presently is in use in other countries, is the burial of wastes in salt beds. This type of storage has the advantage that salt mines are free of water (and have been for millions of years), are self-sealing and so are extremely stable with respect to earthquakes, and will flow (from the waste decay heat) to plastically seal the wastes. Salt is also a good conductor of heat.

Another proposed disposal method uses the ice sheets in the Antarctic. This would be done by placing the waste containers on the ice surface and allowing the heat of the containers to melt shafts into the ice, through which the containers would sink to the ice–rock interface for permanent storage. The remoteness of the area has the advantage of preventing intrusion by people. However, there are questions regarding the monitoring of the wastes and the lifetime of the ice caps.

A third method of storage that removes the danger of accidental human intervention is seabed disposal. This is the controlled placement of the sealed wastes on the floor of the ocean, not the dumping of wastes as has been done by some countries. The deep troughs (20,000 ft under the ocean surface) are thought to offer stable geological regions, far from active tectonic plates. If there is a leak, the ocean water may provide adequate dilution of the radioactive material. Still, such disposal jeopardizes the important resources available to us from the ocean, including food and metals. At the moment, much more study remains on this option, including a study of the effects of radioactive wastes on the surrounding aquatic life and the technology of placing these wastes at such great depths.

Elimination of radioactive wastes in principle can be accomplished by transmutation—that is, changing the nature of the radioisotope by bombarding it with neutrons from a reactor or charged particles from an accelerator. Unfortunately, most of the nuclear reactions have low cross sections, and the separation of certain elements might be necessary to prevent the formation of other radioisotopes of similarly large half-lives. Further, it has been shown that all reactions induced by bombarding particles other than neutrons are impractical, as they will consume more power than was generated in the production of the wastes. The other method of elimination, rocketing the wastes into the sun, has many safety concerns.

Every country seems to be pursuing its own path in nuclear waste management (Fig. 14.17). Most countries with nuclear reactors have interim storage facilities but only ongoing research for permanent disposal, although most involve some form of underground burial. Most programs assume reprocessing. Germany has the only operating geologic disposal facility in western Europe, using the Asse salt mines. Their radioactive wastes after reprocessing

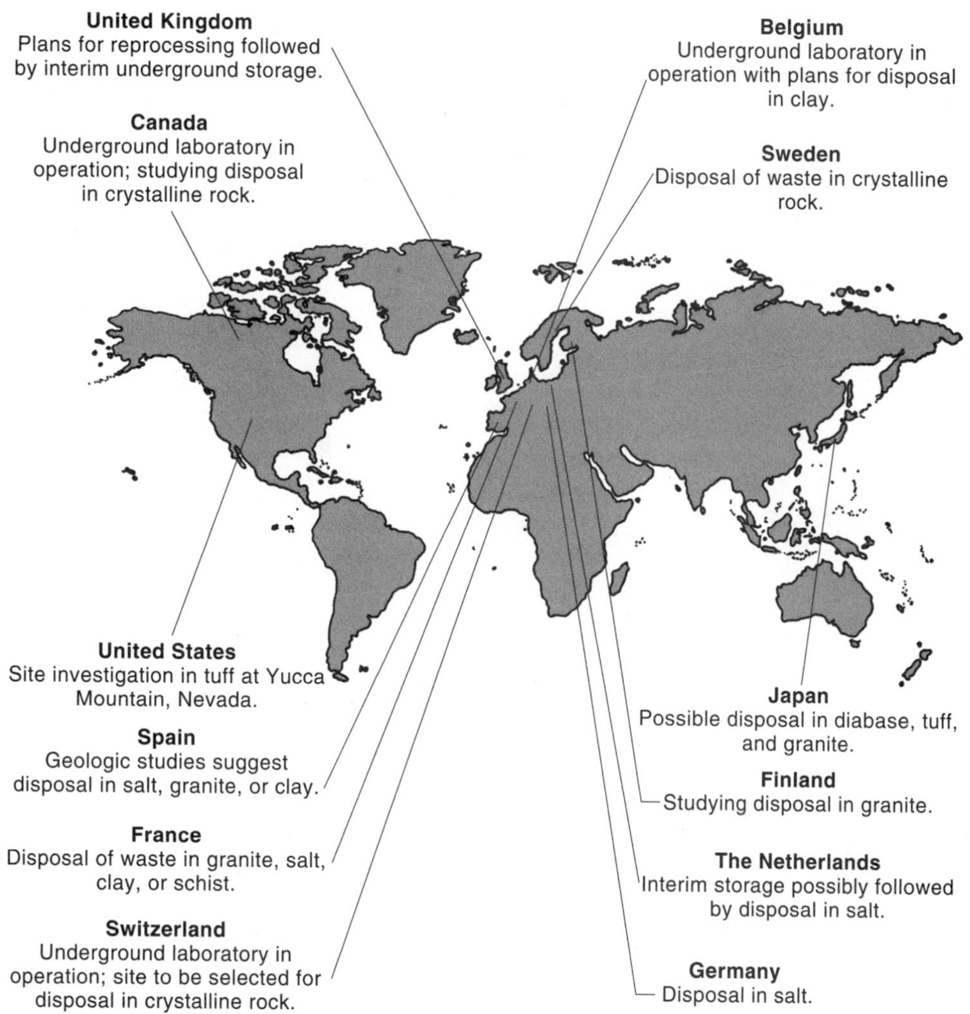

FIGURE 14.17

International nuclear waste disposal programs. (UNITED STATES DEPARTMENT OF ENERGY)

are stored as liquids for a limited period of time, then solidified and chemically bonded into glass. France uses a similar encapsulation process and stores the glass blocks in a surface facility. They are considering disposal in granite rock. In Britain, liquid high-level wastes are stored in tanks, and final burial sites are still being researched. Canada, with its large resources of uranium, will probably not reprocess; granite formations in Manitoba are being examined as a repository.

Focus On 14.2

MONUMENTS FOR THE FUTURE

The high-level radioactive wastes from civilian and military nuclear reactors must be isolated from the public for at least 10,000 years. When a place or places are selected and the burial completed, warning signs must be developed to keep people from disturbing the material. Unknowingly, miners might interfere when looking for water or minerals. But how do you communicate a warning for such a long time—over 300 generations? It's likely that no language in use today will still be spoken. Climatic change might alter the surrounding landscape drastically. What type of monuments can be counted on to last—safe from souvenir takers as well as nature? One suggestion is the construction of a landscape of large stone thorns to mark the area. Another approach is the encapsulation of the waste in a heavy monolith of glass—to prevent entry even if someone uncovered the site. What do you think?

Warnings for future generations.

Whatever the repository selected, the waste material from the spent fuel pools at each power plant site will be transported to the repository in solid form in armored casks. The casks have been designed and tested to withstand abnormal accidents without a release of radioactive material. In one full-scale test, a locomotive was deliberately crashed broadside at 80 mph into a cask; the locomotive was demolished but the cask was only slightly dented. Still, critics fear accidents can spread long-lived radiation over a large area, and that many communities are ill-prepared to handle such a catastrophe. The costs of geologic disposal methods are expected to be relatively small, adding about 1% to the cost of nuclear power. A fee of ⅒¢/kWh is now collected by the federal government from all utilities operating nuclear plants for financing the repository.

G. Decommissioning

Another type of radioactive waste that must be dealt with is the nuclear power plant itself. Eventually the power plant will reach the end of its useful life. (An operating license usually is for 35 to 40 years.) The closing down of a power plant for good is known as "decommissioning." The plant will contain both activated material, that is, material that was made radioactive from neutron bombardment during fission, and contaminated material, that is, radioactive material deposited on a nonradioactive surface. Radioactive contamination can be removed from surfaces by washing, while activated products must be allowed to decay.

In the United States since 1960, about 20 test, demonstration, and power reactors have been or are in the process of being decommissioned. About 10 more worldwide have been decommissioned. Most of these reactors have been small, but many large nuclear plants are due for closure in the next decade, earlier than their predicted 40-year life span. This is due to increased competition brought about by the deregulation of the electric industry. Utilities now are required to set aside funds regularly toward eventual decommissioning. However, early retirement of a reactor may result in a lack of sufficient funds. A recent example is the Trojan reactor in Oregon. After 17 years of operation, this was closed in 1993 by the utility, which deemed it not prudent to spend more money on maintenance. In setting aside funds for decommissioning, the utility had managed to accrue only $40 million for a task that is now estimated to cost 10 times as much. In 1999, the Trojan reactor vessel was shipped to Hanford, Washington.

Three decommissioning options are available: immediate dismantling, mothballing, and entombment. After 25 years of operation, the first commercial nuclear power plant at Shippingport, Pennsylvania (72 MWe) was dismantled in 1985. By mothballing a plant, radioactive levels are allowed to decline to acceptable levels so that dismantling can proceed. Entombment encases the plant in concrete with appropriate surveillance afterwards until radioactivity decays to background levels. The contaminated sections of the plant that are removed in the first two options will be shipped to a low-level waste facility. The radioactive debris will include reactor cores, concrete supports, and metal piping and valves, averaging some 160,000 ft^3. The states are responsible for this low-level waste. The high-level wastes such as the spent fuel await a permanent disposal site.

H. Radioactive Releases

One of the prime concerns of the public, the government, and industry since the beginning of peaceful uses of nuclear energy in the 1950s has been the release of radioactivity associated with the operation of a nuclear power plant during both normal and abnormal conditions. We will consider this question first for the routine operation of a nuclear power plant, and then investigate catastrophic accidents. Since the biological effects of radiation will be covered in the next chapter, only the sources of that radiation will be considered here. This area has been the subject of numerous articles and books arguing both the safety and the dangers of nuclear power plants. Many of the statements made are purely subjective from a social, moral, or economic viewpoint without an understanding of the physical principles. Certainly all these issues must be weighed, since technology does not operate in a vacuum. Eventually a comprehensive view of the entire energy situation will have to be made to form opinions on the merits of nuclear energy.

There are about 10 billion curies of radioactive material within an operating reactor, so reactor safety is a crucial issue. More than $100 million per year in the United States is spent on safety for conventional light water reactors. Many precautions are taken to prevent accidents, including the design, inspection, testing, and maintenance of reactors that would take too much space to discuss here. The plants are designed with multiple safety systems to guard against postulated accidents. The current safety record is excellent: In more than 1000 reactor-years of U.S. operation (plus 2000 reactor-years for U.S. Navy submarines and ships), there

Control room of FitzPatrick Nuclear Power Plant, Oswego, New York. Note the layout of the reactor core under the clock. (NEW YORK POWER AUTHORITY)

have been no direct radiation-related deaths among the general public. However, controversy still exists about the long-term effects of low-level radiation on human health in the normal operation of a reactor (as well as from radioactive waste disposal). These are discussed in more detail in the sections ahead.

Radiation From the Reactor Building: Normal Operation

As shown in Figure 14.5, the steam in a BWR leaves the core and enters the turbine and then the condenser before returning to the reactor vessel. Small quantities of radioactive fission products (solids and gases) can enter the primary coolant through tiny leaks in the fuel-rod cladding. The radioactive fission product gases are primarily krypton, xenon, and vaporized iodine. When the gases leak from the fuel rods, they enter the steam cycle and pass through the turbine to the condenser. (The solids are removed from the water with filters.) The radioisotopes ^{85}Kr, ^{131}I, and ^{133}Xe are gases and cannot be condensed back to liquids, so they are vented into the atmosphere. The gases generally are passed through filters and activated charcoal absorbers, where they are held up for several half-lives before they are released through tall stacks into the environment; there they are subject to normal atmospheric dispersion.

The problem of radioactive emissions is less difficult for a PWR, because the primary coolant never comes into contact with the turbine and condenser. A coolant water purification unit removes the low-level radioactive material from the primary loop and converts it into solids for shipment to a low-level waste disposal facility. The radioactive gases are held up for many half-lives in a large tank and then released to the atmosphere. There are fewer of these gaseous wastes in a PWR so they can be held up longer, resulting in lower radioactive emissions than from a BWR.

Liquid waste emissions from a power plant come from leaks in the coolant piping system of the water purification system. These are collected, chemically treated, temporarily stored, or held up, and then released. The amount of such leakage is usually small, but its presence is one of the reasons for installing the reactor vessel in a special building or containment capable of holding the radioactive material in case of a leak. This was shown in Figure 14.9.

Standards for the maximum allowable radiation exposure for the general public are set by the Environmental Protection Agency. They are based on studies of genetic effects of radiation, although at much higher exposure levels than we would normally encounter. The current limit for the general public is 170 millirem/y above the background radiation level,* which itself averages about 100 mrem/y* (excluding radon). In response to controversy about whether

*A **rem** is a unit of radiation dose; it takes into account the energy deposited per gram of tissue and the biological effectiveness of the type of radiation received (see Chapter 15). A **millirem,** abbreviated mrem, is one thousandth of a rem.

these limits were too high, the Atomic Energy Commission (AEC, the predecessor of today's NRC) proposed in the early 1970s that the routine emission of radioactivity from liquid and gaseous effluents of nuclear power plants should be "as low as practical" and that they should cause exposure no greater than 5 mrem/yr for any individual living as close as the plant boundary. These standards must be met by all operating reactors, and actual exposures are generally less than this.

Catastrophic Accidents

POSSIBILITY OF THE REACTOR EXPLODING LIKE A BOMB. One of the original concerns of the public over the development of nuclear power was the belief that a nuclear reactor might explode like an atomic bomb, with the release of large quantities of radioactivity. This concern is now known to have no basis in fact.

In the activation of a fission bomb, two subcritical masses of very pure ^{235}U or ^{239}Pu are forced together by conventional explosives, and fission is initiated by the addition of neutrons. Since there is such a high concentration of fissionable material, the chain reaction proceeds very rapidly with a multiplication of the number of fissions, and a large amount of energy is generated in a very short time. A vast amount of heat energy is produced, giving rise to a "fireball" as the surrounding gases are ionized. Also, a shock wave or pressure wave is formed, causing a blast that can damage buildings many miles away.

In a nuclear reactor, the fuel is much less rich in fissionable material than in a bomb (3% compared to almost 95% ^{235}U), and the chain reaction proceeds only in a slow, controlled manner. Recall that slow neutrons are better absorbed in the ^{235}U capture process. Any increase in the temperature of the core will decrease the density of the moderator, increase the percentage of fast neutrons, and thus slow down the chain reaction. This is a built-in safety mechanism of the LWR. Even in the unlikely event that the core itself melts, the density of fissionable material will be much less than in the fuel rods because the metal cladding will be added to the "heap." Thus a nuclear explosion is impossible.

LOSS OF COOLANT ACCIDENT. One of the major safety concerns today is the possibility of a meltdown of the reactor core as a result of a loss of coolant accident (LOCA) and the release of radioactive material into the environment. This occurred to some extent at Three Mile Island in 1979. The water that surrounds the fuel elements in the core not only moderates the energy of the neutrons but also carries away the heat produced in the fission process. If there is a rupture or break in the incoming or outgoing waterline, the core will start to increase in temperature because of the decay heat (the energy associated with ongoing radioactive decay) of the fission products, even if the reactor is scrammed by the complete insertion of the control rods. The worst possible accident is labeled a "double-ended guillotine break" of the largest pipe in the coolant loop, which would cause the reactor core to lose its water suddenly. Under normal operating conditions, the UO_2 fuel pellets are at an average temperature of about

1094°C (2000°F), while the Zircoloy cladding is about 343°C (650°F) because of the cooling. With the removal of water in a LOCA, the cladding will rise to an equilibrium temperature with the fuel pellets. Unless an emergency cooling system is activated, the core temperature will continue to rise. At about 1205°C (2200°F), the cladding starts to melt, in the absence of any cooling it might take less than one minute to reach this point. At higher temperatures the cladding will react with the steam to form hydrogen, providing even more heat and explosive gas. This was the source of the "hydrogen bubble" at Three Mile Island, which will be described in the next subsection.

If the fuel rods completely melted because of a LOCA, they would form a "heap" at the bottom of the reactor vessel. Over a period from a few minutes to a few hours, the heap might melt through the 6-in. steel plate of the vessel and fall to the concrete bottom of the containment. The melting point of concrete is 500°C (930°F), so there is a small chance that enough heat could be generated to melt through this 6-ft-thick floor. This might take from hours to days after a pipe rupture. The radioactive heap would then move into the earth below and contaminate the groundwater. This possible chain of events has been called the "China Syndrome," in honor of the country situated on the other side of the earth from the United States. (The impossible scenario in this case is that the radioactive heap would melt through the center of the earth to China.)

A more serious accident with wider effects could occur in the less likely chance that the reactor vessel cracked open as a result of the increased gas pressures within. If the concrete containment buildings were breached also, the radioactive debris would be released to the atmosphere. (This is what occurred at Chernobyl.) Appropriate wind conditions could carry this material into densely populated areas, resulting in more fatalities.

Because of the (very small) possibility of a break in the cooling lines, engineered safety systems have been installed in all reactors to provide emergency cooling of the core in the event of an accident. This **emergency core cooling system** (ECCS) consists of tanks of water under high pressure that are ready to spray water into the core when triggered by an increase in core temperature. The water would turn to steam and carry away the decay heat. Other precautions to prevent a major LOCA are the use of high-quality materials in the piping and reactor vessel and routine inspections to look for small leaks in the cooling system. Tests of the ECCS system have been run at a scale model of a commercial PWR in Idaho. Such loss-of-fluid tests showed that the ECCS performed better than had been estimated.

Nuclear Reactor Accidents: Some Examples

Many incidents have occurred in the past 40 years at civilian nuclear reactor power plants, though very few have injured plant workers and none have caused direct fatalities to the general public. It is important that the lessons learned from these accidents are incorporated into making existing plants safer and in modifying future reactor designs.

THREE MILE ISLAND. On March 28, 1979, the worst accident at a commercial U.S. nuclear power plant occurred at the Three Mile Island (TMI) pressurized water reactor near Harrisburg, Pennsylvania. The incident began at 4 AM (with the reactor running at full power) when a feedwater pump stopped working. According to procedure, an auxiliary pump was started and the reactor was scrammed. However, the pressure in the reactor vessel started to rise because the heat removal in the steam generators was not at the required rate. To compensate for this situation, a pressure relief valve on the vessel was activated to vent some steam. However, this valve failed to close when the pressure returned to normal. Also, in the secondary loop (see Fig. 14.10), no feedwater was reaching the steam generator because a valve between the auxiliary pump and the generator was accidentally closed, and the warning light in the control room was obscured by a tag. The reactor primary coolant circuit continued to blow off radioactive water and steam through the relief valve into the containment building. The ECCS was triggered, but then partially closed by the operator. Not until eight minutes later was it fully opened. Radioactive water from the reactor vessel continued to be emptied into the building and was automatically pumped into an auxiliary building. The decay heat of the fuel rods continued to evaporate the water in the vessel, leading to significant core damage—a core meltdown.

After about two hours, the relief valve was finally closed, but a substantial fraction of the core was uncovered. The high temperature attained in the core before the ECCS was activated caused damage to the fuel rods, which released fission fragments into the vessel and the reactor building. At these high temperatures, the steam reacted with the Zircoloy cladding of the fuel rods to produce hydrogen gas, which formed a hydrogen bubble at the top of the reactor vessel. The bubble remained for several days; it caused a good deal of consternation on the part of the nuclear experts, who feared that a hydrogen gas explosion might occur, rupturing the reactor vessel. (In retrospect this danger was not as bad as had been imagined.) There certainly was a partial meltdown of the core, but safety systems seemed to function. Some radioactive gases were released into the atmosphere in the first few days; one additional death from cancer in the general public is expected from the increased radiation dosage—an estimated 2 mrem per person. The direct cause of the accident was determined to be operator error. As an outgrowth of the TMI accident, many changes were made in operating procedures and training in the nuclear industry.

CHERNOBYL. On April 26, 1986, the worst accident in the history of nuclear power occurred at the Chernobyl Unit 4 Reactor in the southwestern Soviet Union, in what is now Ukraine. The accident resulted in 31 immediate fatalities, the hospitalization of hundreds of people, and contamination of crops and water across western Europe. Although no one from the general public was killed as a direct result of the tragedy (the casualties as a result of acute radiation exposure and thermal burns were among the plant personnel and firefighters), it is estimated that fallout from Chernobyl will cause roughly 47,000 extra cancer deaths worldwide in the next 50 years. Since records are lacking and cancers caused by radiation can take 10 years before they become detectable, this number is speculative and based upon estimates for average radiation doses and their effects. This mortality estimate represents a small percentage increase in natural or spontaneous fatal cancers in the same area. Outside the 30-km zone surrounding Chernobyl, the incremental increase in fatal cancer risk is estimated to be about 0.01%, which is not detectable. What was most observable medically was the major increase in thyroid cancer in children. Certainly one of the largest consequences of Chernobyl has been its effect on the public perception of nuclear power and related safety issues. "The Chernobyl accident demonstrates vividly that nuclear safety is truly a global issue. We would be remiss if we ignored some of the accident's broader issues that transcend the design differences. In a very real sense, we are all hostage to each other's performance." (J. K. Asselstine, NRC member).

The Chernobyl Unit 4 employs an RBMK reactor design, a design significantly different from reactors in any other part of the world. The reactor had an output of 1000 MWe and was completed in 1983. The RBMK's fuel rods are located in separate pressure tubes placed in holes within a graphite moderator block (Fig. 14.18). Water passes through the tubes and then to steam generators

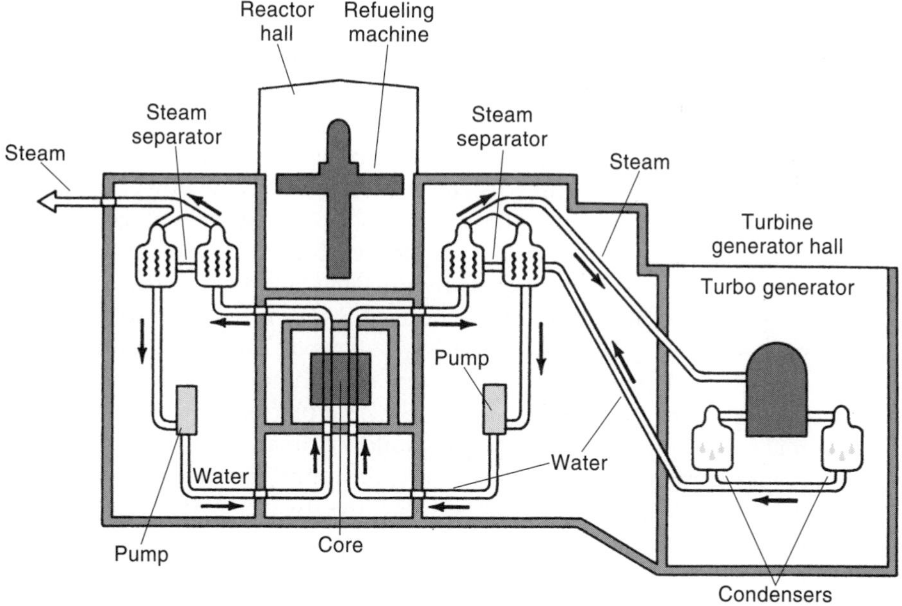

FIGURE 14.18

Schematic of the RBMK reactor at Chernobyl. The core contains bundles of fuel rods placed in a graphite moderator block. Water rising through the tubes in the block is converted into steam. The containment over the top of the core was weak. *(EPRI JOURNAL)*

FIGURE 14.19
Aerial view of Chernobyl's damaged Reactor Number 4. The roof was blown off the plant, unleashing a cloud of radioactive material from the reactor core. (CORBIS/BETTMANN)

(like the PWRs). Although the reactor incorporates a number of barriers between the radioactivity and the environment, there was not an adequate containment vessel to withstand an explosion within the core.

On April 26, the Unit 4 reactor was running at low power to enable the operators to conduct (unauthorized) tests of the electrical generator. Several emergency systems were shut off for the tests. At low power levels, the RBMK reactor design becomes extremely unstable. Suddenly, within seconds, the reactor's power increased to 100 times maximum. The control rods could only be inserted slowly, and, with safety systems off, the temperatures rose rapidly. The fuel rods melted and generated a vast amount of steam, which blew off the top of the reactor (Fig. 14.19). Burning blocks of graphite and radioactive fuel were thrown into the reactor building, finally blowing off its roof and allowing a radioactive plume from the graphite fire to rise 5000 m into the atmosphere, where it was carried northwestward. The explosion released 10 times the amount of radiation of the bomb dropped on Hiroshima, about 100 million curies.

The radioactive cloud from Chernobyl (containing ^{137}Cs and ^{131}I) produced fallout in erratic patterns throughout western Europe (Fig. 14.20). Locations in Sweden measured radiation levels 100 times background, while towns around Chernobyl received doses not much larger. Not until 36 hours after the accident was evacuation ordered for the town of Pripyat (5 km downwind), population 45,000. Altogether, about 160,000 people were evacuated from the surrounding

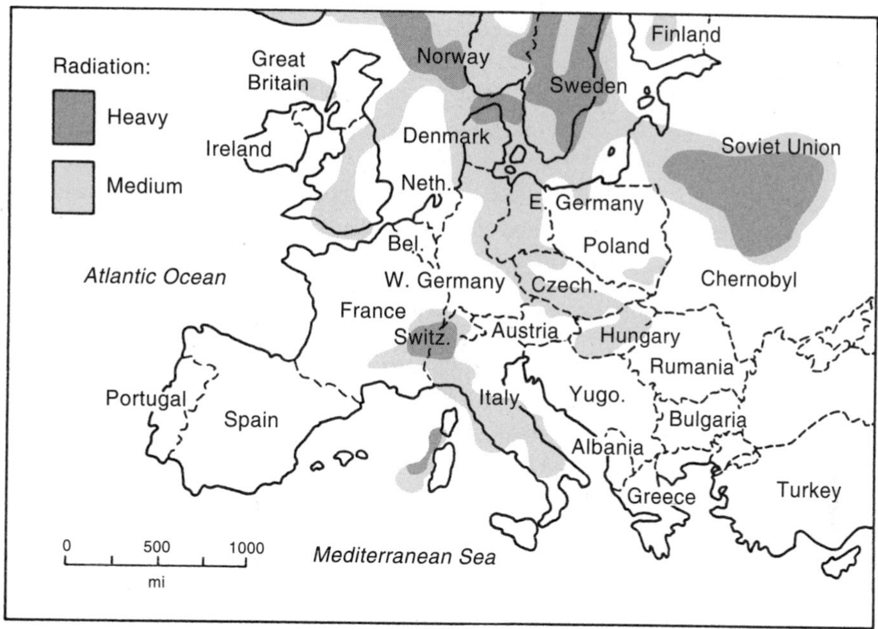

FIGURE 14.20

Radiation fallout pattern in Europe from the Chernobyl accident.

area within a 30-km radius. It might be 10 to 20 years until the residents of Pripyat can return home.

To deal with the accident and quench the raging fire, helicopters dropped tons of sand on the reactor to smother the flames, boron to absorb neutron emissions, and lead for shielding. Today, the reactor is entombed in 300,000 tons of concrete. However, this 10-story tall sarcophagus is unstable and is decaying, and not well sealed to keep out rainwater. (A new foam material, Ekor, might be used to encapsulate this radioactive material to minimize groundwater and air contamination.) The inside remains thermally hot due to the radioactive decay heat. For months after the accident, contaminated produce and milk products from eastern and western Europe were banned from the market. Even cows that were kept indoors, away from contaminated pastures, inhaled enough radioactive material to affect their milk. About 30,000 mi^2 of farmland was contaminated. There is concern for the contamination of the water supply for millions due to the presence of small sites where high-level radioactive wastes were simply dumped.

Lessons are still being learned from Chernobyl. The accident settled the debate over whether a worst-case disaster could actually occur. An assessment of the accident brought out expectedly different opinions within the nuclear community.

- "Differences in design between U.S. and Soviet plants mean that the accident could not happen in precisely the same way here, but what they do not tell the public is that an accident with a release as large as Chernobyl, or worse, could happen." (L. R. Pollard, Union of Concerned Scientists)
- "The risk of a disaster will decrease because of safety improvements that are going to be adopted as a result of Chernobyl. After every major accident, modifications are made." (A. Buhl, Atomic Industrial Forum)

Operator violations and faulty testing procedures proved again that humans are the weak link in reactor safety. Well thought out, sensible, and easy-to-follow rules for reactor operators must be implemented at all facilities. Year-round retraining programs for U.S. reactor operators were an outgrowth of TMI. Chernobyl also raised questions about planning for evacuation in the event of a nuclear accident. United States regulations require a 10-mile evacuation zone around a plant. At Chernobyl, all residents within 18 miles of the plant were evacuated, and children were evacuated from villages as far as 100 miles from the plant.

Concerns over issuing accident warnings to other countries in a timely fashion have been settled since Chernobyl, but there is no wide agreement on economic liability, radiation standards, or the sharing of accident data. International inspection of nuclear reactors remains voluntary. A team of experts from 12 countries that reviewed the Soviet report on Chernobyl said that "an opportunity now exists for the world's safety experts to learn from this tragic event to greatly improve our understanding of nuclear safety. Whether that will occur remains to be seen."

Things in the former Soviet Union are still risky. Fifteen of the RBMK reactors are still operating, providing half of Russia's electricity that is supplied by nuclear energy. Western experts are almost unanimous in the view that these reactors should be phased out. The other reactors at Chernobyl were closed in 2000 due to international pressure. The nuclear situation is even more risky in former Soviet states. Ukraine receives only one fourth of the oil it used to receive from Russia and wants to start up three new reactors that stand near completion. About 25% of its electricity comes from nuclear power.

I. Probabilistic Risk Assessment and Nuclear Safety

An assessment of any technology or any individual action should take into account the benefits and risks involved. There are always risks involved in any activity. For example, ask any coal miner, who has one of the most dangerous jobs. There is clearly a risk in traveling by airplane, but the benefits of time saved and increased safety over other modes of transportation make it an acceptable risk. In considering the risks for various types of accidents, it should be noted that risk is the probability of occurrence of that accident (such as the chance per mile traveled or per reactor-year of operation) multiplied by the consequence per accident (such as the number of fatalities or injuries).

One intensive examination of the risks involved in the use of nuclear power (from mining to electrical generation) was made in a 1975 government-sponsored study on reactor safety, the so-called "Rasmussen Report" (designated WASH-1400). This report attempted to establish the probabilities that accidents of various severities would occur. (Several years after the publication of this report, the NRC withdrew its approval because of questions about the methods used in the analysis and the large uncertainties in some of the numbers, perhaps as large as one or two orders of magnitude. Even though the report cannot prove that reactors are safe or dangerous, it is still helpful to present the conclusions of that study.)

The report "calculated" that the probability of a core meltdown is 1 in 20,000 per reactor per year. (This means that with 100 reactors operating, the chance of a core meltdown would be once in 200 years.) Not every core meltdown would lead to a release of radioactivity (about one in ten LOCA would lead to a breach of containment), and those that did might affect only a small number of people. (The accident at Three Mile Island did not breach containment; its radioactive emissions were released by an operator in the process of controlling rising pressure within the reactor.) The probability of a major reactor accident with a release of radioactivity and about 1000 fatalities—direct (from cancer) and indirect (from genetic effects)—as a result of LOCA was estimated in this study to be 1 in 100 million reactor-years. From the viewpoint of a person living near a reactor, the likelihood of being killed in any one year by a reactor accident is 1 in 5 billion. Compare this with 1 chance in 2 million of death by lightning. Nonnuclear catastrophes were found to be about 10,000 times more likely than nuclear plant accidents to produce equal numbers of fatalities. A summary of these conclusions is given in Table 14.5 and Figure 14.21.

Table 14.5 WASH-1400 ESTIMATES OF FREQUENCY OF AND DAMAGE TO THE PUBLIC FROM THREE TYPES OF MAJOR ACCIDENTS TO LWR

Accident Type	Frequency (Chance per Reactor-Year)	Health Effects Within One Year		Total First Year's Health Effects Plus Delayed Health Effects	
		Deaths	Illnesses	Cancer Deaths	Genetic Defects (for All Generations)
Core meltdown	1 in 20,000	Negligible	Negligible	3	75
Plus above-ground breach of containment	1 in 1,000,000	1	300	5000	3800
Plus adverse weather conditions and population density	1 in 1,000,000,000	3000	45,000	45,000 (~1500/y)	30,000

(WASH-1400, United States Department of Energy)

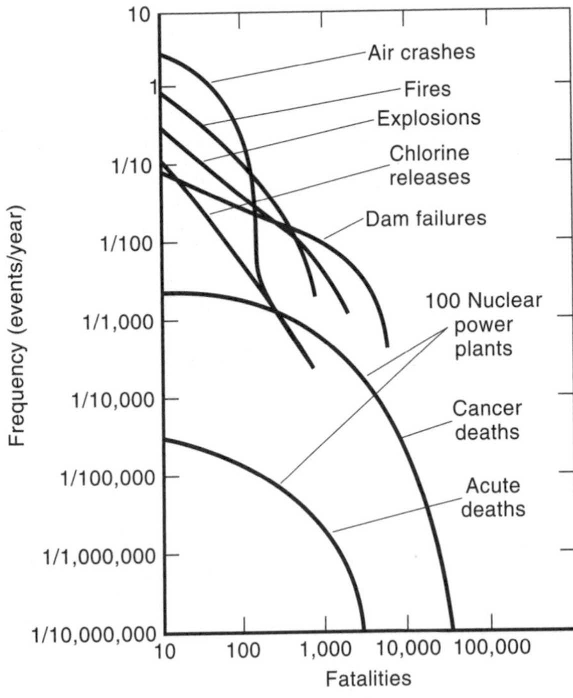

FIGURE 14.21
Probability of an accident producing a certain number of fatalities. (WASH-1250, United States Department of Energy)

The technique of predicting the likelihood and consequences of accidents is called "probabilistic risk assessment." It is used to provide insight into which factors are significant in reactor safety and which are not. For a reactor to experience a severe accident, many essential elements of its safety systems would have to fail simultaneously. We can calculate the probability of such a simultaneous failure of many components (if each failure is independent of all the others) by estimating the probability of failure of one component, multiplying it by the estimated probability of failure of another component, and so on. This product of many probabilities can lead to a very small number. For example, if each of three elements has a 33% chance of failing, and if all three must fail for there to be an accident, then the probability that the accident will occur is $\frac{1}{3} \times \frac{1}{3} \times \frac{1}{3} = \frac{1}{27} = 3.7\%$, if each failure is independent of the others.

The calculations and philosophy leading to the probabilities stated in the Rasmussen Report are not accepted by some critics because they are based on computer models of possible accidents and not on actual experience. However, we can check on these estimates by noting that in 3000 reactor-years of U.S. commercial and naval light water reactor operations, there has never been a core meltdown (although the TMI accident involved a partial meltdown). It appears that the report's estimates might be accurate to within a factor of 10. Note that these calculations are for the LWR, as are all U.S. reactors; the RMBK reactor at Chernobyl does not fall within this study.

In addition to calculating the probability of an accident with a major release of radioactivity, we also have to ask about the consequences of such an accident. To find these numbers, we need to know the effects of radiation on humans, a topic for Chapter 15. In such calculations we need to know the number of cancer deaths or genetic effects caused per person per unit of radiation dose. We do have data for such occurrences from laboratory animal experiments and from the aftermath of the atomic bomb explosions in Japan in 1945. As listed in Table 14.5, one accident scenario—a core meltdown plus aboveground breach of the containment vessel, with a chance of occurring once in a million reactor-years—has been estimated (in WASH-1400) to cause about 5000 cancer fatalities over a 30-year period and nearly 4000 genetic changes for all subsequent generations.

An extremely serious but less likely reactor accident—occurring once in a billion reactor-years—is an explosion of the containment building as a result of a high internal steam pressure, with direct release of radiation into the atmosphere, coupled with adverse weather conditions (such as wind direction toward a population center and a temperature inversion that prevents dispersion of the radioactive gases). There could be many fatalities (3000) in this case because of excessive radiation exposure (over 100 rem) in the first few weeks after the accident, although deaths up to one year are included in the table. The most serious health effect would be fatalities (1500 per year) from latent cancer that would occur over a 30-year period in the exposed population. Most cancer fatalities would occur from increased background radiation from the ground. Ground doses would come primarily from cesium-137 and its 0.66-MeV gamma ray. These consequences could be reduced if the population were evacuated and the land decontaminated.

Nuclear reactor safety continues to be a matter of concern to both the general public and the NRC. It should be noted again that the failure of any one piece of equipment or the mistake of an operator will not necessarily cause a disaster to occur. The nuclear reactor safety system consists of a series of precautionary measures and devices to prevent a malfunction. If one safety system fails, there exists another to back it up. This "defense in depth" approach makes it unlikely that an accident will occur. (These small probabilities were calculated using techniques similar to the multiplication of probabilities example of a previous paragraph.) Continual monitoring and checks of the safety systems are required by the NRC, with stiff fines to the utilities for even small compliance failures.

As a result of the TMI accident, many improvements in instrumentation and operator training were made in nuclear plants across the country. The nuclear industry established committees to study safety problems and conduct evaluations of operating practices at the plant. Major improvements also have been made with respect to emergency preparedness in case of an accident. The NRC requires utilities (in cooperation with local governments) to develop extensive evacuation plans for a 10-mile radius around the plant before an operating license can be received. Both the Shoreham (Long Island, New York) and (initially) the Seabrook (New Hampshire) power plants were not allowed to operate at full power because they did not have evacuation plans that were acceptable to local and state governments. (Shoreham was eventually shut down.)

J. Alternate Reactor Designs

Future reactor designs aim at reducing costs and speeding up approval and construction times by adopting a more generic design. If nuclear power is to be a viable option, then new plants must have simpler and more standardized designs. Over the years, LWRs have become very complicated. A typical 1000-MWe nuclear plant contains between 30,000 and 40,000 valves (10 times more than a similar-sized fossil-fueled plant). Almost half of these could be eliminated by taking advantage of changes in technology and requirements that have occurred since the systems were designed originally. Increased redundancy and implementation of additional safety features have caused the cost of power plants to rise dramatically. Between 1971 and 1982, the amount of material required for construction of a 1000-MWe PWR increased by a factor of 2.8 for concrete and 2.4 for the length of cable, as well as a factor of 6.8 for the work hours spent by craft labor!

The next generation of nuclear power plants will probably be LWRs, since there is so much experience with this design, but they also will be smaller (400 to 600 MWe), which will take less capital to build and require shorter lead times. Traditionally, large power plants have been cheaper to build on a per-kilowatt basis. However, smaller power plants will provide a closer match between growth in electrical demand and generating capacity. The construction of a 1000 MWe plant will create an overcapacity for many more years than would the construction of a 500-MWe facility. Also, smaller plants lend themselves more readily to the use of safer and more reliable safety systems. Several new, standardized LWRs are currently under development in the United States. They should be cost-competitive with other types of electric generating plants. They will employ "passive" safety systems that will ensure there will be no release of radioactive material in the event of an abnormality.

Today, light water reactors depend on the "active" intervention of electromechanical devices and the action of operators to ensure the safety of the plant and to protect the public from radioactive releases. To reduce the likelihood of an accident, more and more safety features have been added over the years, many of which are backup or redundant systems. However, critics claim that not all possibilities or avenues leading to an accident with the release of radioactivity can be foreseen. Consequently, another approach to a safe nuclear reactor is the use of "passive" safety systems, which rely on the inherent characteristics of the reactor itself and the laws of physics. Such reactors depend on gravity feed from a water storage pool to replace the coolant lost from the vessel in the event of a LOCA or depend on natural heat transfer by conduction and convection to dissipate the decay heat from the core to the environment after a shutdown. These ideas have been pursued vigorously since TMI, and many passive features have been incorporated into redesigned, actively safe LWRs, as well as into a completely new generation of inherently safe reactors.

Figure 14.22 is the schematic of an advanced LWR that achieves safety improvements by the use of passive features. These features include the use of gravity feed to provide long-term cooling in the event of a LOCA and increased water in the core to provide smaller power densities. These advanced LWRs also will be much smaller than current reactors, for both safety and financial reasons.

Although light water reactors are by far the dominant reactors used in the United States, several other types of coolants and moderators are in use throughout the world. In Canada, the preferred reactor uses heavy water (in which the more massive isotope deuterium replaces hydrogen) as the coolant moderator. The advantage of using heavy water is that deuterium has a neutron capture cross section only 1/600th as large as that of hydrogen. Thus unenriched, or

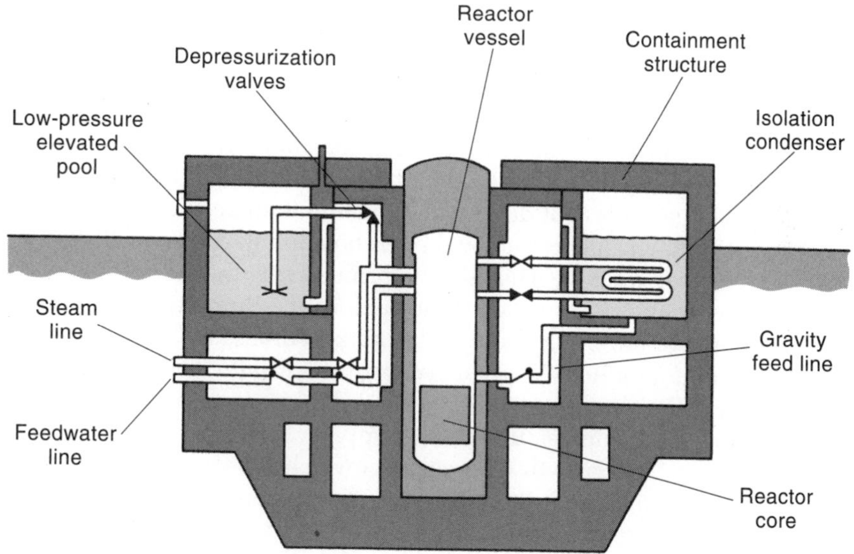

FIGURE 14.22

The passive safety systems for a small BWR are based on an elevated pool of water that encircles the reactor. If the normal heat-removal systems become unavailable, decay heat is removed from the core by circulating coolant from the reactor vessel through an isolation condenser submerged in the pool. If a LOCA occurred, steam from the reactor would be vented into the pool to depressurize the reactor coolant system. Once depressurization is accomplished, water from the elevated pool can flow into the reactor vessel by gravity for long-term cooling. (EPRI JOURNAL)

natural, uranium can be used as the fuel because the chain reaction generates enough neutrons to be self-sustaining. This Canadian reactor, called **CANDU,** has a schematic similar to that of the PWR. The fluid in the secondary loop is regular water. The overall efficiency of the CANDU is about 29%. On-line refueling (that is, while the reactor is operating) is possible. Although the cost of heavy water is high, the Canadian experience to date is that the cost of electricity produced is substantially less than that from a LWR.

Another new type of reactor is the modular high-temperature gas-cooled reactor (HTGR). This new, small reactor (about 150 MWe) is designed to be immune to meltdown in the event of a LOCA. One of its key features is the use of billions of grains of uranium that are individually encased in ceramic shells. These shells cannot melt at the temperatures that would be achieved within the core (1500°C). The HTGR uses helium as the coolant; it is run in the turbine-generator cycle at higher temperatures than steam and so yields a higher efficiency (40 to 45%) than conventional LWRs. The uranium fuel pellets are contained in a fuel region within a carbon moderator. Holes in the moderator allow for the passage of the helium gas. Three or four of these small modular units could be strung together to provide the equivalent output of a larger commercial plant.

Another type of reactor is the **breeder reactor.** This type of reactor is not new, and not used in the United States, but is in use in France, Russia, and Japan, so is worthwhile to consider here. (The first generation of electricity from a nuclear reactor came from a breeder in 1951 at the Argonne [Illinois] National Laboratory.) This type of nuclear reactor is so called because it "breeds" or produces more fuel than it consumes. It's as if you put 3 gallons of gasoline into your car's gas tank, drove around the city, and returned with 4 gallons of gasoline. But don't worry, energy is still conserved! The breeder can produce enough fuel for its own operation and generate additional fissionable fuel for other reactors. This is done primarily by the conversion of the nonfissionable uranium-238 isotope into fissionable plutonium, thus making much greater use of the energy potential in natural uranium. Current nuclear reactors "burn" only uranium-235, which constitutes just 0.7% of natural uranium. The successful development of the breeder would ensure a country with access to uranium of an almost unlimited supply of energy. However, as has been pointed out, the growth of nuclear power has been much less than was expected, and so there is little pressure on uranium supplies, and so interest in breeder reactors has declined.

Basically, a breeder converts some of the (normally) nonfissionable isotope ^{238}U into the fissionable isotope ^{239}Pu. This occurs through the following series of steps:

$$^{238}_{92}U + n \rightarrow {}^{239}_{92}U \xrightarrow{\beta} {}^{239}_{93}Np \xrightarrow{\beta} {}^{239}_{94}Pu$$

A neutron captured by ^{238}U forms ^{239}U, which has a half-life of 23 minutes; this decays into neptunium-239 by beta decay (half-life of 2.4 days), which in turn decays into plutonium-239, with a half-life of 24,400 years.

At this point you might question why this reaction doesn't also happen in current light water reactors (LWRs). The answer is that it does, but not very efficiently. Current 1000-MWe nuclear reactors produce about 200 kg of plutonium per year. However, not enough ^{239}Pu is produced to replace the fissile fuel that is destroyed. Breeders are constructed to maximize the amount of fissionable plutonium produced; unlike LWRs, they produce more fissionable material than is used up.

Breeders use high-energy or "fast" neutrons to accomplish this task. Fast neutrons have a relatively better chance of being captured by ^{238}U than by ^{235}U. When fission occurs with fast neutrons, the number of neutrons produced on average is greater than in slow-neutron fission (2.9 compared to 2.4 per fission event), so the breeding efficiency is increased. The object is to produce enough neutrons to sustain the chain reaction and breed an amount of plutonium that will at least replace the ^{235}U or ^{239}Pu nuclei that are used up. To just maintain the chain reaction and replace the fissioned nuclei, two neutrons per fission must be captured. The surplus of neutrons above two could then go into the production of additional fissionable fuel. The significance of this process is that we are also converting ^{238}U into a useful fuel, ^{239}Pu, rather than just using the ^{235}U. The breeder promises to use 60% of the energy content of natural uranium instead of the present 1% to 2% of the LWR, thus amplifying the reserves of uranium by a factor of about 50.

Since one purpose of the water in the LWR is to slow the fission neutrons so that their chance of capture by ^{235}U is higher, a different coolant must be used in breeder reactors. The coolant most often used today is liquid sodium, a metal. The sodium nucleus has a larger mass than that of the hydrogen or oxygen nuclei of water, so it will not slow the fission neutrons as much. (This is analogous to a tennis ball bouncing off a bowling ball—the tennis ball loses very little energy.) Sodium as a coolant has excellent heat transfer properties. A disadvantage of sodium is that it reacts quite violently with water and (at the temperatures of the breeder reactor) burns spontaneously in air. Thus, extra efforts have to be made to prevent leaks in the cooling system. This type of breeder reactor is called the "liquid metal fast breeder reactor" or LMFBR.

K. Nuclear Proliferation

Another concern with the commercial use of nuclear power is the possible use of its byproducts for the construction of nuclear weapons. In the eyes of many, there has always been a link between nuclear power and nuclear weapons. Indeed, nuclear power development began in the 1950s as an outgrowth of the nuclear weapons program (President Dwight Eisenhower's "Atoms for Peace" program).

Simple nuclear fission bombs are made from very pure (90%) ^{235}U or ^{239}Pu. The critical masses are about 10 kg and 5 kg, respectively. The fuel used in a conventional LWR is enriched to only about 3% ^{235}U. Weapons-grade material can be made from fresh fuel rods only by the expensive, energy-intensive enrichment process. During the operation of a reactor, however, ^{239}Pu is produced at the rate of 200 kg/y for each 1000-MWe plant. The process to separate Pu from spent fuel rods is simpler than uranium enrichment and is well known, although not easy to do. If a country wished to acquire nuclear weapons-grade material, it would probably find it cheaper and quicker to build a "research" reactor fueled with natural uranium designed to yield enough plutonium annually for several weapons. (An example is the Indian reactor that enabled that country to detonate a nuclear device in 1974.)

Focus On 14.3

POST—COLD WAR PROLIFERATION

One of the concerns after the fall of Communism is the funneling of weapons-grade material into the hands of terrorists and nonnuclear countries.

- In 1990, the new republic of Ukraine found itself the third most powerful nuclear nation, having "inherited" 1800 nuclear warheads from the U.S.S.R. Such a possession gives it an excellent bargaining chip to stimulate its weak economy.
- Highly underpaid Russian nuclear scientists are tempted by offers of nations seeking to develop their own bombs or missile systems. There already have been arrests of Russian scientists trying to smuggle highly enriched uranium out of the country—in suitcases.
- North Korea would not allow inspection of its research reactors by the United Nations for many years and might have enough plutonium stockpiled now to build several bombs.
- China has been upgrading its weapons capability by hiring experts from Russia to develop a version of the SS25 rocket with multiple warheads.

During the cold war days, the U.S.S.R. exercised strict control over its nuclear materials and its personnel. Those times have changed.

The Treaty on the Non-Proliferation of Nuclear Weapons (enacted in 1970) states that signers of the treaty are "not to manufacture or otherwise acquire nuclear weapons or other nuclear explosive devices." However, although a majority of countries have endorsed the treaty, some have not, including India, Pakistan, Israel, and North Korea.

The ability of a terrorist group to obtain nuclear weapons-grade material is always a concern. However, the danger of terrorist bombs made from stolen American materials is small. Remember that the spent fuel rods are radioactive, and the chemical separation of the plutonium from the uranium and fission fragments would be difficult for a terrorist group. Still, a simple bomb constructed with reactor-grade plutonium might have an output of several hundred tons of TNT—ten times the power of a conventional World War II bomb.

Plutonium reprocessing and recycling greatly increase the opportunities for theft and nuclear blackmail. A **plutonium economy** with reprocessing and recycling would involve the transportation of thousands of kilograms per year of plutonium from reprocessing to fuel-rod fabrication to power plants. (See Focus On 14.1 Japan and Plutonium.) As a consequence, the system could be vulnerable to theft and diversion. Inspections of reactor facilities, even within countries that have signed the Treaty on Non-Proliferation of Nuclear Weapons, would be complicated and difficult. However, some countries with nuclear power installations are intent on a plutonium economy. Their energy supplies are not as diverse as those of the United States, so extending the amount of nuclear fuel by recycling is important for them. Breeder reactor development would also contribute to increased availability of plutonium since it would be producing plutonium for use in conventional LWRs.

In summary, the connection between commercial nuclear power facilities and nuclear weapons is tenuous. Historically, weapons programs have used research reactors and not commercial power reactors for the production of plutonium. The spread of enrichment facilities could put highly enriched uranium, probably the easiest material from which to make a bomb, within the grasp of nonnuclear nations. Plutonium reprocessing and recycling require strong international efforts to ensure that diversion and theft do not increase the dangers of nuclear war and nuclear terrorism.

L. Environmental and Economic Summary of Nuclear Power

The bulk of this chapter has been a description of the principles behind the operation of nuclear power plants, the nuclear fuel cycle, and the question of reactor safety. Any final assessment of the role of nuclear power in the years ahead must consider the total energy supply-and-demand picture as well as the total environmental impact of energy use. That picture is developed throughout the course of this book, but to aid in a comparison of energy supply alternatives, it is helpful at this point to compare the environmental and economic consequences of nuclear energy and coal in the production of electricity.

The environmental impact of each electrical system includes (1) the land use in extraction, processing, and conversion (to electricity) of the fuel; (2) the air pollution associated with the conversion; (3) the emission of radiation during normal operation or in the event of an accident; (4) the occupational health factors associated with the extraction, processing, and conversion of the fuel; and (5) the storage of solid or stored waste. These items are summarized in Table 14.6 for a 1000-MWe power plant (operating at 75% capacity) using either coal or uranium as the fuel.

If strip mining is used to provide the low-sulfur coal needed by the power plant, then 17,000 acres per year per 1000-MWe plant will have to be mined. The effect on water life because of thermal pollution is a problem for both nuclear and fossil-fueled plants. The nuclear plant emits about 40% more waste heat than an equivalent-sized coal plant because of its lower efficiency (see Chapter 4). Coal-fired plants also emit radiation, as a result of the presence of radioactive uranium and thorium in the coal. The number of curies emitted depends on the type of coal being burned and the air pollution control equipment.

The occupational health hazards of coal mining, especially underground, have been known for years (including black lung disease and cave-ins) but have been reduced because of government safety standards. The health hazard to the general public as a result of the emission of air pollutants is not known very well. In contrast to nuclear power, little government evaluation or assessment has been made of the hazards associated with fossil-fuel based power sys-

Table 14.6 ANNUAL ENVIRONMENTAL IMPACTS ASSOCIATED WITH A 1000-MWe POWER PLANT*

Impact	Coal	Nuclear (LWR)
Land use (acres)	17,000	1900
Water discharges (tons)	40,000	21,000
CO_2 emissions (tons)	6×10^6	0
Air emissions (tons)	380,000	6200
Radioactive emissions (curies)	1	28,000
Occupational Health		
Deaths	0.5–5	0.1–1
Injuries	50	9
Total fatalities (public and worker)	2–100	0.1–1

*Includes extraction, processing, transportation, and conversion. Strip-mined coal.

(WASH-1250 and *Ann. Nuclear Energy,* 13, 173, 1986)

tems. A coal-burning plant emits several hundred thousand tons per year of sulfur dioxide, nitrogen oxides, and particulates. (Chapter 8 deals with air pollution.) It is very difficult to determine the public health effects of these emissions, especially at very low levels. The exposure one receives depends on whether one is indoors or outdoors, the weather, and the type of pollutant. Some elements found in particulate emissions are essential to health at certain concentrations but toxic at higher levels. During transit of the pollutants, their nature (size and composition) may change, causing significant changes in toxicity. For an individual plant (1000 MWe coal fired) that is meeting new source standards with appropriate control devices, the number of premature deaths for the entire operation (mining to generation) is estimated at from 2 to 100 per year for the workers and the public. (On-the-job fatalities account for 0.5 to 5 of this number.) For air pollution one probably also must consider nonfatal illnesses and discomfort caused by respiratory diseases.

The effects of radiation emission (discussed in Chapter 15) are taken from studies of radiation effects on animals and from the Hiroshima and Nagasaki bombings. Much more is known about the effects of radiation than about chemicals at low levels. It is estimated that the number of deaths from radiation from the combined operation of all nuclear plants and reprocessing facilities in this country will total about 10 per year, or about 0.01 to 0.2 premature deaths per 1000-MWe plant per year. (Such calculations also include catastrophic accidents. With the worst possible accident having a probability of occurring once in a billion reactor-years, about 0.02 fatalities per reactor per year would be expected from accidents.) For an individual 1000-MWe plant, 0.1 to 1 fatality per year from occupational accidents is estimated. (Caution should be exercised in using these numbers, because there are large uncertainties in the analyses, as already mentioned.)

In conclusion, the health and environmental effects of operating a coal-fired power plant are apparently greater than those of the nuclear power option, although the uncertainties of catastrophic accidents, high-level radioactive waste disposal, and nuclear proliferation may discolor these conclusions in the minds of many. These environmental impacts do not include other risks that exist. For coal (see Chapter 9), these risks include global climatic effects as a result of CO_2 emissions (the greenhouse effect) and the effect on lakes and vegetation as a result of acid rain.

In many ways, a discussion on the economics of nuclear power today, in the era of deregulation of the utility industry, is superfluous. In the not-too-distant-past history of the electric industry, a utility could build a power plant and then set rates so as to recover its investments, with a fixed profit margin. There was no other game in town. Now, with competition in the generation of electricity, cost is of prime concern. And nuclear plants are much more expensive to build than gas and coal-fired plants. However, a few insights on economics and history might be useful here.

The **operating cost** of a generating plant is made up of plant investment cost (amortization of the original capital cost over the lifetime of the plant), operation and maintenance cost, and fuel cost. (Capital cost accounts for about 70% of the cost of electricity produced by a LWR, but between 35 and 40% for a coal-fired plant.) Fuel prices play a much smaller role in the operating cost of a nuclear plant, accounting for about 5 to 10% of the power cost for nuclear, compared to 50 to 60% for coal plants.

In the mid-1970s, construction costs were \$600/kW for a coal plant and \$730/kW for a nuclear plant. In 1986, costs for a nuclear plant were as high as \$4000/kW! The costs escalated quite rapidly because of increased labor and material costs, high interest and inflation rates, construction delays because of environmental and legal constraints, and added complexity in the plants because of safety concerns. Cost is the primary reason why no nuclear reactors have been ordered in the United States since 1979. There are also hidden costs in the construction of nuclear plants that are not included in these figures—for example, the research that has been done by the federal government over the years on nuclear safety and waste management. There is also a break in the special liability insurance required by the companies that own the nuclear plants, which is limited by law (Price-Anderson Act) to a certain maximum.

While construction of new nuclear power plants in the United States seems unlikely in the near future due to capital costs, there have been "fire sales" of many existing nuclear plants beginning in 1999 and 2000. Utilities have begun to divest themselves of their generating facilities. Nuclear plants have been selling to large national corporations for 20¢ on the dollar in some cases. The loss ("stranded costs") that a utility would take in this sale is passed on to the ratepayer in most states. Thus, in new hands, with inexpensive fuel costs and good operating records, nuclear has become very competitive with fossil fuels.

Nuclear plants were constructed in the 1970s because utilities wanted a mix of fuels in their generation portfolio. In several parts of the country, experience by utilities in the 1970s using both coal and nuclear units found that coal-fired electricity costs were slightly greater per kilowatt-hour than nuclear. This number turned around in the 1980s for the reasons listed above. However, some countries, such as Japan, with no natural energy resources, continue to build nuclear power plants. Concern also for the environment (including global warming) has prompted some countries to continue to build nuclear plants. (Refer back to Table 14.1.)

M. Summary

Nuclear power supplies about 19% of the electricity generated in the United States. A nuclear plant uses uranium enriched to 3% ^{235}U. The heat given off during the fission of ^{235}U is removed by water in the reactor core. In a boiling water reactor (BWR), this water turns to steam, which is used to drive a turbine generator. In a pressurized water reactor (PWR), the water in contact with the fuel remains in the liquid phase and transfers energy through a heat exchanger to boil water, which is used in the turbine generator. If this water circulating through the core is interrupted, as in a loss of coolant accident (LOCA), core overheating could result. An emergency core cooling system (ECCS) is designed to provide makeup water in case of this emergency.

One concern about nuclear energy is the disposal of high-level radioactive waste. High activity and long half-lives make the need for isolated burial for thousands of years very important. Burial in stable geologic formations seems to be the choice of most nations. In addition to this issue, and that of the economics of nuclear power, the perceived risks of nuclear power are ever present. Human errors as well as technical factors are involved in this question. Concerns about global warming, and reduced construction costs made possible by adopting more generic designs and streamlining the licensing procedures, might allow nuclear energy to expand again, but more than likely, economics and retirement of aging plants will cut nuclear power's contribution to electricity generation in half in the next two decades.

Internet Sites

For an up-to-date list of Internet resources related to the material in this chapter, go to the Harcourt College Publishers website at **http://www.harcourtcollege.com**. The links are in the *Energy: Its Use and the Environment* site on the Physics page. General energy related sites and some guidelines for using the World Wide Web in your class are on the inside front cover of this book.

References

Chapter 14

Beck, P. 1999. Nuclear Energy in the Twenty-First Century: Examination of a Contentious Subject. *Annual Review of Energy*, 24.

Bethe, H. A. 1976. The Necessity of Fission Power. *Scientific American*, 234 (January).

Blair, B., and H. Kendall. 1990. Accidental Nuclear War. *Scientific American*, 263 (December).

Bodansky, D. 1996. *Nuclear Energy: Principles, Practices, Prospects*. New York, American Institute of Physics.

Cohen, B. L. 1977. The Disposal of Radioactive Wastes from Fission Reactors. *Scientific American*, 236 (June).

Cohen, B. L. 1990. *The Nuclear Energy Option—An Alternative for the '90s.* New York, Plenum Press.

Cohrssen, J., and V. Covellow. 1989. *Risk Analysis: A Guide to Principles and Methods for Analyzing Health and Environmental Risks.* Washington, D.C., U.S. Council on Environmental Quality.

Craig, P. 1999. High Level Radioactive Waste: The Status of Yucca Mountain. *Annual Review of Energy*, 24.

Gould, J., and B. Goldman. 1991. *Deadly Deceit: Low Level Radiation.* New York, Four Walls Eight Windows.

Holdren, J. P. 1992. Radioactive Waste Management in the U.S. *Annual Review of Energy*, 17.

Hoyle, F., and G. Hoyle. 1980. *Commonsense in Nuclear Energy.* San Francisco, W. H. Freeman.
League of Women Voters. 1996. *The Nuclear Waste Primer*. New York, Nick Lyons.
Lewis, H. W. 1980. The Safety of Fission Reactors. *Scientific American*, 242 (March).
Morgan, G. 1993. Risk Analysis and Management. *Scientific American*, 269 (July).
Nero, A. 1979. *A Guidebook to Nuclear Reactors.* Berkeley, University of California Press.
Roberts, J. T., R. Shaw, and K. Stahlkopf. 1985. Decommissioning of Commercial Nuclear Power Plants. *Annual Review of Energy*, 10.
Rose, D., and R. Lester. 1978. Nuclear Power, Nuclear Weapons, and International Stability. *Scientific American*, 239 (April).
Schmidt, F., and D. Bodansky. 1976. *The Fight over Nuclear Power.* San Francisco, Albion.

QUESTIONS

1. Why doesn't uranium undergo spontaneous chain reactions in nature?
2. What are the two major roles of water in the operation of a reactor?
3. How is radiation emitted to the environment in the normal operation of a BWR?
4. Why is the efficiency of a nuclear plant less than that of a fossil-fuel electric generating plant? What does this say about the heat emitted from two such plants with the same electric output?
5. If the temperature of the steam in a PWR is 315°C (600°F), what is the maximum efficiency obtainable from the power plant?
6. If the water in a BWR core gets too hot, what will happen to the power output of the reactor? Why?
7. Why is a conventional nuclear reactor not able to explode as a bomb?
8. Why do radioactive wastes remain thermally hot for thousands of years?
9. What is a passive reactor safety system?
10. Why is there such a controversy over exporting reprocessing-plant technology?
11. What are the differences between high-level and low-level radioactive wastes?
12. What disadvantages to the public would occur if activities that produced low-level radioactive wastes were severely restricted? Why might this be popular today? Discuss.
13. Give arguments for and against the statement that the burial of radioactive wastes is an unjust burden to be given to future generations.
14. What are arguments for and against the permanent storage of radioactive wastes in geologic formations such as rock or salt beds?
15. What safety features would be used if the cooling water circulation should stop (e.g., from a broken pipe) in a BWR?

16. If the probability of being killed in an airplane crash is many times greater than that of being killed as a result of a nuclear reactor accident, then why aren't all planes grounded until their safety is improved drastically? Discuss.
17. The net energy analysis of any energy technology is important. You don't want to put more energy into the system than you get out. Write a general expression for the net energy ratio for nuclear reactors, that is, the ratio of the electrical energy output to the total energy input needed to obtain that output (e.g., the steps in the fuel cycle, mining, reactor construction). This ratio should be greater than one. (Unlike the definition for plant efficiency, this ratio does not include the thermal energy content of the uranium fuel itself.)
18. Analogous to the risk assessment given with respect to nuclear power, list risks that are faced on your campus. What are the chances you take given those risks?
19. Investigate the change in ownership of nuclear power plants in your state. What was the price paid for the power plant? What fraction of the original price was this?

FURTHER ACTIVITIES

1. The moderation of fission neutrons by water within the reactor core can be simulated by collisions between objects of roughly the same mass. Try this out for two marbles, two steel balls, or two billiard balls. Observe the transfer of velocity between the moving ball (the neutron) and the stationary one (the hydrogen nucleus). If the moving ball is a marble, what happens in a collision with a more massive object, such as a steel ball?
2. Using the U.S. map of Figure 14.1, or more current information, locate the nuclear power plant nearest to you. Is it a PWR or a BWR? Inquire about its size, cost, date of completion, storage of spent fuel rods, and the cost of electricity generated (cents per kWh).

15

Effects and Uses of Radiation

A. Introduction

Mention of the subject of "radiation" might bring to mind nuclear power plants, nuclear bombs, or X-ray tubes. Such topics certainly are associated with "radiation," but they are not the only ones. Radiation is present everywhere, from sunlight to microwaves to nuclear radiation. Radio and TV signals are transmitted using electromagnetic waves. It is impossible to avoid some exposure to radiation. Radiation comes in many forms, as both waves and particles, and can have both beneficial and harmful effects on living things. Our earth has been subject to radiation from cosmic rays since it was formed. In fact, this so-called "background" radiation probably has played an important role in the development of living systems, from simple cells on up. Only since the discovery of X-rays in 1895 by Wilhelm Roentgen has artificial radiation come to play a role in our society.

For our purposes we will study only those types of radiation that can *produce harmful biological effects in animals.* Such forms of radiation as radiowaves and visible light carry small concentrations of energy and are unlikely to harm biological material. When absorbed by living tissue, the energy in these radiations is converted into heat, which causes the molecules in the tissue to vibrate more rapidly, but does not pull them apart or cause any chemical reactions. (An exception to this is photosynthesis in plants where, with the help of visible light, carbon dioxide and water combine to produce oxygen and simple sugars.) The ultraviolet component of solar radiation carries more energy and can cause skin cancer if overexposure occurs. Even more energetic electromagnetic and particle radiations can provide the energy necessary to break up a molecule or knock an electron loose from an atom. Such energetic radiation is called **ionizing radiation,** and it is the most important type to consider in our study of radiation effects.

B. Radiation Dose

Ionizing radiation comes from gamma rays, X-rays, and charged particles such as electrons and alpha particles. Table 13.1 listed the properties of these radiations, including their abilities to penetrate matter. Gamma rays and X-rays are not charged, so they cannot ionize an atom directly through electrical interactions. However, they produce electrons indirectly through other interactions with matter.

To understand the effects of radiation on human tissue, we must have some way of measuring the amount delivered. For many years after the discovery of X-rays, there was no adequate way to measure the amount received. Early X-ray workers noticed that hair fell off or that their skin became ulcerated in areas exposed to X-rays. Skin cancer developed, in some cases, years later. A "suitable" exposure in the early 1900s was that amount of X-rays that would just produce a detectable reddening of the skin. By today's standards this is much too great, especially when observable effects do not develop for weeks to years. Many of the first radiation workers suffered radiation burns or died prematurely of cancer.

There are several radiation dose units used today. The **roentgen** is a measurement of radiation exposure and is a unit proportional to the amount of ionization produced in air by X-rays or γ-rays. To relate this to biological dose, the unit **rad** (for "radiation absorbed dose") is used; it is that quantity of radiation that delivers 100 ergs (1 erg $= 10^{-7}$ J) of energy to 1 gram of tissue. (For X-rays 1 rad is about equal to 1 roentgen.) In SI, the unit for radiation dose is the gray (Gy), which is equal to 1 J of energy deposited per 1 kg of material: 1 Gy $=$ 100 rad.

It is known that some radiations are more effective than others at producing ionization. For example, 1 rad of α particles produces about 10 to 20 times as much biological effect as 1 rad of X-rays or electrons of the same energy. This is because the α particles have a much shorter range and so concentrate their damage within a much smaller volume. This is analogous to the differences in damage that would occur if a person were struck by a 20-g sponge or a 20-g razor blade, both traveling at the same speed. To take these differences into

Table 15.1 QUALITY FACTORS (QF) OF SEVERAL RADIATIONS

Radiation Type	QF
γ- and β- rays	1
Low-energy neutrons and protons	5
α Particles, high-energy neutrons, and protons	10–20

Note: Dose in rems = absorbed dose in rads × *QF*

account, we introduce the biological unit of radiation damage, the **rem** (for "roentgen equivalent man"). The dose in rems is numerically equal to the absorbed dose in rads multiplied by a quality factor, which depends on the type of radiation.

$$\text{Dose in rems} = \text{absorbed dose in rads} \times \text{quality factor } (QF)$$

The ratio of rems to rads is called the "quality factor" (QF). QF values for several radiations of the same energy are given in Table 15.1. Most radiation doses are fairly small, so the unit "millirem" (1/1000 rem, abbreviated mrem) is commonly used. The equivalent dose unit in SI is the sievert (Sv); 1 Sv = 100 rem. Table 15.2 summarizes both sets of radiation units.

Table 15.2 RADIATION UNITS

Unit	Definition
For radioactivity	
curie (Ci)	3.7×10^{10} nuclear disintegration/s
becquerel (Bq)	1 nuclear disintegration/s (SI unit)
For absorbed dose	
rad	100 erg/g deposited
gray (Gy)	1 J/kg (SI unit); 1 Gy = 100 rad
rem	rad × *QF* (dose equivalent)
sievert (Sv)	1 Sv = 100 rem (dose equivalent in SI)

C. Biological Effects of Radiation

The cell is the fundamental unit of life. The average adult human has about 50 trillion cells. Just as the fundamental unit of matter, the atom, is itself composed of smaller building blocks, so the cell has a complicated structure. A simplified diagram of a cell is shown in Figure 15.1. The nucleus is the control center of the cell. Here reside the chromosomes, which carry the "instructions" for cell development. A normal human cell contains 23 pairs of chromosomes. In cell division (mitosis), each chromosome duplicates itself so that each new cell will be an exact copy of the first, with an identical set of chromosomes. The "instructions" for cell development and duplication are found in the long molecular chains of DNA (deoxyribonucleic acid) located within the nucleus.

Ionizing radiation is energetic enough to displace atomic electrons and break the bonds that hold molecules together. Ionizing radiation causes a number of chemical reactions to take place that precede the biological effects. Some of the most important chemical events occur with water (which makes up 70% of the cell). The water molecule can be ionized or be excited, forming what is called a "free radical." A free radical is an electrically neutral molecule with a highly excited unpaired electron in one of the outer shells. It is extremely reactive chemically and can attack molecules in the cell, leading to biological damage.

Damage or alteration of a cell by radiation can induce **somatic effects** (those that affect the individual receiving the radiation) or **genetic effects** (those that affect the individual's offspring), depending on the type of cell exposed. Radiation damage to a cell (such as the breaking of bonds within a DNA molecule) can inhibit cell repair because the cell has to "read" the instructions contained in the DNA coding. Radiation damage also can inhibit the proper division of a cell. A damaged cell might grow in a new uncontrolled manner and invade and destroy the surrounding host cells, becoming a malignant cancer. Cancer as a disease is probably caused by a combination of several factors, so simple explanations are not possible. One factor in the production of cancers might be a virus that attacks normal cells, causing them to reproduce wildly. Radiation and other carcinogenic agents (chemical and

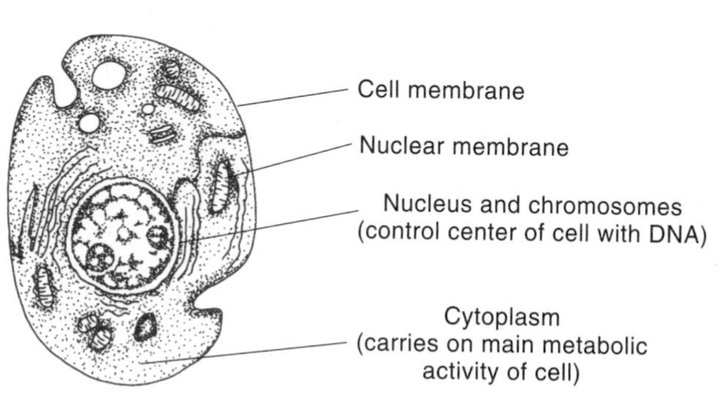

FIGURE 15.1
Diagram of a cell.

physical) might interfere with a healthy cell's resistance to this virus. Genetic damage in gender cells produces mutations that will be passed along to offspring, although in many cases it does not become visible for several generations.

The effects of high radiation doses may appear within hours or days in such ways as nausea, vomiting, and death. However, for small radiation doses, effects like cancer are not seen for 20 years or more, so it is difficult to establish good quantitative relationships between cause and effect. Experiments on fruit flies and mice have been used to quantify the effects of large radiation doses. Figure 15.2 shows a curve of the percentage of survivors as a function of radiation dose for mice. The death rate increases with increasing radiation dose until finally 100% of the mice are affected by the large doses. The point at which 50% of the population is affected and dies (in weeks) is called the "lethal dose–50%" or LD-50. Figure 15.2 indicates that the LD-50 for mice is about 1000 rads.

For humans, the LD-50 is about 300 to 500 rads dose over the whole body. Such a dose will produce an alteration or reduction in the blood cells and blood-forming organs, leading to nausea, vomiting, and susceptibility to infection, and death to 50% of the people within 30 days. Doses greater than 2000 rads will damage the central nervous system and lead to death within a few days (Table 15.3).

Biological effects of radiation depend on what part of the body is irradiated. Hands and feet can take larger doses than most any other part of the body. The most sensitive tissues are the intestinal wall, the spleen, the gonads, and bone marrow. The intestinal wall is important because it is the chief barrier to infection from bacteria that reside in the gut. If it is adversely affected, intestinal bacteria can enter the bloodstream directly. In the likely case that the white blood cells (the infection-fighting cells) of the irradiated person also have been reduced, then infection might be fatal.

To set standards governing our exposure to artificial radiation, it is important to know the effects on human populations of low as well as high levels of radiation. The health effects of high levels of radiation are well known, but information on low-level radiation is difficult to obtain. The only complete study of the effects of radiation on humans comes from the Hiroshima and Nagasaki

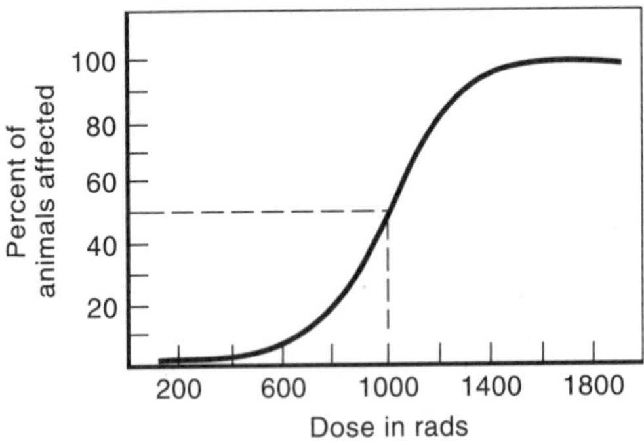

FIGURE 15.2

Dose-effect distribution curve for mice. The point at which 50% of the population dies is called the lethal dose–50, or LD–50. It is about 1000 rads in this example.

Table 15.3 EFFECTS OF WHOLE-BODY EXPOSURE IN HUMANS

Dose (rads)	Effect
1	No detectable change
10	Blood changes detectable
100	Some injury
200	Injury and some disability
400	50% deaths in 30 days
600	100% deaths in 30 days
2000	50% deaths in 4 days

tragedies. Here 24,000 survivors received an average exposure of 130 rem, with about 120 additional cancers developing among them in the next 27 years. The 40-year study of the A-bomb survivors and their offspring have yielded many somatic and genetic dose-response curves. One such radiation effect curve (Fig. 15.3) shows the annual incidence rate of leukemia as a function of total dose. Based on these data, the incidence rate of leukemia is determined to be about two cases per million people per year per rem of radiation. Other evidences of the carcinogenic nature of radiation are found in studies of uranium miners (with high incidences of lung cancer), early medical and dental users of X-rays, and workers who painted luminous watch dials. In the latter case,

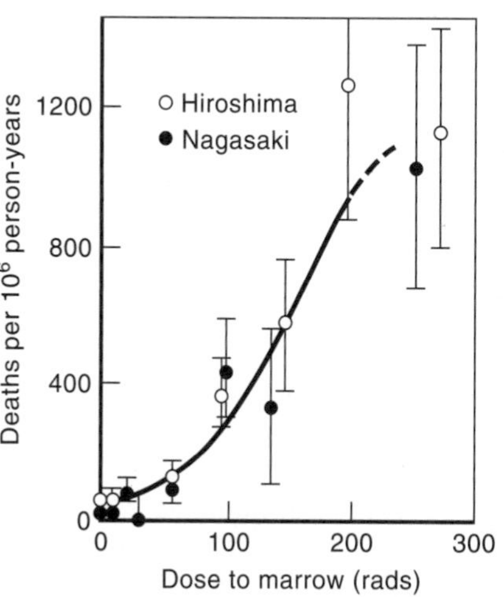

FIGURE 15.3
Leukemia mortality dose-response curves for Hiroshima and Nagasaki.
(American Institue of Physics, c 1981. Used with permission.)

Focus On 15.1

AMERICA'S NUCLEAR SECRETS: "WAS IT JUSTIFIED?"

In the years since the demise of the cold war and the fall of communism, a good deal of information has surfaced on U.S. experiments with ionizing radiation during the 1940s to the 1970s. We have learned that experiments were conducted on the effects of nuclear radiation on humans, hundreds of undisclosed (underground) nuclear bomb tests were carried out, and safety at nuclear weapons facilities (e.g., Rocky Flats in Colorado) was lacking.

- Unknowing or ill-informed people were subjected to irradiation from direct exposure during A-bomb tests in Nevada or from injection of radioisotopes or X-ray exposures.
- Some of the above-ground nuclear tests in the 1940s were viewed by people (civilian and military) up close.
- Patients were injected with plutonium to see how quickly the body rids itself of this toxic element.

While the number of people involved in these tests was admittedly small (fewer than 1000), did scientists have a right to treat people as guinea pigs? The United States was locked into a cold war with the Soviet Union and there was the perceived need to learn as much as possible about the effects of radiation on people. In the midst of the secrecy of the times, doctors and scientists were reluctant to disclose fully what was being done, even if they knew about the possible consequences (which many times they did not). Did the ends justify the means?

radium-containing paint was put on watch numerals with a small brush that was often wetted in the mouth. The painters showed a marked increase in bone cancer in later years.

Such studies of radiation dose effects generally assume that a **linear response** to radiation occurs, meaning that a dose-response curve can be extrapolated from experimental data on high doses back to zero dose and zero effect. This assumes that no threshold exists for radiation damage: Any radiation will have a harmful effect, no matter how little (Fig. 15.4). There is debate on the validity of this linear hypothesis, especially for somatic damage. It is known that

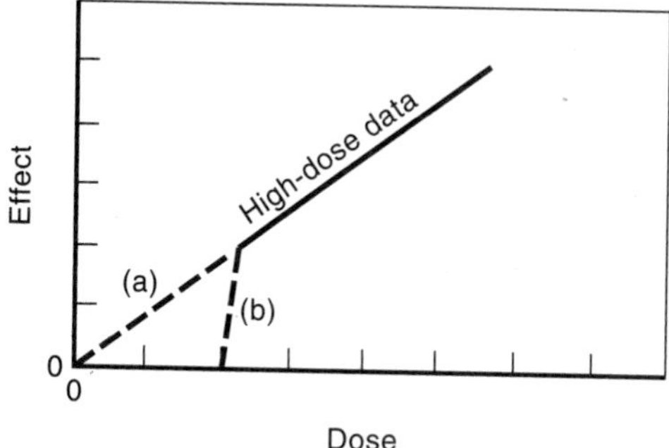

FIGURE 15.4
Dose-response curves for low doses of radiation are based on the assumption of a linear response—that is, that the data from high doses (*solid curve*) can be extrapolated back to low doses (as in curve *a*). No threshold exists (as is assumed for curve *b*), so that any radiation is assumed to have a harmful effect.

small damage in a localized region can be repaired by the replacement of dead cells by new ones. A dose of 10 rem at one time might cause the same cellular damage as 40 rem over a longer period of time, as a result of the body's repair mechanism. Clearly it is difficult to gather data on the effects of low levels of radiation in a population. One would have to separate out the effects from naturally occurring radiation and other environmental effects. Also, such low-level effects might not be observable for years, or generations. Thus data for high doses must be used. The use of these data to establish radiation standards will be outlined in a later section, after an examination of contributions from background radiation.

D. Background Radiation, Including Radon

Before examining the radiation emitted from nuclear power plants, it is necessary to know the amount of radiation the average person receives from other sources. This background radiation can be divided into two types: that from natural sources and that from human-made devices, primarily for medical exposures.

The most important source of natural radiation comes from **radon** and its decay products (called radon daughters). Radon gas is a naturally occurring radioactive element found in soils and rocks that make up the earth's crust. It comes from the decay of radium, which originally came from the decay of uranium. Figure 15.5 shows the decay scheme of radon. Radon, with a half-life of 3.8 days, decays into polonium-218 by emitting an α particle. This decays by α emission into lead-214. The decay of radon and its daughters occurs within a relatively short time, with the release of α radiation that is very harmful in tissue because of its short range. Because they are solids, the radon decay products can become trapped in lungs. Medical studies of uranium mine workers have shown that long-term exposure to radon gas and its daughters can cause cancer. It's generally agreed that radon is a major cause of lung cancer, with estimates ranging from 5000 to 30,000 deaths each year; this is about 10 to 15% of

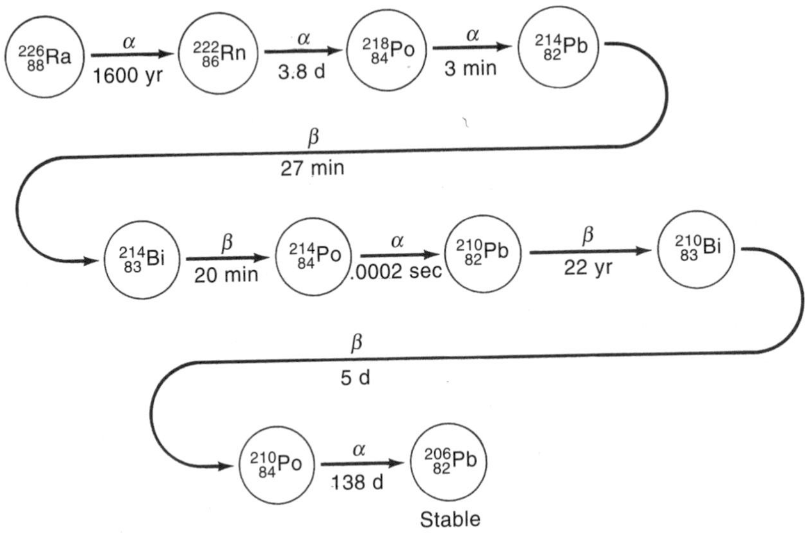

FIGURE 15.5

Radon (Rn) decay scheme, starting with radioactive radium-226. The half-life of the parent, and type of radiation, is noted.

all lung cancers! Figure 15.6 shows the risk of lung cancer from radon exposure compared with the risk incurred by smoking. An indoor concentration of 10 picocuries per liter of air (10×10^{-12} curies per liter, abbreviated pCi/L) over a lifetime is equivalent in health effects to smoking one pack of cigarettes per day!

Radon gas can enter a house through cracks in concrete floors and walls of a basement, floor drains, sump pumps, and the water supply within private wells. The main cause of radon entry from soils is pressure-driven air flow, because the air in a house is generally at a slightly lower pressure than the surrounding environment. In some unusual situations, radon may be released from the materials used in the construction of a home, such as stone fireplaces. These problems are especially great if the exchange of indoor air with the outside is poor, as might occur in the winter in an energy-efficient house with excellent infiltration barriers. Thus radon concentrations can vary drastically from house to house in the same neighborhood.

Indoor levels of radon are measured in units of the activity per standard volume of air. The units used are picocuries (pCi) per liter. The EPA has set a standard of 4.0 pCi/L; this is equivalent to about two disintegrations of radon per minute per liter of air. House levels above this should call for immediate attention. The national average is about 1.3 pCi/L. At a level of 4 pCi/L, people would receive a radiation dose of about 7700 mrem to the sensitive cells in the lung, or about 1000 mrem whole-body equivalent dose each year if they spent

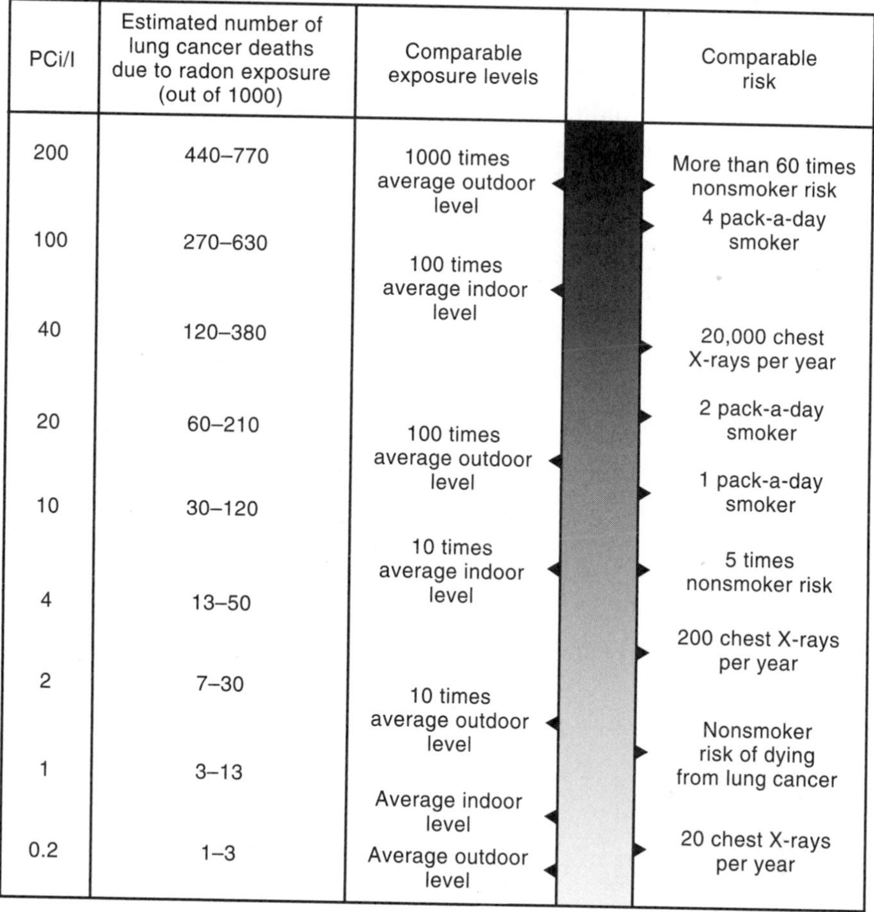

FIGURE 15.6

Comparable risks from exposure to radon gas. (United States Environmental Protection Agency)

75% of their time in the house. A committee of the National Academy of Sciences has studied the biological effects of ionizing radiation, and in a 1998 report (BEIR VI) it uses an average effective dose equivalent for the United States from radon of 200 mrem/y. Remember, though, that radon levels vary dramatically from location to location. The level within a particular home seems to depend as much on the microgeology of the area as on the airtightness of the home. Inexpensive radon detection kits can be purchased from most hardware stores. These charcoal containing canisters are placed in the basement for about a week and then analyzed by the manufacturer. Many local laws require such a test before a home can be sold.

Radon concentrations in a house can be reduced inexpensively by sealing the cracks in the basement floor and walls, and introducing an air-to-air heat exchanger. Figure 15.7 shows major radon entry routes into a house. You can also reduce the infiltration of radon gas from the soil by locating a ventilation drain around the house's perimeter or under a basement slab that is ventilated to the outside. This can be an expensive task.

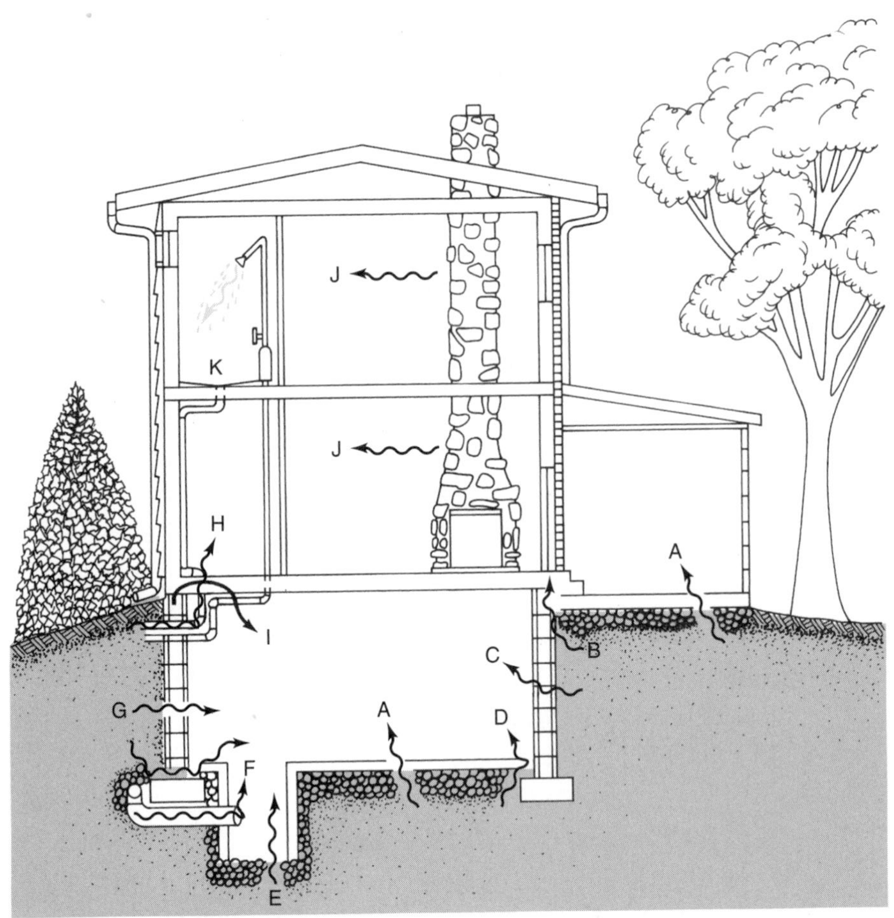

A. Cracks in concrete slabs
B. Spaces behind brick veneer walls that rest on uncapped hollow-block foundation
C. Pores and cracks in concrete blocks
D. Floor–wall joints
E. Exposed soil, as in a sump
F. Weeping (drain) tile, if drained to open sump
G. Mortar joints
H. Loose-fitting pipe penetrations
I. Open tops of block walls
J. Building materials, such as some rock
K. Water (from some wells)

FIGURE 15.7

Key to major radon entry routes. (UNITED STATES ENVIRONMENTAL PROTECTION AGENCY)

Another important type of external natural radiation comes from **cosmic rays,** which are high-energy particles (primarily protons) and γ-rays from outer space. The dose rate from this source depends on altitude and latitude. Recall from Chapter 11 that the earth's magnetic field deflects charged particle radiations best at the equator and least at the poles (Fig. 11.5). Air acts as an attenuator (reduces the intensity) for cosmic rays, so people living in mountainous regions will receive more radiation than those at sea level. There is an increase by a factor of about 3 in cosmic radiation between sea level and 10,000 feet. By state, the dose varies from 30 mrem/y in such states as Alabama, California, and Massachusetts to 120 mrem/y in Colorado. The average whole-body dose for the United States from cosmic rays is about 30 mrem/y. Radiation exposure has been measured on commercial airplane flights, where there are increased cosmic ray intensities. A dose rate of ⅓ mrem/h has been found. Thus, for a transcontinental flight, one could expect to receive about 2 mrem.

Another source of natural radiation is from the foods and liquids we ingest. Fertilizers containing phosphates have potassium (which contains the radioisotope ^{40}K) and uranium. Some cereals and teas have radiation levels of almost 1 pCi/g. Brazil nuts have high concentrations of radium (14 pCi/g), radioactive enough to qualify as low-level radioactive waste! On the average, our **internal radiation** dose is 40 mrem/y.

External doses from the **earth** itself vary greatly depending on location, ranging from 30 mrem/y in Texas to 100 mrem/y in Colorado and South Dakota. Some places have such great natural concentrations of thorium and uranium that the background radiation is 10 times as high as the average value. Because of radiation from the earth and its materials, the external radiation you receive also depends on the type of house in which you live. A brick house will expose the occupant to 40 mrem/y radiation more than if the house were made of wood. The high uranium content of the granite walls in the U.S. Capitol Building in Washington, D.C., provides an exposure rate much above background. The average dose from the earth's external radiation is about 30 mrem/y.

The **medical and dental** use of X-rays and radioisotopes is widespread and important. Such radiation is used for diagnostic and therapeutic purposes, and will be discussed in the next section. More than 60% of the U.S. population have some medical or dental X-rays every year. Medical exposures are the largest component of artificial radiation. The dose received depends on the specific type of examination. Genetic effects may occur with any radiation that reaches the reproductive organs (the gonads). Since X-rays are scattered by matter, even radiation to the head (such as dental X-rays) will result in some dose to the gonads. Lower-body examinations have increased gonad doses. To formulate an average dose for the entire U.S. population from medical diagnostic radiology and nuclear medicine, a "genetically significant dose" (GSD) is used. This is derived using gonad dose data, weighted and corrected for the proportion of examinations done and the fraction of the population irradiated that is expected to have children. The average GSD for the entire United States is about 55 mrem/y per person. (It is estimated that this number could be reduced by 50% by proper adjustment of equipment and shielding.)

Table 15.4 MEAN EXPOSURE BY TYPE OF X-RAY EXAMINATION

Type	Dose at Skin Entrance (mrem)
Head and neck	300
Chest	45
Abdomen and gonads	950
GI series (Gastrointestinal tract)	1000–8000
Arms and legs	100
Dental film	1000

Table 15.5 AVERAGE ANNUAL RADIATION DOSE RECEIVED BY INDIVIDUALS IN THE UNITED STATES*

Source	Effective Dose Equivalent (mrem/y)
Natural sources	
Inhaled radon daughters	200
Cosmic rays	30
Terrestrial	30
Internal natural radionuclides	40
Artificial Sources	
Medical/dental X-rays	39
Nuclear medicine	14
Consumer products	9
All other sources (including occupational, fallout, nuclear fuel cycle)	< 3
Rounded Total	**360**

*From "Ionizing Radiation Exposure of the Population of the United States." (National Council on Radiation Protection, 1987)

It should be remembered that these values for medical doses are averages for the entire population. Even though many people never have an X-ray, the average for Americans is four dental X-rays and one medical X-ray every 10 years. Mean exposures from some common X-ray examinations are given in Table 15.4. The average skin or entrance exposure is given. While the skin exposure from a dental X-ray is about 1000 mrem, its abdominal (somatic) dose is only 2 mrem.

The average components for all forms of radiation exposure in the United States are summarized in Table 15.5. The sum is about 360 mrem/y per person. With the decline in the above-ground testing of nuclear weapons, the contribution from fallout decreased from 13 mrem/y in 1963 to less than 1 mrem/y now. The contribution of nuclear power plants (and the fuel cycle) is very small, less than 1 mrem/y as an average to the entire population. Exposure from these plants comes primarily from the gaseous effluents ^{85}Kr, ^{131}Xe, and ^{133}Xe. The dose to someone living at the power plant boundary is about 1 mrem/y. You can calculate your own dose from all sources by using Table 15.7 at the end of this chapter.

E. Radiation Standards

Establishing and maintaining radiation standards should be carried out with the attitude that no exposure to ionizing radiation should be permitted without an evaluation of expected benefits. A comparison should be made between the potential risks and attainable benefits, and risks of alternative options if no radiation is allowed. This is not always done. The present radiation guide for the general population is based on genetic effect risks. The (now extinct) Federal Radiation Council recommended, in 1960, that a guide or maximum exposure limit of 5 rem (5000 mrem) per year (above background radiation—both natural and medical) be adopted. If such an exposure occurs over 30 years, this corresponds to 170 mrem/y per person (i.e., 5000 mrem/30 y = 170 mrem/y). Such a standard is now felt to be much too high by most experts.

To reexamine the basis for these standards and to make risk–benefit comparisons, we must work with data that relate risk to radiation exposure. This task was carried out by the National Academy of Sciences committee on the Biological Effects of Ionizing Radiation. The BEIR Report was first published in 1972 and was updated in following years; the latest (BEIR VI) came out in 1998. The committee used data from mouse and fruit-fly exposures and Hiroshima–Nagasaki studies to arrive at risk estimates for genetic effects. It estimated the genetic doubling dose (the dose required to produce a number of mutations equal to those that occur naturally) to be 100 rem. For somatic effects, risk estimates were made from Hiroshima–Nagasaki survivors and occupationally exposed groups of people. Extrapolating from the data, the number of cancer deaths annually per rem could be estimated. It was estimated that the excess mortality rate is 560 cancer deaths per million people per rem of annual dose to the whole body. (This number includes all cancers. Normally, there are about 18,000 cancer deaths per year per million people.) Using these numbers, an exposure of 170 mrem/y above the background level to the entire U.S. population would cause about 24,000 extra cancer deaths per year, an increase of about 6% in the spontaneous cancer death rate. Fortunately, the general public does not receive an additional 170 mrem/y exposure.

In 1994, after evaluation of new dose-effect data, the NRC set the limit on the maximum permissible dose a member of the public could receive in 1 year above background at 100 mrem. In other activities, the public should receive no more than 25 mrem/y from all phases of the nuclear cycle. The dose limit for a person living at the boundary of a nuclear power plant is 5 mrem/y. (The doses actually received are much lower.) Radiation workers are allowed to receive up to 5000 mrem/y whole-body exposure, although many nuclear plants have set lower limits.

F. Medical and Industrial Uses of Radiation

To describe the many beneficial uses of ionizing radiation would take an entire book in itself. In this section, a brief overview of the medical and industrial uses of radiation and the principles involved will be covered. As we have seen, the effects of radiation depend not only on the dose received, but the type of radiation, the organ irradiated, and the time between doses. One of the most important therapeutic uses of radiation is the treatment of cancer. Chances are that you have had a friend or relative who has received cobalt-60 treatment to kill cancer cells. Cancer cells usually divide rapidly, so that they are expected to be more sensitive to radiation than normal cells. However, other factors make this radio-sensitivity difference less distinct, so one has to be careful about inducing harmful effects in the surrounding good tissue when irradiating the bad tissue.

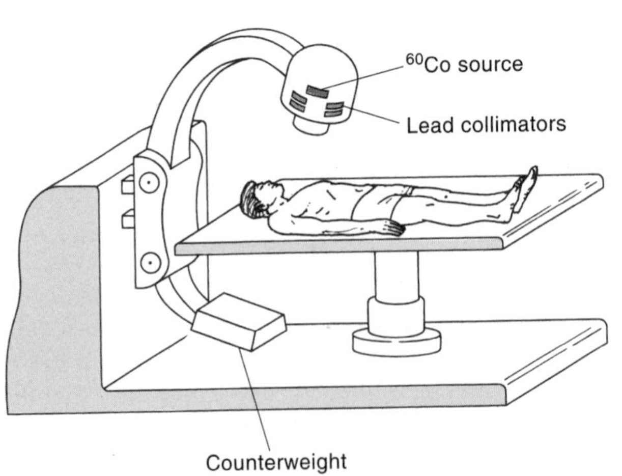

FIGURE 15.8
A cobalt-60 unit for radiation therapy. The machine allows the radioactive source to rotate around the stationary patient. The tumor being treated is placed exactly at the center of rotation, so that damage to good tissue is minimized.

One procedure to reduce damaging effects of radiation treatment is illustrated in Figure 15.8. Here the patient receives exposures from all sides via a rotating γ-ray source. The rotation is around the bad tissue, so it is always receiving radiation. The good tissue only receives radiation periodically. Such sources are usually quite intense (10,000 Ci), yielding a dose rate of about 100 rem/min at 1 m. Consequently, a large amount of lead shielding is needed around the source.

Another method for tumor treatment uses particles from an **accelerator.** At high enough energies, the particles are able to pass into the center of the body. Charged particles have the advantage that cell destruction can be localized at a particular region inside the body, thus killing an internal tumor with little damage to the surrounding cells.

Radioisotopes have many uses today, from clinical medicine to industry. In medicine, radioisotopes are used for both diagnostic and therapeutic purposes (Table 15.6). They have the advantages that only small amounts are usually needed, and they can be directed to specific parts of the body. In diagnosis,

Table 15.6 SOME RADIOISOTOPES AND THEIR USES*

Americium-241 (458 y) Used in many smoke detectors; to check uniform thickness in sheet metal, paper; to measure toxic lead levels in paint	**Krypton-85** (10.7 y) Used in indicator lights in appliances; measures dust and pollutant levels
Californium-252 (2.6 y) Used to inspect airline luggage for explosives; to measure moisture content of soil	**Phosphorus-32** (14 days) Tracer in molecular biology and genetics research
Carbon-14 (5730 y) Tracer in biomedical research; radioactive dating	**Technetium-99** (6 h) Used in imaging brain, bone, liver, kidneys
Cesium-137 (30 y) Treatment of cancerous tumors; check fill-level for packaged products; thickness gauge	**Thallium-204** (3.8 y) Measures dust and pollutant levels on filter paper
Cobalt-60 (5.2 y) Treatment of cancerous tumors; sterilizing surgical instruments	**Xenon-133** (2.3 days) Lung ventilation and blood flow studies
Iodine-131 (8 days) Used to diagnose and treat thyroid disorders	**Tritium (H-3)** (12 y) Tracer for drug metabolism; lumination for aircraft exit signs

*Half-lives appear in parentheses.

radioisotopes are used as "tracer atoms" to determine flow and absorption of elements in animals and plants. The chemical and biological properties of elements are determined by their electron structure, so the body does not differentiate between isotopes of the same element: The radioisotope's chemical behavior will be the same as that of other nonradioactive atoms of the same element. To trace the flow of a particular element, some of the atoms can be "tagged" by using a radioactive isotope of that element and then using a radiation detector to determine its location in the body at intervals of time after injection.

One of the most commonly used tracer radioisotopes is ^{131}I. This isotope emits 0.36-MeV γ-rays and has a half-life of eight days. Because iodine becomes concentrated in the thyroid gland, it is particularly useful in evaluating the functioning of this gland. (The thyroid uses iodine in the production of hormones to control the body's metabolism.) Iodine is usually taken by mouth, and the γ-rays from ^{131}I are counted days later with a meter. Another tracer is ^{51}Cr. Since the chromium atom attaches itself to red blood cells, doctors can study blood flow by injecting radioactive chromium into the blood. Nitrogen-15 can be used as a tracer to study proteins in the body and the functioning of liver and kidneys.

Not only can we observe the rate of uptake or flow of a radioisotope in the body with a radiation detector, but we can take "pictures" of how a radioisotope is distributed within an organ. The organ is scanned with a γ-ray camera (see the Special Topic section) to provide an image. One radioisotope commonly used for this purpose is technetium-99, with a 6-hour half-life. It is used for imaging bone, heart, kidneys, and brain. A tumor can be identified because some elements are taken up better by tumor tissue than by the surrounding normal tissue. Whole-body scans are also common today. Bone scans are useful for showing any portion of the skeleton that is cancerous. The bone being destroyed by cancer takes up more of certain elements (^{99}Tc in this case) than normal bone as it tries to rebuild itself, so a scan will show increased radioactivity in the area of the bone tumor.

In selecting radioisotopes to be used for diagnostic purposes, the half-life of the isotope is important; it should not be too short (less than a few seconds) or too long (more than a few weeks). Also, the radiation emitted must be observable by a detector outside the body. Since γ-rays are very penetrating, γ-ray emitters are the most useful radioisotopes for medical diagnosis. An example of the therapeutic use of radioisotopes is the implantation of radioactive "seeds" (usually an iodine isotope) into the prostate gland. This procedure may

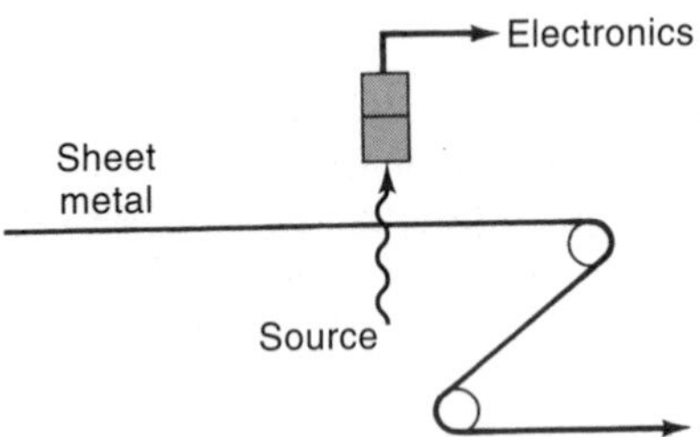

FIGURE 15.9

Radioisotope use in industry. An increase or reduction in the intensity of γ-rays transmitted through the sheet metal indicates a thinner or thicker sheet, respectively. This is used as a thickness gauge.

be used as an alternative to surgery or external radiation for the treatment of prostate cancer, and has proven effective in several studies.

There are many industrial uses of radiation and radioisotopes as well. In the production of sheets of metal, radiation can be used to measure the thickness of the stock (Fig. 15.9). The transmission of radiation through the material will

Focus On 15.2

FOOD IRRADIATION

The radiation of food with γ-rays, X-rays, or electron beams is known to rid foods of a variety of bacteria, molds, and fungi, and will increase its shelf life. In 1963, the USDA approved irradiation to control insects in wheat and wheat flour; in 1985 they approved its use for pork to control trichinosis, in 1986 for certain fruits to delay ripening and infestation, in 1990 to control salmonella in poultry, and in 1999 to control E-coli bacteria in red meat. Yet, many people are afraid of irradiated foods and few markets carry a large menu of such products. The international symbol for irradiated foods that must appear on all such products is called the **radura** as shown. Its color is green.

Irradiation will not make food radioactive, since the energy of the radiation is not enough to do anything but ionize the atom. The nutritional value may be altered slightly, but not the taste. The food is usually passed under a radioactive source (cobalt-60) and then packaged. (It must still be refrigerated.) The dose permitted varies according to the type of food, but most doses are less than 1 kiloGray (100,000 rad). (This number is equivalent to about 10 million chest X-rays.) Some chemical changes will produce "radiolytic products," but such products are also found naturally at much higher levels. While most food organizations have endorsed food irradiation as a foolproof and safe way to rid our food supply of disease-causing microbes, public acceptance is lacking. It may take an outbreak of food poisoning that results in many deaths before the public accepts food irradiation.

change if the thickness varies. A change in the detected signal can be fed back to control the manufacturing equipment.

Radioisotopes can be used to measure piston ring wear in an automobile engine. The piston ring can be made radioactive by neutron irradiation, and then the wear can be measured by observing the increasing concentration of radioactive material in the lubricating oil. An energy-related use of radioisotopes is the identification of oil from different companies that is sent through the same pipeline. There are more than 225,000 miles of pipeline in the United States. Crude oil from wells goes into large, centralized lines that carry it to some 250 refineries. Pipelines that cross state boundaries are "common carriers"; they must accept oil from any company that meets the tariff conditions. Different shipments can be separated by adding a radioactive tracer to each batch and detecting it at the end of the pipeline.

G. Radiation Protection

It is generally believed that the best radiation dose is the least dose, preferably zero. This is especially true with regard to the gender cells, where every bit of radiation adds to the number of mutated cells. However, when radiation is present, there are some procedures that can be followed to minimize exposure. The first is **distance.** As radiation expands outward from a point source, its inten-

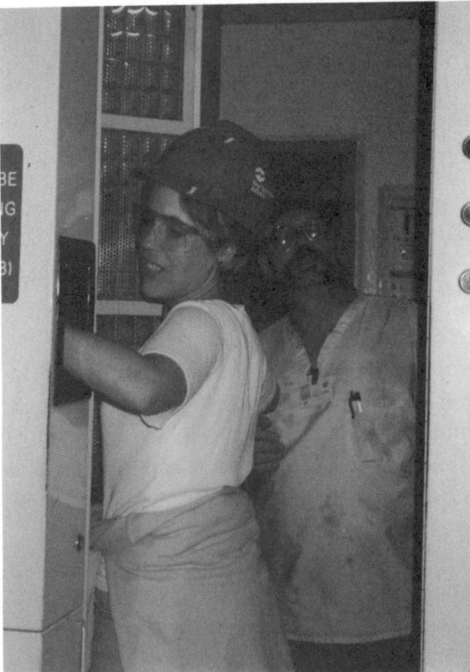

Being checked for contamination by whole-body scan (using NaI detectors), upon exiting nuclear power plant.

sity at any one point is inversely proportional to the square of the distance from the source. The radiation received at 2 meters will be four times less than that received at 1 meter, and nine times less at 3 meters than at 1 meter.

The use of appropriate **shielding** (concrete blocks, lead walls) is also important in minimizing exposure. For α-radiation, air itself usually provides sufficient shielding. Shielding for radiation is expressed in terms of **half-value thicknesses** for a particular radiation, specifically, the thickness that will reduce the incident radiation intensity by a factor of 2. For X-rays and γ-rays, the best type of shielding is lead. About 1 cm of lead will attenuate (reduce the intensity of) the 1.3-MeV γ-rays from ^{60}Co by a factor of 2. Since the radiation intensity is reduced exponentially with increased shielding thickness, each half-value thickness will reduce the intensity by another factor of 2. Thus, 3 cm of lead will reduce the number of γ-rays by a factor of 8. For neutrons, the best shield is material with a small atomic number; the neutron will lose a good deal of its energy in its collisions with the hydrogen nuclei in the shielding. Water, paraffin, and concrete are good shields. About 5 cm of water or paraffin will attenuate 4-MeV neutrons by a factor of 2. (It would take 17 cm of lead to do the same job.)

Finally, **time of exposure** is important. It is the cumulative dose that counts, so exposure should be kept to a minimum. Radiation workers wear film badges (Fig. 15.10), which record the radiation received during a particular time interval. The blackening of the film emulsion is an indication of the

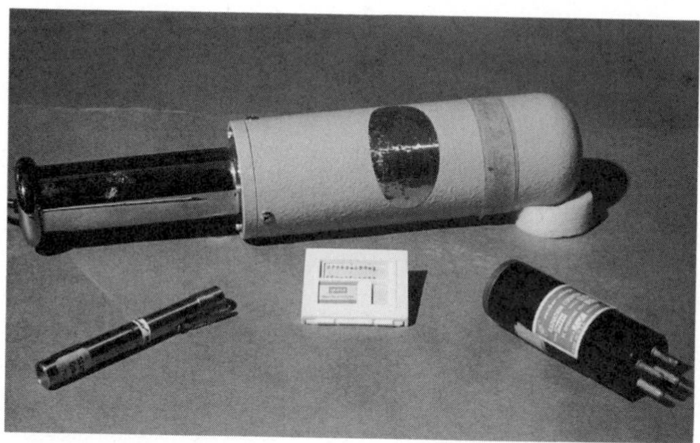

FIGURE 15.10

Radiation detection devices. The film badge in the center consists of photographic film, which darkens as a function of the amount of ionizing radiation received. The pocket dosimeter on the left can be read at any time to see how much radiation has been accumulated. The Geiger tube on the right is used with a counter to measure radiation intensity, while the large device in the background is a detector to monitor neutron activity. (GARY BURGESS)

person's exposure to β and γ radiation, usually given in millirems. The individual's cumulative dose is kept by the employer on permanent file, even if the job position changes. Another device for the measurement of personal exposure is a thermoluminescent dosimeter (TLD), which is more sensitive than film badges and provides for periodic evaluation of cumulative exposure. TLDs use crystals in which electrons are excited by the incident radiation and become "trapped" in excited states within the crystal. The electrons are freed on the application of heat, giving up energy in the form of light, which can be measured with a photomultiplier tube. Another device worn by workers in radiation areas to measure total dose (not dose rate) is a pocket dosimeter. This instrument is an ionization chamber that responds to electrons created by the interaction of γ-rays with the metal case. (See the Special Topic section.)

H. Summary

Radiation is defined as the process by which energy is emitted from a body as waves or particles. Radiation from a radioisotope is energetic enough to cause damage to biological tissue, and is called ionizing radiation. The activity of a radioactive source (expressed in disintegrations per second or curies) is related to the amount of source material present and its half-life. The radiation dose one might receive from exposure to a source is related to the energy deposited in tissue, expressed in units of rads. Different types of radiation (α, β, γ) can cause varying degrees of damage, and this is taken into account by using a biological dose unit, expressed in rems.

The biological effects of radiation are of somatic (physical) and genetic types. Somatic effects, such as cancer, are often not observed for many years in the exposed individual. Radiation effects are a function of dose received and the part of the body irradiated. (The extremities are less sensitive.) For ingested radioactivity, the effects are a function of the type of radiation, the biological and physical half-lives of the element, and the organ affected.

Present radiation standards permit doses of 100 mrem/y above background. Background radiation from natural sources includes indoor radon (200 mrem/y average), cosmic rays (30 mrem/y average), the earth (30 mrem/y average), and internally ingested radioisotopes (40 mrem/y). Americans will receive on the average about 55 mrem/y from medical and dental diagnostic and therapeutic irradiation.

Radiation doses can be reduced by increasing your distance from the source, minimizing your time near the source, or having appropriate shielding between yourself and the source.

Industrial and medical uses of radiation are numerous. Such applications make use of (1) radioisotopes that are injected into a material to trace fluid flow or treat cancerous tissue, or (2) direct radiation from a source (such as ^{60}Co) or radiation-producing equipment (such as an accelerator).

Internet Sites

For an up-to-date list of Internet resources related to the material in this chapter, go to the Harcourt College Publishers website at **http://www.harcourtcollege.com**. The links are in the *Energy: Its Use and the Environment* site on the Physics page. General energy related sites and some guidelines for using the World Wide Web in your class are on the inside front cover of this book.

References

Committee on Health Effects of Exposure to Low Levels of Ionizing Radiation. 1990. *Biological Effects of Ionizing Radiation.* Washington, D.C., National Academy Press (BEIR VI, 1998).

Inglis, D. 1973. *Nuclear Energy—Its Physics and Its Social Challenge.* Reading, MA, Addison-Wesley.

Macklis, R. 1993. The Great Radium Scandal. *Scientific American,* 269 (August).

Nero, A. 1988. Controlling Indoor Air Pollution. *Scientific American,* 256 (May).

Upton, A. C. 1982. The Biological Effects of Low-Level Radiation. *Scientific American,* 246 (May).

QUESTIONS

1. What is the most penetrating radiation at the same energy: α, β, or γ radiation?
2. What are the differences between activity, dose, and dose rate?
3. Is there always damage to a cell if it is hit by radiation?
4. What are the advantages of using heavy charged particles (such as nitrogen nuclei) rather than γ-rays for the treatment of tumors within the body?
5. Which is worse: a 1-rad dose from α particles or 5 rads from X-rays?
6. Suppose a friend swallows some radioactive material by accident. What information would you want to know to be able to assess possible damage?
7. A person who worked around radiation-producing equipment (e.g., TV tubes) was worried that radiation had accumulated in his body to such an extent that he could not get rid of it. What would you say?
8. There has been talk about the development of a "neutron bomb" that would inflict damage to people but not buildings. What shielding would be best for protection against neutrons?
9. Why do chest X-rays contribute to the gonad dose?
10. If everyone in the United States had an extra 5 mrem/y of radiation, what would be the increase in the number of cancer deaths per year (after time had elapsed for the effects to become apparent)? Assume that there are 200×10^{-6} deaths per rem of radiation per year.

11. A piece of equipment in a nuclear power plant is reading 900 mrem/h at a distance of 1 meter. How far would you need to back up to reduce the radiation at your position to 100 mrem/h?
12. How long can a worker remain in an area where the dose rate is 150 mrem/h if her allowable dose is 3 rem?
13. As you wander about a dump site, your survey meter begins to read a radiation field that would give you an exposure of 5000 mrem in 20 minutes. What dose rate are you reading and what types of radiation are you detecting?
14. How would you use a radioisotope to study particular effluents in a water system?
15. Calculate your average annual radiation dose by filling in Table 15.7.

Table 15.7 AVERAGE ANNUAL PERSONAL RADIATION DOSE*

Common Sources of Radiation		Your Annual Dose (mrem)
Where You Live	**Location:** Cosmic radiation at sea level	26
	For your elevation (in feet), add this number of mrem Elevation mrem Elevation mrem Elevation mrem 1000 2 4000 15 7000 40 2000 5 5000 21 8000 53 3000 9 6000 29 9000 70	______
	Ground: U.S. average	26
	House Construction: For stone, concrete, or masonry building, add 7	______
What You Eat, Drink, Drink, and Breathe	**Food, Water, Air:** U.S. average	24
	Weapons test fallout	1
How You Live	**X-ray and Radiopharmaceutical Diagnosis:** Number of chest X-rays ______ × 10	______
	Number of lower gastrointestinal tract X-rays ______ × 500	______
	Number of radiopharmaceutical examinations ______ × 300 (Average dose to total U.S. population = 92 mrem)	______
	Jet Plane Travel: For each 2500 miles add 1 mrem	______
	TV Viewing (or Computer Usage): Number of hours per day ______ × 0.15	______

How Close You Live to a Nuclear Plant	**At Site Boundary:** Average number of hours per day _____ × 0.2	_____
	One Mile Away: Average number of hours per day _____ × 0.02	_____
	Five Miles Away: Average number of hours per day _____ × 0.002	_____
	Over 5 Miles Away: NONE *Note:* Maximum allowable dose determined by "as low as reasonably achievable" (ALARA) criteria established by the U.S. Nuclear Regulatory Commission. Experience shows that your actual dose is substantially less than these limits.	_____
	Your total annual mrem dose	_____

Compare your annual dose to the U.S. annual average of 160 mrem. One mrem per year is equal to increasing your diet by 4%, or taking a 5-day vacation in the Sierra Nevada (California) mountains.

*Based on the "BEIR Report III," National Academy of Sciences, Committee on Biological Effects of Ionizing Radiation. The Effects on Populations of Exposure to Low Levels of Ionizing Radiation, National Academy of Sciences, Washington, D.C., 1980. Dose from radon is excluded.

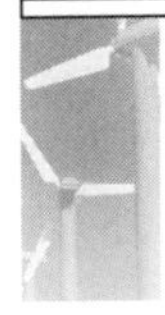

SPECIAL TOPIC

Radiation Detection Instruments

Instruments for monitoring radiation exposure or determining the type and energy of radiation make use of the ions produced by the radiation within a material. These detectors can be grouped into gas-filled counters, scintillators, and solid-state detectors.

Gas-Filled Detectors

These detectors are used primarily for counting radiation or determining dose rate. The detector consists basically of a volume of gas in a container (usually cylindrical in shape) where there are positive and negative electrodes to collect the ion pairs that are produced by the incident radiation (Fig. 15.11). An external voltage is applied between the electrodes. A meter is placed in the circuit to measure the current produced by the electrons (produced by ionization) collected at the positive electrode. The walls or window of the detector must allow the radiation to enter the gaseous region; α radiation thus requires a very thin window. Gas-filled detectors used as portable survey instruments use batteries to provide external voltages. These survey instruments are called Geiger-Mueller (GM) detectors. These detectors count individual radiation events and can provide a calibrated output in mrem/h or counts per minute. To measure a total radiation exposure over a period of time, a pocket dosimeter (see Fig. 15.10) can be used, which is a small gas-filled detector measuring total ionization, not ionization per hour.

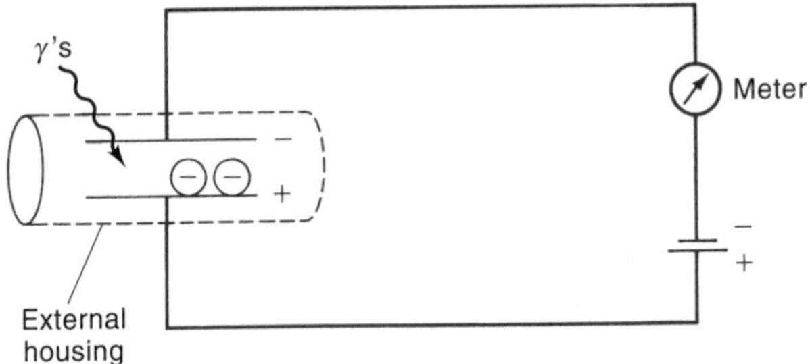

FIGURE 15.11
A gas-filled radiation detector.

Scintillation Counters

Substances that are capable of emitting light when exposed to ionizing radiation are called scintillators. Ernest Rutherford observed the scattered α particles of his famous experiment (see Chapter 13) by using a ZnS scintillator. Scintillators can be made of organic or inorganic materials in solid, liquid, or gaseous form. The most widely used inorganic scintillator is sodium iodide, NaI, which has the largest light output. The total amount of light produced is a function of the incident radiation energy, so that a measurement of light intensity can be used to identify different radioisotopes present in a sample. The flashes of light in a scintillator are converted into an electrical signal by a photomultiplier tube. A photosensitive material between the scintillator and the photomultiplier tube converts light into electrons (the photoelectric effect). The current is converted to a voltage for amplification and analysis in an external circuit (Fig. 15.12). One can use this technique to obtain a large picture of body organs with a γ-ray camera. This device positions many photomultiplier tubes above a large NaI crystal. Each tube's output can be analyzed to piece together a two-dimensional image.

Semiconductor Detectors

Semiconductor detectors are similar in principle to gas-filled detectors, except that the gas is replaced with a semiconductor material such as silicon or germanium. The current collected at the electrode is proportional to the energy of the

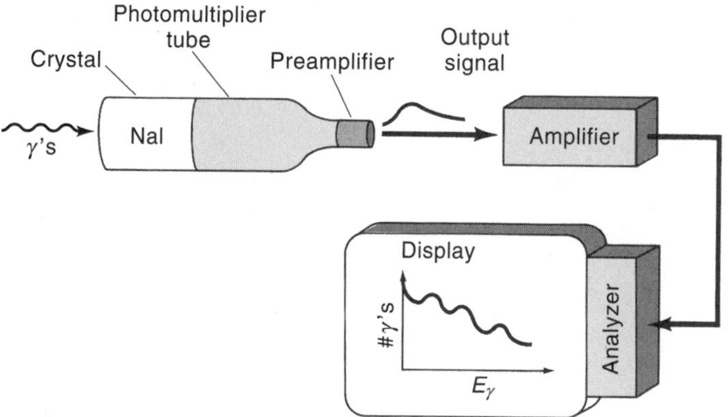

FIGURE 15.12

Analysis of γ-rays from a radioactive source, using a NaI scintillation detector. The NaI crystal is coupled to a photomultiplier tube, whose output is amplified and directed to an analyzer. The analyzer determines the number of γ-rays as a function of their energy, and the results are displayed on the monitor.

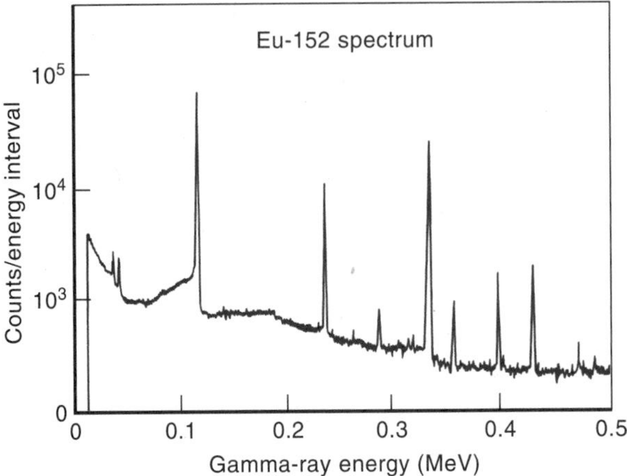

FIGURE 15.13
Gamma-ray spectrum of Eu-152, obtained with a germanium detector. The spectrum shows the number of counts (on the y-axis) as a function of γ-ray energy.

incident radiation. Germanium detectors are unsurpassed for γ-ray detection with fine energy resolution, although they are not as sensitive as NaI detectors since their volume is much smaller than that of NaI. They are used to identify and quantify radioactive material (Fig. 15.13).

By examining a spectrum of γ-rays from a sample, one can determine the amount and type of radioisotope present. Such methods are used in trace metal analysis of environmental samples. The sample can be irradiated in a reactor and then examined with a germanium detector. Trace impurities of different elements (as Se, Pb, Hg, etc.) are made radioactive and give off distinctive γ-rays of known energies, so identification is possible after the γ-ray energy of the observed radiation is determined.

16

Future Energy Alternatives: Fusion

A. Potential for Fusion Power

Since Chapter 1, we have emphasized the finiteness of many of our resources, and the environmental and economic constraints on their use. Worldwide reserves of oil and natural gas, in that order, will certainly undergo large declines in this century at present rates of use. Coal and uranium have longer lifetimes, but environmental and economic matters weigh heavily upon their increased use. There are essentially only two long-term options. One is solar energy, which we have separated into the renewable energies of radiant solar, wind, hydroelectric, and biomass. The other is nuclear fusion, which some consider to be our ultimate energy source. Recall that **fusion** is the joining together of two small nuclei to form a larger nucleus, while **fission** is the splitting of a very large nucleus (such as uranium), usually by the addition of a neutron, into two smaller nuclei. In both cases, the mass of the end products is less than the mass of the original reacting nuclei. This lost mass is converted into energy.

Enthusiasm for fusion as a future energy source is based on several facts. One is that the fuel that could be used—deuterium (D)—is found in ordinary water, in which about 1 out of every 6500 hydrogen atoms is this isotope. (A deuterium nucleus contains one proton and one neutron.) The complete fusion of just 1 g of deuterium (obtained from 8 gallons of water) will release energy equivalent to burning 2400 gallons of gasoline. The fusion of all the deuterium in an Olympic-sized swimming pool could provide enough electricity for a city of 100,000 people for 1 year! The energy released in the complete fusion of the deuterium found in a 1 km^3 of water is equivalent to about 2 trillion barrels of oil, approximately twice the earth's estimated total oil reserves. The extraction of deuterium from water is not very difficult or expensive, so the fuel for deuterium fusion is essentially infinite and extremely inexpensive.

Another advantage of fusion is the potential reduction in environmental pollution. The end products of the fusion reaction are hydrogen, helium, and neutrons, so we do not have to worry about the long-lived radioactive wastes of fission reactors, although there will be some radioactive reactor parts with which to contend. Also no bomb materials will be made in fusion reactors, and global warming will not be a factor.

The prospect for fusion power is not all roses, however. The technological feasibility of fusion reactors is very much open to question today. Billions of dollars have been spent since World War II trying to achieve controlled fusion. So far, the energy output obtained from demonstration reactors is less than the total energy input, although scientists feel they are close to the break-even point. There are also strong economic concerns and large price tags for fusion research in many labs around the country. In the last five years, federal funding for fusion research has declined by 40%. The role that fusion may play in this new century probably will not be known for several decades. What will make fusion a winner will be determined as much by environmental factors as by economic ones.

B. Energy From the Stars: The Fusion Process

The best example of fusion reactions are those that occur in our sun. The sun's energy output is derived from the transformation (nuclear "burning") of 4 million tons of matter into energy *each second.* In several steps, four hydrogen nuclei fuse together to form a helium nucleus plus two positrons (particles identical to electrons except that they have positive electrical charge—the electron's "antiparticle"). In stars that are hotter than our sun, energy is also released from fusion reactions of heavier nuclei (such as fusion between carbon nuclei). The most common fusion reactions carried out in experiments on earth use deuterium, D, and tritium, T (consisting of one proton p and two neutrons n) nuclei:

$$\mathrm{D + D \rightarrow p + T + 3.3\ MeV}$$
$$\mathrm{D + D \rightarrow n + {}^{3}He + 4.0\ MeV}$$
$$\mathrm{D + T \rightarrow n + {}^{4}He + 17.6\ MeV}$$
$$\mathrm{D + {}^{3}He \rightarrow p + {}^{4}He + 18.3\ MeV}$$

(The energy released in each reaction is expressed in millions of electron volts, abbreviated MeV.)

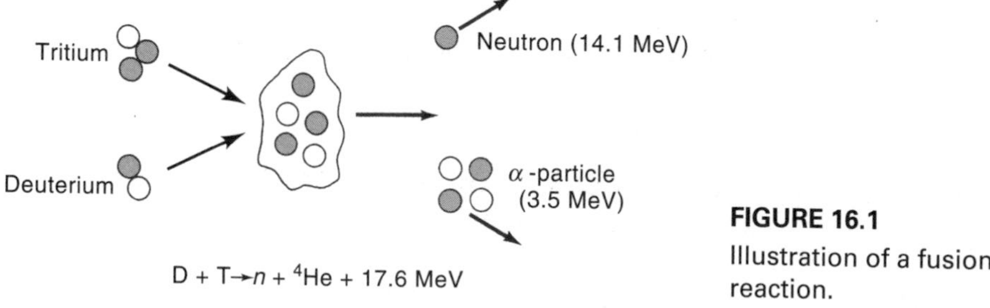

FIGURE 16.1
Illustration of a fusion reaction.

The D–T reaction is the one on which most current work is focusing (Fig. 16.1). This is because it has the lowest "ignition" temperature of any of these reactions—about 50 million °C—and it is less demanding in the conditions necessary to achieve a net power output. Tritium (T), the radioactive isotope of hydrogen with a 12-year half-life, is not found in nature except in trace amounts, but can be produced in a reactor by

$$\text{n} + {}^{6}\text{Li} \rightarrow {}^{4}\text{He} + \text{T} + 4.8 \text{ MeV}$$

Although the world supply of lithium is much less than that of deuterium, we have sufficient reserves to use for D–T technology, at least until the technology for the D–D reaction becomes available.

C. Conditions for Fusion

One of the obstacles to nuclear fusion is the electrical force of repulsion between the positive nuclei. This force is proportional to the product of the charges of the interacting nuclei and inversely proportional to their separation distance squared. To be able to overcome this repulsion, the nuclei must have high kinetic energies, which is possible only if the fuel has temperatures on the order of 50 to 100 million °C.

While the fusion reaction might appear to be simple, its use for the controlled release of energy poses some serious scientific and technological problems. Consider the requirements:

- Very high temperatures must be attained so the kinetic energies of the nuclei will be enough to overcome their electric repulsion. At high temperatures, a gas will decompose into free electrons and positive nuclei. This ionized gas, with an equal number of positive and negative charges, is called a **plasma.** Our sun is one example of a plasma; the gas in a fluorescent light bulb is another.
- Confinement of the plasma is difficult. As the temperature of a gas increases, its volume or its pressure (or both) must increase. So the gas must be confined to a fixed volume to allow the nuclei to come close enough together to fuse.

- High densities are needed because, for a sufficient release of energy through fusion, a large number of nuclei must react.
- The fusion energy released must be converted into a useful form, such as electricity.

One of the immediate goals of fusion research is "scientific breakeven," in which the energy released in the fusion reaction is equal to that required to heat the fuel and contain it. The next goal will be "ignition," in which the fusion reaction becomes *self-sustaining* with no need for external sources to heat the fuel. The heat comes from the collisions of the alpha particles (^{4}He), generated by the reaction $D + T \rightarrow n + {}^4He$, and the other particles in the plasma.

There are two key parameters that determine whether a fusion reaction will produce a net positive energy output. The first one, known as the Lawson number, characterizes the quality of the plasma confinement. To achieve substantial power output from a fusion reactor, enough deuterium and tritium ions must be kept together for a long enough time for them to react. This relationship is expressed by the **Lawson criterion**; this states that a net release of fusion energy occurs only when the product of the plasma density (in particles per cubic centimeter) and the confinement time (in seconds) exceeds 10^{14} s/cm^3. Some test reactors have achieved this condition on a limited basis. The other factor is the plasma temperature. Temperatures above 100,000,000°C are needed for ignition. This condition also has been obtained in some reactors.

There are several schemes for plasma confinement. A material box cannot be used as a container, because the hot plasma ions will lose energy by colliding with the walls. In the sun, confinement is achieved by the force of gravity. However, this won't work on the earth because of the much smaller mass of plasma. Scientists are trying two methods of confinement. One makes use of the electrical properties of the plasma and is called magnetic confinement. Another method does not invoke any physical confinement of the plasma, but makes use of such rapid heating that the nuclei fuse before the plasma has a chance to expand and decrease in density. This method, called inertial confinement, is made possible using lasers.

D. Magnetic Confinement Fusion Reactors

The most established method of plasma confinement uses magnetic fields. Several designs are being pursued for such "magnetic bottles." Recall from Chapter 11 that a charged particle experiences a force when it moves in a magnetic field. We represent the magnetic field by a set of field lines. A charged particle that has a velocity perpendicular to the magnetic field will be forced to move in a circular path around a field line. If the particle has a velocity along the field line as well, its trajectory will be helical, like a corkscrew (Fig. 16.2). The motion of the particle is tied to or confined to the field line. This container is known as a "magnetic bottle."

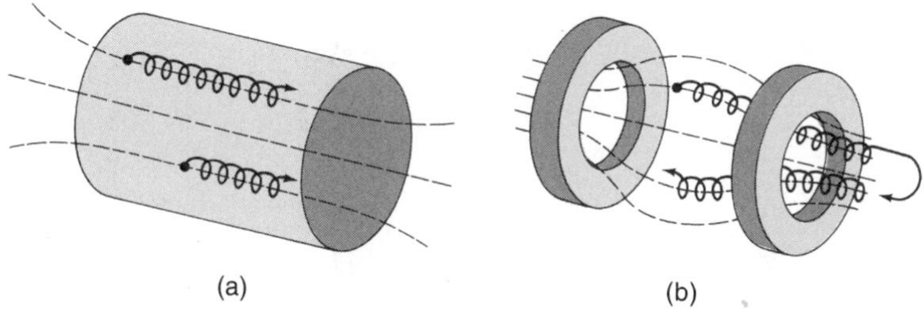

FIGURE 16.2

(*a*) Helical trajectory of a charged particle along magnetic field lines. (*b*) Mirror-confinement type of magnetic "bottle." The charged particle is reflected in the region of concentrated magnetic field lines.

The plasma must be kept within a fixed region, not just confined to a magnetic field line with no limits. This is done in one case by shaping the magnetic field with coils so that near the end of each region the magnetic field becomes much stronger, causing the particle to experience a force backward and so to be reflected from each end (Fig. 16.2b). This device is called a magnetic mirror.

Leakage of the plasma through the ends of the mirror machine can be eliminated by bringing one end of the cylindrical bottle around to join with the other end, thus making a toroid shape, like a doughnut (Fig. 16.3). In this scheme, the magnetic-field lines form closed loops. The most popular toroidal or closed confinement system is the "Tokamak," a reactor initially developed in Russia and brought into this country in 1969. Figure 11.6 showed the (now-closed) Tokamak Fusion Test Reactor at Princeton University. Figure 16.4 shows an interior view of its toroidal vacuum vessel in which plasma is confined. The **Tokamak** (Russian for "toroidal magnetic chamber") uses current windings around the toroid to provide a magnetic field around the perimeter of the

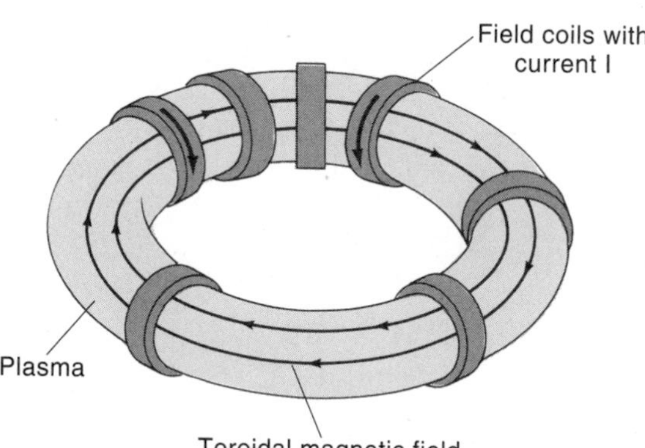

FIGURE 16.3

Toroidal or doughnut-shaped magnetic field for confinement of plasma. Only a few of the field coils are shown.

FIGURE 16.4
Interior view of the Tokamak Fusion Test Reactor vacuum vessel. (PRINCETON PLASMA PHYSICS LABORATORY)

toroid. The Tokamak also uses a current through the plasma itself to provide another magnetic field that encircles the plasma. This secondary field provides a great deal of stability to keep the plasma ions from drifting to the outside walls of the toroid. This secondary current is attained by using the plasma as a secondary coil of a transformer (Fig. 16.5). A pulse of current through the primary coil will induce a current in the secondary coil, the secondary "coil" being the plasma in this case. This current (thousands of amps) is also used to heat the plasma, much like resistive heating.

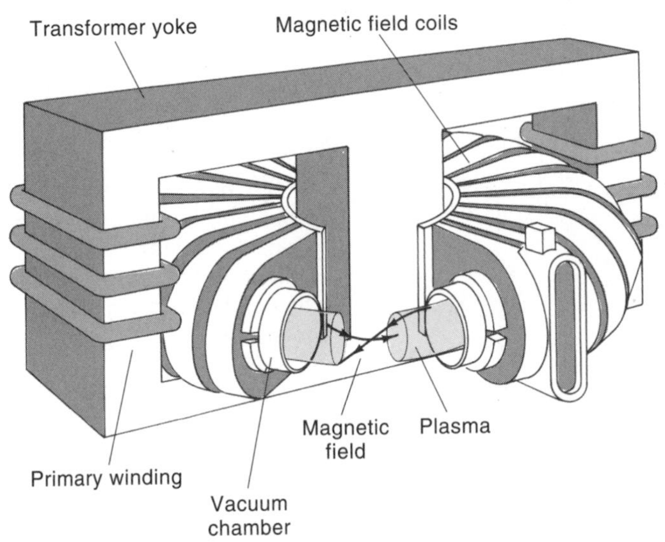

FIGURE 16.5
A Tokamak machine. Plasma is confined by a toroidal magnetic field. The toroid forms the secondary winding of a transformer. The current induced in the toroid is used for heating the plasma.

One of the problems in fusion reactors is plasma instability. If the magnetic fields develop any irregularities, such as a "kink," the plasma can leak out of the bottle. To contain the plasma at high temperatures, where the average ion speed is 1000 mph, the density of the plasma must be kept small, about 100,000 times smaller than atmospheric density, almost a vacuum. (Even at high temperatures, the plasma is not dense enough to be able to melt metal objects placed in it.) More powerful magnetic fields are being developed (using superconducting materials) to allow for an increase in the plasma density.

How might we heat our plasma to temperatures of 50 to 100 million °C? It would take a pretty good sized Bunsen burner! One method we have already mentioned: resistive heating with the secondary current in Tokamaks. However, high enough temperatures cannot be reached this way, as the resistance of the plasma decreases with higher temperatures. One promising method of plasma heating is by adding "hot" or energetic particles to the plasma from the outside. This is done by directing into the plasma a beam of deuterium atoms that have been given kinetic energies on the order of 5 to 10 keV by an accelerator.

Significant advances were made at the Princeton Plasma Physics Laboratory in 1994, when a fuel of tritium and deuterium was heated inside the Tokamak reactor. (Only deuterium had been used before.) Temperatures well above 100 million °C were achieved and a record 10 million watts of power was produced—all in the 4-second experiment. Work was completed with this reactor in 1997, and it is now in the process of being dismantled and decontaminated. A new type of magnetic confinement design (a spherical torus) began operation at Princeton in 1999 to study plasma performance.

E. Laser-Induced Fusion

The other confinement method being used in fusion is **inertial confinement.** In this case, there is no external confinement of the plasma at all. We know that as a gas increases in temperature, it usually expands. However, if the heating is done fast enough, fusion can take place before the fuel particles can move apart. The fuel is confined during this brief time by the inertia of its own mass. The heating can be done with an intense laser beam, giving us what is called "laser-induced fusion." The hydrogen bomb, which also releases energy through fusion, is triggered by a fission "atomic" bomb. This is a case of "inertial confinement" in which the fuel is driven together by the explosion of the fission device. The fusion fuel is heated by the radiation released in fission. The first H-bomb explosion in 1952 marked the first time fusion (on a large scale) had been achieved on earth. (In fact, much of the original money for U.S. laser fusion research was from the weapons program of the Department of Energy, where applications to H-bombs were foreseen.)

The number of fusion reactions in a given time will increase as the plasma's density increases, so increased output from the fuel can be achieved by compressing it. The fuel (a mixture of deuterium and tritium) is contained within a

Tiny gold microshells, similar to those containing high-pressure gaseous D–T fuel for use in laser fusion, on a U.S. quarter. (United States Department of Energy)

glass or metal pellet, a "microballoon" about 1 mm in diameter. The heating and compression is usually accomplished by subjecting this pellet to intense uniform laser pulses from all sides (Fig. 16.6). Beam splitters are used to divide the original laser beam into many components (e.g., 60 beams at the University of Rochester, New York), which are then amplified further. Mirrors are then used to direct all the beams simultaneously onto the fuel pellet.

The pellet is compressed through the process of "implosion." The outer pellet layer is heated by the laser beams so that it evaporates. This happens so rapidly that the material leaving the surface compresses the inner layers, increasing the density of the remaining core by a factor of 1000 as well as increasing its temperature. The pressure rises to a trillion atmospheres. This compression of the inner

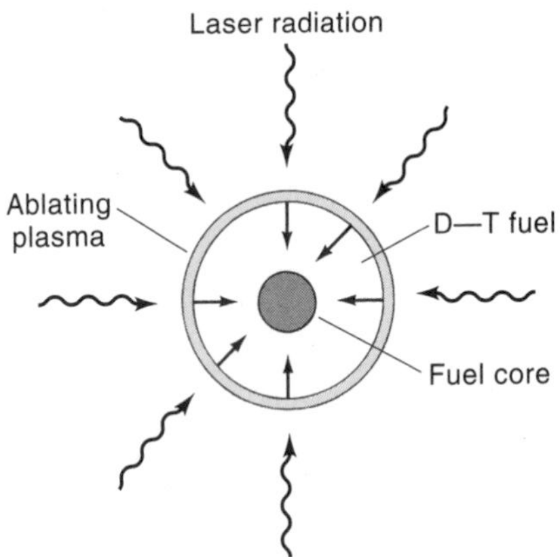

FIGURE 16.6
A fuel pellet of D–T is heated by laser radiation from all sides. It implodes as the surface explodes outward. Pressures on the order of one trillion atmospheres are achieved.

material, implosion, is an example of Newton's third law: "for every action force there is an equal and opposite reaction force." It is analogous to jumping off a sled that is on the ice: You go forward and the sled goes backward.

To reach the temperatures and densities necessary for fusion (and to meet the Lawson criterion), the energy delivered by the laser must be large (on the order of one million joules) and of a very short duration (one billionth of a second). This corresponds to a power rating of 10^{15} W, 1000 times larger than the power of *all* human-made machinery. (However, the energy output is small because of the short time period.) To meet the Lawson criterion, high densities (about 10^{26} ions/cm^3) must be attained for short times (about 10^{-12} s).

The apparatus for laser-induced fusion is enormous. One example is the high-energy laser facility at the University of Rochester. (Other examples are at the Lawrence Livermore Laboratory, California, and at Los Alamos, New Mexico.) The laser system for irradiation of the fusion target at Rochester consists of 60 separate laser amplifier chains. A master oscillator provides a low-energy infrared laser pulse that is split to provide 60 input pulses, one for each of the laser amplifier chains. In each of these chains, a crystal is used to generate higher harmonics of the incident laser frequency. The results are 60 beams of wavelength 0.35 microns—ultraviolet waves. It is important that all pulses arrive at the target at the same time and stay focused on the pellet while it undergoes compression.

The laser-target setup is about the size of a football field (Fig. 16.7). As each laser pulse moves down a chain through many amplifiers, its energy and size increases. As one of the scientists on the project has said, "it's like throwing pies." The frequency of the pulses will be quite high. To produce a steady output of about 1000 MWe, the "pies" will have to come at a rate of about 10 per second.

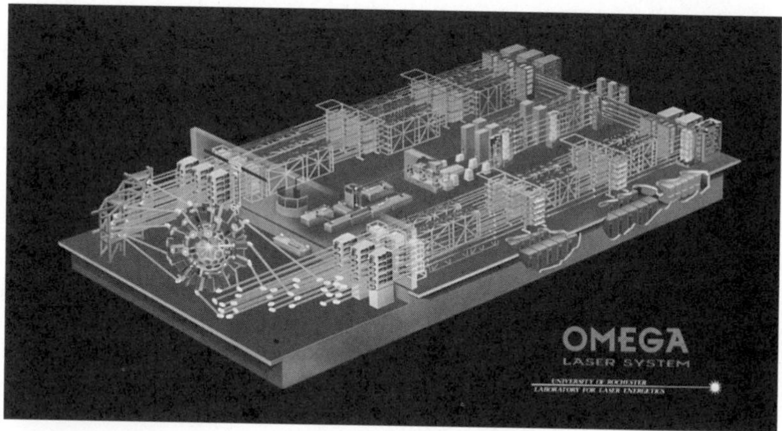

FIGURE 16.7

The 60-beam OMEGA laser system is a 30,000-J ultraviolet laser. The fusion target chamber is to the left. (LASER ENERGETICS LABORATORY, UNIVERSITY OF ROCHESTER)

Technically, the lasers needed to yield useful fusion energy do not exist today. The primary laser in use now for fusion experiments is a neodymium-glass laser, which produces infrared light of about 1 micron in wavelength. Lasers being built will produce an output of more than 10,000 J with a pulse of less than one billionth of a second. This will produce an energy output within a few percent of scientific breakeven. To achieve breakeven, more powerful types of lasers must be built, and their efficiencies must also increase. The neodymium-glass laser has an efficiency of only about 1%, which means that 99% of the energy pumped into the laser does not emerge as useful light. Also, lasers in use today can be fired, at most, a few times an hour—a long way from the 10 times per second that is needed for successful commercial operation.

A laser-fusion power plant might look like the one shown in Fig. 16.8. The laser beams are focused to the center of the combustion chamber. A D–T pellet is dropped so that it arrives in the center of the chamber at the same time as the laser pulses. The subsequent explosion is equivalent to about 10 lb of TNT, not enough to damage the parts of the combustion chamber. The energy from each explosion is carried away primarily by the escape of neutrons and X-rays from the target. Outside the chamber, in a configuration similar to that of the magnetic confinement machines, is a blanket of lithium to capture this energy. Heat exchangers throughout this blanket carry water, which is turned into steam for driving turbines. The lithium is also a source of tritium, to be recycled back to the "pellet factory."

Another inertial confinement scheme uses X-rays rather than laser pulses. The X-rays are produced by a pulsed power facility that uses currents of magnitude 20 million amps! The X-ray bursts are used to implode the fuel pellet and initiate fusion. The "Z-Pinch" machine is being developed at Sandia National Laboratory, New Mexico.

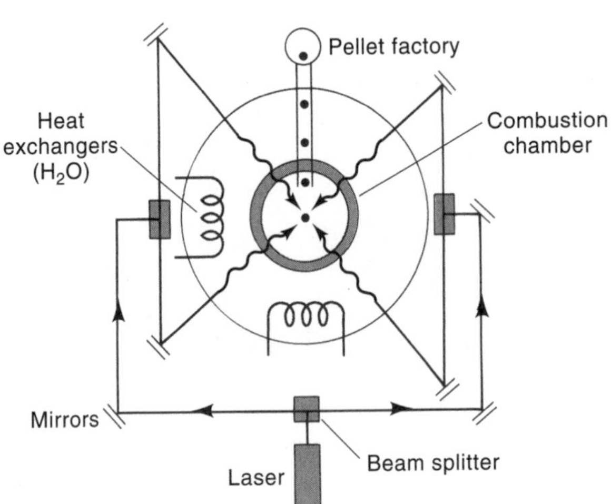

FIGURE 16.8
Schematic of a laser-fusion power plant. Pellets of frozen D–T enter from the top. They are irradiated by laser beams. Lithium in the outer layer absorbs the energy released in fusion, and then turns water into steam in the heat exchangers.

F. Cold Fusion

In 1989, the scientific community was startled by the announcement of what some thought to be one of the major breakthroughs of the century: the possible achievement of the fusion of deuterium at room temperatures. Two chemists from the University of Utah had carried out an experiment that could be done in a test tube with equipment costing only thousands of dollars and with much simpler technology than that used in the types of fusion programs we have been describing.

The apparatus (Fig. 16.9) consisted of an ordinary test tube that contained a platinum wire wrapped around a palladium metal rod. Both metals were immersed in heavy water—water enriched with deuterium. A battery was used to establish an electric current between the two metals, causing the heavy water to break up into deuterium and oxygen. This is the process of electrolysis of water, which is usually done with other metals. Usually one sees bubbles emerging from each electrode as the oxygen and hydrogen migrate toward each electrode and bubble away. But using palladium instead of a more common metal apparently made a difference. Palladium has a great affinity for hydrogen, and it soaked up the hydrogen atoms before they could bubble out of the solution. So much hydrogen was stored in the palladium rod that fusion between the nuclei reportedly started to take place, and up to 50 times as much energy output (in the form of heat) as input was reportedly liberated. Apparently the reaction taking place was $D + D \rightarrow {}^3He + n$, and some neutrons were apparently observed.

These results were first disclosed to the world via the *Wall Street Journal* and London's *Financial Times,* as well as a news conference that made it onto the national evening news as the lead story—not the normal ways that scientists report their results. Usually research is reported in refereed journals, which provide the opportunity for each new result to be examined critically by experts to assure its accuracy.

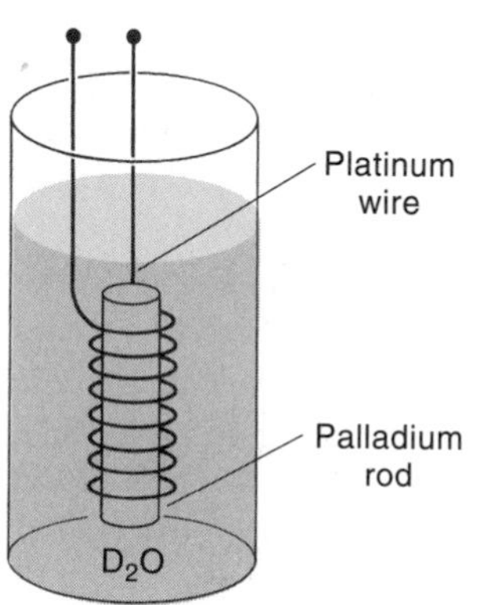

FIGURE 16.9

Apparatus set up to test for cold fusion. Electric voltage exists between the platinum wire and the palladium electrodes. The water in the tube is enriched with deuterium.

Since the original claim in March 1989, all but a handful of other laboratories have declared the results negative. Most laboratories did not find the excess heat originally reported or the fusion by-product of neutrons. It was suggested later that contamination of tritium on the electrodes or problems with the calibration of the neutron detectors yielded spurious results. Possibly fusion was taking place in the test tube, but at rates far below what the original experiment had indicated. Excitement for cold fusion has since waned, but questions still remain about the phenomena behind the sporadic results reported by several laboratories in 1989. A handful of labs around the world still continue to find results that would only be possible through cold nuclear fusion. Replication of these results is difficult, so skepticism remains while some work continues.

The phenomenon of cold fusion has caused some people to ask about the procedures and prevalent atmosphere for doing science. "It is now virtually impossible to publish positive cold fusion results in certain journals because the editors or their peer reviewers are already convinced that the effect is bogus. This creates a 'catch-22': The journals will not accept papers until more papers published in journals show additional evidence for the effect." (E. Storms, Los Alamos National Lab., 1994.) A Department of Energy report issued late in 1989 concluded that "nuclear fusion at room temperature, of the type discussed in this report, would be contrary to all understanding gained of nuclear reactions in the last half century. It would require the invention of an entirely new nuclear process."

G. The Outlook for Fusion

The atmosphere at most fusion centers today is not "if" but "when" fusion reactor feasibility will be attained. Many of the problems of instabilities in magnetic confinement systems are being solved, and larger magnets and upgraded lasers should be able to achieve the fuel densities and confinement times necessary to satisfy the Lawson criterion. However, large engineering obstacles still remain. Just consider the conditions that must be met: temperatures several times that at the center of the sun; uniform magnetic fields some 100,000 times stronger than the earth's; densities for laser fusion pellets a thousand times that of an ordinary solid; laser beams focused to within millionths of an inch, and all arriving within a trillionth of a second of one another, while pulsing many times per second; and many more challenges.

Another difficulty is the radiation and heat damage to the chamber wall. In D–T fusion, there are four times as many neutrons produced as in a comparably sized fission reactor. The fast neutrons from fusion can lead to metal fatigue and embrittlement. Replacing the chamber wall during the reactor's lifetime would be very expensive.

The economics are certainly a crucial point. Economic break-even conditions clearly will be larger than those for scientific breakeven. The Tokamaks will have to be large (1000 MWe) for efficient operation—possibly too large for private industry to invest in. For laser-induced fusion, the microballoon pellets will have to cost about 1¢ each for economical operation.

The advantages of fusion power are many: an essentially infinite fuel supply, higher thermal efficiencies, few radioactive waste problems, no runaway reactions, no global warming. However, the prospect for controlled fusion still has many uncertainties. Many of the requirements for achieving breakeven have been met individually in several labs, but not all at once. Yet the outlook continues to be optimistic, and both types of confinement programs insist that the first commercial fusion plants could be brought on line within 40 years.

Plans are underway for a $2 billion International Thermonuclear Experimental Reactor, a joint effort of Japan, Russia, and several European countries. It is hoped that this gigantic machine will reach the break-even point and produce more energy than it requires to operate. The new design will be twice as large as the present English fusion test reactor and is expected to run for weeks, compared to runs of several seconds in Tokamaks today. It will use a combination of tritium and deuterium fuel and will need new superconducting magnets to confine the plasma. If all goes well, a similar but scaled-up commercial reactor could be on line by the year 2040.

In the area of laser fusion, plans are underway to build a giant laser complex that would be able to pump enough energy into the fuel pellet to achieve breakeven. There, 192 laser beams will be fired at a chamber containing the D–T fuel pellet, heating it up to 50 million °C. The $1.2 billion project is the size of a football stadium and is known as the National Ignition Facility and is being built at the Lawrence Livermore National Laboratory in California. If successful, the National Ignition Facility will produce at least as much energy from fusion as the laser delivers to the pellet. It will still lack the 100 times more energy needed to power these lasers. When it begins operating in 2003, its scientists hope that progress toward sustained power-producing fusion will be accelerated.

H. Summary

Fusion is the combining of two small nuclei to form a larger nucleus, with the release of energy. The reaction most commonly studied is $D + T \rightarrow {}^4He + n + 17.6$ MeV. For a fusion reactor to produce a net energy output, the D–T plasma must achieve the Lawson criterion—defined as the product of the particle density times the confinement time of the plasma—of about 10^{14} s/cm^3. It must also reach a high plasma temperature (about 100 million °C). Confinement of the plasma is achieved with magnetic bottles of doughnut shape (Tokamaks) or with inertial confinement, as done in laser-induced fusion. In the first case, relatively low plasma densities with long confinement times can be used to achieve the Lawson criteria; for the latter case, high densities and short times are achieved with high-energy pulsed lasers.

In the generation of electricity in fusion reactors, neutrons produced in the D–T reaction heat up surrounding lithium, which is used to turn water into steam via a heat exchanger. It is hoped that within the next few years a scientific break-even point will be achieved, in which as much useful energy will be produced in the fusion reactor as is put into it. However, many serious technical difficulties must still be overcome before fusion can make any significant contribution to our energy needs.

Internet Sites

For an up-to-date list of Internet resources related to the material in this chapter, go to the Harcourt College Publishers' website at **http://www.harcourtcollege.com**. The links are in the *Energy: Its Use and the Environment* site on the Physics page. General energy related sites and some guidelines for using the World Wide Web in your class are on the inside front cover of this book.

References

Chapter 16

Colombo, U., and U. Farinelli. 1992. Progress in Fusion Energy. *Annual Review of Energy,* 171:123-59.
Conn, R. W., V. A. Chuyanov, N. Inove, and D. R. Sweetman. 1992. The International Thermonuclear Experimental Reactor. *Scientific American,* 266 (April).
Huizenga, J. R. 1993. Cold Fusion: The Scientific Fiasco of the Century. New York, Oxford University Press.
Maniscalco, J. A. 1980. Inertial Confinement Fusion. *Annual Review of Energy,* 5.
Post, R. F. 1976. Nuclear Fusion. *Annual Review of Energy,* 1.

QUESTIONS

1. Where does the energy come from that is released in a fusion reaction?
2. Why is fusion such a difficult task to achieve? Why is our sun so successful in providing us energy via fusion reactions?
3. In words, explain the Lawson criterion and its significance.
4. Verify the statement that "the fusion of the deuterium in the water in an Olympic-sized swimming pool could provide enough electricity for a city of 100,000 people for one year." State the assumptions in your calculations.
5. What is the advantage of a "magnetic bottle" for fusion?
6. How can we heat a plasma up to 50 million °C?

7. How does the use of lasers provide the temperatures and densities necessary for fusion to occur?
8. What is meant by "scientific breakeven"?
9. What power output does one achieve with a 10,000-J laser that provides a pulse of duration one billionth of a second?
10. React to the statement made by E. Storms on page 536. Can you think of other types of scientific studies that cannot get funding because of the nature or thrust of the research?

17

Biomass: From Plants to Garbage

A. Introduction

Because we eventually expect to run out of conventional fuels such as oil and natural gas, alternatives that provide the usefulness, flexibility, cleanliness, and economy of these resources have been sought for many years. One of these alternatives is as close as the kitchen garbage pail or the plants outside. This is "biomass": a source of energy that is both as old as humankind and as new as the morning paper.

Biomass energy is that energy derived from living matter such as field crops (corn, wheat), trees, and water plants; it is also agricultural and forestry wastes (including crop residues and manure), and municipal solid wastes. Biomass can be used as fuel in three forms: solid biomass fuels or feedstocks such as wood chips; liquid fuels produced from solid biomass through chemical or biological action and/or conversion of plant sugars to ethanol or methanol; and gaseous fuels produced by high temperature and high pressure processing.

Biomass resources cover a lot of ground. They have the potential of providing anywhere from 4 to 25% of the energy needed in the United States. Biomass presently provides 3.6% of U.S. energy needs and can provide several times the output expected from wind and photovoltaics. Sweden and Ireland use biomass for 13% of their energy needs, and Finland provides 14% in this manner. In addition, this resource has particular usefulness in nations of the developing world, where high oil prices have slowed economic growth. As a stored form of solar energy, biomass has the advantage that collector costs are small and energy storage is built in. Biomass can be converted into liquid and gaseous fuels in several steps, yet direct combustion to produce steam or electricity is also very popular. Biomass sources are under strong consideration as alternative fuels for transportation, especially because of new air pollution standards. Figure 17.1 illustrates some ways in which biomass is converted into other fuels.

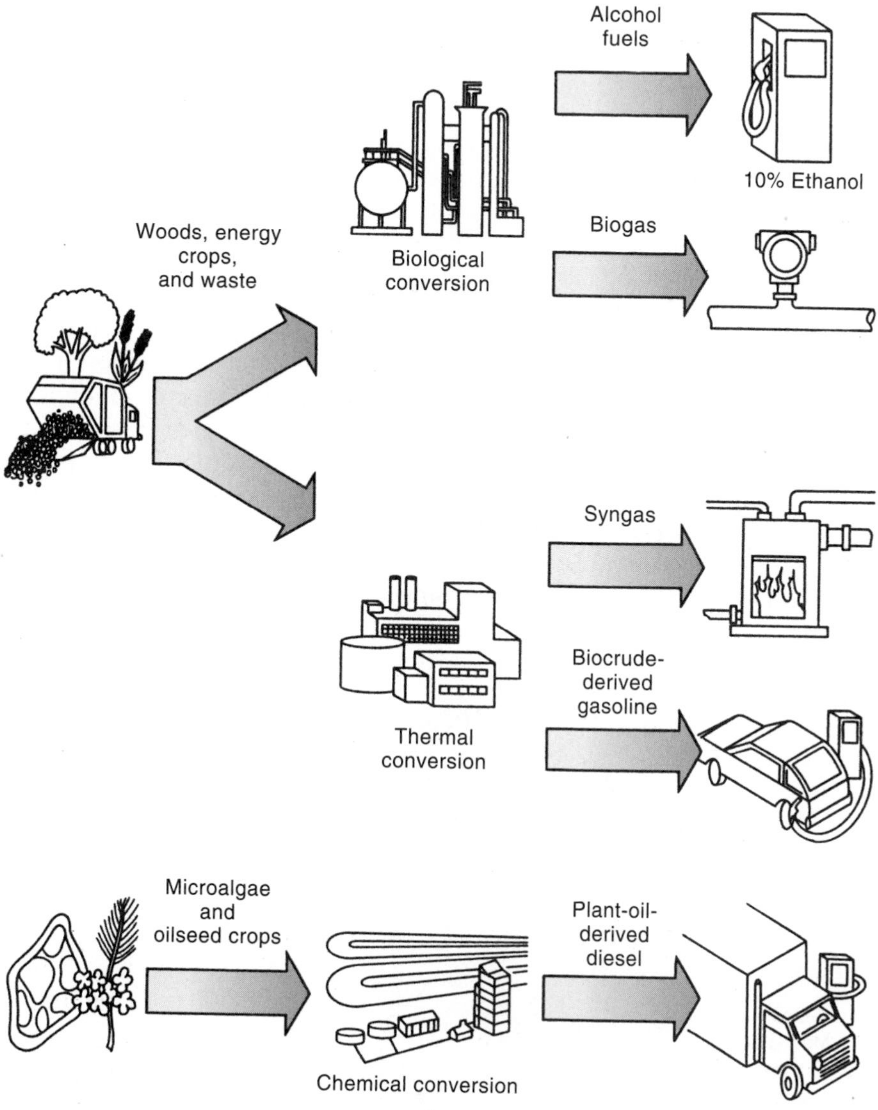

FIGURE 17.1
Conversion of biomass into useful fuels.

B. Municipal Solid Waste

Introduction

One of the hottest issues today is the use and disposal of municipal solid waste (residential and commercial garbage). Town meetings to discuss (and protest!) the proposed location of a new landfill or the construction of a waste-to-energy incineration plant or the organization of a recycling program for the community occur regularly.

Every year, Americans toss 81 million tons of paper, 44 million tons of food and yard wastes, 15 million tons of metal, 13 million tons of glass, and 19 million tons of plastics into landfills. Figure 17.2 characterizes our municipal solid waste situation. Our garbage also contains hazardous wastes—cancer-causing chemicals, lead, insecticides, household cleaning chemicals, and much more. Our solid waste amounts to about 4 lb per day per person, twice what the Europeans and Japanese generate. In New York State, with 18 million people, the trash generated per year could cover a football field to a height of 3 miles.

While the amount of solid wastes generated is growing at a rate of 2.3% per year, the landfill space available is drastically shrinking. More than two thirds of the nation's landfills have closed since the late 1970s, and one third of the remaining 6000 landfills will be full in the next five years. As landfills close and disposal fees soar, many garbage haulers are taking their trash out of state, or even out of the country, if they can. Many of the landfills that were in use in the 1970s were nothing more than open pits. They received normal garbage as well as hazardous chemicals. Serious groundwater contamination occurred in many locations throughout the country because of chemicals and pollutants from garbage from the site. Even the "sanitary landfills" of the 1970s and early 1980s

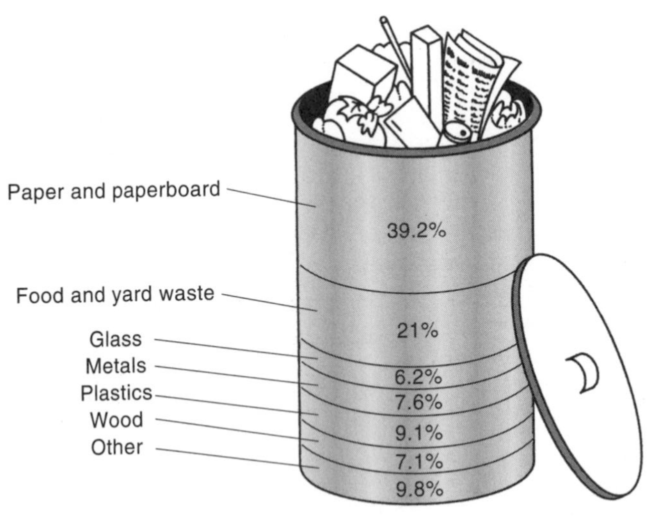

FIGURE 17.2
National waste profile. (EPA *Municipal Solid Waste Fact Book*, 1998)

Municipal dump in Chile.
(R. SCRUDATO)

only covered the garbage with a layer of dirt every day. Strong environmental regulations came into effect in the 1980s, which dictated that landfills must be "secure" to prevent leachate contamination. (**Leachate,** also called "garbage juice," is the liquid that results when water percolates through the landfill.) The Fresh Kills landfill near New York City is the world's largest city dump, receiving up to 24,000 tons of garbage each day at its peak. It is the tallest "mountain" on the East Coast, and it leaches 2 million gallons of contaminated junk into the groundwater each day. It is scheduled to be closed in 2001.

There are four or five nonexclusive options for solving our municipal solid waste crisis:

- reduce the amount of trash generated,
- institute vigorous recycling programs,
- incinerate garbage in "waste-to-energy" plants,
- construct more "secure" landfills,
- anaerobic digestion (see Section C).

The first option is obvious but difficult to achieve. We are a throw-away society. Think of the common items used every day and disposed of in the garbage: Styrofoam cups, juice boxes, razors, plastic spoons, diapers, soft drink bottles, etc. Americans will not give up their disposable lifestyle easily. But reducing garbage at the source would do much to alleviate the burden on recycling programs, incinerators, and landfills. More than 10% of the cost of food goes just to packaging; packaging accounts for 50% of the volume and 30% of the weight of household waste.

Recycling

Today about 30% of the U.S. waste stream is recovered and returned to use; some countries have achieved 50%. In most recycling efforts, trash is separated into newspapers, glass bottles, aluminum and steel cans, light plastics, and hazardous waste. Many states already have laws requiring the recycling of aluminum cans, glass bottles, and plastic containers that hold our beverages and foods. About 65% of the aluminum cans used today are recycled as well as 61% of the 50 billion glass beverage containers that are used each year. Glass bottles and jars can be remelted to produce new containers, while plastic containers find other uses, for example as fiberfill stuffing for pillows, sleeping bags, and automobile seats. About 40% of today's plastic soft-drink bottles are recycled.

Thousands of communities today have some sort of recycling program. Recyclables can be separated in the home, at the curb during collection, or at a processing facility. Recycling in the home seems to have the biggest impact on the waste stream, especially if the separated materials are picked up the same day as other trash; drop-off facilities have been the center of many recycling programs but achieve only a fraction of the participation achieved in home collection programs. In community recycling programs, newspapers are the most common material collected. Nearly 30% of all paper is now reused to make insulation, building materials, or other paper products. Recovery of the paper used to produce only the Sunday edition of a major newspaper, such as the *New York Times*, would preserve 75,000 trees per year!

For a recycling program to succeed, there must be a market for the materials that are reclaimed. This can be a problem. Requirements for the use of recycled materials will have to be strengthened. This is particularly true in the area of paper products, which are 40% by weight of our trash. Not many U.S. paper mills are equipped to turn old newspapers into new newsprint.

- In the United States, we throw away enough glass bottles to fill both towers of the New York World Trade Center every two weeks.

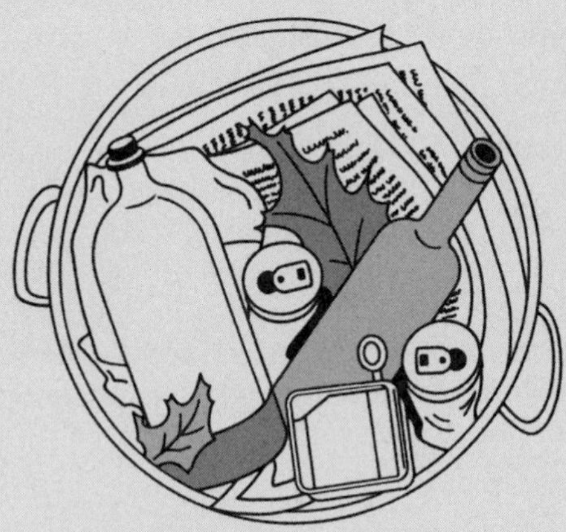

- Americans go through 2.5 million plastic bottles every hour, only a small percentage of which are now recycled.

- If you drink two aluminum cans of soft drinks per day and fail to recycle the cans, you waste more energy than is used daily by each of one billion humans in less developed countries.
- We throw away enough paper annually to build a wall 12 ft high from Los Angeles to New York.

Recycled aluminum cans require only 6% of the energy needed to make the can in the first place. Recycling one aluminum can will save enough energy to run a TV for 3 hours. Reusing old glass will cost less than forming it from new material. Only about 10% of our glass is recycled today. It is cheaper to recycle than to put garbage in landfills, because the landfill option requires transportation and tipping fees (the charge imposed on trucks entering the landfill), and doesn't gain any revenues from selling the material reclaimed. There is also a greater energy savings in recycling rather than incineration of the material. The energy derived from incineration is usually less than the energy saved by not producing that product from raw materials.

Composting is another method of dealing with some of our garbage. About 20% of municipal waste is organic kitchen and yard wastes. These can be collected, shredded, and arranged in piles to undergo decomposition. The product is rich composted materials that can be used in agriculture, nursery, and landscaping. Many communities today forbid the disposal of leaves in a landfill. About half of the states in the United States restrict the inflow of grass clippings and other yard wastes to landfills. The material is composted or used as mulch. Well over one third of U.S. and European sewage is now applied to land, often with minimal precautions for safety with regard to heavy metals and other potentially harmful components that might be present due to industrial wastes.

In the last few years the demand for recycled materials has climbed dramatically, and even surpassed supply in some cases. In just two years (1993–1995), the prices paid by processors for old newspaper climbed about 1300%. In the same period, prices paid for clear glass rose 75%. Prices will fluctuate rapidly with season, but concern for the environment, a perceived shortage of landfill space, and high demand for recycled products should keep the price of trash high.

Japan serves as an example of a premiere recycling country. Half of its waste stream is recycled. This includes 95% of all newspapers, 50% of all other paper,

ACTIVITY 17.1

Keep track of the amount (total) and type of garbage that your family generates in a week's time. Separate the garbage into glass, paper, metal, waste food, plastics, and miscellaneous, and weigh each component.

and 55% of all metal. The sorting is done at home or in centralized centers. Of the nonrecyclable waste, two thirds is sent to 1 of 2000 incinerators and one third directly to landfills. The ash from these incinerators is buried, sometimes after it has been embedded into concrete.

Incineration

Even with 30% of the U.S. waste stream recycled, 70% must still be disposed of. For many communities, the best approach to these residuals is incineration. This process reduces the volume by a factor of 10; the ash can be sent to a landfill. The heat produced in incineration can be used to generate steam for use in making electricity or industrial process steam.

A schematic of an incineration plant is shown in Figure 17.3. Basically, the sorted garbage is brought into the plant by truck from local transfer stations and dumped onto the floor. From there it is moved into large burners for combustion at temperatures of about 1000°C. Steam is generated in the boilers. The combustion gases are treated in state-of-the-art emission control pathways, using cyclone collectors and electrostatic precipitators or fabric filters to reduce the emission of particulates. (See Chapter 8.) The bottom ash from the boiler and the much finer particles in the fly ash from precipitators are collected by truck and hauled off to the local landfill. Financially, many waste-energy plants are breaking even, with the income from the sale of process steam and electricity and the tipping fee at the facility (for each truck) equaling the expense of running the plants.

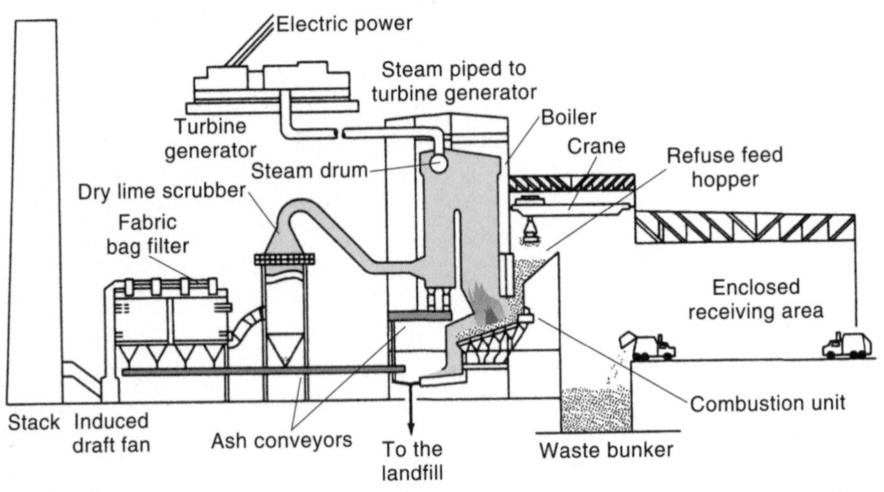

FIGURE 17.3
Waste-to-energy facility. The steam produced can be used to drive a turbine generator or to provide process heat to a nearby customer.

Table 17.1 TYPICAL METAL CONCENTRATIONS IN INCINERATOR ASH

Element	Fly Ash (mg/kg)	Bottom Ash (mg/kg)
Calcium	54,500	50,500
Cadmium	470	100
Lead	5200	900
Aluminum	70,000	33,000
Iron	17,000	132,000

("Energy from Wastes," *Power Magazine,* March 1988, The McGraw-Hill Companies, New York, NY)

Strong concerns exist about the air pollutants emitted during combustion. There will be heavy metals and dioxins (from chlorine compounds formed during combustion). **Dioxins** are considered some of the most potent carcinogens known. Dioxin played a significant role in the contamination of Love Canal, New York, and in the exposure of soldiers to Agent Orange in Vietnam. It has been proven to cause cancer in animals and is linked with a variety of other health problems, including genetic effects. Dioxin is formed either in the combustion chamber or after combustion when the gases cool in the exhaust stack. Disagreements continue over establishing safe levels of dioxin emissions from any incinerator. It has been discovered at various levels at every incinerator tested. Also of concern is the residual ash, which has high metal concentrations. This is usually disposed of in local landfills; the leachate must be well contained. Table 17.1 lists typical metal concentrations (in parts per million—ppM) found in fly ash (at the bottom of the stacks) and bottom ash (from the boiler).

Not all household waste goes to municipal incinerators. The Environmental Protection Agency (EPA) estimates 20 million people get rid of their trash by burning it outdoors in open barrels. Items such as paper, cardboard, plastic, and food waste as well as tin cans and glass go into the fire. This practice may release as many dioxins per year as that released by all municipal incinerators in 1995—before stiffer regulations were put in place. Open burning is regulated only by state or local regulations—if at all.

Secure Landfills

The point has already been made that we are running out of places to put our garbage. Today one third as many landfills exist as in the late 1980s, and this number continues to decrease. Not everything can be recycled or incinerated, and landfills will continue to be needed.

One of the problems of many present-day landfills is the contamination of groundwater as a result of leaching of the garbage by rains and underground

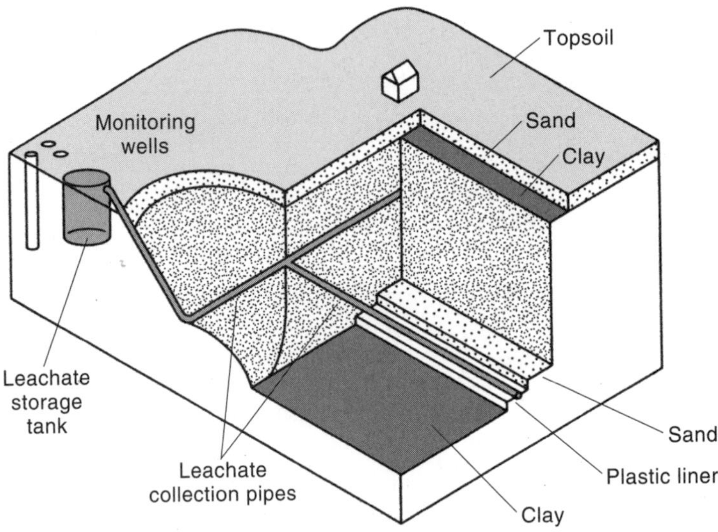

FIGURE 17.4
Cross section of a secure landfill. Note the multibarrier liner of plastic and clay to protect the groundwater from leachate.

water sources. To prevent this problem, "secure" landfills today are required to be sited on land distant from active groundwater locations and to be lined with both clay and synthetic (plastic) liners that will trap the leachate (Fig. 17.4). Some landfills have a double-liner system: plastic, clay, plastic, clay. The site will also have a leachate collection system and continuous monitoring to make sure groundwater contamination and runoff are avoided. The leachate collected is carried away to a sewage-treatment facility.

When the landfill is filled and ready for closure, a plastic liner is placed over the top and then covered with soil. However, there is a debate between the EPA and others on such closure methods. The EPA wants dry storage to minimize water contamination. Others want to keep the waste wet to speed up decomposition. In this method, the leachate would be recycled through the landfill to accelerate decomposition. One benefit is the production of methane gas from the landfill by the bacteria breaking down the complex organics in the garbage. Active landfills are already producing methane gas. You can sometimes see the white plastic pipes in the ground that vent the methane to a holding tank.

C. Biomass Conversion

The fundamental conversion process in green plants is **photosynthesis,** which is the combining of CO_2 from the atmosphere with water plus light energy to produce oxygen and carbohydrates (such as sugars and starches):

$$CO_2 + H_2O + \text{light energy} \rightarrow O_2 + \text{carbohydrates}$$

The reverse of this process is called **respiration,** in which CO_2 and H_2O and heat are produced in the combustion of carbohydrates and oxygen. This takes place in plant leaves and roots and in animals, as well as in decaying organic matter. The efficiency of photosynthesis is only about 1% for the process of changing solar energy into stored chemical energy. In green plants, photosynthesis and respiration occur during the day and respiration at night, leading to daily changes in atmospheric CO_2 concentration. Carbon dioxide concentration varies seasonally, being highest at the beginning of spring and lowest in the fall, at the end of the growing season.

Table 17.2 gives the potential energy available from agricultural and commercial forest production for the United States. Twenty percent of our land area (in the 48 contiguous states) is cropland and 30% is commercial forest and woodlands. In terms of its energy value, corn is the largest crop.

Processes for the conversion of biomass into other energy forms are numerous, but can be classified into three types:

1. Direct combustion—the burning of biomass to produce heat for space heating or for the production of electricity through a steam turbine. Anything from solid wastes to crop residues to wood can serve as feedstock for this process.
2. Pyrolysis—the thermal decomposition of wastes into a gas or liquid (with a relatively low heating value) under high temperatures (500°C to 900°C) in a low-oxygen atmosphere.
3. Biochemical processes—the decomposition of organic wastes in an oxygen-deficient atmosphere with the production of methane gas (anaerobic digestion), or controlled fermentation for production of the alcohols ethanol and methanol.

Anaerobic digestion is a decomposition process by which bacteria convert organic material into methane and carbon dioxide gases in the absence of oxygen. It

Cattle pens, with fermentation tanks in the background for converting animal waste into methane gas. (UNITED STATES DEPARTMENT OF ENERGY)

Table 17.2 POTENTIAL ENERGY FROM CROP AND FOREST RESIDUES

Source	Hectares* (ha) Harvested ($\times 10^6$)	Crop Yield (ton/ha)	Residues (ton/ha)	Total Yield (tons $\times$ 10^6)
Barley	3.8	0.24	3.5	13.3
Corn	28.3	5.6	5.6	158.5
Cotton	5.4	0.6	0.5	2.7
Oats	5.4	2.0	4.0	21.6
Rice	0.9	4.9	7.4	6.7
Rye	0.3	1.5	2.3	0.7
Sorghum	5.7	3.5	1.2	7.4
Soybeans	23.4	2.0	3.0	70.2
Wheat, winter	19.6	2.1	3.5	68.6
Wheat, spring	7.2	1.8	2.3	16.6
Other	56.0		1.1	62.0
Total crops	156.0			428.3
Total forest	4.5		24.7	111.2
Grand total				539.5

*1 ha = 2.47 acres

(D. Pimental, *Science,* vol. 212, p. 1110, 1981. Copyright 1981, American Association for the Advancement of Science)

is a method used today in municipal sewage treatment plants for removing organics and destroying harmful microorganisms. The process can use agricultural waste—such as manure—to generate a gas with a heating value of 500 to 600 Btu/ft^3. (Natural gas of pure methane has a heating value of about 1000 Btu/ft.^{3}) In sewage treatment plants this gas is burned to provide the temperatures necessary for rapid decomposition of the waste.

Anaerobic digestion usually takes place in **digesters** built to hold a slurry of waste and water. If the raw material is cow manure, the output is a gas that is about 60% methane and 40% carbon dioxide. One pound of cow manure will yield about 1 ft^3 of gas, while one cow's annual output is equivalent to the energy in about 50 gallons of gasoline. This gas (termed **biogas**) can be used for heating, cooking, gas refrigerators, electricity generation, and other energy needs. Digesters in Kenya with volumes of 1.5 m^3 (16 ft^3) yield about 0.75 m^3 (8 ft^3) of biogas per day from the daily manure of two cows—enough for 5 hours

Table 17.2 POTENTIAL ENERGY FROM CROP AND FOREST RESIDUES

Amount Readily Useable (tons × 10^6)	Potential Net Heat Energy (kcal × 10^6)	Potential Net Electrical Energy (kcal × 10^9)	Potential Net Ethanol Energy (kcal × 10^9)
3.6	6940	2652	758
39.6	71,874	19,800	9308
0	0	0	0
6.5	12,513	4789	1368
5.2	10,010	3832	1095
0	0	0	0
0	0	0	0
0	0	0	0
18.6	35,805	13,705	3916
0	0	0	0
0	0	0	0
73.5	137,132	47,480	16,445
44.0	71,032	20,307	8372
117.5	208,164	65,085	24,817

of lighting or 1 hour of cooking. The use of biogas plants has been dramatic in China and India, with more than one million plants in operation. Two thirds of China's rural families use biogas as their primary fuel.

Figure 17.5 shows a small-scale methane digester. The collection tank has a floating cover to allow gas expansion. Raw waste and water are mixed with microorganisms that convert the waste into biogas. The temperature at which the digesting bacteria work is about 35°C to 43°C (95°F to 110°F), so an external heat source (such as a solar collector) is needed. Besides biogas, another useful product of the digester is fertilizer. Household solid waste and sewage sludge can be combined in a process with heat to provide another useful product—compost. A plant in Massachusetts uses previously landfilled materials to produce a marketable product. Air and heat are added to a mix of household waste and sewage sludge to make the compost. The product is suitable for agriculture and landscaping.

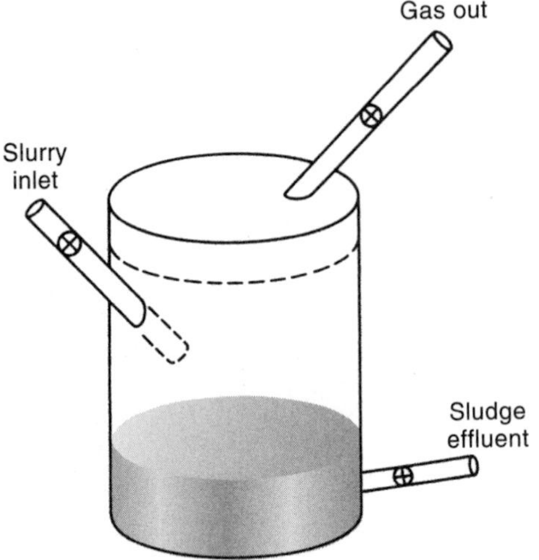

FIGURE 17.5
Fixed-container methane digester. Family-sized digesters using manure are relatively common in countries such as India and China. In China, an estimated 4.5 million biogas digesters are currently in use.

Alcohols are other products of biomass conversion that have received considerable attention in recent years as a substitute for our petroleum-based liquids. The fermentation of plant materials for the conversion of their sugars into alcohol dates back more than 4000 years: The Egyptians brewed beer from grains and grapes. More than 100 years ago, Louis Pasteur identified yeast as a catalyst in the fermentation process.

The two alcohols of primary importance are ethanol and methanol. **Ethanol** is a colorless liquid with a boiling point of 78°C. Its formula is C_2H_5OH; it is also referred to as grain alcohol or ethyl alcohol. It can be made from a wide range of raw materials, but the most common are sugarcane, corn, and wood. Figure 17.6 is a simple flow diagram for ethanol production from corn, starting with complex carbohydrates.

Ethanol as produced from crops uses large amounts of energy for planting, fertilizing, and harvesting and so is an expensive fuel both money-wise and energy-wise. (Some calculations show that more energy is used to produce ethanol from corn than one can get out of it!) Ethanol can improve vehicle performance and produces fewer emissions than gasoline vehicles. Ethanol made from wood can reduce greenhouse gas emissions, since plants absorb CO_2 as they grow. However, ethanol made from corn results in no reduction in greenhouse gas emissions because of the petroleum used for cultivation and fertilizers.

Ethanol was used as a motor fuel in the Ford Model-T in the early part of the 20th century and found increased use during the Depression (as an alcohol–gasoline mixture) when efforts were made to stimulate the nation's agricultural economy. During the oil crises of the 1970s and 1980s, gasohol (5% to 10% alcohol added to gasoline mixture) was promoted to replace some of the petroleum used for automobile fuel. The cost of producing alcohol for this mix-

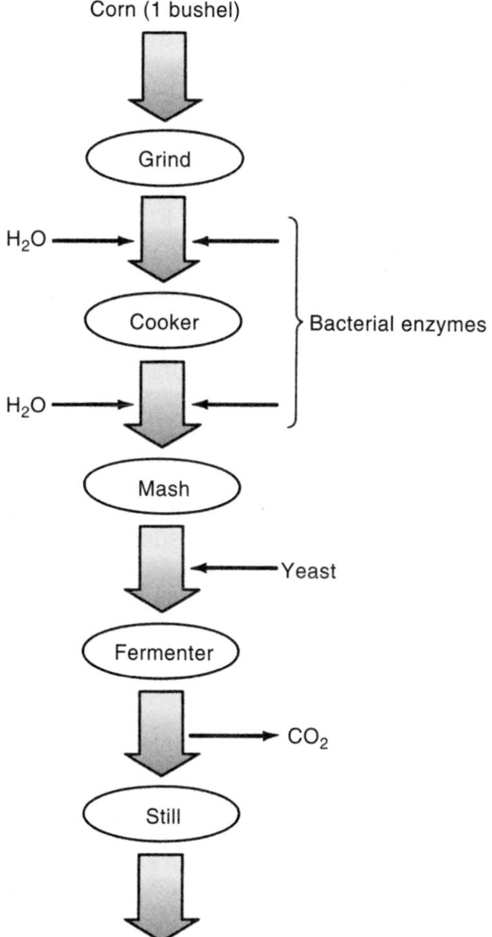

FIGURE 17.6
Flow diagram for production of ethanol from corn.

ture was often greater than the cost of the petroleum product it replaced. However, research is continuing and some gasoline companies do sell gasohol at the pump. No conversion of the engine is necessary for this blend. If pure ethanol is burned, modifications would have to be made since ethanol is corrosive and could degrade some metals and polymers used in gasoline systems. Also, ethanol's low energy density means a car using ethanol would have a range about one-third less than that of gasoline. Brazil has had a long history of using ethanol from sugarcane for its cars. About half the fleet runs on pure ethanol.

Methanol is a colorless liquid with a boiling point of 65°C. Its formula is CH_3OH; it is also called wood alcohol or methyl alcohol. It can be produced from virtually any carbon-containing material, but originally was produced in

The Illinois "Corn Bus" in Peoria runs on E95, a fuel made of 95% ethanol and 5% unleaded gasoline. (S. TARTAR)

the United States as a by-product of wood distillation. In addition to biomass, coal or natural gas can be used as the raw material. Germany made extensive use of methanol (from coal) during World War II for submarine and aircraft fuel. Methanol is a superior fuel for internal combustion engines and is used today as the fuel for racing cars.

Focus On 17.1

BRAZIL'S ETHANOL PROGRAM

Brazil is the third largest energy consumer in the Western Hemisphere. Its Amazon rainforest is about 30% of the world's remaining tropical forests. About 95% of its electricity comes from hydroelectric, including the Itaipu facility with 12,000 MW. Brazil imports about 35% of its petroleum.

In 1975, an ethanol program began in response to the oil price increases of 1973 that guaranteed that all gasoline sold would be blended with 22% ethanol from sugarcane. However, over the years as gasoline prices decreased, it was difficult to sustain this program; less than 1% of new cars sold in 1997 ran on ethanol.

Today this program is being revived since it reduces carbon emissions, burns cleaner (air pollution is a large problem in the major cities), creates jobs in the sugarcane industry, and reduces petroleum imports. The new initiative seeks to raise annual production of cars running on 100% ethanol. Currently, 41% of Brazil's transportation fuel demand is met with ethanol. However, prices of ethanol are still high.

Methanol derived from natural gas is the cheapest form, costing not much more than regular gasoline. It can be up to twice as expensive when derived from wood or coal. Using methanol might slightly reduce tailpipe emissions, especially NO_X, and would result in about the same greenhouse gas emissions as regular gasoline (if derived from natural gas). Methanol mixed with gasoline can be used in a vehicle with no major modifications.

Other alternative fuels for use in the transportation sector have been mentioned elsewhere in the book and are listed in Table 17.3. **Hydrogen fuel** is a clean-burning fuel that can be made from natural gas (methane + steam $\rightarrow$ $CO_2 + H_2$), coal, or the electrolysis of water. Presently the cheapest method is from natural gas. Using solar cells to provide electricity for electrolysis of water is an ideal way to store the sun's energy and cut down on greenhouse gas emissions. Hydrogen can be used in a conventional gasoline engine, but is much more efficient in a modified engine. It can be stored as a gas under high pressure, as liquid hydrogen (−253°C) or inside a metal hydride, to be released with the addition of heat. It also can be used in a fuel cell. **Fuel cells** (Chapter 10) are twice as efficient as the standard internal combustion engine and directly provide electrical energy by combining hydrogen and oxygen. At present, fuel cells are very heavy and so have found most of their uses in buses, although new advances are being made in lighter cells for cars. Fuel cells emit almost no pollution. Battery-powered **electric vehicles** (EVs) are more developed than fuel cells. However, large costs, short ranges, long recharging times, and battery replacement costs continue to be crucial drawbacks to EVs. Becoming popular now are hybrid vehicles—cars that use the internal combustion engine and electric motor with a single battery. In these vehicles, 70 mpg performances are common. **Compressed natural gas** (CNG) has the lowest cost of any alternative fuel, and has a very stable fuel supply. Conventional gasoline engines require only minor modifications to run on natural gas exclusively. The fuel is stored in pressure cylinders in the vehicle. Today, most CNG vehicles in use are dual-fuel operated and are converted from conventional vehicles using conversion kits, which include a storage tank, pressure regulator, mixing device, and various valves and fuel lines. Some electric utilities have moved into the transportation area by selling CNG at refueling stations.

Table 17.3 ALTERNATIVE FUEL FOR VEHICLES

Methanol from natural gas	Battery power
Ethanol from biomass	Methanol from biomass
Propane (liquefied petroleum gas)	Hydrogen gas
Fuel cell using hydrogen from natural gas	Compressed natural gas
Fuel cell using hydrogen from PV electrolysis	

D. Wood Combustion

Perhaps the most obvious source of biomass energy is wood, a renewable resource that is certainly not a new form of fuel. In 1860, three quarters of the energy needs of the United States were met by wood. In 1900 this figure was down to 25%, and it reached a low of about 2% in 1973. Higher energy prices and a desire for greater self-reliance have seen this percentage rise somewhat through increased use of wood-burning stoves in the home and industrial use of wood and its waste products. In Scandinavian countries, wood accounts for about 10% of total energy use. In developing countries, wood is often times *the* source of energy, accounting for upward of 90% of the energy used by villages in Africa and Asia. However, many problems occur with this practice because the gathering of wood without reforestation brings about erosion, floods, and a loss of nutrients for crop production.

Environmentally, emissions of sulfur and nitrogen oxides from wood are low. However, emissions of carbon monoxide and organic and inorganic particulates, including carcinogenic molecules of polycyclic organic compounds, are higher than from oil or gas furnaces. Adverse effects on air quality have been reported in New England and in the Pacific Northwest, attributable to the use of wood-burning stoves. Many cities have suffered through decreased visibility during the winter. Oregon was the first state to require standards for wood-burning-stove emissions, which include the use of catalytic combusters to improve the combustion of gases and smoke from the wood. Many other states have followed.

Residential Wood Combustion and Stove Design

Wood burning in homes is different from commercial or utility operations. Wood is burned in chunks at home, while industry uses chips or scrap pieces obtained from forest residues or logging waste products. The pulp and paper industry obtains about 40% of its energy needs from forest wastes. In developed countries, wood is used primarily for space heating (and some domestic hot water), while in developing countries it is used primarily for cooking. In larger cities in the Third World, charcoal is used for cooking because it is lighter and easier to transport.

To support a fire, three things are needed: fuel, oxygen, and high temperatures. When wood burns, it goes through three stages of combustion:

1. Moisture evaporates from the wood. The greener or more freshly cut the wood, the less energy per piece is available for heating.
2. The wood catches fire at about 315°C (600°F) as some of the solid wood and volatile compounds (released as gases) burn.
3. The heated wood decomposes into charcoal and smoke. Charcoal burns at about 540°C (1000°F), and the volatile compounds ignite at about 600°C (1100°F). This step is unlikely to occur in a wood stove unless the entry of excess air is limited to prevent cooling and the smoke released in step 2 is routed over an area of high temperature.

The wood to be burned must be stacked in the stove in such a way that it allows enough air to get in, and that the burning pieces heat each other to sustain good combustion. Wood that is too close together defeats the first criterion, while wood that is too far apart makes the second part unlikely. The overall heating efficiency of the unit is the product of the combustion efficiency of the wood and the heat-transfer efficiency of the stove.

For space heating, wood is burned in fireplaces, stoves, and larger central furnaces. Since the early 1970s, energy-efficient wood burning stoves have been very popular. Figure 17.7 shows a conventional fireplace. As wood is burned in the hearth, radiant energy is emitted into the room while the hot combustion gases, which still contain a lot of chemical energy, rise up the chimney flue. Combustion air is taken from the room itself, which must then be replenished by cold outside air. However, much more air is taken from the room than is needed for the combustion of the wood, and all of it goes up the chimney with the hot exhaust gases. For these reasons, the efficiency of a conventional fireplace might be close to zero, or even negative! To increase the efficiency, we first need to reduce the excess air intake. This can be done by using glass fireplace doors with small vents at the base of the hearth. Even better efficiencies are obtained by drawing combustion air from outside the room through pipes. The efficiency is also improved by arranging the logs being burned into a configuration that allows more rapid burning (and not smoldering) and that concentrates the emission of radiant heat energy toward the room.

Two other problems with fireplace heating can be addressed with the use of a wood-burning stove: The heat of combustion can be absorbed in all directions by capturing it in the metal walls of a stove, and then allowing these walls to

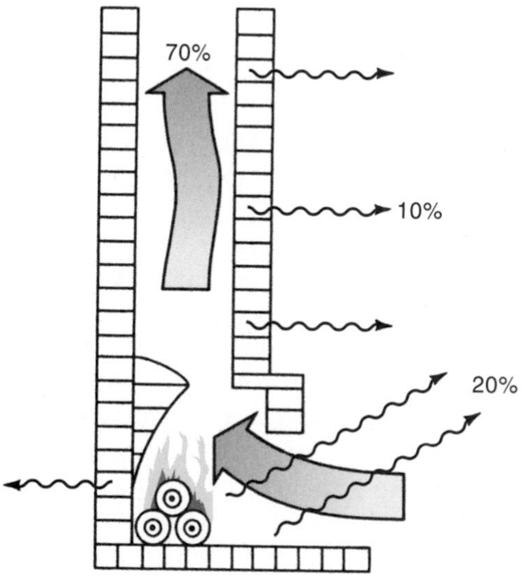

FIGURE 17.7

Heat transfer from a conventional fireplace. Heat output to house is offset by heat loss as a result of room airflow up the chimney.

High-efficiency wood-burning stove. (VERMONT CASTINGS)

transfer the heat energy into the room. Also, in a fireplace the combustion gases themselves still carry away a great deal of energy. If they can be burned further, the stove becomes more efficient and emissions will be reduced. This is the principle behind secondary combustion in a stove. Figure 17.8 shows the inside of a stove with secondary combustion capabilities. Secondary combustion takes place in the chamber above the primary combustion chamber. Secondary air is needed for complete combustion of the smoke, since there is usually little oxygen left from the primary air. This region must be kept hot because the combustion gases need temperatures of at least 550°C. The "S-shape" of the stove design illustrated in Figure 17.8 allows a longer exit path for the combustion gases and so a more complete burn.

Two thirds of a stove's output is via radiant heat, one third by convection. Therefore, best heat transfer efficiencies are gained from a stove with a large surface area. Black is a preferred color, but most other colors give equally good results. (Shiny metallic paints or clean metal surfaces give poor emittances.) Since the radiant energy is not in the visible spectrum, the color seen by the eye is not relevant here.

Wood varies significantly with respect to its energy content per cubic foot or per cord. (A cord of wood is a stack of wood measuring 4 ft × 4 ft × 8 ft.) Table 17.4 lists such values for a variety of woods. Each *pound* of wood has about the same heating value—8600 Btu/lb. In all cases the wood is assumed to be dry. Wet or green wood has a high water content and so a considerably lower heating value, because energy must be used to evaporate the water. Green wood also has the disadvantage that it will cause more creosote buildup in the chimney. The creosote is formed as combustion gases condense out onto the cool chimney surface to form a black, tarry substance. If layers of creosote build up

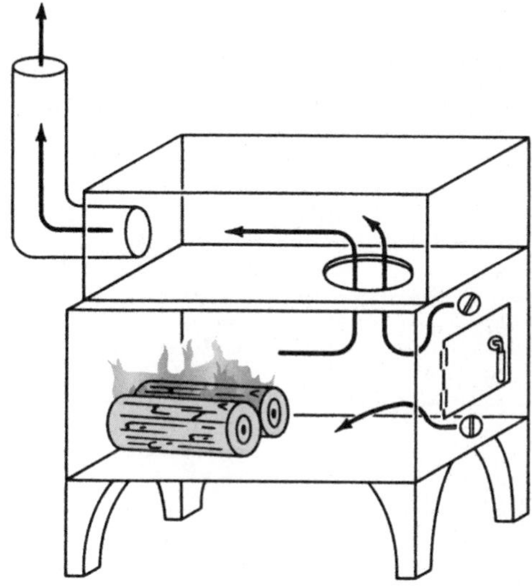

FIGURE 17.8
Airtight stove, with secondary combustion in the upper chamber.

on the inside of the chimney, the draft will be restricted. At high enough temperatures creosote will burn, causing a chimney fire.

Wood-Burning Economics

The economic grounds for switching to wood as a means of providing space heating depend on the cost of competing fuels. Natural gas is the most competitive. If gas costs \$0.65 per therm (100,000 Btu), a cord of hardwood would have to cost less than \$100 to provide comparable Btu's for the same price (Table 17.5). This assumes that the efficiency of the wood stove is 50% and that of the gas furnace is 75%. (Newer models of gas furnaces have efficiencies up to 96%.) Such calculations neglect the attractiveness of a wood-burning stove to provide the coziness that comes from its use. It also neglects the disadvantages of carrying wood into the house on a cold winter night, the removal of the ash, and the threat of chimney fires.

EXAMPLE

If natural gas costs 55¢ per therm, what should be the price per cord of mixed hardwoods to deliver the same Btu?

Solution

Using Table 17.5, we see that natural gas costs 13.33 × \$0.55 = \$7.33/MBtu. This should equal 0.083 × \$/cord. Solving finds the price per cord should be \$88.

Table 17.4 DENSITY AND HEATING POTENTIAL OF AIR-DRIED WOOD

Type of Wood	Air Dried Weight (lb)		Millions of Btu per cord
	Per ft^3	Per cord	
Shagbark, pignut hickory	53	4505	31.5
White oak, sweet birch, black locust	48	4080	28.5
Chestnut (rock) oak, bitternut hickory	47	3995	27.9
Red oak, Pacific madrone	46	3910	27.3
American beech	45	3825	26.7
Sugar maple	44	3740	26.1
Black oak, southern red oak	43	3655	25.5
White ash	42	3570	24.9
Yellow birch, California white oak	40	3400	23.8
California laurel	39	3315	23.2
Red maple	38	3230	22.6
Paper birch	37	3145	22.0
Black cherry	36	3060	21.4
American elm, sycamore	34	2890	20.2
Sassafrass	32	2720	19.0
Virginia pine	30	2550	20.6
Douglas fir	30	2540	20.6
Eastern red cedar	29	2490	20.1
Tulip tree	26	2210	15.4
Quaking aspen	25	2125	14.8
Redwood	25	2120	17.1
White spruce	25	2120	17.1
Balsam fir	22	1905	15.4
Eastern white pine	22	1855	15.0

(*Knowing Your Trees*, American Forestry Association, 1974)

Table 17.5 HEATING FUEL COST COMPARISON

Fuel	Unit Conversions	Conversion Efficiency (%)	$/MBtu =
Natural gas	100,000 Btu/therm	75	13.33 × $/therm
Fuel oil	138,000 Btu/gal	65	11.15 × $/gal
LP gas	93,000 Btu/gal	75	14.34 × $/gal
Mixed hardwoods	24 MBtu/cord	50	0.083 × $/cord
Mixed softwoods	15 MBtu/cord	50	0.13 × $/cord
Coal	12,500 Btu/ton	60	0.067 × $/ton
Electricity	3412 Btu/kWh	100	293 × $/kWh

Cooking Stoves in Developing Countries

Household energy demands, especially those for cooking and heating, dominate the energy needs for developing countries. The majority of the population of developing countries live in rural settings, where cooking uses about 80% of basic energy demand. India has 75% of its population living in 570,000 villages. Rural energy demand there is met primarily by wood (60%) and dung and crop residues (30%). It has been estimated that 56% of the world's population eats food cooked with such fuels. Women in many villages spend two hours per day gathering firewood for cooking purposes, and this takes time away from other household tasks such as getting water. Less water often means increased health problems, while less wood means less hot water.

Upward of five hours per day can be spent gathering firewood, as this Indian man can testify. (A. ODDIE/ PHOTOEDIT)

Laying out dung cakes to dry in the sun before use as fuel for cooking. (N. PIERCE)

Firewood supplies are becoming a problem in many developing countries. More than one billion people have identifiable shortages. The problem lies not only in wood supplies but also in the process of deforestation itself. Without forest management, the removal of trees, bush, and leaves leads to soil erosion and desertification. This problem is increased because many villagers remove bark and strip away live branches to provide the smaller wood pieces needed for cooking. The use of animal dung—formed into dung cakes for cooking—robs farmland of needed nutrients. At least 1 million hectares of forest disappear every year, especially in tropical regions. What effect this will have on global climate changes is unknown.

The most common stove in rural areas of the Third World is the "three-stone" open fire (Fig. 17.9). This stove is easy to build, provides light, space heating, and insect reduction, and is a good social gathering place. However, it is very inefficient (about 10%). Many new stove designs have been developed over the past decade or two, but most have been unsuccessful because they failed to include the multiuse characteristics of the three-stone cooker. Solar cookers (Chapter 6) also have received a lot of attention because many of the developing countries have high insolation. Yet it's hard to cook this way at night.

As developing countries become increasingly urbanized, the demand for charcoal increases. Charcoal is easier to transport than wood, has twice as much energy content per pound, and can be used in stoves that are more efficient than open wood fires. However, about 60 to 80% of the energy stored in wood is lost when it is converted to charcoal. (Charcoal is prepared by covering a stack of wood with earth and allowing it to smolder in a low-oxygen atmosphere for several days.) The increased demand for charcoal is having a sharp impact on the forest reserves.

FIGURE 17.9
Three-stone stove is used for cooking in many developing countries.

The heat energy transferred to foods during cooking is divided into three parts:

Raising temperature of the liquids to their boiling point	20%
Losses from the cooking vessel	45%
Vaporization of water	35%

Since most foods contain a high percentage of water, the time for cooking is independent of the heat input rate once the boiling point is reached, since the temperature of the water remains at 100°C (in an uncovered pot). Figure 17.10 shows three types of stoves used for cooking with firewood. Notice the tight-fitting holes for the pots and their closeness to the heat source. (Radiant heat is inversely proportional to the square of the distance.) Dampers control the rate of combustion and direct the air to the fuel where it is best used. Air entering a stove without a damper may cool the flames or the bottom of the pot. Most of the heat transfer takes place through radiation from the flames and convective heat transfer from the circulating hot gases. Conductive heat transfer to the pots is small but conduction is a major cause of heat loss through the stove walls. Not all improved stoves need a chimney. A chimney is used where

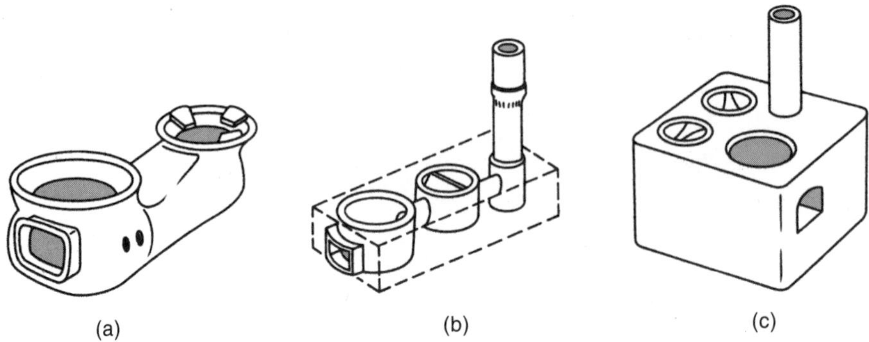

FIGURE 17.10
Low-cost cooking stoves with increased efficiency. From (*a*) India, (*b*) Nepal, (*c*) Guatemala. (BILL STEWART, *IMPROVED WOOD, WASTE, AND CHARCOAL BURNING STOVES*, INTERMEDIATE TECHNOLOGY PUBLICATIONS, 1987)

smoke in the cooking area is a problem or when a draft is required. Chimneyless stoves leave a gap around the second pot to enable the hot waste gases to leave. (Note stove in Figure 17.10a.)

E. Energy Plantations

Growing biomass for energy provides a multitude of advantages over nonrenewable fuels. Biomass can reduce our dependence on fossil fuels, thereby reducing expenditures on fuel imports. It can provide a boost to rural economies by creating a varied crop basis and thereby reducing federal subsidies for farming. A more varied crop can result in reduced soil loss, better water quality, and better habitat for wildlife. Increased job growth in new energy and environmental technologies can also occur.

A biomass energy plantation is a farm that is devoted to converting sunlight into energy. A standard farm already does this as it routinely converts sunlight into food, and a wood lot achieves the same results as it grows trees for sale as firewood. Yet there are many other plants that can be raised for the purpose of converting sunlight into energy. This is usually accomplished by conversion of biomass into a liquid or gaseous fuel. Marine plants and grains have shown much potential as sources of methane and ethanol. Important factors that must be considered in evaluating energy plantations are yields (in tons per acre per year), energy content per pound, ease of harvesting, and crop maintenance needs (weather, water, soil conditions). The chemical energy obtained from the biomass source must be greater than the energy expended in growing and obtaining that resource. Because of the different types of fuels that are available, biomass offers an important alternative to the fossil-fuel-based liquid fuels, especially for the transportation sector.

FIGURE 17.11
Eucalyptus tree—a prime candidate as a biomass energy crop. These trees grow rapidly, have a high heat content, require little water, and are quite hardy.
(STANFORD UNIVERSITY)

One species receiving attention as an energy crop is the fast-growing hybrid poplar. Research is showing that the high growth rates of these trees also makes them effective pollution fighters. They seem to be able to clean up groundwater and surface flows moving downslope from agricultural lands. Another plant species is the eucalyptus tree (Fig. 17.11). These plants have a high energy content, grow very quickly (as much as 12 ft/yr), require little water, use land that is poor for agriculture, and are quite hardy. Other species being used are sugarcane, sunflowers, water hyacinth, and freshwater algae.

F. Food, Fuel, Famine

When we consider the use of grains such as corn to produce alcohol fuels, we should ask whether food should be used to produce fuel, especially in light of the malnutrition of about one half of the world's people. Before addressing this issue, we will first look at the energy required to produce food and some of the other questions that concern land use.

Today, the average dietary intake in the world is about 2100 kcal per day. (Remember that one food Calorie = 1000 cal = 1 kcal.) Protein is a requirement in the diet, and its recommended intake is about 40 g per day (for average body weight). In the United States, the average diet includes about

3300 kcal per day and about 100 g of protein, two thirds of which comes from products of animal origin. While the majority of the world's population obtains its protein intake from plant foods—grains and legumes—Americans consume about 110 kg of meat per year per person. More than half of the feed for livestock consists of cereal grains that humans could eat. (The other half is forage and grasses not suitable for human consumption.) More than 90% of our corn crop is used as feed for animals! The efficiency of conversion from plant protein to animal protein is poor, less than 20%. In the United States about 25 million tons of plant protein per year are fed to animals to produce 6 million tons of animal protein. In essence, plant protein is being recycled to provide higher quality animal protein. As we shall see next, the energy input per kilogram of protein output is much higher for animals than for grain.

It's interesting to note the total use of energy in the production of animal and plant protein. Overall—from production to transportation to processing to preparation—about 13% of the total energy used in the United States goes into food. In developing countries, 60 to 80% of the total energy consumption is used for food. However, in absolute terms, three to four times as much energy is used for food production per person in the United States than in developing countries.

The uses of this energy for food production are different, too. In the United States, about half of the food energy input goes into tractors and other machines used in crop planting, tilling, and harvesting; the other half is used in making and applying fertilizers and in irrigation. In developing countries, most of the energy goes into harvesting and into gathering firewood, both of which require much humanpower. Only 1% of the energy used in developed countries is provided by humans and animals.

Clearly, more energy is required for the mechanized management of a farm (mostly because it uses petroleum) than for one in which all operations are carried out by hand. However, the modern farm is more economical to run. If all operations were done by hand, only about 1 hectare (ha) could be successfully managed per person during the growing season. With machinery, a farmer can manage about 100 ha (1 ha = 10,000 m^2 = 2.47 acres) for grain production, allowing more produce to be sold to support a higher standard of living. The fossil-fuel input replaces human input, and it is generally much cheaper.

EXAMPLE

A can of corn costs $0.58 and contains 370 kcal of food energy. This is 58/370 = 0.157¢/kcal. A gallon of gasoline costs $1.50. This is equivalent to 31,500 kcal, or a cost of 0.005¢/kcal.

A great deal of energy is used to produce a product, since fossil-fuel inputs are required for the manufacture of fertilizers and the operation of tractors. For beef raised in a feedlot, about 800 kg of feed protein is fed to them for a yield of about 50 kg of beef protein. This is a 6% conversion. For milk, about 30% of the plant protein fed to the cattle produces milk protein. When fossil-fuel inputs are included to produce the beef, the ratio of fossil-fuel-energy input to protein-energy output is about 75 to 1. Table 17.6 lists the energy used in the production

Table 17.6 ENERGY IN U.S. CORN PRODUCTION

	Needed per Hectare (appropriate units)	Energy per Hectare (kcal)
Inputs		
Labor	12 h	5580
Machinery	31 kg	558,000
Diesel fuel	112 L	1,278,368
Nitrogen	128 kg	1,881,600
Phosphorus	72 kg	216,000
Potassium	80 kg	128,000
Limestone	100 kg	31,500
Seeds	21 kg	525,000
Irrigation	780,000 kcal	780,000
Insecticides	1 kg	86,910
Herbicides	2 kg	199,820
Drying	426,341 kcal	426,341
Electricity	380,000 kcal	380,000
Transportation	136 kg	34,952
Total		6,532,071
Outputs		
Corn yield	5394 kg	19,148,700
kcal output/kcal input		2.93
Protein yield	485 kg	

(D. and M. Pimental, *Food, Energy, and Society*, Edward Arnold, 1979)

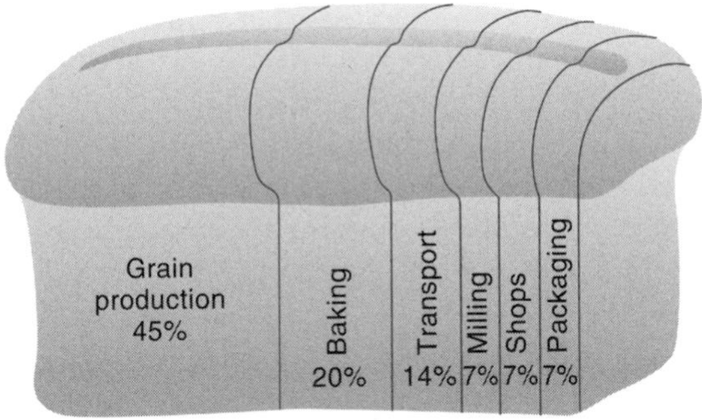

FIGURE 17.12
Energy used in the production of a loaf of bread. (D. AND M. PIMENTAL, *FOOD, ENERGY, AND SOCIETY*, EDWARD ARNOLD, 1979)

of corn, one of our most productive crops. Even with the use of fertilizer, corn is less energy-intensive than livestock. The ratio of corn energy output to energy input is about 2.9 to 1, compared to 0.01 for feedlot beef.

One of the most energy-intensive steps in our own use of food is in its processing and preparation in the home. Much more energy is used to cook the food than is found in the food itself. (But it's hard to eat wood.) A breakdown of the energy used in the production of a loaf of bread is shown in Figure 17.12.

Some basic issues emerge at this point. As we continue to deplete our supply of finite fossil fuels, especially oil and natural gas, how might our diets change? The United States uses about 600 kg of grains to produce about 115 kg of animal foods that we eat per year per person. In developing countries, with primarily a vegetarian diet, about 180 kg of grain products per year per person are consumed. If the United States were to provide the increased amounts of animal protein necessary to meet our projected population in ten years, an increase in crop production of about 10% would be necessary. This increase would be equal to 600 kg $\times$ 0.10 $\times$ 250 million people = 150 $\times$ 10^8 kg of grain. This is enough to feed 83 million more people in developing countries at their present consumption rate. As the world's population grows, a larger percent increase in fossil fuels will be required to meet the need for food. Some people suggest that an easier way to increase food supplies is to reduce the intake of animal protein, since about 4 kg of plant protein is needed for the production of 1 kg of animal protein.

G. Summary

Biomass resources include field crops, wood, agricultural wastes, and solid waste. They can be converted into gaseous and liquid fuels or burned directly. In general, biomass can be converted into other energy forms by anaerobic digestion (conversion of organic material into methane, in the absence of oxygen) and fermentation processes (producing alcohols such as ethanol and methanol). While Americans generate 0.7 tons of garbage per year per person, landfill space traditionally used to dispose of our waste is diminishing. The ever-increasing volume of solid wastes can be dealt with by reducing waste generated, by recycling, or by incinerating.

The use of food crops as fuel raises issues that touch on our dietary habits. Agricultural products such as livestock are energy intensive, with more energy used to produce the product than is obtained from it. Designs for wood-burning stoves for heating and cooking stoves for developing countries must be concerned with energy efficiency and so must use the concepts of Chapter 4 on appropriate heat transfer. Controlling the amount of air intake and employing secondary combustion are important energy-conserving principles for such stoves.

Internet Sites

For an up-to-date list of Internet resources related to the material in this chapter, go to the Harcourt College Publishers' website at **http://www.harcourtcollege.com**. The links are in the *Energy: Its Use and the Environment* site on the Physics page. General energy related sites and some guidelines for using the World Wide Web in your class are on the inside front cover of this book.

References

Chapter 17

Blumberg, L., and R. Gottlieb. 1989. *War on Waste.* Washington, D.C., Island Press.

Cointreau, S. 1984. *Recycling from Municipal Refuse.* Washington, D.C., World Bank.

Darrow, K., and M. Saxenian. 1993. *Appropriate Technology Sourcebook.* Boulder, CO, Appropriate Technology Institute.

Durning, A. 1993. *Saving the Forests: What Will It take?.* Washington, D.C., Worldwatch Institute.

Garner, G. 1997. *Recycling Organic Waste: From Urban Pollutant to Farm Resource.* Washington, D.C., Worldwatch Institute.

Geller, H. S. 1985. Ethanol Fuel from Sugar Cane in Brazil. *Annual Review of Energy,* 10.

Gordon, D. 1991. *Steering a New Course.* Cambridge, MA, Union of Concerned Scientists.

Harrison, P. 1993. *Inside the Third World: The Anatomy of Poverty.* London, Penguin.

Kammen, D. 1995. Cookstoves for the Developing World. *Scientific American,* 272 (July).

Kleinbach, M., and C. Salvagin. 1986. *Energy Technologies and Conversion Systems.* Englewood Cliffs, NJ, Prentice-Hall.

Kozloff, K., and R. Dower. 1993. *A New Power Base: Renewable Energy Policies for the Nineties and Beyond.* Washington, D.C., World Resources Institute.

Lappé, F. M. 1991. 20th ed. *Diet for a Small Planet.* New York, Ballantine.

MacKenzie, J. 1994. *The Keys to the Car: Electric and Hydrogen Vehicles for the 21st Century.* Washington, D.C., World Resources Institute.
Manibog, F. R. 1984. Improved Cooking Stoves in Developing Countries. *Annual Review of Energy,* 9.
Ogden, J., and R. Williams. 1989. *Solar Hydrogen.* Washington, D.C., World Resources Institute.
Penner, S. S., D. F. Wiesenhahn, and C. P. Li. 1987. Mass Burning of Municipal Wastes. *Annual Review of Energy,* 12.
Pimentel, D., and C. Hall, eds. 1984. *Food and Energy Resources.* Orlando, FL, Academic Press.
Pimentel, D., and M. Pimental. 1979. *Food, Energy, and Society.* London, Edward Arnold.
Ross, B., and C. Ross. 1978. *Modern and Classic Woodburning Stoves.* Woodstock, NY, Overlook Press.
Schramm, G., and J. Warford, eds. 1989. *Environmental Management and Economic Development.* Washington, D.C., World Bank.
Shelton, J. 1983. *Solid Fuels Encyclopedia.* Charlotte, VT, Garden Way.
Stewart, B. 1987. *Improved Wood, Waste, and Charcoal Burning Stoves.* London, Intermediate Technology Publications.
Wood, T. S., and S. Baldwin. 1985. Fuelwood and Charcoal Use in Developing Countries. *Annual Review of Energy,* 10.

QUESTIONS

1. Make a table listing biomass fuels and their possible uses.
2. What are some of the concerns about the incineration of solid waste?
3. What steps are taken in modern landfills to prevent groundwater contamination?
4. What is secondary combustion in a wood-burning stove? Why does one need to control the amount of combustion air?
5. Why is the efficiency of a fireplace sometimes close to 0%?
6. What are some of the difficulties with the use of traditional cooking stoves in developing countries?
7. At the present time, what is the least expensive alternative fuel that could be used for the automobile? What limitations (including market factors) are there to expanded use of this fuel for the automobile?

PROBLEMS

1. The following table is a handy guide to compare the quantity of fuel needed to deliver equivalent amounts of heat. Using current costs, fill in the last column. (You will have to make some calls or use the Internet.)

AMOUNTS OF FUEL NEEDED TO PRODUCE 20 MBtu OF HEAT

Fuel	Quantity	MBtu Available	Efficiency (%)	MBtu Delivered	Fuel Cost
Medium-density Wood	1 cord	29	70	20	
Number 2 Fuel Oil	222 gal	31	65	20	
Electricity	5876 kWh	20	100	20	
Natural gas	26,800 ft^3	27	75	20	
Bottled gas	292 gal	27	75	20	
Coal	2440 lb	31	65	20	

2. Using information on recycling, estimate the energy (in Btu's) needed to produce one aluminum can. Compare that with the energy needed to run a TV set for 5 hours and the energy used per day by a person from India (see Fig. 2.7).
3. A kitchen stove uses about 50 ft^3 of natural gas per day. The daily manure from how many cows would have to be put into a digester to meet this need? (Use the Kenyan example numbers in the discussion of anaerobic digestion.)
4. Using Table 17.5, a cord of hardwood at $120 is equivalent to using electric heat at how many cents per kilowatt-hour?
5. Find the price per gallon of heating oil (fuel oil) in your area. Using Table 17.5 (and the example in the same section of the text), calculate the cost to heat a space requiring 100 MBtu per season (typical of a medium-sized home in a northern U.S. climate).
6. Calculate the cost of electric heat to provide 100 MBtu to heat a typical medium-sized home for one heating season.
7. Using the information from Table 17.2, calculate the potential heat energy of rice versus wheat per acre.

FURTHER ACTIVITIES

1. Outline the recycling program that your community has (or will have). What is mandatory?
2. Investigate the market for recycled newspaper and plastic products in your area.

18

Tapping the Earth's Heat: Geothermal Energy

A. Introduction

Geothermal energy is produced from heat originating in the earth's interior. Volcanoes, geysers, hot springs, and boiling mud pots are visible evidence of the great reservoirs of heat that lie within and beneath the earth's crust. Although the amount of thermal energy within the earth is very large, useful geothermal energy is limited to certain sites. These resources are not infinite and can be depleted at a particular site under intensive exploitation. Nevertheless, geothermal energy is a resource that can be further developed in favorable locations. Currently, 4% of the electricity generated in the United States from so-called renewable sources comes from geothermal energy. (This is almost four times the contribution from wind and solar.) While growth in the total energy used from geothermal resources in the United States has shown little growth in the last 10 years, globally, geothermal power has been growing steadily at a rate of about 8.5% per year.

Table 18.1 GEOTHERMAL POWER PLANTS: 1998

Site	Installed Capacity (MWe)
United States	2850
Philippines	1848
Mexico	743
Italy	769
Indonesia	590
Japan	530
New Zealand	345
Iceland	140
Costa Rica	120
El Salvador	105

(U.S. EIA)

Electricity was first produced from naturally occurring steam in Italy in 1904. Today many geothermal power stations are operating worldwide. Table 18.1 lists world geothermal generating capacity in 1998 for those nations with substantial geothermal facilities. The world's total capacity is about 8000 MWe. The Geysers in northern California, the largest such facility in the world, has a total installed capacity of 1224 MWe—enough to power the cities of San Francisco and Oakland, located 90 miles to the south; this is about 7% of California's electricity needs. The island of Hawaii provides 25% of its electricity from geothermal resources. The Philippines, Indonesia and Mexico have enjoyed rapid growth in generating capacity in the past decade. El Salvador generates the majority of its electricity with steam from geothermal resources.

Nonelectric applications of geothermal energy have been developed extensively in some countries. Hot water from underground sources provides direct heating for the majority of homes in Iceland's capital, Reykjavik, and has done so for six decades. Budapest, Hungary, has been partly heated by geothermal steam since the times of the Roman Empire. Heated greenhouses there provide vegetables and flowers the year around. Space or district heating via geothermal energy in the United States is on a small scale. The main developments have been for residences and businesses in the cities of Klamath Falls, Oregon; Boise, Idaho; and San Bernadino, California.

In this chapter we will investigate the sources of geothermal energy and their potential yields, the use of this resource, and obstacles to its development.

B. Origin and Nature of Geothermal Energy

Geothermal energy has its origin in the molten core of the earth, where temperatures are about 4000°C (7200°F). It's Mother Nature's own boiler. This thermal energy is produced primarily by the decay of radioactive materials within the interior, leading some people to refer to geothermal energy as a form of "fossil nuclear energy." The interior of the earth is thought to consist of a central molten core surrounded by a region of semi-fluid material called the **mantle** (Fig. 18.1). This is covered by the **crust,** which has a thickness between 30 and 90 km. The temperature in the crust increases proportionally with depth at a rate of about 30°C/km. The temperature at the base of the crust (the top of the mantle) levels off at a value of about 1000°C and then increases slowly into the earth's core. If these average conditions for the temperature of the crust were all we had to work with, we'd be out of luck since the earth's heat energy that could be used effectively would be much too deep for tapping. Fortunately, there are regions in which hot molten rock of the mantle (called **magma**) has pushed up through faults and cracks to near the surface, creating "hot spots" within 2 to 3 km of the surface. We see evidence of such activity in volcanic eruptions, geysers, and bubbling mud holes. In fact, the zone of likely geothermal sites corresponds roughly to the region of earthquakes and volcanic activity, as shown in Figure 18.2. These regions are at the junctions of **tectonic plates** that make up the earth's crust (Fig. 18.3). These plates are in a state of constant relative motion (at rates of several centimeters per year). Where they collide or grind, there are very strong forces that can build mountains or cause earthquakes. It is near the junctions of these plates that heat travels most rapidly from the interior via subsurface magma to surface volcanoes. Most of the world's geothermal sites today are located near the edges of the Pacific plate, the so-called "ring of fire."

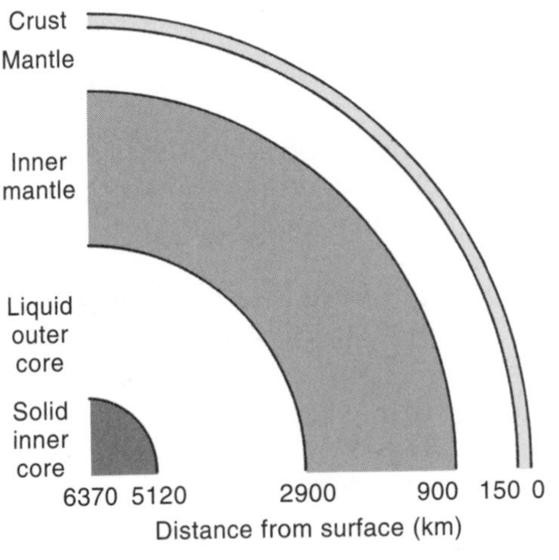

FIGURE 18.1
Cross section of the earth, showing the layered structure.

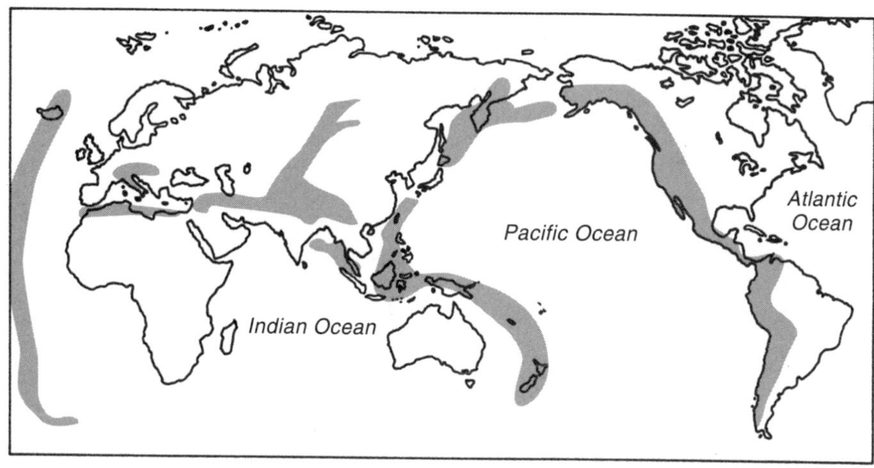

FIGURE 18.2

Regions of potential geothermal sites, the "ring of fire." These regions correspond roughly to the zones of earthquakes and volcanic activity.

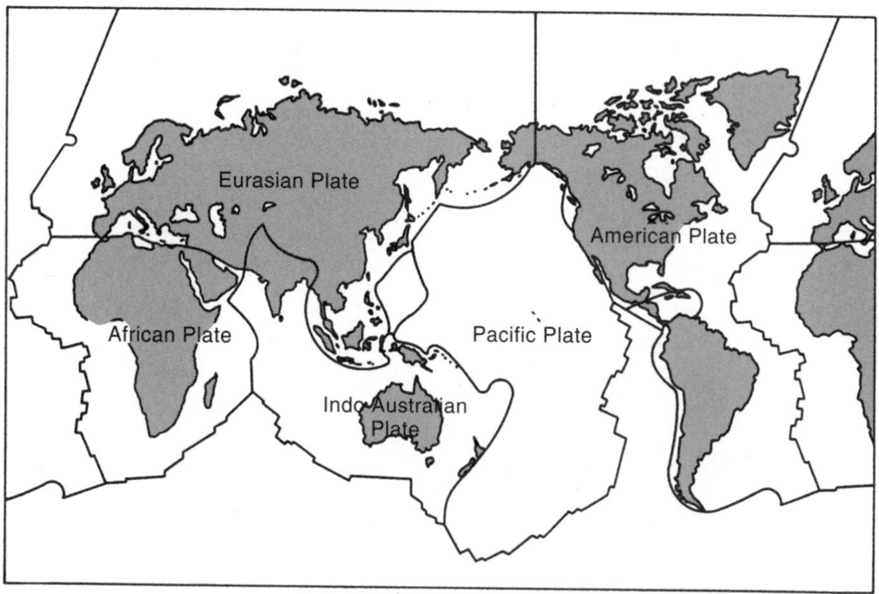

FIGURE 18.3

Tectonic plate boundaries. The earth's crust is made up of six major plates and a number of smaller ones. These plates are in a state of constant relative motion (at rates of several centimeters per year). Near the junctions of the plates, heat travels most rapidly from the interior. (NEW ZEALAND MINISTRY OF ENERGY)

From these hot spots, geothermal energy is most easily extracted, technologically and economically. In the most common types of geothermal reservoirs, called **hydrothermal systems,** the thermal energy of the magma is stored in water or steam that fills the pores and fractures in the rock. These reservoirs can be classified into wet steam (or hot water) and dry steam systems. Although the wet steam systems are 10 to 20 times more abundant, the dry steam systems have been used more often in the generation of electricity because of their convenience. An example is The Geysers in California. There are other types of geothermal energy resources that have as much if not more potential, but they will have to await new extraction technologies before they can be developed. These other resources—hot dry rock systems and geopressured regions—will be discussed later in the chapter.

C. Hydrothermal Systems

Wet Steam Systems

When water is trapped in an underground reservoir and is heated by the surrounding rocks, it is under high pressures and can reach temperatures as high as 370°C (700°F) without boiling. If this hot water is released to the surface, it will "flash" into steam as the external pressure falls below that necessary to keep it a liquid. Places where steam escapes through cracks in the surface are

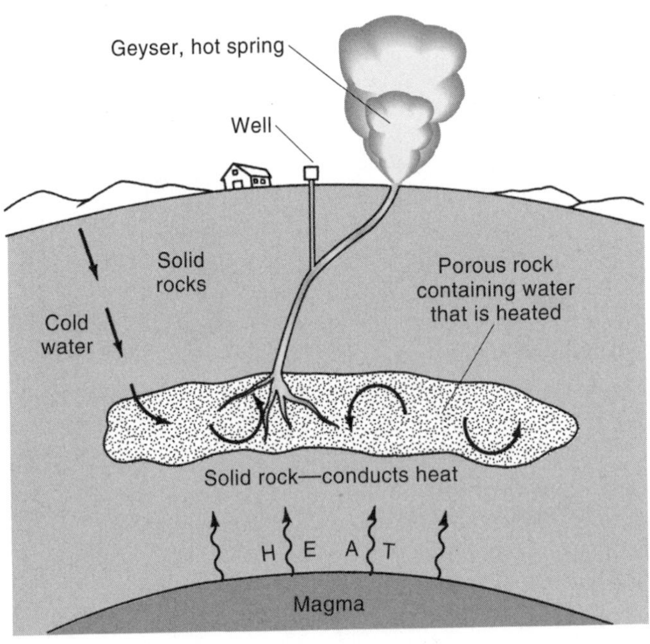

FIGURE 18.4
Model of a high-temperature hot water geothermal system. The water within the porous rock is heated by conduction from the magma. The hot water escapes through fissures to the surface, boiling near the top. A well to tap the steam within the fissure is also shown.

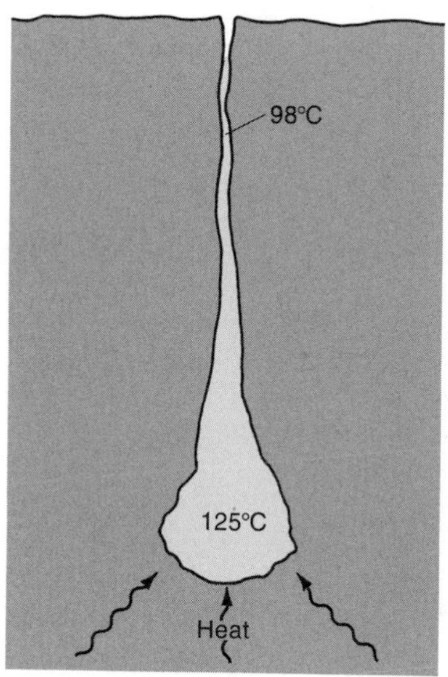

FIGURE 18.5

Model of a geyser. Water at the bottom of the container is under great pressure and will not boil until temperatures above 100°C are reached. When boiling begins, the pressure is released, causing the water to boil very rapidly. The steam-driven water rushes up the neck and is sprayed into the air in the form of steam.

called **fumaroles.** In some geothermal reservoirs, hot water leaks to the surface, forming hot springs or geysers. Geothermal wells tap these wet steam fields (Fig. 18.4). As the hot water rises in the well, it flashes into a mixture of about one part steam and four parts hot water. The steam is separated from the water and used to run turbines to generate electricity. The hot water can be used for direct heating or for a desalination plant.

It might be good to consider at this point the principles behind a periodic geyser like "Old Faithful" in Yellowstone National Park. Suppose water is stored in an underground reservoir with a small vent to the surface (Fig. 18.5). The pressure on the water at the bottom of the reservoir is greater than atmospheric pressure because of the overlying column of water in the reservoir's neck, and so the temperature in the base can rise above 100°C without boiling occurring as the water is heated by the surrounding rocks. As the water continues to increase in temperature, the boiling point at that pressure is reached. The beginning of boiling releases the pressure in the base rapidly, increasing the boiling rate and causing the water to gush out of the ground—yielding a geyser. Cooler water from the surface replaces the hot water. The time between eruptions will be equal to the time it takes the water in the reservoir to be heated up to reach the boiling temperature at that pressure. A percolator coffeepot works on this same principle.

Dry Steam Systems: The Geysers Power Plant

Dry steam geothermal fields occur when the pressure is not much above atmospheric pressure and the temperature is high. In this situation, water boils underground and generates steam at temperatures of about 165°C (350°F) and

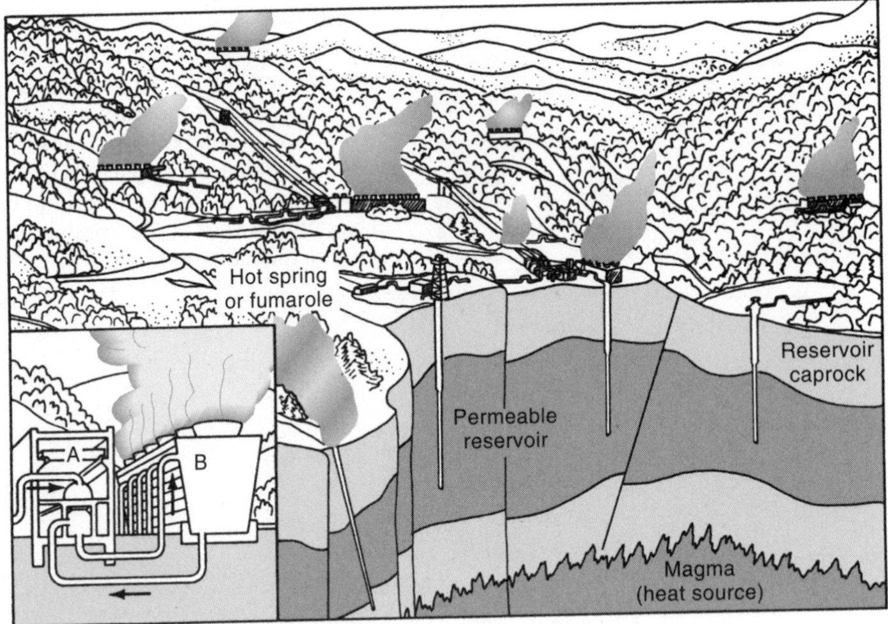

Schematic of the geothermal field of The Geysers. Insert shows a power plant unit (A) with condenser (B). (PACIFIC GAS AND ELECTRIC)

pressures of about 100 psi. The steam taken from such a geothermal well can be used directly to drive a turbine. Geothermal plants at Larderello, Italy, and The Geysers, California, use dry steam.

The dry steam fields of The Geysers were discovered in 1847 by a hunter looking for grizzly bear. Seeing the natural steam venting into the air, he later told his friends he thought he had found the "gates of Hell." A resort was built in the area in the 1860s, featuring therapeutic hot springs. Electric power for the resort was provided in the 1920s by using the steam from several wells to run a steam turbine generator. Thus the first geothermal power plant in the United

Table 18.2 THE GEYSERS (CALIFORNIA)

Capacity	1224 MWe
Well depth	800–5000 ft
Steam pressure	115 psi
Steam temperature	350°F (165°C)

(Pacific Gas and Electric)

FIGURE 18.6
The Geysers Geothermal Power Plant in northern California. Most of the steam seen here comes from the condensers of the individual units. (PACIFIC GAS AND ELECTRIC)

States came into existence. Large-scale operations did not begin for another 30 years, when two small private power companies invited Pacific Gas and Electric (PG&E) to come into the area and test the dry steam wells that they had drilled. Engineering and economic conditions seemed right, so in 1960 the first unit of The Geysers went into operation, with an output of 11 MWe. Today the output is more than 1224 MWe, with 14 power plants (Fig. 18.6). Although several of the units have been abandoned in recent years, reducing capacity by about 30%, The Geysers remains the largest complex of geothermal generating facilities in the world. In 1999, under divestiture, The Geysers was sold by PG&E.

During operation, the dry steam used for the turbines is at a temperature of 165°C (350°F) and a pressure of 115 psi. The steam leaving the turbine goes to a condenser and then to a cooling tower. (The evaporated water from the cooling tower is seen as steam in the figure.) Some of the water leaving the cooling tower is returned to the condenser to cool the incoming steam, while the rest of the water is reinjected into the ground. Further statistics on this facility are given in Table 18.2.

D. Geothermal Exploration and Resources

The size of geothermal resources is estimated by some to be very high. The U.S. Geological Survey (USGS) defines the resource to be the entire heat content of the earth's crust above 15°C to a depth of 10 km. With this definition, the USGS estimates that more than 2×10^{22} Btu of thermal energy exists within the crust. This is equivalent to 900 trillion tons of coal, enough to supply our energy needs at current rates for 350,000 years! However, one must be careful with these numbers. Geothermal energy is quite *low grade* because the temperature of the steam or hot water used is only usually between 150°C and 250°C (at 100 psi). This compares with the steam in a conventional fossil-fuel plant at 550°C and 1000 psi.

The earth acts like a large heat engine for the generation of electricity as it provides hot water and/or steam at temperatures of 150°C to 250°C. However, these relatively low temperatures mean lower efficiencies. Recall that the maximum efficiency of a heat engine operating between a hot temperature T_H and a cold temperature T_C is given by the Carnot efficiency:

$$\text{Maximum (Carnot) efficiency} = \left(1 - \frac{T_C}{T_H}\right) \times 100\%$$

For a geothermal plant, if $T_H = 200°\text{C} = 473\text{ K}$ and $T_C = 27°\text{C} = 300\text{ K}$, then the Carnot efficiency is $(1 - 300/473) \times 100\% = 36\%$. If the real efficiency is half of the Carnot efficiency, then an 18% efficiency is achieved, compared to the 35 to 40% efficiency of a fossil-fuel plant or the 50 to 55% efficiency of a gas turbine combined cycle plant.

Geothermal hot spots are sparsely distributed and usually found some distance from the markets needing energy. The minimum temperature of steam required for the economic production of electricity is about 110°C. As a consequence, many reservoirs of hot water can be used only for space heating (as in Iceland). Since thermal energy cannot be efficiently transported very far, the point of use must be close to the source.

Some of the more obvious evidences of geothermal resources are the surface heat displays due to leakage from deeper reservoirs, such as geysers, hot springs, and steam vents. These displays are analogous to oil seeps, and the same care must be taken in evaluating the evidence. Oil sometimes will seep to the surface from an underground reservoir by strange paths, so drilling around such seeps will not always be productive. The present method for proving the commercial viability of a geothermal resource is to drill deep holes and perform long-term flow tests. One must also study the local geologic environment, including the types and properties of rock formations.

Other exploratory techniques use (1) aerial and surface observations to look for faults and volcanic activity, (2) seismic methods to look for underground reservoirs, (3) geochemical methods, such as analyzing water from springs, and (4) electrical measurements to indicate the presence of deep water with high temperatures and salt content. Using infrared aerial photography to look for temperature gradients has not been very successful because of interference from the local terrain.

In addition to hydrothermal, there are two other types of geothermal resources: geopressurized reservoirs and hot dry rocks. **Geopressurized reservoirs** consist of hot, liquid brines that lie in large, deep (3000 to 6000 m) areas (not in small, localized hot spots near the surface) and are usually under pressures as high as 10,000 psi. The energy contained in these geopressurized zones is not only thermal but also mechanical (hydraulic) and chemical (as a result of dissolved methane gas). In the United States, these reservoirs are thought to lie primarily along the Gulf Coast. The energy potentially recoverable from these reservoirs is immense, but the technology to exploit it is still being developed. Experimental wells are already in place in Texas and Louisiana.

Hot dry rocks are located underground, but lack the aquifers or fractures (cracks) to convey fluid to the surface, as in hydrothermal reservoirs. This source can be exploited by circulating water through the cracks to extract the heat energy. Artificial reservoirs can be made by hydraulically fracturing these rocks and then circulating water through the cracks. Some fractured reservoirs at depths of 3000 to 4000 m (10,000 to 13,000 ft) currently are being tested in New Mexico. Hot dry rocks are much more common than hydrothermal reservoirs and more accessible, so their potential is quite high.

Electricity is extracted from hot dry rock systems by using a secondary or binary fluid cycle (Fig. 18.7). In this cycle, the hot water is passed through a heat exchanger to transfer heat to a liquid with a low boiling point, such as a refrigerant or isobutane, whose vapors then are used to drive the turbine. The original geothermal water is then reinjected into the ground. In this technique, one doesn't

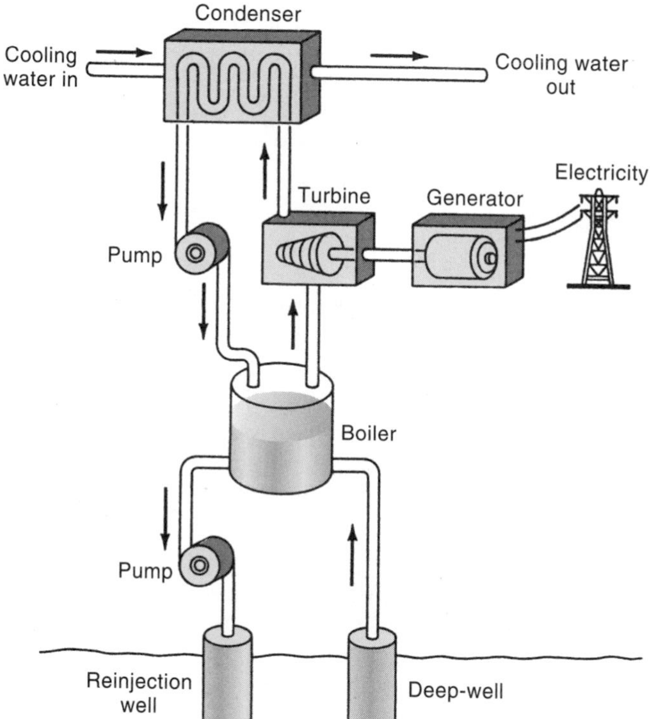

FIGURE 18.7

Vapor-turbine cycle. Water is circulated through a hot dry rock system. The heated water is pumped through a heat exchanger, boiling the working fluid, which is isobutane. The cooled water is reinjected into the ground. The vapor expands through a turbine and then is condensed back to a liquid in the water-cooled condenser.

Table 18.3 U.S. GEOTHERMAL RESOURCES (ASSUMED RECOVERABLE)

Resource	Heat at Wellhead* (Quads†)	Electricity (MW-30††)
Hydrothermal convection systems		
Vapor dominated (dry steam)	26	
Hot water (>150°C)	350	11,500
Intermediate temperature hot water (90°–150°C)	340§	
Hot igneous magma systems		
Regions of hot, dry rock, 0–10 km deep	26,000	
Geopressured regions		
Thermal energy	900	80,000
Methane	500	
Mechanical energy	40	33,000

*Assumes a 25% recovery rate for convection resources at the wellhead.

†1 Quad = 10^{15} Btu.

††1 MW-30 is equivalent to 1000 kW (1 MW) produced continuously for 30 years.

§For thermal (nonelectrical) uses close to the source.

(U.S. Geological Survey, Circular 790, 1979)

need to use high-pressure steam from a conventional hydrothermal reservoir to drive the turbine. Low-temperature steam can be used. A binary fluid plant in Nevada makes electricity economically from 103°C (218°F) geothermal fluid.

The estimated recoverable energy from all these types of resources in the United States and the estimated resource base are listed in Table 18.3. The potential for energy from hot dry rocks could be much larger than this if technology improves and costs decline. Historically it should be noted that with any relatively new resource, projections of its potential tend to be high. For example, estimated contributions for the generation of electricity for 1985 in the United States from geothermal energy were made in the following reports (report date is shown in parentheses): Department of the Interior (1972): 19,000 MWe; Project Independence (1975): 7000 to 15,000 MWe; Energy Research and Development Administration (ERDA-48) (1975): 10,000 to 15,000 MWe; Hickel U.S. Geological Survey (USGS) Report (1972): 132,000 MWe! Actual capacity in 1985 in the United States was about 1300 MWe.

E. Low-Temperature Geothermal Resources

Geothermal reservoirs of low to moderate temperatures (20°C to 150°C) can be used to provide direct heat for residential and industrial uses. These reservoirs are usually hot water under pressure in an underground reservoir. To be used, the hot water is brought to the surface where a heat exchanger system transfers the thermal energy to another fluid. The cooled geothermal fluid is then pumped through an injection well back into the ground. The primary uses of the heated fluid are in district and space heating (sometimes individual structures), greenhouses, and aquaculture (fish farming).

Recent surveys have identified a large potential for new direct-use geothermal applications in the western United States. More than 9000 thermal wells and springs are now in use. These applications are saving the energy equivalent of almost 2 million barrels of oil per year.

F. Environmental Impacts

The expanded use of geothermal energy has run into some opposition from environmental groups, which protest that such plants are dangerous, dirty, noisy, and unsightly. One of the problems of a geothermal plant is emission of noxious gases, such as hydrogen sulfide (H_2S). Hydrogen sulfide has the smell of rotten eggs. Carbon dioxide is also emitted in geothermal processes, although significantly less is emitted than from any other fossil-fuel plants with the same output. The steam from dry steam fields contains minerals that can contaminate the groundwater and poison fish and other aquatic life after it condenses. In wet steam fields, the mineral and salt content of the hot water (so-called brines) can

Pipes carrying steam to a generator at The Geysers. (PACIFIC GAS AND ELECTRIC)

be as high as 20 to 30% dissolved solids. Damage from corrosion can occur to the turbine blades, and the piping can become clogged. There are additional problems with liquid waste disposal. Procedures have been developed in which the hot water would be evaporated after use so that the minerals could be extracted. Another problem in some geothermal regions is that the removal of steam from the reservoirs can cause subsidence (setting or slumping) of the land above. One of Mexico's steam plants reported a subsidence of 13 cm. This problem might be remedied by reinjecting the waste water from the fields into the area via injection wells. In hot dry rock processes, the fluid from the hot reservoir is reinjected into the ground, making this process environmentally attractive.

G. Summary

The prospects for geothermal power are certainly promising, although restricted to specific geographical areas. The different forms of geothermal resources are

- hydrothermal resources containing steam or hot water reservoirs that can be tapped by drilling.
- hot dry rock resources (heat stored in largely impermeable rocks). These can be used by injecting cold water into the well.
- geopressured resources (deeply buried brines that contain energy in thermal , mechanical, and chemical forms).

For accelerated exploitation of this resource to occur, several problems have to be overcome:

- lack of reliable information about geothermal resources, such as location, lifetime, and energy available. This is especially true for hydrothermal systems for which much of the technology is available. Investors won't back power plant construction unless they can be assured of a reliable supply of geothermal fluids over the lifetime of the plant.
- lack of proven technology for extraction and use of the resource. This is especially true for geopressurized systems.
- lack of complete knowledge about environmental impacts, such as air pollution and subsidence.
- regulatory requirements and complexity of leasing regulations. Most of the currently exploitable geothermal resources are on public land, and leases have been available for only about 25 years. (Federal law prohibits the exploitation of geothermal resources in National Parks.)
- lack of substantial funding to research these concerns.

The absence of information in these areas will hinder private industry from committing capital for the development of these large resources. The economics certainly appear to be good. The costs of geothermal energy are one half to three quarters of those for fossil-fueled plants in similar locations.

Internet Sites

For an up-to-date list of Internet resources related to the material in this chapter, go to the Harcourt College Publishers' website at **http://www.harcourtcollege.com**. The links are in the *Energy: Its Use and the Environment* site on the Physics page. General energy related sites and some guidelines for using the World Wide Web in your class are on the inside front cover of this book.

References

Chapter 18

Cuff, D., and W. Young. 1986. *The United States Energy Atlas*. 2nd ed. New York, Macmillan.

Kruger, P. 1976. Geothermal Energy. *Annual Review of Energy*, 1.

Mock, J. E., J.W. Tester, and P.M. Wright. 1997. Geothermal Energy for the Earth: Its Potential Impact as an Environmentally Sustainable Resource. *Annual Review of Energy*, 22.

Pollack, H., and D. S. Chapman. 1977. The Flow of Heat from the Earth's Interior. *Scientific American*, 237 (August).

QUESTIONS

1. List the different types of sources of geothermal power.
2. Why is the vast amount of thermal energy lying beneath the surface of the earth not a very useful resource?
3. Describe the similarity between the geysers discussed in this chapter and a percolator coffeepot.
4. Why are the most likely geothermal sites located in regions of high volcanic activity?
5. What are the environmental impacts of geothermal energy?
6. If a geothermal plant runs at one third its Carnot efficiency, how much *more* heat will be added to the atmosphere than from a fossil-fuel plant operating at two-thirds Carnot efficiency? Take the condenser temperature to be 27°C (300 K) for both systems; for the geothermal plant the boiler temperature is at 150°C, and for the fossil plant it is 550°C. Assume the electrical output from both plants is the same.
7. Investigate the manner in which geothermal energy is used to heat homes in Reykjavik, Iceland, or businesses in San Bernadino, California.

19

A National and Personal Commitment

Few will have the greatness to bend history itself, but each of us can work to change a small portion of events, and in the total of all these acts will be written the history of this generation. Robert F. Kennedy

The first 18 chapters of this book have shown the complexities, the pros and cons, and some of the technical principles of many energy technologies. You can now better appreciate that energy is one of the building blocks of modern society and that the availability and cost of energy resources are key factors in a country's economic growth. However, knowing the parts does not necessarily mean that we understand the whole—or that we can even see it. A national energy policy, if that is possible, must be built on more than simplistic statements such as "All we need to succeed is solar power, or nuclear power, or . . . ." We must first determine what goals we want to achieve for our country, and then decide how energy resources can better help us meet those ends. What long-range constraints (economic, environmental, political, resource availability) will there be? The 1991 war in the Persian Gulf brought many of these issues to the attention of the public—and Congress—once again. Unfortunately, it seems that only in times of crises, real or imagined, do people pay much attention to energy matters. Will this be the case as we enter an era of competition, especially in the electric power business? A strong economy also makes us less sensitive to higher energy prices.

Figure 19.1 tries to capture the complexities in the use of energy. As we seek to meet certain needs (light, heat, etc.), we must use our energy resources. We have to connect the resource with the need: A lump of coal doesn't provide light by itself. Carriers or converters (such as high-voltage lines, gas furnaces, etc.) are needed to go from resource to need. However, there are barriers in this process: economic, environmental, political, and so on. These must be considered and dealt with.

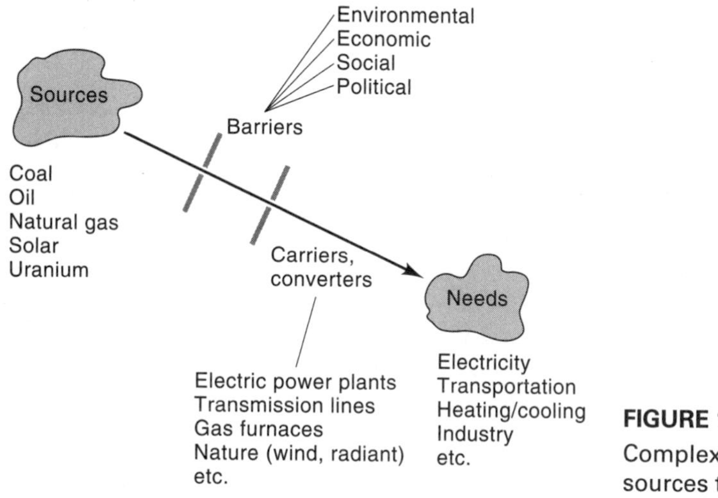

FIGURE 19.1
Complex path from energy sources to material needs.

The U.S. government has wrestled with energy policies since the 1920s, when it tried to encourage the production of then-scarce liquid fuels from oil shale for use in a blossoming automobile industry. The discovery of vast resources of oil in Texas and Oklahoma in the late 1920s ushered in an era of cheap oil. It also brought about an almost insatiable demand for petroleum products that has been the cornerstone of U.S. energy policy and its economy ever since.

Perceived shortages of petroleum after World War II brought about U.S. federal energy policies that encouraged the production of oil from our vast resources of coal. However, the discovery of large quantities of low-cost oil in the Middle East in the 1950s made such synthetic fuels relatively too expensive and so began an era of increasing oil imports and foreign trade deficits. The energy-use patterns discussed in Chapter 1 showed that oil has fueled most of the increase in global energy consumption over the past five decades. Because more than half of the world's oil reserves are in the Middle East, oil prices over the past three decades have strongly reflected international events in that region of the world. The 1991 Persian Gulf crisis was no exception.

Estimates of the magnitude of the world's oil and gas reserves are still uncertain. With trillions of barrels of oil reserves available, many groups feel that the United States does not have to rush to a transition to other fuels. Should an emphasis be placed on producing more oil from our own reserves? What damages will be done to our environment and economy by extending the lifetime of our supply of oil by only a few years by drilling in new areas?

Patterns of fuel use are very important. (You might wish to review Chapter 1 at this point.) About 63% of the 19 million barrels per day of oil used in the United States goes to transportation. Eighty-five percent of our oil imports are used for transportation. Without significant changes in this pattern, mandated or not, many of the fundamental problems of fuel supply for our 160 million vehicles will only grow. And problems of air pollution in urban areas will not improve either.

In this book we have discussed many alternatives to fossil fuels. In the 1960s and 1970s, many energy experts saw nuclear power as the technology of choice to provide the transition away from fossil fuels. The Three Mile Island scare in 1979 and the explosion at Chernobyl in 1986, plus huge construction cost overruns and lack of a policy for nuclear waste disposal, have placed nuclear power development on the back burner. However, the design of smaller nuclear power plants to include passive safety features, and increased concerns about global climatic changes, might cause nuclear power to once again be considered a viable option, especially in countries other than the United States.

Renewable energy options have been suggested for years as being able to provide a sizeable contribution to our energy supply. While public interest in solar energy has always been large, there has been a resurgence of development in wind energy and photovoltaics. Worldwide, wind energy is today's fastest growing energy technology, although it still provides less than 1% of the energy consumed in the United States. A sharp drop in solar cell prices has made photovoltaics more attractive. Solar-thermal, solar-electric, and biomass-energy conversion technologies are also able to make contributions to energy supply. However, they provide less than 5% of the energy consumed in this country. This is not true in developing countries, where biomass makes a substantial contribution.

Factors other than supply must also be examined. Environmental values are now regarded politically as having the same importance as energy security, resulting in the Clean Air Act of 1990 and the Kyoto Protocol of 1997. Air pollu-

Six-kW PV unit at Strang Middle School, Yorktown Heights, New York. (NEW YORK POWER AUTHORITY)

tion and potentially catastrophic climatic changes as a result of global warming might bring an end to the fossil-fuel era not in centuries but in decades. To what extent will market forces alone (with no intervention by the federal government to change fuel use patterns) provide a shift away from fossil fuels? Many energy analysts argue that such a transition, especially in light of our position as part of a global village, requires a blend of free-market forces and strong government policy and leadership. A free market often doesn't take into account the cost to the environment. Does the public recognize the relationship between energy choices and environmental quality?

One of the most important planks of any energy policy must be energy conservation and the efficient use of energy. These ideas were interwoven into each chapter. Recall Figure 1.15, depicting intensity of use versus frequency of activity. We can have a growing economy with the use of less energy if we make more efficient use of our limited resources. Indeed, the amounts of energy used in 1973 and 1984 were the same even through our Gross Domestic Product (GDP) grew by 34% during this period. The energy consumption per dollar of GDP fell 41% between 1973 and 1999. Reducing demand is much less expensive than increasing energy supplies.

Currently, quite large changes are occurring in the electric power industry due to deregulation in almost every state. Independent power producers grew strongly in the 1990s; large utilities have found it profitable to reduce demand by encouraging energy efficiency among their customers. Fueled by new rules and an apparent abundance of coal and natural gas, the market in the United States has become quite competitive. However, the California electric power crisis of 2001 has brought about a rethinking of deregulation across the country.

One of the basic problems in developing an energy strategy is that politics makes it difficult to formulate long-term plans. Compounding this problem is that we seem to be a nation with a short memory. Are periodic energy crises required to make us sit up and take notice of the crucial role that energy plays in our country's well being? We should recognize that long periods of time are necessary for significant changes in energy technology and policy. For these things to occur, citizens must realize that everyone will have to make trade-offs: higher energy prices, opening up certain geographical areas for energy exploration, acceptance of the benefits as well as the risks of technology. Energy is not an end in itself but a means to achieve the goals of a healthy economy and a healthy environment. Each of us has an important role to play; each of us can have a positive effect on the outcome. **But to succeed, we must play. We must take on a proactive role. We must be an informed and energetic citizenry.**

APPENDIX A

Units of Measurement and Powers of Ten Notation

Physics is an experimental science, and measurements must be made. A measurement is expressed as a number and a unit, such as 3 meters, so comparisons can be made with other measurements. There are many systems of units in use, but the basic set is the Système International, or SI. The unit of length in SI is the meter. Besides length, the other fundamental units are time (in seconds) and mass, expressed in kilograms (kg). Any measurable quantity can be expressed in terms of length, mass, and time. The English (or American) system is based on the foot as a unit of length; we will use both SI and English units, as dictated by common usage. A list of the commonly used units in SI and English systems and conversions between them, is given in Table 2.4.

Many of the distances and other quantities encountered in science are very large or very small. The diameter of an oxygen nucleus is approximately 0.0000000000000035 m, while the distance from the earth to the sun is 155,000,000,000 m. It is cumbersome to write such quantities in this form. As a result, it is customary to use powers of ten notation, in which the quantity is written in decimal form as a number between 1 and 10 multiplied by the appropriate power of ten. The number of times that 10 is multiplied together appears as a superscript—the "power" to which 10 is raised. For example, $100 = 10 \times 10 = 10^2$, $1000 = 10 \times 10 \times 10 = 10^3$, and $0.01 = 1/100 = 10^{-2}$. The two distances just mentioned now can be written as 3.5×10^{-15} m and 1.55×10^{11} m. For products of powers of ten, the superscripts are added, as $10^2 \times 10^3 = (10 \times 10) \times (10 \times 10 \times 10) = 10^{(2+3)} = 10^5$ and $10^4 \times 10^{-2} = 10^{(4-2)} = 10^2$.

An illustration of the large range in distances found in nature is shown in Table A.1. If we start with the size of a person being about 1 meter (10^0 m), the sizes of known objects go from 10^{-15} m (diameter of the hydrogen nucleus) to 10^{+27} m (diameter of the known universe), a range of 42 orders of magnitude.

Table A.1 DISTANCES AND FACTORS OF 10

Size (m)	Object	Size (m)	Object
10^{-14}	Nucleus	10^{3} (1 km)	Several long street blocks
10^{-9}	Atom	10^{6}	Several states
10^{-8}	Virus cells	10^{9}	Earth–Moon distance
10^{-6}	Dust particles	10^{12}	Sun–Mars distance
10^{-4} (0.01 cm)	Human hair (thickness)	10^{15}	Our solar system (no other stars)
10^{-3} (0.1 cm)	Fingernail (thickness)	10^{18}	Many stars observed
10^{-2} (1 cm)	Human finger (thickness)	10^{21}	Our entire galaxy
10^{0} (1 m)	Average person's size	10^{24}	Clusters of galaxies
		10^{27}	$>$ Size of known universe

APPENDIX B

Conversions and Equivalencies

CONVERSIONS AND EQUIVALENCIES *Also available in Table 3.4*	
Energy Units	1 Btu = 1055 J = 778 ft-lb = 252 cal 1 ft-lb = 1.356 J = 0.33 cal 1 calorie = 4.184 J 1 food Calorie = 1000 cal = 1 kcal 1 hp-hr = 2.68×10^6 J = 0.746 kWh 1kWh = 3.61×10^6 J = 3413 Btu = 2.65×10^6 ft-lb 1 Quad = 10^{15} Btu 1 GJ = 10^9 J = 948,000 Btu
Power Units	1 watt = 1 J/s = 3.41 Btu/h 1 hp = 550 ft-lb/s = 2545 Btu/h = 746 W
Fuel Relationships	1 barrel (bbl) crude oil = 42 gal = 5.8×10^6 Btu = 6.12×10^9 J 1 standard ft^3 natural gas (SCF) = 1000 Btu 1 therm = 100,000 Btu 1 gal gasoline = 1.24×10^5 Btu 10^6 ft^3 natural gas = 172 bbl crude oil 1 ton bituminous coal = 25×10^6 Btu 1 ton ^{235}U = 70×10^{12} Btu 1000 bbl/day of oil = 2.117×10^{12} Btu/y 1 million bbl/day of oil (1 MBPD) = 5.8×10^{12} Btu/d = 80 million tons per year of coal = ⅕ ton per year of uranium oxide
Fuel requirements for a 1000-MWe Power Plant (2.4×10^{11} Btu/d Input)	Coal: 9000 tons/d or 1 unit train load (100–90 ton cars)/d Oil: 40,000 bbl/d or 1 tanker /wk Natural gas: 2.4×10^8 SCF/d Uranium (as ^{235}U): 3 kg/d

continued

CONVERSIONS AND EQUIVALENCIES *continued*

Also available in Table 3.4

Energy Needs	U.S. Total Energy Consumption (1999) $= 97 \times 10^{15}$ Btu (97 Quads) = 45 MBPD oil equivalent $= 102 \times 10^{9}$ GJ
Everyday Usage and Energy Equivalencies	1 barrel (bbl) of oil = driving 1400 km (840 mi) in average car Electricity for city of 100,000 takes 4000 bbl/d of oil State of California energy needs for 8 h = 1 million bbl oil 1 gal gasoline = 11 kWh electricity (@ 30% generation efficiency) = 5 hours of operation of standard air conditioner = 200 days for electric clock = 48 hours for color TV = average summer days' solar energy incident on 2 m^2 (22 ft^2)
One Million Btu Equals Approximately	90 lb of coal 125 lb of oven-dried wood 8 gal of motor gasoline 10 therms of natural gas 1 day energy consumption *per capita* in the United States 100 kWh of electricity produced at a power plant
Power Data:	1000 MWe utility, at 60% load factor, generates 5.3×10^{9} kWh/y, enough for a city of about 1 million people. U.S. *per capita* power use = 12 kW Human, sitting = 60 W Human, running = 400 W Automobile at 65 mph = 33 kW

APPENDIX C

Home Heating Analysis

Introduction

This home energy audit will guide you in calculating the heat loss of your residence. This audit includes heat gains from various sources; Item H (on the summary page) will allow you to compare the energy use of your home with that of others. You will also be able to compare costs of heating your home with various fuels by using Section II and Item F. Analysis of your findings will indicate which features in your house are wasting the greatest amounts of energy.

Directions

Page 1
- A. Fill in each of the blanks in Section I.
- B. From your utility companies, find out the *local* cost per unit measure for each of the fuel sources in Section II.
- C. Calculate the cost per MBtu delivered to the living space.
- D. Make a sketch of your home, including the area of each type of exterior surface.

Page 2
- A. You may wish to make two copies of page 2, allowing calculations of heat losses for up to 12 rooms.
- B. For *each* room of your home with exterior heat-dissipating surfaces, fill in the data requested:
    1. Provide the name of the room.
    2. Provide the surface areas of all heat-dissipating surfaces.
    3. Referring to the table of heat-loss factors at the end of this appendix, note the surface type at ① and the corresponding factor at ② for each surface.
    4. Multiply the surface area by the factor ② to determine the Btu/h loss.
    5. Add the column of Btu/h losses for each room to determine the total room loss.

Page 3 A(1). Add the Btu/h losses of all windows from all rooms and record under Sum of Rows.

A(2). Add the Btu/h losses of each other feature and record under Sum of Rows.

A(3). Calculate infiltration losses.

1. Select the K factor from Item 7 in the table of heat-loss factors.
2. Transfer the volume of heated space from page 1.
3. Multiply $0.018 \times 65 \times K \times$ volume of heated space.

A(4). Add all losses (Sum of Rows) to determine the total peak hourly loss (Btu/h).

A(5). Calculate the percent total heat loss for each surface type (Total loss of each type $\div$ total peak hourly loss $\times$ 100).

B. Calculate the losses as a result of heat transfer from the central heating system.

C. Calculate the heat gains.

D. Calculate the peak net heat loss.

E. Calculate the seasonal heating needs.

F. Determine the total heating cost by multiplying Item E by the cost per MBtu for the fuel used to heat your home (see Section II on page 1).

G. Compare your calculated heating season cost with your actual heating season cost.

H. Calculate the Btu/degree-day/ft^2 and compare with the heating requirements of other homes.

I. If there is a discrepancy between the *calculated* and *actual* total heating season costs (Items F and G), try to explain why.

Name: ________________________________ Page 1

Home Heating Analysis

General Information:

1. All dimensions should be measured and not estimated.
2. The surface heat loss is equal to the factor you choose for that surface multiplied by the surface area.
3. Selection of a living unit with a self-contained heating system is preferred.
4. Prepare a drawing illustrating floor plan(s); room names, dimensions and areas; locations and sizes of exterior doors and windows. Calculate total area by floor level. Show wall thickness. Indicate north compass direction.
5. Round all mathematics to the nearest tenth.

I. Location of building: ________________ (city); ________________ (state)

Latitude: ________ (to nearest whole degree) Degree-days: ________

Type of structure: ___ house number of stories ___

___ apartment/duplex ___ mobile home

Type of construction (Example: standard height concrete block foundation; wood frame with 2″ × 4″ studs, 3½″ fiberglass insulation, cedar shiplap siding, 12″ fiberglass ceiling insulation, ceiling heights: 8′ 1st floor, 7′ 6″ 2nd floor)

____

____

Total amount of heated space: ________ ft^2 (floor area); ________ ft^3 (volume)

Type of heat source(s): ________________________________

(oil, natural gas, electric, LP, wood, etc.)

Type of heating system: ________________________________

(forced hot air, baseboard hot water, electric baseboard, radiant, etc.)

Rating of heating system: ________________________ Btu/h (output)

(see specification plate on unit)

II. Fuel/energy costs in your area × units/MBtu ÷ Conversion Efficiency = $/MBtu delivered

Energy Source	Cost	Conversion Efficiency	Cost/MBtu Delivered	Home Heating Cost If This Fuel Were Used
Electricity	$ ___ /kWh × 293 kWh/MBtu	÷ 1.00 =	________	________
Oil (#2)	$ ___ /gal × 7.14 gal/MBtu	÷ 0.70 =	________	________
Natural gas	$ ___ /therm × 10 therm/MBtu	÷ 0.80 =	________	________
LP gas	$ ___ /gal × 11 gal/MBtu	÷ 0.80 =	________	________
Coal (bit.)	$ ___ /ton × 0.039 ton/MBtu	÷ 0.65 =	________	________
Wood (128 ft^3)	$ ___ /cord × 0.035 cord/MBtu	÷ 0.60 =	________	________
________ (other)	$ ___ / × /MBtu	÷ =	________	________

Name: ______________________________ Page 2

HOME HEATING ANALYSIS

	Room ________			Room ________			Room ________		
Conduction Losses from Heat-Dissipating (Exterior) Surfaces	*Area*	*Surface Type Factor*	*Btu/h Loss*	*Area*	*Surface Type Factor*	*Btu/h Loss*	*Area*	*Surface Type Factor*	*Btu/h Loss*
Windows—Conduction Loss		① ____ ②			① ____ ②			① ____ ②	
Doors—Conduction Loss		① ____ ②			① ____ ②			① ____ ②	
Walls—Conduction Loss (minus window and door area)		① ____ ②			① ____ ②			① ____ ②	
Roof or Ceiling		① ____ ②			① ____ ②			① ____ ②	
First Floor/Slab/ Crawl Space/Basement Floor		① ____ ②			① ____ ②			① ____ ②	
		Total Room Loss			*Total Room Loss*			*Total Room Loss*	

	Room ________			Room ________			Room ________		
Conduction Losses from Heat-Dissipating (Exterior) Surfaces	*Area*	*Surface Type Factor*	*Btu/h Loss*	*Area*	*Surface Type Factor*	*Btu/h Loss*	*Area*	*Surface Type Factor*	*Btu/h Loss*
Windows—Conduction Loss		① ____ ②			① ____ ②			① ____ ②	
Doors—Conduction Loss		① ____ ②			① ____ ②			① ____ ②	
Walls—Conduction Loss (minus window and door area)		① ____ ②			① ____ ②			① ____ ②	
Roof or Ceiling		① ____ ②			① ____ ②			① ____ ②	
First Floor/Slab/ Crawl Space/Basement Floor		① ____ ②			① ____ ②			① ____ ②	
		Total Room Loss			*Total Room Loss*			*Total Room Loss*	

Name: ______________________________

Summary Page

III. **A.** For all rooms, total the heat loss per surface type (sum of rows):

Conduction Losses	*Sum of Rows*	*% of Total Heat Loss*
Windows		
Doors		
Outside Walls		
Roof/Ceiling		
Basement Floor/Slab		
Infiltration Losses (See at right)		
Total Peak Hourly Loss (Btu/h)		100%

Air exchanges per hour (see Item 7 on factor sheet)
Factor chosen: K = ____________
Volume of heated space = ______ ft^3 (from p. 1)
Infiltration loss = heat capacity of air $\times \Delta T \times$ air exchanges/h $\times$ volume of heated space = $0.018 \times 65 \times K \times$ volume of heated space

B. Losses as a result of heat transfer from central heating systems: If hot air ducts or hot water baseboard supply pipes are *not* insulated, multiply Total Peak Hourly Heating Loss by 1.20. If ducts/pipes *are* insulated, multiply by 1.10. Multiply by 1.0 if electric heat is used. ________ Btu/h

C. Heat gains (yearly averages):

- Appliances and lighting—Enter 2500 (2000 if water heater is in unheated space) ________ Btu/h
- People—Enter 400 for each occupant (averaged) ________ Btu/h
- Solar Gain—Enter 35 × area of south-facing windows ________ Btu/h

Total Gains ________ Btu/h

D. Peak net heat loss:
(Subtract total gains [C] from total losses [B].) = ________ Btu/h

E. Seasonal heating needs:
Divide item D by 1,000,000 and multiply by the number of degree-days × 24⁄65 = ________ MBtu

F. Total heating season *costs* based on previous calculations using fuel/energy costs in *your* area (Part II, p. 1)
Calculation based on $ ________ /MBtu for ____________ kind(s) of fuel or energy used = ________ dollars

G. Actual heating season costs taken from your bills. (Subtract domestic water heating costs if you use the same type of energy source for both space and water heating—approximately 20% of the total bill. You can also find the cost by comparing winter and summer energy bills.) = ________ dollars

H. For comparison purposes, divide item D by the *area* of heated space and the number of degree-days and multiply by $24/65$. = ________ Btu/degree-day/ft^2

I. Comment on any discrepancy between your results in F and G.

HEAT LOSS FACTORS FOR HOME ENERGY AUDIT CALCULATIONS (INCLUDES COMPOSITE CONSTRUCTION; CONSIDERS CONDUCTION LOSSES ONLY; $\Delta T = 65°F$)

Surface Type	Factor (Btu/ft²/h)			
			12 Hour Use of:	
1. *Windows and Sliding Glass Doors*	*Standard Factor*	*Low Emmitance*	*Drapes*	*Quilts*
Single pane	71.4	59.1	46.8	20.6
w/storm window	32.5	—	26.2	15.3
Double pane w/¼″ air space	37.8	—	29.6	16.4
w/½″ air space	31.9	22.4	25.8	15.2
Heat mirror	—	14.4	—	—
Triple pane w/¼″ air space	25.4	—	21.4	13.5
w/½″ air space	20.1	17.6	17.5	11.9

2. *Exterior Doors (Excluding Sliding Glass Doors)*	*Factor (Btu/ft²/h)*
(Calculate glass area in doors as windows)	
Wood: 1¼″ no storm door	27.0
1¼″ w/1″ storm door	17.0
1½″ no storm door	24.0
1½″ w/1″ storm door	15.0
1⅔″ solid-core door	21.0
Steel with foam core:	
1¾″ Pella	5.1
1¾″ Therma-Tru	4.1

3. *Exterior Walls (with Siding)*	*Factor (Btu/ft²/h)*
Concrete block (8″)	33.0
w/Vermiculite insulated cores	5.1
w/foam-insulated cores	3.3
w/4″ on uninsulated stud wall	15.0
w/4″ insulated stud wall	4.5
w/1″ air space and ½″ drywall	24.0

See Item 8 for special conditions of structures with infiltration barriers, adjacent unheated areas, and earth sheltering.

(continued)

3. ***Exterior Walls (with Siding)*** *(continued)*	***Factor (Btu/ft²/h)***	
Brick (4″)		
w/4″ uninsulated stud wall	16.0	
w/4″ insulated stud wall	4.6	
Logs (6″)	7.8	
(8″)	6.0	See item 8 for special conditions of structures with infiltration barriers, adjacent unheated areas, and earth sheltering.
Uninsulated wood frame w/2″ × 4″ construction	14.0	
Insulated wood frame		
w/1½″ fiberglass	7.3	
w/3½″ fiberglass; studs 16″ on-center	5.3	
w/3½″ fiberglass & 1″ foam	3.2	
w/6″ fiberglass; studs 24″ on-center	3.4	
w/6″ fiberglass and 1″ foam	2.5	
w/6″ cellulose	3.0	
w/6″ cellulose and 1″ foam	2.3	
Insulated double-wall frame		
11½″ fiberglass	1.8	
11½″ cellulose	1.6	
13½″ fiberglass	1.6	
13½″ cellulose	1.4	
Solid polystyrene walls 6″	2.5	
8″	1.9	
For heated underground basement, use above factors × 0.5		

4. ***Roof/Ceiling***	***Factor (Btu/ft²/h)***
No insulation	20.0
3½″ fiberglass	5.1
6″ fiberglass	3.2
6″ cellulose	2.8
12″ fiberglass	1.5
12″ cellulose	1.4
14″ cellulose	1.2

(continued)

5. Floor	*Factor (Btu/ft²/h)*
Over unheated basement or crawl space vented to outside	
Uninsulated floor	15.0
6″ fiberglass floor insulation	2.6
Over sealed, unheated, completely underground basement[a]	
Uninsulated floor	8.0
Uninsulated floor	
w/1″ foam on basement walls	3.4
w/3½″ fiberglass on basement walls	3.2
6″ fiberglass floor insulation	1.5
On concrete slab	
No insulation	6.0
1″ foam perimeter insulation	1.4
2″ foam perimeter insulation	1.0

6. *Other*

To determine factors for construction not listed on these sheets, use $65/R_{total}$

7. Air exchanges per hour = K (Btu/h/ft³)

Old, uninsulated, not maintained	$K = 4.0$
Old, uninsulated house, maintained	$K = 2.0$
Average insulated house, well maintained	$K = 1.0$
New, well-insulated house	$K = 0.5$
New, superinsulated (12″ walls)	$K = 0.2$

8. Notes

a. Multiply percentage of exposed basement wall above frostline (3 ft below ground level) by 4.0 for a 6″ fiberglass insulated floor or 20 for an uninsulated floor. (Omit window areas.)
b. If walls are earth sheltered, multiply factor by 0.5.
c. If walls are wrapped with Tyvek or Tu-Tuff infiltration barrier, multiply factor by 0.66.
d. If a closed, unheated space (such as an unused room, hall, or garage) is adjacent to the exterior wall, multiply the factor by 0.66.

APPENDIX D

Insolation and Temperature Data for Selected U.S. Cities

Monthly average daily radiation on a horizontal surface and on a surface tilted at an angle equal to the latitude, Btu/ft^2/d; monthly mean temperatures in °F.

	January	February	March	April	May	June	July	August	September	October	November	December
Albuquerque, NM 35° 03′	1151 1970 37.3	1454 2110 43.3	1925 2330 50.1	2344 2390 59.6	2560 2280 69.4	2757 2315 7901	2561 2250 82.8	2387 2290 80.6	2120 2480 73.6	1640 2230 62.1	1274 2060 47.8	1052 1870 39.4
Atlanta, GA 33° 39′	848 1290 47.2	1080 1410 49.6	1427 1660 55.9	1807 1810 65.0	2018 1860 73.2	2103 1790 80.9	2003 1740 82.4	1898 1820 81.4	1519 1640 77.4	1291 1650 66.5	998 1460 54.8	752 1160 47.7
Bismarck, ND 46° 47′	587 1370 12.4	934 1730 15.9	1328 1850 29.7	1668 1720 46.6	2056 1850 58.6	2174 1840 67.9	2305 1940 76.1	1929 1910 73.5	1441 1770 61.6	1018 1660 49.6	600 1310 31.4	464 1160 18.4
Boston, MA 42° 22′	505 890 31.4	738 1100 31.4	1067 1310 39.9	1355 1370 47.5	1769 1570 60.4	1864 1570 69.8	1860 1640 74.5	1570 1520 73.8	1268 1460 66.8	897 1260 57.4	636 970 46.6	443 830 34.9
Columbus, OH 40° 00′	486 760 32.1	747 1030 33.7	1112 1330 42.7	1480 1510 53.5	1839 1660 64.4	2110 1770 74.2	2041 1760 78.0	1573 1510 75.9	1189 1320 70.1	920 1220 58.0	479 680 44.5	430 700 34.0
Davis, CA 38° 33′	599 964 47.6	945 1350 52.1	1504 1880 56.8	1959 2020 63.1	2368 2130 69.6	2619 2230 75.7	2565 2210 81.0	2287 2200 79.4	1856 2130 76.7	1288 1800 67.8	795 1280 57.0	550 920 48.7
Denver, CO 39° 10′	848 1590 26.9	1210 1900 35.0	1622 2060 45.0	2002 2080 55.8	2300 2070 66.3	2645 2220 75.7	2517 2190 82.5	2157 2110 79.6	1957 2330 71.4	1394 2050 58.3	970 1720 42.0	793 1590 31.4
Dodge City, KA 37° 46′	953 1710 33.8	1186 1740 38.7	1565 1940 46.5	1975 2010 57.7	2126 1910 66.7	2459 2090 77.2	2400 2064 83.8	2210 2140 82.4	1841 2110 73.7	1421 1960 61.7	1065 1790 46.5	873 1650 36.8

Ithaca, NY 42° 27′	434 720 27.2	755 1140 26.5	1074 1320 36.0	1322 1340 48.4	1779 1600 59.6	2025 1760 68.9	2031 1750 73.9	1736 1680 71.9	1320 1540 64.2	918 10,300 53.6	466 720 41.5	370 640 29.6
Los Angeles, CA 34° 03′	911 1410 57.9	1223 1680 59.2	1640 1930 61.8	1866 1880 64.3	2031 1880 67.6	2259 1920 70.7	2428 2110 75.8	2198 2130 76.1	1891 2120 74.2	1362 1780 69.6	1053 1590 65.4	977 1440 60.2
Madison, WI 43° 08′	564 1090 21.8	812 1270 24.6	1232 1340 35.3	1455 1470 49.0	1745 1570 61.0	2031 1750 70.9	2046 1740 76.8	1740 1760 74.4	1443 1690 65.6	993 1450 53.7	556 920 37.8	496 1020 25.4
Miami, FL 25° 47′	1292 1720 71.6	1554 1890 72.0	1828 2010 73.8	2026 1990 77.0	2068 1880 79.9	1991 1750 82.9	1992 1770 84.1	1890 1790 84.5	1646 1690 83.3	1436 1670 80.2	1321 1690 75.6	1183 1710 72.6
Nashville, TN 36° 07′	589 850 42.6	907 1224 45.1	1246 1450 52.9	1662 1680 63.0	1997 1820 71.4	2149 1830 80.1	2079 1930 80.1	1862 1780 81.0	1601 1780 76.6	1223 1610 65.4	823 1220 52.3	614 960 44.3
St. Cloud, MN 45° 35′	632 1410 13.6	976 1750 16.9	1383 1880 29.8	1598 1680 46.2	1859 1640 58.8	2003 1700 68.5	2087 1820 74.4	1828 1810 71.9	1369 1630 62.5	890 1340 50.2	545 1010 32.1	463 1040 18.3
San Antonio, TX 29° 32′	1045 1490 53.7	1299 1660 58.4	1560 1680 65	1664 1610 72.2	2024 1860 93.2	2250 1940 85.0	2364 2060 87.4	2185 2080 87.8	1844 1610 82.6	1487 1930 74.7	1104 1500 63.3	954 1400 56.5
Seattle, WA 47° 36′	252 420 38.9	471 710 42.9	917 1170 46.9	1375 1400 51.9	1664 1514 58.1	1724 1470 62.8	1805 1550 67.2	1617 1600 66.7	1129 1330 61.6	638 930 54	325 520 45.7	218 400 41.5
Washington, DC 38° 51′	632 1000 38.4	901 1280 39.6	1255 1500 48.1	1600 1620 57.5	1846 1640 67.7	2080 1790 76.2	1929 1720 79.9	1712 1660 77.9	1446 1630 72.2	1083 1460 60.9	763 1210 50.2	594 1030 40.2

Horizontal values and tilt correction factors taken from "Solar Heating and Cooling of Buildings: Design of Systems," U.S. Department of Commerce, 1980.

APPENDIX E

World Energy Consumption, 1997
U.S. Consumption of Energy by Source, 1949–1998
U.S. Energy Efficiency, 1970–1998

WORLD ENERGY CONSUMPTION, 1997

	Population (Millions)	GDP (Billion $ U.S.)	Total Energy (Trillion Btu)	Energy per capita (MBtu)	Energy per GDP (Thousand Btu/$)	GDP per capita/($)
North America						
Canada	30.0	658.0	12,070	402.199	18.34	21,700
Mexico	98.4	694.3	5,680	57.724	8.18	7,700
United States	267.74	8,080.0	94,380	352.506	11.68	30,200
Central and South America						
Argentina	35.67	348.2	2,610	73.17	7.5	9,700
Brazil	159.64	1,040.0	7,690	48.17	7.39	6,300
Chile	14.62	168.5	890	60.88	5.28	11,600
Columbia	40.06	231.1	1,210	30.2	5.24	6,200
Cuba	11.07	16.9	400	36.13	23.67	1,540
Ecuador	11.94	53.4	310	25.96	5.81	4,400
Peru	24.37	110.2	510	20.93	4.63	8,300
Venezuela	22.78	185.0	2,660	117	14.38	8,670

continued

WORLD ENERGY CONSUMPTION, 1997 *continued*

	Population (Millions)	GDP (Billion $ U.S.)	Total Energy (Trillion Btu)	Energy per capita (MBtu)	Energy per GDP (Thousand Btu/$)	GDP per capita/($)
Europe						
Albania	3.73	4.5	35	9.38	7.78	1,370
Austria	8.07	174.1	1,320	163.57	7.58	21,400
Belarus	10.40	50.4	1,060	101.92	21.03	4,800
Belgium	10.19	236.3	2,600	255.15	11.0	23,200
Czech Republic	10.30	111.9	1,690	164.08	15.1	10,800
Denmark	5.28	122.5	890	168.56	7.27	23,200
Finland	5.14	102.1	1,260	245.14	12.34	20,000
France	58.61	1,320.0	9,780	166.87	7.41	22,700
Germany	82.05	1,740.0	14,100	171.85	8.1	20,800
Greece	10.50	137.4	1,200	114.29	8.73	13,000
Hungary	10.15	73.2	1,040	102.46	14.21	7,400
Ireland	3.66	59.9	480	131.15	8.01	16,800
Italy	57.37	1,240.0	7,700	134.22	6.21	21,500
Kazakhstan	16.82	55.2	1,860	110.55	33.7	900
Netherlands	15.60	343.9	3,810	244.23	11.08	22,000
Norway	4.40	120.5	1,800	408.16	14.94	27,400
Poland	38.65	280.7	3,860	99.87	13.75	7,250
Portugal	9.93	149.5	930	93.66	6.22	15,200
Romania	22.55	114.2	2,050	90.91	17.95	5,300
Russia	147.10	692.0	2,611	17.7	3.77	4,700
Spain	39.32	642.4	4,700	119.53	7.32	16,400
Sweden	8.85	176.2	2,210	249.72	12.54	19,700
Switzerland	7.09	172.4	1,230	173.48	7.13	23,800
Uzbekistan	24.10	60.7	1,890	78.42	31.14	2,500

continued

WORLD ENERGY CONSUMPTION, 1997 *continued*

	Population (Millions)	GDP (Billion $ U.S.)	Total Energy (Trillion Btu)	Energy per capita (MBtu)	Energy per GDP (Thousand Btu/$)	GDP per capita/($)
Ukraine	50.89	124.9	6,520	126.12	52.2	2,500
United Kingdom	58.82	1,242.0	9,740	165.59	7.84	21,200
Africa						
Algeria	29.05	120.4	1,260	43.37	10.47	4,000
Congo Republic	48.04	5.25	49	1.02	9.33	2,000
Cote d'Ivoire	15.30	25.8	215	14.05	8.33	1,700
Egypt	62.01	267.1	1,800	29.03	6.74	4,400
Ethiopia	58.12	29.0	490	8.43	16.9	530
Gabon	1.14	6.0	60	52.63	10.0	5,000
Ghana	18.66	36.2	310	16.61	8.56	2,000
Kenya	33.14	45.3	510	15.39	11.26	1,600
Libya	5.78	38.0	590	102.08	15.53	6,700
Morocco	27.31	107.0	410	15.01	3.83	3,500
Nigeria	103.90	132.7	970	9.34	7.31	1,300
South Africa	43.34	270.0	4,310	99.45	15.96	6,200
Sudan	30.90	26.6	290	9.39	10.9	875
Tunisia	9.21	565.0	240	26.06	0.42	6,100
Congo (Brazzaville)	2.75	18.0	490	178.18	27.22	400
Zimbabwe	10.92	24.9	250	22.89	10.04	2,200
Asia						
Bangladesh	122.01	167.0	380	3.11	2.28	1,330
China	1,244.20	4,250.0	35,470	28.51	8.35	3,460
India	955.12	1,534.0	12,100	12.67	7.89	1,600

continued

WORLD ENERGY CONSUMPTION, 1997 *continued*

	Population (Millions)	GDP (Billion $ U.S.)	Total Energy (Trillion Btu)	Energy per capita (MBtu)	Energy per GDP (Thousand Btu/$)	GDP per capita/($)
Indonesia	201.39	960.0	3,830	19.02	3.99	4,600
Iran	61.69	371.2	4,440	71.97	11.96	5,500
Iraq	22.40	42.8	1,030	45.98	24.07	2,000
Israel	5.83	96.7	720	123.5	7.45	17,500
Japan	126.07	3,080.0	21,570	171.1	7.0	24,500
North Korea	21.40	21.8	1,820	85.05	83.49	900
South Korea	22.98	631.2	7,620	331.59	12.07	13,700
Malaysia	21.00	227.0	1,650	78.57	7.27	11,100
Oman	2.26	17.4	270	119.47	15.52	8,000
Pakistan	138.16	344.0	1,680	12.16	4.88	2,600
Philippines	53.53	244.0	1,090	20.36	4.47	3,200
Saudi Arabia	19.48	206.5	4,080	209.45	19.76	10,300
Singapore	3.50	84.6	1,420	405.71	16.78	24,600
Taiwan	22.10	308.0	3,240	146.61	10.52	14,200
United Arab Emirates	2.30	54.2	1,790	778.26	33.03	24,000
Thailand	60.60	525.0	2,520	41.58	4.8	8,800
Turkey	62.51	388.3	2,850	45.59	7.34	6,100
Australia						
	18.52	394.0	4,070	219.76	10.33	21,400

Source: World Almanac and Book of Facts, 2000

U.S. CONSUMPTION OF ENERGY BY SOURCE, 1949–1998 (Quadrillion Btu, Except As Noted)

Year	Coal	Natural Gas	Petroleum	Hydroelectric Power	Nuclear Electric Power	Renewables	Total	Percent Change
1949	11.98	5.15	11.88	1.45	0	0	30.46	—
1950	12.35	5.97	13.32	1.44	0	0	33.08	8.6
1951	12.55	7.05	14.43	1.45	0	0	35.47	7.2
1952	11.31	7.55	14.96	1.50	0	0	35.30	−0.5
1953	11.37	7.91	15.56	1.44	0	0	36.27	2.7
1954	9.71	8.33	15.84	1.39	0	0	35.27	−2.8
1955	11.17	9.00	17.25	1.41	0	0	38.82	10.1
1956	11.35	9.61	17.94	1.49	0	0	40.38	4.0
1957	10.82	10.19	17.93	1.56	0	0	40.48	0.3
1958	9.53	10.66	18.53	1.63	0	0	40.35	−0.3
1959	9.52	11.72	19.32	1.59	0	0	42.12	4.4
1960	9.84	12.39	19.92	1.66	0.01	0	43.80	3.9
1961	9.62	12.93	20.22	1.68	0.02	0	44.46	1.5
1962	9.91	13.73	21.05	1.82	0.03	0	46.53	4.7
1963	10.41	14.40	21.70	1.77	0.04	0	48.52	3.9
1964	10.96	15.29	22.30	1.91	0.04	0	50.50	4.5
1965	11.58	15.77	23.25	2.06	0.04	0	52.68	4.3

continued

U.S. CONSUMPTION OF ENERGY BY SOURCE, 1949–1998 (Quadrillion Btu, Except As Noted) ***continued***

Year	Coal	Natural Gas	Petroleum	Hydroelectric Power	Nuclear Electric Power	Renewables	Total	Percent Change
1966	12.14	17.00	24.40	2.07	0.06	0	55.66	5.6
1967	11.91	17.94	25.28	2.34	0.09	0.01	57.57	3.4
1968	12.33	19.21	26.98	2.34	0.14	0.01	61.00	6.0
1969	12.38	20.68	28.34	2.66	0.15	0.01	64.19	5.2
1970	12.26	21.79	29.52	2.65	0.24	0.01	66.43	3.5
1971	11.60	22.47	30.56	2.86	0.41	0.01	67.89	2.2
1972	12.08	22.70	32.95	2.94	0.58	0.03	71.26	5.0
1973	12.97	22.51	34.84	3.01	0.91	0.04	74.28	4.2
1974	12.66	21.73	33.45	3.31	1.27	0.05	72.54	−2.3
1975	12.66	19.95	32.73	3.22	1.90	0.07	70.55	−2.8
1976	13.58	20.35	35.17	3.07	2.11	0.08	74.36	5.4
1977	13.92	19.93	37.12	2.51	2.70	0.08	76.29	2.6
1978	13.77	20.00	37.97	3.14	3.02	0.06	78.09	2.4
1979	15.04	20.67	37.12	3.14	2.78	0.08	78.90	1.0
1980	15.42	20.39	34.20	3.12	2.74	0.11	75.96	−3.7
1981	15.91	19.93	31.93	3.11	3.01	0.12	73.99	−2.6
1982	15.32	18.51	30.23	3.57	3.13	0.10	70.85	−4.2

1983	15.89	17.36	30.05	3.90	3.20	0.13	70.52	−0.5
1984	17.07	18.51	31.05	3.76	3.55	0.16	74.10	5.1
1985	17.48	17.83	30.92	3.36	4.15	0.20	73.95	−0.2
1986	17.26	16.71	32.20	3.39	4.47	0.22	74.24	0.4
1987	18.01	17.74	32.87	3.07	4.91	0.23	76.84	3.5
1988	18.85	18.55	34.23	2.64	5.66	0.22	80.20	4.4
1989	18.92	19.50	34.02	2.85	5.69	0.20	81.23	1.3
1990	19.1	19.3	33.55	3.03	6.16	2.98*	84.12	3.6
1991	18.77	19.61	32.85	3.21	6.58	3.01	84.03	−0.1
1992	18.87	20.13	33.53	2.9	6.61	3.15	85.19	1.4
1993	19.43	20.84	33.84	3.18	6.52	3.11	86.92	2
1994	19.54	21.16	34.65	3.11	6.83	3.18	88.47	1.8
1995	21.98	19.11	34.66	3.21	7.18	6.48	92.61	4.7
1996	22.61	19.53	35.72	3.59	7.17	7.06	95.68	3.3
1997	23.21	19.39	36.38	3.71	6.68	7.12	96.48	0.1
1998	23.81	19.47	36.57	3.39	7.16	6.87	97.27	0.1

Source: U.S. Energy Information Administration

U.S. ENERGY EFFICIENCY, 1970–1998

Year	GNP* (Billion 1982 Dollars)	Energy Consumption (Trillion Btu)	Energy Consumption per GNP* 1982 $U.S. (Thousand Btu)	Oil Energy Consumption (Trillion Btu)	Oil Energy Consumption per GNP* 1982 $U.S. (Thousand Btu)
1970	2,416	67,143	27.8	29,537	12.2
1971	2,485	68,348	27.5	30,570	12.3
1972	2,609	71,643	27.5	32,966	12.6
1973	2,744	74,282	27.1	34,840	12.7
1974	2,729	72,543	26.6	33,455	12.3
1975	2,695	70,545	26.2	32,731	12.1
1976	2,827	74,362	26.3	35,175	12.4
1977	2,959	76,289	25.8	37,122	12.5
1978	3,115	78,089	25.1	37,965	12.2
1979	3,192	78,897	24.7	37,123	11.6
1980	3,187	75,955	23.8	34,202	10.7
1981	3,249	73,991	22.8	31,931	9.8
1982	3,166	70,848	22.4	30,231	9.5
1983	3,279	70,524	21.5	30,054	9.2
1984	3,501	74,101	21.2	31,051	8.9
1985	3,619	73,945	20.4	30,922	8.5
1986	3,718	74,237	20.0	32,196	8.7
1987	3,854	76,845	19.9	32,865	8.5
1988	4,024	80,069	19.9	34,209	8.5
1989	4,143	81,070	19.6	34,290	8.3
1990	4,226	84,070	19.8	33,550	7.9
1991	4,186	84,020	20.1	32,850	7.8
1992	4,282	85,190	19.9	33,530	7.8

U.S. ENERGY EFFICIENCY, 1970–1998 *continued*

Year	GNP* (Billion 1982 Dollars)	Energy Consumption (Trillion Btu)	Energy Consumption per GNP* 1982 $U.S. (Thousand Btu)	Oil Energy Consumption (Trillion Btu)	Oil Energy Consumption per GNP* (Thousand Btu)
1993	4,416	86,990	19.7	33,840	7.7
1994	4,541	88,420	19.5	34,650	7.6
1995	7,565	90,860	12.1	34,660	4.6
1996	7,636	93,810	12.3	35,720	4.7
1997	8,110	94,370	11.6	36,380	4.5
1998	8,511	94,230	11.1	36,570	4.3

*Data after 1994 uses GDP in place of GNP.

Source: U.S. Energy Information Administration

Glossary

absolute zero: the lowest achievable temperature, 0 K = −273°C.

acid deposition: the depositing of sulfuric or nitric acid in the environment through snow, rain, or dry sediment.

active solar heating: heating a house by using a solar collector with pumps or fans to transfer the heat into the house.

adiabatic process: a process in which no heat is transferred into or out of the system.

aerosols: solid or liquid matter in the atmosphere, usually smaller than 10 microns.

alcohol fuel: fuel made by distilling grain, wood, or other plant products into alcohol.

alpha ray: the nucleus of a helium atom; one of the natural radiations.

altitude: elevation or angle of the sun above the horizon.

amorphous silicon: type of solar cell that uses disordered and noncrystalline silicon.

ampere: the unit of measurement of electric current. Amount of current produced by 1 V through a resistance of 1 ohm.

anaerobic digestion: decomposition process by which bacteria convert organic material into methane in the absence of oxygen.

Archimedes' principle: the buoyant force on an object is equal to the weight of the fluid displaced by that object.

armature: part of an electric motor or generator that revolves between the poles of a magnet. It is made of wire wound around an iron core.

atomic number: the number of protons in the nucleus.

available energy: the maximum amount of work that can be extracted in a particular process.

avoided cost: cost to a utility to generate electricity if it buys the same amount from another source.

azimuth: angle of the sun from true south.

becquerel (Bq): a measure of radioisotope activity. 1 Bq = 1 disintegration per second.
beta ray: a positive or negative electron; a natural component of radioactivity.
binding energy: the energy that holds a nucleus together; the difference between the sum of the masses of the individual nucleons and the actual mass of the nucleus.
biomass: organic material in any form—wood, crop residue, animal manure, and so forth. Biomass contains energy stored in chemical form.
breeding: the process whereby a fissile nucleus is produced from a nonfissile nucleus in a reactor by neutron absorption.
British thermal unit (Btu): the energy required to raise the temperature of 1 lb of water by 1°F.
BWR: boiling water reactor; a nuclear reactor in which the reactor water is allowed to boil to produce steam.

CANDU: the Canadian natural-uranium, heavy water reactor.
capacity: the maximum amount of electricity that would be available from a generating unit. In units of megawatts for electrical power.
Carnot cycle: a particular cycle for a heat engine; the cycle that gives the maximum efficiency when operating between two temperatures.
Carnot efficiency: maximum efficiency for converting heat energy into work; given by $(1 - T_c/T_h)$, with the temperatures in absolute units.
caulking: a soft material that can be squeezed into the cracks of a building to reduce the air flow into or out of the building.
CFC (chlorofluorocarbons): chlorinated compounds contributing to the destruction of stratospheric ozone.
China syndrome: reference to a nuclear reactor core melting its way into the ground beneath the reactor in the event of a loss of coolant accident.
cogeneration: production of both electricity and useful heat from the same fuel source.
competitive market: an environment that allows many sellers and buyers to buy and sell goods or services from each other. Customers have the choice of buying their energy from more than one provider.
conduction: the process by which heat travels through substances.
convection: the process by which heat is transferred by the movement of fluids.
COP (coefficient of performance): ratio of heat transferred by a heat pump to its electricity input.
cross section: probability for a nuclear reaction to occur.
curie: a measure of radioisotope activity; 1 Ci = 3.7×10^{10} disintegrations per second.

Darrieus rotor: a vertical eggbeater-type wind turbine machine.
degree-days: the difference between the mean daily temperature and 65°F.
demand: the amount of electricity that must be generated to meet the needs of all customers. Sometimes called the load.

Demand-side management (DSM): utility programs used to reduce peak electrical demand and help customers use electricity more efficiently.

deregulation: the act or process of removing regulations or other barriers that may restrict an industry.

DHW: domestic hot water.

diffuse radiation: solar radiation received under cloudy skies; cannot be focused by mirrors or lenses.

direct radiation: solar radiation received under clear skies. Can be focused.

divestiture: the separation of a utility's generation and transmission functions into smaller, individually owned businesses.

dosimeter: a device to measure the amount of radiation absorbed over time.

doubling time (DT): the time required for a given quantity to double in value.

$$\text{Approximately, } DT = \frac{70 \text{ years}}{\text{percent growth rate}}$$

ECCS (emergency core cooling system): a high-pressure water spray system that is used to cool the nuclear reactor core in the event of a loss of coolant accident.

efficiency: the ratio of the useful work or energy output to the total energy input or energy converted.

EGR (exhaust gas recirculation): a system in an auto engine to reduce nitrogen-oxide formation by adding exhaust gases to the incoming air to lower the combustion temperature in the engine.

electrolyte: a chemical that, when dissolved in water, will conduct an electrical current.

electrostatic precipitator: a device for removing particulates from the combustion gases in a power plant.

EMF: electromagnetic field, usually used in association with high-voltage electrical transmission lines.

emissivity: measure of a material's ability to give off thermal radiation.

enrichment: the process in which the abundance of the fissile isotope, ^{235}U, is increased from 0.7% to 2 to 3% in the uranium oxide fuel.

entropy: a term used in thermodynamics to measure the disorder of a system. The total entropy of a system always increases in an isolated process.

eutrophication: enrichment of a body of water by the addition of extra nutrients, stimulating algae growth.

eV (electron volt): an amount of energy equal to 1.6×10^{-19} J.

exponential growth: a process in which the amount of a substance has a fixed growth rate, or the change in amount of something is proportional to the amount present.

fertile material: nuclei that do not fission but can be used to create fissionable nuclei through neutron absorption, for example, ^{238}U.

first law: conservation of energy in a thermodynamic system; the heat added plus the work done equals the change in the total energy of a system.

fissile material: nuclei that will undergo fission when it absorbs a neutron (e.g., ^{235}U, ^{239}Pu).

fission products: unstable radioactive isotopes produced when a uranium nucleus is split after capturing a neutron.

fluidized bed combustion: process by which coal is burned on a moving bed of air; limestone is added to remove SO_2.

fly ash: minute particles of ash resulting from the burning of coal.

fuel cell: device that produces electricity from a chemical reaction between hydrogen and oxygen.

fusion: the process of bringing two nuclei together to form one nucleus; energy is released through the loss of mass in the product nucleus.

gamma ray: a high-energy photon of electromagnetic energy released in some radioactive decays.

gas desulfurization: process to remove SO_2 from exhaust gases in a power plant.

gas turbine: a heat engine that produces electricity by using the force of hot, expanding gases to make a turbine revolve. The gases are produced by burning a fuel.

gaseous diffusion: the enrichment of ^{235}U by diffusion of uranium hexafluoride gas through porous barriers.

gauss: a measure of magnetic field. Earth's magnetic field is about 0.5 gauss.

generator: a device that changes mechanical energy into electrical energy. It consists of a magnet and a coil of wire rotating relative to the magnet.

geothermal energy: heat energy available in rocks, hot water, and steam beneath the earth's surface.

gray: a unit of absorbed radiation dose; 1 Gy = 1 J/kg.

greenhouse effect: the trapping in the atmosphere, primarily by CO_2, of long-wavelength radiation emitted by the earth.

greenhouse gases: gases in the atmosphere that absorb infrared radiation from the earth; primarily CO_2, methane, CFCs, ozone, nitrous oxides.

green energy: electricity produced by renewable resources.

grid: a system of power lines and generators that are coordinated to provide electricity to customers at various points.

half-life: time it takes for one-half of the amount of a radioactive material to decay into another element.

head: the height of water behind a dam to the turbine below.

heat capacity: amount of energy needed to raise the temperature of 1 ft^3 of material by 1° F; equal to specific heat multiplied by the object's density.

heat engine: device operating between two temperatures that converts fraction of heat energy into useful work.

heat exchanger: device that is used to transfer heat between two fluids—one loses heat energy and the other gains it.

heat of fusion: heat energy needed to change unit mass of object from solid to liquid at the same temperature.

heat of vaporization: heat energy needed to change unit mass of object from liquid to vapor at the same temperature.

heat pump: a device that operates as a refrigerator and moves heat into the house in the winter from the outside; in the summer it cools the house, moving the heat to the outside.

HTGC (high-temperature gas-cooled reactor): a reactor that uses graphite as the moderator, and gas, rather than water, for heat transfer.

hydrocarbons: compounds containing only hydrogen and carbon. Also called volatile organic compounds. For example, benzene and methane.

hydrothermal: the type of geothermal energy in which water heated within the earth flows to the surface as hot water or steam.

inertial confinement: laser-induced fusion in which fuel is heated rapidly before expansion can begin.

independent power producer (IPP): a company, other than a utility, that generates electricity.

independent system operator (ISO): an entity that monitors the reliability of the power system and coordinates the supply of electricity around the state.

infiltration: process by which cold air leaks into a house through cracks.

insolation: incident solar radiation; measured in $Btu/ft^2/d$ or W/m^2.

insulation: a material that slows heat loss or heat gain.

isotopes: nuclei that have the same atomic number but different numbers of neutrons.

kcal: the energy required to raise the temperature of 1 kg of water by 1° C.

kinetic energy: energy of motion.

Kyoto Protocol: International conference held in Japan in 1997 that dealt with reductions in greenhouse gas emissions.

latent heat: heat necessary to cause a phase change in a material, such as going from ice to liquid, or liquid to gas, without a change in temperature.

Lawson criteria: conditions that must be achieved in fusion for energy breakeven (energy output at least as large as energy input).

leachate: liquid resulting from water moving through a landfill ("garbage juice").

leaching: the process by which water moves through a material (e.g., garbage), picking up substances in the material.

liquefaction: the process of converting coal into useable liquid fuel.

LMFBR (liquid metal fast breeder reactor): a reactor using liquid sodium as the coolant.

LOCA (loss of coolant accident): a nuclear accident in which a major water coolant pipe breaks, possibly leading to a core meltdown.

MAGLEV: magnetic levitation—suspension of an object using repulsive force between two magnets. Used for high-speed trains.
magma: hot material under the earth's crust and mantle.
magnetic confinement: technique used in fusion reactors to contain the plasma that is being heated.
methane: colorless, odorless gas—CH_4. Main component of natural gas.
micron: 1 μm = 10^{-6} m.
moderator: substance used in nuclear reactors to slow down neutrons so they can be more readily captured by ^{235}U and produce fission. Usually water.
mrem: millirem: 1/1000 of a rem (radiation dose).

NRC (Nuclear Regulatory Commission): the federal agency responsible for the licensing and operation of nuclear reactors.
nucleon: general term for the proton and neutron.

off-peak period: hours of the day when demand for electricity is low. Usually has lower prices for electricity.
Ohm's law: the empirical relationship between the current, potential difference, and resistance in an electrical circuit: $V = IR$.
oil shale: sedimentary rock containing solid organic material that can be converted to crude oil.
overburden: soil and rock that overlie a buried deposit of coal. In surface mining, the overburden is first removed.

PANs: peroxyacyl nitrates—reactive organic radicals that contribute to formation of smog.
passive solar heating: using the building to collect and store incoming solar energy. Does not use fans or pumps to distribute heat through the building.
penstock: a large pipe that carries water to the turbine in a hydroelectric plant.
photoelectric effect: the release of electrons from a metal by the absorption of light.
photon: a massless particle of electromagnetic energy, moving at the speed of light.
photosynthesis: the production of carbohydrates in a plant from water and carbon dioxide using solar radiation.
pH: a measure of how acidic or basic a substance is. Distilled water has a pH of 7. A pH of 6 to 0 is increasingly acidic; 8 to 14 increasingly alkaline.
plasma: ionized gas, usually at a high temperature.
p–n junction: the region of contact between p-type and n-type semiconductor materials. A depletion region exists at this junction, as well as a potential difference.
potential energy: energy stored; a function of the object's position.
pumped storage: a technique for storing energy by using excess electricity to pump water to a high reservoir, from which it can be used in a hydroelectric facility to generate electricity during periods of peak demand.

PURPA (Public Utilities Regulatory Policy Act): 1978 federal law requiring competition in the electrical generating industry. Requires utilities to buy power from eligible cogeneration sources, small hydro, or waste-fueled facilities, under contracts at an avoided cost rate.

PWR (pressurized water reactor): a nuclear power reactor in which the cooling water is kept under a high pressure and not allowed to boil.

quad: an amount of energy equal to 10^{15} Btu.

quantized value: a discrete value possessed by a physical quantity.

radiation: energy emitted from atoms or molecules in form of rays, waves, or particles. Radiation from the nucleus can have enough energy to produce ionization and harm living tissue.

radon: a radioactive gas which emits alpha rays; atomic weight 222, atomic number 86.

Rankine cycle: a particular cycle in a turbine power system; the working fluid is both a liquid and vapor in the cycle.

rated speed: the speed for which a given wind turbine is designed to produce maximum power.

rem: a unit for measuring absorbed doses of radiation.

renewable resource: an alternative energy source to oil, gas, coal, or uranium used to produce electricity.

reserve: the amount of a resource that is recoverable at current prices and with current technology.

retrofitting: installation of equipment or materials after initial construction.

reversible: a process that can have its direction changed or reversed with no effect upon the environment.

roentgen: an old unit of radiation exposure.

R-value: the "resistance" of a substance to heat transfer by conduction.

sequestration, of carbon: capture of CO_2 to nonatmospheric sinks.

shale oil: a form of oil trapped within a rock called shale.

solar constant: incident solar radiation at top of earth's atmosphere, per unit area. Measured in W/m^2 or $Btu/ft^2/h$.

somatic effect: an effect on the health of an individual receiving radiation.

specific heat: amount of heat added per unit mass per degree of temperature increase.

stranded costs: costs that a utility has an obligation to pay for, but may not be able to recover from a customer because the customer no longer uses the utility's service.

sustainable development: development that meets needs of the present while protecting the environment for the future.

synergism: action of two or more things to achieve an effect of which each is individually incapable.

temperature inversion: event occurring when the temperature of a layer of air increases, trapping rising air pollutants.

therm: a measure of heat energy content; 1 therm equals 100,000 Btu.

thermal mass: a heat storage material, such as water or masonry, used in passive solar heating systems.

thermal neutrons: neutrons in a reactor that have very small energies, about 1/40th of an eV.

thermosiphoning: the circulation of water by natural convection.

time-of-use pricing: rates that are designed to reflect changes in a utility's cost of providing service that change by time of day or season.

Tokamak: a particular type of fusion reactor using magnetic confinement; doughnut shaped.

total energy: sum of a body's potential, kinetic, and thermal energies.

volatile organic compounds (VOC): compounds consisting of carbon and hydrogen atoms. Air pollutants associated with petroleum use.

Index